Lecture Notes in Artificial Intelligence   2036

Subseries of Lecture Notes in Computer Science
Edited by J. G. Carbonell and J. Siekmann

Lecture Notes in Computer Science

Edited by G. Goos, J. Hartmanis and J. van Leeuwen

T0189658

Springer
Berlin
Heidelberg
New York
Barcelona
Hong Kong
London
Milan
Paris
Singapore
Tokyo

Stefan Wermter   Jim Austin
David Willshaw (Eds.)

Emergent Neural Computational Architectures Based on Neuroscience

Towards Neuroscience-Inspired Computing

 Springer

Series Editors

Jaime G. Carbonell,Carnegie Mellon University, Pittsburgh, PA, USA
Jörg Siekmann, University of Saarland, Saarbrücken, Germany

Volume Editors

Stefan Wermter
University of Sunderland, Centre of Informatics, SCET
St Peters Way, Sunderland SR6 0DD, UK
E-mail: stefan.wermter@sunderland.ac.uk

Jim Austin
University of York, Department of Computer Science
York, Y010 5DD, UK
E-mail: austin@minstor.york.ac

David Willshaw
University of Edinburgh, Institute for Adaptive and Neural Computation
5 Forrest Hill, Edinburgh, Scotland, UK
E-mail: D.Willshaw@anc.ed.ac.uk

Cataloging-in-Publication Data applied for

Die Deutsche Bibliothek - CIP-Einheitsaufnahme

Emergent neural computational architectures based on neuroscience : towards
neuroscience inspired computing / Stefan Wermter ... (ed.). - Berlin ;
Heidelberg ; New York ; Barcelona ; Hong Kong ; London ; Milan ; Paris ;
Singapore ; Tokyo : Springer, 2001
    (Lecture notes in computer science ; 2036 : Lecture notes in artificial
    intelligence)
    ISBN 3-540-42363-X

CR Subject Classification (1998): I.2, F.1, F.2.2, I.5, F.4.1, J.3

ISBN 3-540-42363-X Springer-Verlag Berlin Heidelberg New York

Springer-Verlag Berlin Heidelberg New York
a member of BertelsmannSpringer Science+Business Media GmbH

http://www.springer.de

© Springer-Verlag Berlin Heidelberg 2001
Printed in Germany

Typesetting: Camera-ready by author, data conversion by PTP Berlin, Stefan Sossna
Printed on acid-free paper   SPIN 10782442   06/3142   5 4 3 2 1 0

Preface

This book is the result of a series of international workshops organized by the EmerNet project on Emergent Neural Computational Architectures based on Neuroscience sponsored by the Engineering and Physical Sciences Research Council (EPSRC). The overall aim of the book is to present a broad spectrum of current research into biologically inspired computational systems and hence encourage the emergence of new computational approaches based on neuroscience. It is generally understood that the present approaches to computing do not have the performance, flexibility, and reliability of biological information processing systems. Although there is a massive body of knowledge regarding how processing occurs in the brain and central nervous system this has had little impact on mainstream computing so far.

The process of developing biologically inspired computerized systems involves the examination of the functionality and architecture of the brain with an emphasis on the information processing activities. Biologically inspired computerized systems address neural computation from the position of both neuroscience and computing by using experimental evidence to create general neuroscience-inspired systems.

The book focuses on the main research areas of modular organization and robustness, timing and synchronization, and learning and memory storage. The issues considered as part of these include: How can the modularity in the brain be used to produce large scale computational architectures? How does the human memory manage to continue to operate despite failure of its components? How does the brain synchronize its processing? How does the brain compute with relatively slow computing elements but still achieve rapid and real-time performance? How can we build computational models of these processes and architectures? How can we design incremental learning algorithms and dynamic memory architectures? How can the natural information processing systems be exploited for artificial computational methods?

We hope that this book stimulates and encourages new research in this area. We would like to thank all contributors to this book and the few hundred participants of the various workshops. Especially we would like to express our thanks to Mark Elshaw, network assistant in the EmerNet network who put in tremendous effort during the process of publishing this book.

Finally, we would like to thank EPSRC and James Fleming for their support and Alfred Hofmann and his staff at Springer-Verlag for their continuing assistance.

March 2001

Stefan Wermter
Jim Austin
David Willshaw

Table of Contents

Timing and Synchronisation ....

Towards Novel Neuroscience-Inspired Computing

Stefan Wermter[1], Jim Austin[2], David Willshaw[3], and Mark Elshaw[1]

[1] Hybrid Intelligent Systems Group
University of Sunderland,
Centre for Informatics, SCET
St Peter's Way, Sunderland, SR6 0DD, UK
[Stefan.Wermter][Mark.Elshaw]@sunderland.ac.uk
www.his.sunderland.ac.uk
[2] Department of Computer Science
University of York, York YO10 5DD, UK
austin@cs.york.ac.uk
[3] Institute for Adaptive and Neural Computation
University of Edinburgh, 5 Forrest Hill, Edinburgh
david@anc.ed.ac.uk

Abstract. Present approaches for computing do not have the performance, flexibility and reliability of neural information processing systems. In order to overcome this, conventional computing systems could
benefit from various characteristics of the brain such as modular organisation, robustness, timing and synchronisation, and learning and memory
storage in the central nervous system. This overview incorporates some
of the key research issues in the field of biologically inspired computing
systems.

1   Introduction

It is generally understood that the present approaches for computing do not have
the performance, flexibility and reliability of biological information processing
systems. Although there is a massive body of knowledge regarding how processing occurs in the brain this has had little impact on mainstream computing. As a
response the EPSRC[1] sponsored the project entitled Emergent Neural Computational Architectures based on Neuroscience (EmerNet) which was initiated by
the Universities of Sunderland, York and Edinburgh. Four workshops were held
in the USA, Scotland and England. This book is a response to the workshops
and explores how computational systems might benefit from the inclusion of the
architecture and processing characteristics of the brain.

The process of developing biologically inspired computerised systems involves
the examination of the functionality and architecture of the brain with an emphasis on the information processing activities. Biologically inspired computerised

---

[1] Engineering and Physical Sciences Research Council.

S. Wermter et al. (Eds.): Emergent Neural Computational Architectures, LNAI 2036, pp. 1–19, 2001.
© Springer-Verlag Berlin Heidelberg 2001

systems examine the basics of neural computation from the position of both neuroscience and computing by using experimental evidence to create general neuroscience-inspired systems.

Various restrictions have limited the degree of progress made in using biological inspiration to improve computerised systems. Most of the biologically realistic models have been very limited in terms of what they attempt to achieve compared to the brain. Despite the advances made in understanding the neuronal processing level and the connectivity of the brain, there is still much that is not known about what happens at the various systems levels [26]. There is disagreement over what the large amount of information provided on the brain imaging techniques means for computational systems [51].

Nevertheless, the last decade has seen a significant growth in interest in studying the brain. The likely reason for this is the expectation that it is possible to exploit inspiration from the brain to improve the performance of computerised systems [11]. Furthermore, we observe the benefits of biological neural systems since even a child's brain can currently outperform the most powerful computing algorithms. Within biologically inspired computerised systems there is a growing belief that one key factor to unlocking the performance capabilities of the brain is its architecture and processing [47], and that this will lead to new forms of computation.

There are several architectural and information processing characteristics of the brain that could be included in computing systems to enable them to achieve novel forms of performance, including modular organisation, robustness, information processing and encoding approaches based on timing and synchronisation, and learning and memory storage.

2   Some Key Research Issues

In this chapter and based on the EmerNet workshops we look at various key research issues: For biologically inspired computerised systems it is critical to consider what is offered by computer science when researching biological computation and by biological and neural computation for computer science. By considering four architectural and information processing forms of inspiration it is possible to identify some research issues associated with each of them.

Modular Organisation: There is good knowledge of how to build artificial neural networks to do real world tasks, but little knowledge of how we bring these together in systems to solve larger tasks (such as in associative retrieval and memory). There may be hints from studying the brain to give us ideas on how to solve these problems.

Robustness: How does human memory manage to continue to operate despite failure of its components? What are its properties? Current computers use a fast but brittle memory, brains are slow but robust. Can we learn more about the properties that can be used in conventional computers.

Sychronisation and Timing: How does the brain synchronise its processing? How does the brain prevent the well known race conditions found in computers?

How does the brain schedule its processing? The brain operates without a central clock (possibly). How is the asynchronous operation achieved? How does the brain compute with relatively slow computing elements but still achieve rapid and real-time performance? How does the brain deal with real-time? Do they exploit any-time properties, do they use special scheduling methods. How well do natural systems achieve this and can we learn from any methods they may use?

Learning and Memory Storage: There is evidence from neuron, network and brain levels that the internal state of such a neurobiological system has an influence on processing, learning and memory. However, how can we build computational models of these processes and states? How can we design incremental learning algorithms and dynamic memory architectures?

3 Modular Organisation

Modularity in the brain developed over many thousands of years of evolution to perform cognitive functions using a compact structure [47] and takes various forms such as neurons, columns, regions or hemispheres [58].

3.1 Regional Modularity

The brain is viewed as various distributed neural networks in diverse regions which carry out processing in a parallel fashion to perform specific cognitive functions [21,42,58]. The brain is sometimes described as a group of collaborating specialists that achieve the overall cognitive function by splitting the task into smaller elements [59]. The cerebral cortex which is the biggest part of the human brain is highly organised into many regions that are responsible for higher level functionality that would not be possible without regional modularity [72,58]. A feature of regional modularity is the division of the activities required to perform a cognitive function between different hemispheres of the brain. For instances, in this volume, Hicks and Monaghan (2001) [38] show that the split character of the visual processing between different brain hemispheres improves visual word identification by producing a modular architecture.

Brain imaging techniques have successfully provided a great deal of information on the regions associated with cognitive functions [27]. The oldest of these techniques mainly involves the examination of the brain for lesions that are held responsible for an observed cognitive deficit [29]. The lesion approach has been criticised since it does not identify all the regions involved in a cognitive function, produces misleading results due to naturally occurring lesions and alternative imaging techniques contradict its findings [9]. Due to the difficulties observed with the lesion approach and technical developments, four alternative techniques known as positron emission tomography (PET), functional magnetic resonance imaging (fMRI), electronencephalogram (EEG) and magnetoencephalogram (MEG) have received more attention. PET and fMRI both examine precisely the neural activity within the brain in an indirect manner and so create an image of the regions associated with a cognitive task [60,9]. For PET this is done by identifying the regions with the greatest blood flow, while for fMRI the brain map is the blood oxygen levels. Although PET and fMRI have good spatial attainment, their temporal competence is limited [68]. In contrast, EEG measures voltage fluctuations produced by regional brain activity through electrodes position on the surface of the scalp. MEG uses variations in the magnetic field to establish brain activity by exploiting sophisticated superconducting quantum devices. The temporal properties of EEG and MEG are significantly better that PET and fMRI with a sensitivity of a millisecond [68].

A major issue that is currently being investigated by biological inspired computer system researchers is the manner that modules in the brain interact [66]. In this volume Taylor (2001) [68] establishes an approach to examine the degree of association between those regions identified as responsible for a subtask by considering the correlation coefficients. This approach incorporates structural modelling where linear associations among the active regions are accepted and the path strengths are established via the correlation matrix. When bridging the gap between the brain image information and underlying neural network operations, activity is described by coupled neural equations using basic neurons. The outcomes from brain imaging, as well as from single cell examinations lead to the identification of new conceptions for neural networks.

A related cortical approach is taken by Érdi and Kiss in this volume. Érdi and Kiss (2001) [24] develop a model of the interaction between cortical regions. Using sixty-five cortical regions, connection strengths and delay levels a connection matrix was devised of a dynamically outlined model of the cortex. Finally, Reilly (2001) [59] identified both feedforward and feedback routes linking the modules performing a particular cognitive function.

The concept of regional modularity in the brain has been used to develop various computing systems. For example, Bryson and Stein (2001) [12] point out in this volume that robotics used modularity for some time and has produced means of developing and coordinating modular systems. These authors also show that these means can be used to make functioning models of brain-inspired modular decomposition. Deco (2001) [18] in this volume also devises a regional modular approach to visual attention for object recognition and visual search. The system is based on three modules that match the two principal visual pathways of the visual cortex and performs in two modes: the learning and recognition modes.

A biological inspired computer model of contour extraction processes was devised by Hansen et al. (2001) [34] and Hansen and Neumann (2001) [33] that is based on a modular approach. This approach involves long-range links, feedback and feedforward processing, lateral competitive interaction and horizontal long-range integration, and localised receptive fields for oriented contract processing. The model depends on a simplified representation of visual cortex regions V1 and V2, the interaction between these regions and two layers of V1. Because there is a large number of cortical regions, a description of their mutual connectivity is complex. Weber and Obermayer (2001) [72] have devised computational models for learning the relationships between simplified cortical areas. Based on a paradigm of maximum likelihood reconstruction of artificial data, the architecture adapts to the data to represent it best.

3.2 Columnar Modularity of Cerebral Cortex

Turning to a more detailed interpretation of the brain's modular construction, Érdi and Kiss (2001) [24], Guigon et al. (1994) [31] and Fulvi Mari (2000) [28] built on the fact that the cerebral cortex is composed completely of blocks of repetitive modules known as cortical columns with basically the same six layer structure [24,28]. Variations in cognitive functionality are achieved as the columnar organisations have diverse start and end connections, and the cortical neurons have regional specific integrative and registering features [31]. According to Reilly (2001) [59] the columns can be 0.03mm in diameter and include around 100 neurons. These columns are used to provide reliable distributed representations of cognitive functions by creating a spatio-temporal pattern of activation and at any particular time millions are active. Their development was the result of the evolutionary need for better functionality and the bandwidth of the sensory system. There are two extreme views on the form that representation takes in the cerebral cortex. The older view sees representation as context independent and compositionality of the kind linked with formal linguistics and logical depiction. The new view holds that the brain is a dynamic system and that predicate calculus is relevant for describing brain functionality.

A model of the cortex and its columns designed by Doya points to a layered structure and a very recurrent processing approach [22]. The system provides both inhibitory and excitatory synaptic connections among three types of neurons (pyramidal neurons, spiny stellate neurons and inhibitory neurons). The pyramidal and spiny stellate neurons are responsible for passing an excitatory signal to various other cells in the column including cells of the same kind. Inhibitory neurons restrict the spiking of the pyramidal and spiny stellate neurons that are near by, while the pyramidal and spiny stellate neurons use the inhibitory neurons to control themselves and other cells in the column. In this volume, a related columnar model is devised by Bartsch et al. (2001) [7] when considering the visual cortex. A prominent character of the neurons in the primary visual cortex is the preference input in their classical receptive field. The model combines various structured orientation columns to produce a full hypercolumn. Orientation columns are mutually coupled by lateral links with Gaussian profiles and are driven by weakly orientation-biased inputs.

There has been some research to create these multi-cellar models using cells made from many linked compartments and so a higher degree of biological plausibility. However, there is the difficulty of high processing time due to the ionic channel processing elements within the compartments [39]. To a certain extent this can be overcome by using Lie series solutions and Lie algebra to create a restricted Hodgkin-Huxley type model [30].

In general, the brain consists of a distributed and recurrent interaction of billions of neurons. However, a lot of insight and inspiration for computational architectures can be gained from areas, regions or column organisation.

4 Robustness

A second important feature of the brain is its robustness. Robustness in the human brain can be achieved through recovery of certain functions following a defect. The brain has to compensate for the loss of neurons or even neuron areas and whole functional networks on a constant basis. The degree of recovery and hence of robustness is dependent on various factors such as the level of the injury, the location and size of the lesion, and the age of the patient. Recovery is felt to be best when the patient is younger and still in the maturation period, but the approaches for recovery are complicated and variable [50,43,8].

Two approaches to recovery are: i) the repair of the damaged neural networks and the reactivation of those networks which although not damaged due to their close proximity to the injury stopped functioning; and ii) redistribution of functionality to new regions of the brain [14]. There is mixed evidence about the time it normally takes for repair of injured tissue. However, researchers have found that the redistribution of functionality to new regions of the brain can take longer and repair of the left superior temporal gyrus occurs over numerous months following the injury [50]. Restoration of the cortex regions is critical to good recovery of the functionality of the region and is known to inhibit the degree of reallocation of functionality to new regions [71,73]. According to Reggia *et al.* (2001) [58] in this volume the reorganisation of the brain regions responsible for a cognitive function explains the remarkable capacity to recover from injury and robust, fault-tolerant processing.

4.1 Computerised Models of Recovery through Regeneration

It is possible to model recovery through tissue regeneration by considering the neural network's performance at various degrees of recovery. For instance, Martin *et al.* (1996) [46] examined recovery through the regeneration of tissue in a deep dysphasia by considering the attainment of a subject on naming and repetition tests. The model used to examine robustness is associated with the interaction activation and competition neural network and recovery comes from the decay rate returning to more normal levels. Wright and Ahmad (1997) [81] have also developed a modular neural network model that can be trained to perform the naming function and then damaged to varying degrees to examining recovery. A model that incorporates a method to achieve robustness through recovery that is closer to the technique employed in the brain is that of Rust *et al.* (2001) [61] in this volume, which considers the creation of neural systems that are dynamic and adaptive. This computational model produces recovery by allowing adaptability and so achieving self-repair of axons and dendrites to produce new links.

4.2 Computerised Model of Robustness through Functional Reallocation

A second form of robustness is reallocation. When considering the recovery of functionality through reallocation, Reggia *et al.* (2001) [58] in this volume devise biologically plausible models of the regions of the cerebral cortex responsible for the two functions of phoneme sequence creation and letter identification. The former model is based on a recurrent unsupervised learning and the latter on both unsupervised and supervised learning. When the sections of the models that represent one hemisphere of the cerebral cortex were left undamaged they contributed to the recovery of functionality, particularly when the level of injury to the other hemisphere was significant. In general, such graded forms of dynamic robustness go beyond current computing systems.

5 Timing and Synchronisation

Although the neurophysiological activity of the brain seems complicated, diverse and random experimental data indicates the importance of temporal associations in the activities of neurons, neural populations and brain regions [11]. Hence, timing and synchronisation are features of the brain that are considered critical in achieving high levels of performance [17]. According to Denham (2001) [19] in this volume, the alterations of synaptic efficacy coming from pairing of pre- and postsynaptic activity can significantly alter the synaptic links. The induction of long-term alterations in synaptic efficacy through such pairing relies significantly on the relative timing of the onset of excitatory post-synaptic potential (EPSP) produced by the pre-synaptic action potential.

There is disagreement over the importance of the information encoding role played by the interaction between the individual neurons in the form of synchronisation. Schultz *et al.* (2001) [62] consider synchronisation as only secondary to firing rates. However, other research has questioned this based on the temporal organisation of spiking trains [11].

Another critical feature of timing in the brain is how it performs real-time and fast processing despite relatively slow processing elements. For instance Bugmann (2001) [13] points to the role of the cerebellum in off-line planning to achieve real-time processing. According to Panzeri *et al.* (2001) a commonly held view is that fast processing speed in the cerebral cortex comes from an entirely feedforward-oriented approach. However, Panzeri *et al.* (2001) [55] were able to contradict this view by producing a model made up of three layers of excitatory and inhibitory integrate-and-fire neurons that included within-layer recurrent processing.

Given the importance of timing and synchronisation in the brain, computational modelling is used in several architectures to achieve various cognitive functions including vision and language. For instance, Sterratt (2001) [67] examined how the brain synchronises and schedules its processing by considering the locust olfactory system. The desert locust olfactory system's neural activity has interesting spatiotemporal and synchronisation coding features. In the olfactory system the receptor cells connect to both the projection neurons and inhibitory local neurons in the Antennal Lobe, as well as the projection neurons and inhibitory local neuron groups being interconnected. The projection neurons appear to depict the odour via a spatiotemporal code in around one second, which is made up of three principal elements: the slow spatiotemporal activity, fast global oscillations and transient synchronisation. Synchronisation in this system is used to refine the spatiotemporal depiction of the odours.

A biologically inspired computerised model of attention that considers the role played by sychronisation was formulated by Borisyuk *et al.* (2001) [11] with a central oscillator linked to peripheral oscillators via feedforward and feedback links. In this approach the septo-hippocampal area acts like the central oscillator and the peripheral oscillators are the cortical columns that are sensitive to particular characteristics. Attention is produced in the network via synchronisation of the central oscillator with certain peripheral oscillators.

Henderson (2001) [37] devised a biologically inspired computing model of synchronisation to segment patterns according to entities using simple synchrony networks. Simple synchronisation networks are an enlargement of simple recurrent networks by using pulsing units. During each period pulsing units have diverse activation levels for the phrases in the period.

A related biologically inspired model addresses the effects of axonal and dendritic conduction time delays on temporal coding in neural populations, Halliday (2001) [32]. The model uses two cells with common and independent synaptic input based on morphologically detailed models of the dendritic tree typical of spinal a motoneurones. Temporal coding in the inputs is carried by weakly correlated components present in the common input spike trains. Temporal coding in the outputs is manifest as a tendency for synchronized discharge between the two output spike trains. Dendritic and axonal conduction delays of several ms do not alter the sensitivity of the cells to the temporal coding present in the input spike trains.

There is growing support for chaotic dynamics in biological neural activity and that individual neurons create chaotic firing in certain conditions [52]. In a new approach to brain chaos András (2001) [1] states in this volume that the stimuli to the brain are represented as chaotic neural objects. Chaotic neural objects provide stability characteristics as well as superior information representation. Such neural objects are dynamic activity patterns that can be described by mathematical chaos.

Assadollahi and Pulvermüller (2001) [2] were able to identify the importance of a spatio-temporal depiction of information in the brain. This was performed by looking at the representations of single words by using a Kohonen network to classify the words. Sixteen words from four lexico-semantic classes were used and brain responses that represent the diverse characteristics of the words such as their length, frequency and meaning measured using MEG.

6   Learning and Memory Storage

An additional structural characteristic of the brain and central nervous system is the manner it learns and stores memories. Denham (2001) [19] argues that the character of neural connections and the approach to learning and memory storage in the brain currently does not have a major impact on computational neural architectures despite the significant benefits that are available. A school of though known as neo-constructivism lead by Elman (1999) [23] argue that learning and its underlying brain structure does not come from a particular organisation that is available at birth, but from modifications that results from the many experiences that are faced over time. Although this model does have a certain appeal, Marcus (2001) [45] points to various limitations with it, learning mechanism have a certain degree of innateness as infants a few months old often have the ability to learn 'abstract rules', developmental flexibility does not necessarily entail learning and it relies too greatly on learning and neural activity. Marcus (2001) [45] holds that neo-constructivists lack a toolkit of developmental biology and has put forward his own approach to developing neural networks that grow and offer self-organising without experience. This toolkit includes cell division, migration and death, gene expression, cell-to-cell interaction and gene hierarchies.

For many years computational scientists have attempted to incorporate learning and memory storage into artificial intelligent computer systems typically as artificial neural networks. However, in most systems the computational elements are still a gross simplification of biological neurons. There is too little biological plausibility or indication of how the brain constrains can incorporated in a better way [66,12,25]. Nevertheless, Hanson *et al.* (2001) [35] in this volume outlines that artificial neural network such as recurrent ones can perform emergent behaviour close to human cognitive performance. These networks are able to produces an abstract structure that is situation sensitive, hierarchical and extensible. When performing the activity of learning a grammar from valid set of examples the recurrent network is able to recode the input to defer symbol binding until it has received sufficient string sequences.

6.1   Synaptic Alteration to Achieve Learning and Memory Storage

Two regions of the brain that are fundamental in learning and memory storage are the cortex and the hippocampus. However, these are not the only areas involved as shown below by Pérez-Uribe (2001) [57] who describes a basal ganglion model and its role in trial-and-error learning. The hippocampus system is a cortical subsystem found in the temporal lobe and has a fundamental role in short-term memory storage and transferring of short-term memories to longer-term ones. The cortex is the final location of such memories [48].

One of the first accounts of how learning occurs is that of Hebb (1949) [36] who devised a model of how the brain stores memories through a simple synaptic approach based on cell assemblies for cortical processing. Alterations in synaptic strengths is the approach for learning, the persistence of memories and repeated co-activation is used for memory retrieval. The determinant of an assembly is the connectivity structure between neurons that lends support to one another's firing and hence have a greater probability of being co-activated in a reliable fashion. Cell assemblies are found in working and long-term memory storage and interact with other cell assemblies. There has been a substantial amount of work on learning and memory [66,54,79,74,69,49,65].

Long-term potentiation (LTP) is a growth in synaptic strength that is caused rapidly by short periods of synaptic stimulation and is close to the Hebb's notion of activity-reliant alterable synapses. Given that there is an approach like LTP for strengthening links between synapses, it is likely that there is a device for reducing the synaptic strength which is known as long-term depression (LTD). Shastri (2001) [64] in this volume devises a computational abstraction of LTP and LTD which is a greatly simplified representation of the processes involved in the creation of LTP and LTD. A cell is represented as an idealised integrate-and-fire neuron with spatio-temporal integration of activity arriving at a cell. Certain cell-kinds have two firing modes: supra-active and normal. Neurally, the supra-active model relates to a high-frequency burst reaction and the normal mode relates to a basic spiking reaction made up of isolated spikes. LTP and LTD are identified by Shastri (2001) [64] as critical in episodic memory through their role in binding-detection. In Shastri's model a structure for the fast production of cell responses to binding matches is made up of three areas: role, entity and bind. Areas role and entity are felt to have 750,000 primary cells each, and bind 15 million cells. The role and entity areas match the subareas of the entorhinal cortex, and the bind area the dentrate gyrus.

Huyck (2001) [40] devised a biologically inspired model of cell assemblies known as the CANT system. The CANT system is made up of a network of neurons that may contain many cell assemblies that are unidirectionally linked to other neurons. As with many neural network models connection strengths are altered by the local Hebbian rule and learning through a Hebbian-based unsupervised approach.

6.2   Models of Learning

There have been some recent models of learning in artificial systems which are particularly interesting since they are based on neuroscience learning methods. For instance, McClelland and Goddard (1996) [48] examined the role of the hippocampal system in learning by devising a biologically inspired model. Forward pathways from the association regions of the neocortex to the entorhinal cortex create a pattern of activation on the entorhinal cortex that maximises preservation of knowledge about the neocortical pattern. The entorhinal cortex gives inputs to the hippocampal memory system, which is recoded in the dentate gyrus and CA3 in a manner that is suitable for storage. The hippocampus computerised model is split into three main subsystems: i) structure-preserving invertible encoder subsystem; ii) memory separation, storage and retrieval subsystem; and iii) memory decoding system.

The learning process is outlined by Denham (2001) [19] in this volume as a simple biologically inspired computational model. The model requires the determination of the EPSP at the synapse and the back-propagating action potential. A learning rule is then produced that relies on the integration of the product of these two potentials. The EPSP at the synapse is determine by the effective synapse current using the equation for the passive membrane mechanism.

Two biologically inspired computerised systems of learning are included in robots, which shows that these systems can improve on existing technology. Kazer and Sharkey (2001) [41] developed a model of how the hippocampus combines memory and anxiety to produce novelty detection in a robot. The robot offers knowledge for learning and an approach for making any alterations in anxiety behaviourally explicit. A learning robot was devised by Pérez-Uribe (2001) [57] that uses a biologically inspired approach based on the basal ganglion to learn by trial-and-error.

Bogacz et al. (2001) [10] devised a biologically plausible algorithm of familiarity discrimination based on energy. This is based on the information processing of the perirhinal cortex of the hippocampus system. This approach does not need assumptions related to the distribution of patterns and discriminates if a certain pattern was presented before and keeps knowledge on the familiar patterns in the weights of Hopfield Networks.

A related biologically inspired computerised system was devised by Chady (2001) [16] for compositionality and context-sensitive learning founded on a group of Hopfield Networks. The inspiration comes from the cortical column by using a two-dimensional grid of networks and basing interaction on the nearest neighbour approach. In the model the individual network states are discrete and their transitions synchronous. The state alteration of the grid is carried out in an asynchronous fashion.

When considering a biological inspired computerised systems for natural language understanding Moisl (2001) [51] proposes sequential processing using Freeman's work on brain intentionality and meaning. Moisl (2001) [51] proposed approach will include: i) Processing components that output pulse trains as a nonlinear reaction to input; ii) Modules of excitatory and inhibitory neurons that create oscillatory actions; iii) Feedforward and feedback links between modules to foster chaotic behaviour; and iv) A local learning mechanism such as Hebbian learning to achieve self-organising in modules.

Pearce et al. (2001) [56] argues that the olfactory system offers an ideal model for examining the issues of robust sensory signal transmission and efficient information representation in a neural system. A critical feature of mammalian olfactory system is the large scale convergence of spiking receptor stimulus from thousands of olfactory receptors, which seems fundamental for information representation and greater sensitivity. Typically the information representation approaches used in the olfactory cortex are action and grading potentials, rate codes and particular temporal codings. The study considered whether the rate-coded depiction of the input restricts the quality of the signal that can be recovered in the glomerulus of the olfactory bulb. This was done by looking at the outcomes from two models, one that uses probabilistic spike trains and another which uses graded receptor inputs.

Auditory perception has various characteristics with the brain having the capability to detect growths in loudness as well as differentiating between two clicks that are very close together. Based on the findings of Denham and Denham (2001) [20] this is the result of the manner of information representation in the primary auditory cortex through cortical synaptic dynamics. When synapses are repeatedly activated they do not react in the same manner to every incoming impulse and synapses might produce a short-term depression or facilitation. When there is a great deal of activity in the synapse, the amount of resources that are available is reduced, which is likely to be followed by a period of recovery for the synapse. A leaky integrate-and-fire neuron model is then used, with the input to the neuron model gained through summing the synaptic EPSPs. In the examination of the model the reaction features of the neuron that incorporated a dynamic synapse are close to those of the primary auditory cortex.

Caragea et al. (2001) [15] have proposed a set of biologically inspired approaches for knowledge discovery operations. The databases in this domain are normally large, distributed and constantly growing in size. There is a need for computerised approaches to achieve learning from distributed data that do not reprocess already processed data. The techniques devised by Caragea et al. (2001) [15] for distributed or incremental algorithms attempt to determine information needs of the learner and devising effective approaches to providing this in an distributed or incremental setting. By splitting the learning activity into information extraction and hypothesis production stages this allows the enhancement of current learning approaches to perform in a distributed context. The hypothesis production element is the control part that causes the information extraction component to occur. The long-term aim of the research is to develop well-founded multi-agent systems that are able to learn through interaction with open-ended dynamic systems from knowledge discovery activities.

6.3   Models of Memory Storage

In this section we provide an outline of the various biologically inspired computerised models connected with memory storage in the brain. Knoblauch and Palm (2001) [42] took a similar approach to autoassociative networks as by Willshaw [79,75,78,76,80,77] and extend it based on biological neurons and synapses. In particular Knoblauch and Palm added characteristics that represent the spiking actions of real neurons in addition to the characteristics of spatio-temporal integration on dendrites. Individual cells are modelled like 'spiking neurons': each time the potential level is at a particular threshold a pulse-like action potential is created. The Knoblauch and Palm (2001) [42] model of associative memory is included in a model of reciprocally connected visual areas comprising three areas (R, P and C) each made up of various neuron populations. In region R (retina) input patterns matching input objects in the visual field and are depicted in a 100 x 100 bitmap. Area P (primary visual cortex) is made up of 100 x 100 excitatory spike neurons and 100 x 100 inhibitory gradual neurons. Area C (central visual area) is modelled as the SSI-variant of the spiking associative memory.

A biologically inspired model of episodic memory by Shastri (2000) [63] known as SMRITI outlines how a transient pattern of rhythmic activity depicting an event can be altered swiftly into a persistent and robust memory trace. Creation of such a memory trace matches the recruitment of a complicated circuit in the hippocampal system that includes the required elements. In order to analysis characteristics of the model its performance is examined using plausible values of system variables. The outcomes display that robust memory traces can be produced if one assumes there is an episodic memory capacity of 75,000 events made up of 300,000 bindings.

Forcada and Carrasco (2001) [25] argue that, although a finite-state machine could be modelled by any discrete-time recurrent neural network (DTRNN) with discrete-time processing elements, biological networks perform in continuous time and so methods for synchronisation and memory should be postulated. It ought to be possible to produce a more natural and biologically plausible approach to finite-state computation founded on continuous-time recurrent neural networks (CTRNN). CTRNN have inputs and outputs that are functions of a continuous-time variable and neurons that have a temporal reaction. It is possible to encode in an indirect manner finite-state machines using a sigmoid DTRNN as a CTRNN and then changing the CTRNN into an integrate-and-fire network.

A considerable amount of research has been carried out by Austin and his associates at York into memory storage [6,3,5,53,4,70,82]. An example of this work in this volume is the Correlation Matrix Memories (CMMs) which according to Kustrin and Austin (2001) [44] are simple binary weighted feedforward neural networks that are used for various tasks that offers an indication of how memories are stored in the human cerebral cortex. CMMs are close to single layer, binary weighted neural networks, but use much less complex learning and recall algorithms.

7   Conclusion

There is no doubt that current computer systems are not able to perform many of the cognitive functions such as vision, motion, language processing to the level associated with the brain. It seems there is a strong need to use new architectural and information processing characteristics to improve computerised systems. Although interest in biologically inspired computerised system has grown significantly recently, the approaches currently available are simplistic and understanding of the brain at the system level is still limited. Key research issues for biologically inspired computer systems relate to the fundamental architectural features and processing-related associated with the brain, and what computer science can learn from biological and neural computing.

The characteristics of the brain that potentially could benefit computerised systems include modular organisation, robustness, timing and synchronisation and learning and memory storage. Modularity in the brain takes various forms of abstraction including regional, columnar and cellular, and is central to many biologically inspired computerised systems. Robustness comes from the brain's ability to recover functionality despite injury through tissue repair and re-allocation of functionality to other brain regions. While conventional computer systems are presently based on synchronised or clocked processing these systems could potentially be enriched by basing information processing and encoding on the timing and synchronisation approaches of the brain. Furthermore, as seen in this chapter and volume a particular fertile research area is the development of biological inspired computing models of learning and memory storage to perform various cognitive functions [42,20,63,41,25,20].

To move the field of biological inspired computing systems forward, consideration should be given to how the architectural features of the brain such as modular organisation, robustness, timing and sychronisation, and learning and memory could benefit the performance of computering systems. The most suitable level of neuroscience-inspired abstraction for producing such systems should be identified. Although the findings offered by neuroscience research should be taken serious, there is a need to understand the constraints offered by computer hardware and software. Greater concentration should be given to dynamic network architectures that can alter their structure based on experience by either using elaborate circuitry or constant network modification. Finally, a more comprehensive understanding of the brain and the central nervous system is critical to achieve better biologically inspired adaptive computing systems.

References

1. P. András. The role of brain chaos. In S. Wermter, J. Austin, and D. Willshaw, editors, *Emergent Neural Computational Architectures based on Neuroscience (this volume)*. Springer-Verlag, Heidelberg, Germany, 2001.
2. R. Assadollahi and F. Pulvermüller. Neural network classification of word evoked neuromagnetic brain activity. In S. Wermter, J. Austin, and D. Willshaw, editors, *Emergent Neural Computational Architectures based on Neuroscience (this volume)*. Springer-Verlag, Heidelberg, Germany, 2001.
3. J. Austin. ADAM: A distributed associative memory for scene analysis. In *Proceedings of First International Conference on Neural Networks*, page 287, San Diego, 1987.
4. J. Austin. Matching performance of binary correlation matrix memories. In R. Sun and F. Alexandre, editors, *Connectionist-Symbolic Integration: From Unified to Hybrid Approaches*. Lawrence Erlbaum Associates Inc, New Jersey, 1997.
5. J. Austin and S. O'Keefe. Application of an associative memory to the analysis of document fax images. *The British Machine Vision Conference*, pages 315–325, 1994.
6. J. Austin and T. Stonham. An associative memory for use in image recognition and occlusion analysis. *Image and Vision Computing*, 5(4):251–261, 1987.
7. H. Bartsch, M. Stetter, and K. Obermayer. On the influence of threshold variability in a mean-field model of the visual cortex. In S. Wermter, J. Austin, and D. Willshaw, editors, *Emergent Neural Computational Architectures based on Neuroscience (this volume)*. Springer-Verlag, Heidelberg, Germany, 2001.

8. A. Basso, M. Gardelli, M. Grassi, and M. Mariotti. The role of the right hemisphere in recovery from aphasia: Two case studies. *Cortex*, 25:555–566, 1989.

9. J. Binder. Functional magnetic resonance imaging of language cortex. *International Journal of Imaging Systems and Technology*, 6:280–288, 1995.

10. R. Bogacz, M. Brown, and C. Giraud-Carrier. A familiarity discrimination algorithm inspired by computations of the perirhinal cortex. In S. Wermter, J. Austin, and D. Willshaw, editors, *Emergent Neural Computational Architectures based on Neuroscience (this volume)*. Springer-Verlag, Heidelberg, Germany, 2001.

11. R. Borisyuk, G. Borisyuk, and Y. Kazanovich. Temporal structure of neural activity and modelling of information processing in the brain. In S. Wermter, J. Austin, and D. Willshaw, editors, *Emergent Neural Computational Architectures based on Neuroscience (this volume)*. Springer-Verlag, Heidelberg, Germany, 2001.

12. J. Bryson and L. Stein. Modularity and specialized learning: Mapping between agent architectures and brain organization. In S. Wermter, J. Austin, and D. Willshaw, editors, *Emergent Neural Computational Architectures based on Neuroscience (this volume)*. Springer-Verlag, Heidelberg, Germany, 2001.

13. G. Bugmann. Role of the cerebellum in time-critical goal-oriented behaviour: Anatomical basis and control principle. In S. Wermter, J. Austin, and D. Willshaw, editors, *Emergent Neural Computational Architectures based on Neuroscience (this volume)*. Springer-Verlag, Heidelberg, Germany, 2001.

14. S. Capp, D. Perani, F. Grassi, S. Bressi, M. Alberoni, M. Franceschi, V. Bettinardi, S. Todde, and F. Frazio. A PET follow-up study of recovery after stroke in acute aphasics. *Brain and Language*, 56:55–67, 1997.

15. D. Caragea, A. Silvescu, and V. Honavar. Analysis and synthesis of agents that learn from distributed dynamic data sources. In S. Wermter, J. Austin, and D. Willshaw, editors, *Emergent Neural Computational Architectures based on Neuroscience (this volume)*. Springer-Verlag, Heidelberg, Germany, 2001.

16. M. Chady. Modelling higher cognitive functions with Hebbian cell assemblies. In S. Wermter, J. Austin, and D. Willshaw, editors, *Emergent Neural Computational Architectures based on Neuroscience (this volume)*. Springer-Verlag, Heidelberg, Germany, 2001.

17. D. Chawla, E. Lumer, and K. Friston. The relationship between synchronization among neuronal populations and their mean activity levels. *Neural Computation*, 11:319–328, 1999.

18. G. Deco. Biased competition mechanisms for visual attention in a multimodular neurodynamical system. In S. Wermter, J. Austin, and D. Willshaw, editors, *Emergent Neural Computational Architectures based on Neuroscience (this volume)*. Springer-Verlag, Heidelberg, Germany, 2001.

19. M. Denham. The dynamics of learning and memory: Lessons from neuroscience. In S. Wermter, J. Austin, and D. Willshaw, editors, *Emergent Neural Computational Architectures based on Neuroscience (this volume)*. Springer-Verlag, Heidelberg, Germany, 2001.

20. S. Denham and M. Denham. An investigation into the role of cortical synaptic depression in auditory processing. In S. Wermter, J. Austin, and D. Willshaw, editors, *Emergent Neural Computational Architectures based on Neuroscience (this volume)*. Springer-Verlag, Heidelberg, Germany, 2001.

21. S. Dodel, J.M. Herrmann, and T. Geisel. Stimulus-independent data analysis for fMRI. In S. Wermter, J. Austin, and D. Willshaw, editors, *Emergent Neural Computational Architectures based on Neuroscience (this volume)*. Springer-Verlag, Heidelberg, Germany, 2001.

22. K. Doya. What are the computations of the cerebellum, the basal ganglia and the cerebral cortex? *Neural Networks*, 12(7-8):961–974, 1999.

23. J. Elman. Origins of language: A conspiracy theory. In B. MacWhinney, editor, *The Emergence of Language*, pages 1–27. Lawrence Earlbaum Associates, Hillsdale, NJ, 1999.

24. P. Érdi and T. Kiss. The complexity of the brain: Structural, functional and dynamic modules. In S. Wermter, J. Austin, and D. Willshaw, editors, *Emergent Neural Computational Architectures based on Neuroscience (this volume)*. Springer-Verlag, Heidelberg, Germany, 2001.

25. M. Forcada and R. Carrasco. Finite-state computation in analog neural networks: Steps towards biologically plausible models? In S. Wermter, J. Austin, and D. Willshaw, editors, *Emergent Neural Computational Architectures based on Neuroscience (this volume)*. Springer-Verlag, Heidelberg, Germany, 2001.

26. A. Friederici. The developmental cognitive neuroscience of language: A new research domain. *Brain and Language*, 71:65–68, 2000.

27. L. Friedman, J. Kenny, A. Wise, D. Wu, T. Stuve, D. Miller, J. Jesberger, and J. Lewin. Brain activation during silent word generation evaluated with functional MRI. *Brain and Language*, 64:943–959, 1998.

28. C. Fulvi Mari. Modular auto-associators: Achieving proper memory retrieval. In S. Wermter, J. Austin, and D. Willshaw, editors, *EmerNet: Third International Workshop on Current Computational Architectures Integrating Neural Networks and Neuroscience*, pages 15–18. EmerNet, 2000.

29. M. Gazzaniga, R. Ivry, and G. Mangun. *Cognitive Neuroscience: The Biology of the Mind*. W.W. Norton & Company Ltd, New York, 1998.

30. G. Green, W. Woods, and S. Manchanda. Representations of neuronal models using minimal and bilinear realisations. In S. Wermter, J. Austin, and D. Willshaw, editors, *Emergent Neural Computational Architectures based on Neuroscience (this volume)*. Springer-Verlag, Heidelberg, Germany, 2001.

31. E. Guigon, P. Gramdguillaume, I. Otto, L. Boutkhil, and Y. Burnod. Neural network models of cortical functions based on the computational properties of the cerebral cortex. *Journal of Physiology (Paris)*, 88:291–308, 1994.

32. D. Halliday. Temporal coding in neuronal populations in the presence of axonal and dendritic conduction time delays. In S. Wermter, J. Austin, and D. Willshaw, editors, *Emergent Neural Computational Architectures based on Neuroscience (this volume)*. Springer-Verlag, Heidelberg, Germany, 2001.

33. T. Hansen and H. Neumann. Neural mechanisms for representing surface and contour features. In S. Wermter, J. Austin, and D. Willshaw, editors, *Emergent Neural Computational Architectures based on Neuroscience (this volume)*. Springer-Verlag, Heidelberg, Germany, 2001.

34. T. Hansen, W. Sepp, and H. Neumann. Recurrent long-range interactions in early vision. In S. Wermter, J. Austin, and D. Willshaw, editors, *Emergent Neural Computational Architectures based on Neuroscience (this volume)*. Springer-Verlag, Heidelberg, Germany, 2001.

35. S. Hanson, M. Negishi, and C. Hanson. Connectionist neuroimaging. In S. Wermter, J. Austin, and D. Willshaw, editors, *Emergent Neural Computational Architectures based on Neuroscience (this volume)*. Springer-Verlag, Heidelberg, Germany, 2001.

36. D. Hebb. *The Organization of Behaviour*. Wiley, New York, 1949.
37. J. Henderson. Segmenting state into entities and its implication for learning. In S. Wermter, J. Austin, and D. Willshaw, editors, *Emergent Neural Computational Architectures based on Neuroscience (this volume)*. Springer-Verlag, Heidelberg, Germany, 2001.
38. J. Hicks and P. Monaghan. Explorations of the interaction between split processing and stimulus types. In S. Wermter, J. Austin, and D. Willshaw, editors, *Emergent Neural Computational Architectures based on Neuroscience (this volume)*. Springer-Verlag, Heidelberg, Germany, 2001.
39. A. Hodgkin and A. Huxley. Current carried by sodium and potassium ions through the membrane of the giant axon of *Loligo*. *J. Phyiol.*, 116:449–472, 1952.
40. C. Huyuk. Cell assemblies as an intermediate level model of cognition. In S. Wermter, J. Austin, and D. Willshaw, editors, *Emergent Neural Computational Architectures based on Neuroscience (this volume)*. Springer-Verlag, Heidelberg, Germany, 2001.
41. J. Kazer and A. Sharkey. The role of memory, anxiety and Hebbian learning in hippocampal function: Novel explorations in computational neuroscience and robotics. In S. Wermter, J. Austin, and D. Willshaw, editors, *Emergent Neural Computational Architectures based on Neuroscience (this volume)*. Springer-Verlag, Heidelberg, Germany, 2001.
42. A. Knoblauch and G. Palm. Spiking associative memory and scene segmentation by synchronization of cortical activity. In S. Wermter, J. Austin, and D. Willshaw, editors, *Emergent Neural Computational Architectures based on Neuroscience (this volume)*. Springer-Verlag, Heidelberg, Germany, 2001.
43. H. Krabe, A. Thiel, G. Weber-Luxenburger, K. Herholz, and W. Heiss. Brain plasticity in poststroke aphasia: What is the contribution of the right hemisphere? *Brain and Language*, 64(2):215–230, 1998.
44. D. Kustrin and J. Austin. Connectionist propositional logic a simple correlation matrix memory based reasoning system. In S. Wermter, J. Austin, and D. Willshaw, editors, *Emergent Neural Computational Architectures based on Neuroscience (this volume)*. Springer-Verlag, Heidelberg, Germany, 2001.
45. G. Marcus. Plasticity and nativism: Towards a resolution of an apparent paradox. In S. Wermter, J. Austin, and D. Willshaw, editors, *Emergent Neural Computational Architectures based on Neuroscience (this volume)*. Springer-Verlag, Heidelberg, Germany, 2001.
46. G. Martin, N., E. Saffran, and N. Dell. Recovery in deep dysphasia: Evidence for a relation between auditory - verbal STM capacity and lexical error in repetition. *Brain and Language*, 52:83–113, 1996.
47. G. Matsumoto, E. Körner, and M. Kawato. Organisation of computation in brain-like systems. *Neural Networks*, 26-27:v–vi, 1999.
48. J. McClelland and N. Goddard. Considerations arising from a complementary learning systems perspective on hippocampus and neocortex. *Hippocampus*, 6(6):655–665, 1996.
49. J. McClelland, B. McNaughton, and R. O'Reilly. Why there are complementary learning systems in the hippocampus and neocortex: Insights from the successes and failures of connectionist models of learning and memory. *Pyschological Review*, 102(3):419–457, 1995.
50. M. Mimura, M. Kato, M. Kato, Y. Santo, T. Kojima, M. Naeser, and T. Kashima. Prospective and retrospective studies of recovery in aphasia: Changes in cerebral blood flow and language functions. *Brain*, 121:2083–2094, 1998.

51. H. Moisl. Linguistic computation with state space trajectories. In S. Wermter, J. Austin, and D. Willshaw, editors, *Emergent Neural Computational Architectures based on Neuroscience (this volume)*. Springer-Verlag, Heidelberg, Germany, 2001.

52. G. Mpitsos, R. Burton, H. Creech, and S. Soinilla. Evidence for chaos in spiking trains of neurons that generate rhythmic motor patterns. *Brain Research Bulletin*, 21:529–538, 1988.

53. S. O'Keefe and J. Austin. An application of the ADAM associative memory to the analysis of document images. *The British Machine Vision Conference*, pages 315–325, 1995.

54. G. Palm, F. Schwenker, F.T. Sommer, and A. Strey. Neural associative memory. In A. Krikelis and C. Weems, editors, *Associative Processing and Processors*. IEEE Computer Society, Los Alamitos, CA, 1997.

55. S. Panzeri, E. Rolls, F. Battaglia, and R. Lavis. Simulation studies of the speed of recurrent processing. In S. Wermter, J. Austin, and D. Willshaw, editors, *Emergent Neural Computational Architectures based on Neuroscience (this volume)*. Springer-Verlag, Heidelberg, Germany, 2001.

56. T. Pearce, P. Verschure, J. White, and J. Kauer. Robust stimulus encoding in olfactory processing: Hyperacuity and efficient signal transmission. In S. Wermter, J. Austin, and D. Willshaw, editors, *Emergent Neural Computational Architectures based on Neuroscience (this volume)*. Springer-Verlag, Heidelberg, Germany, 2001.

57. A. Pérez-Uribe. Using a time-delay actor-critic neural architecture with dopamine-like reinforcement signal for learning in autonomous robots. In S. Wermter, J. Austin, and D. Willshaw, editors, *Emergent Neural Computational Architectures based on Neuroscience (this volume)*. Springer-Verlag, Heidelberg, Germany, 2001.

58. J. Reggia, Y. Shkuro, and N. Shevtsova. Computational investigation of hemispheric specialization and interactions. In S. Wermter, J. Austin, and D. Willshaw, editors, *Emergent Neural Computational Architectures based on Neuroscience (this volume)*. Springer-Verlag, Heidelberg, Germany, 2001.

59. R. Reilly. Collaborative cell assemblies: Building blocks of cortical computation. In S. Wermter, J. Austin, and D. Willshaw, editors, *Emergent Neural Computational Architectures based on Neuroscience (this volume)*. Springer-Verlag, Heidelberg, Germany, 2001.

60. M. Rugg. Introduction. In M. Rugg, editor, *Cognitive Neuroscience*, pages 1–10. Psychology Press, Hove East Sussex, 1997.

61. A. Rust, R. Adams, S. George, and H. Bolouri. Towards computational neural systems through developmental evolution. In S. Wermter, J. Austin, and D. Willshaw, editors, *Emergent Neural Computational Architectures based on Neuroscience (this volume)*. Springer-Verlag, Heidelberg, Germany, 2001.

62. S. Schultz, H. Golledge, and S. Panzeri. Sychronisation, binding and the role of correlated firing in fast information transmission. In S. Wermter, J. Austin, and D. Willshaw, editors, *Emergent Neural Computational Architectures based on Neuroscience (this volume)*. Springer-Verlag, Heidelberg, Germany, 2001.

63. L. Shastri. SMRITI: A computational model of episodic memory formation inspired by the hippocampus system. In S. Wermter, J. Austin, and D. Willshaw, editors, *EmerNet: Third International Workshop on Current Computational Architectures Integrating Neural Networks and Neuroscience*, pages 7–10. EmerNet, 2000.

64. L. Shastri. Biological grounding of recruitment learning and vicinal algorithms in long-term potentiation. In S. Wermter, J. Austin, and D. Willshaw, editors, *Emergent Neural Computational Architectures based on Neuroscience (this volume)*. Springer-Verlag, Heidelberg, Germany, 2001.

65. S. Song, K. Miller, and L. Abbott. Competitive Hebbian learning through spike-timing dependent synaptic plasticity. *Nature Neuroscience*, 3:919–926, 2000.
66. M. Spitzer. *The Mind Within the Net: Models of Learning, Thinking and Acting*. MIT Press, Cambridge, MA, 1999.
67. D. Sterratt. Locus olfaction synchronous oscillations in excitatory and inhibitory groups of spiking neurons. In S. Wermter, J. Austin, and D. Willshaw, editors, *Emergent Neural Computational Architectures based on Neuroscience (this volume)*. Springer-Verlag, Heidelberg, Germany, 2001.
68. J. Taylor. Images of the mind: Brain images and neural networks. In S. Wermter, J. Austin, and D. Willshaw, editors, *Emergent Neural Computational Architectures based on Neuroscience (this volume)*. Springer-Verlag, Heidelberg, Germany, 2001.
69. A. Treves and E. Rolls. Computational analysis of the role of the hippocampus in memory. *Hippocampus*, 4(3):374–391, 1994.
70. M. Turner and J. Austin. Matching perfromance of binary correlation matrix memories. *Neural Networks*, 1997.
71. E. Warburton, C. Price, K. Swinnburn, and R. Wise. Mechanisms of recovery from aphasia: Evidence from positron emission tomography studies. *Journal Neurol Neurosurg Psychiatry*, 66:151–161, 1999.
72. C. Weber and K. Obermayer. Emergence of modularity within one sheet of neurons: A model comparison. In S. Wermter, J. Austin, and D. Willshaw, editors, *Emergent Neural Computational Architectures based on Neuroscience (this volume)*. Springer-Verlag, Heidelberg, Germany, 2001.
73. C. Weiller, C. Isensee, M. Rijntjes, W. Huber, S. Muller, D. Bier, K. Dutschka, R. Woods, J. Noth, and H. Diener. Recovery from Wernicke's aphasia: A positron emission tomographic study. *Ann Neurol*, 37:723–732, 1995.
74. T. Wennekers and G. Palm. Cell assemblies, associative memory and temporal structure in brain signals. In R. Miller, editor, *Time and the Brain: Conceptual Advances in Brain Research, Vol. 2*. Academic Publisher, 2000.
75. D. Willshaw. Non-symbolic approaches to artificial intelligence and the mind. *Philos. Trans. R. Soc. A*, 349:87–102, 1995.
76. D. Willshaw and J. Buckingham. An assessment of Marr's theory of the hippocampus as a temporary memory store. *Philos. Trans. R. Soc. Lond. [B]*, 329:205–215, 1990.
77. D. Willshaw, O. Buneman, and H. Longuet-Higgins. Nonholographic associative memory. *Nature*, 222:960–962, 1969.
78. D. Willshaw and P. Dayan. Optimal plasticity from matrix memories: What goes up must come down. *Neural Computation*, 2:85–93, 1990.
79. D. Willshaw, J. Hallam, S. Gingell, and S. Lau. Marr's theory of the neocortex as a self-organising neural network. *Neural Computation*, 9:911–936, 1997.
80. D. Willshaw and C. von der Malsburg. How patterned neural connections can be set up by self-organization. *Philos. Trans. R. Soc. Lond. [B]*, 194:431–445, 1976.
81. J. Wright and K. Ahmad. The connectionist simulation of aphasic naming. *Brain and language*, 59:367–389, 1997.
82. P. Zhou, J. Austin, and J. Kennedy. A high performance k-NN classifier using a binary correlation matrix memory. In *Advances in Neural Information Processing Systems, Vol. 11*. MIT, 1999.

Images of the Mind: Brain Images and Neural Networks

John G. Taylor

Department of Mathematics, King's College, Strand, London WC2R2LS, UK
john.g.taylor@kcl.ac.uk

Abstract. An overview is given of recent results coming from non-invasive brain imaging (PET, fMRI, EEG & MEG), and how these relate to, and illuminate, the underpinning neural networks. The main techniques are briefly surveyed and data analysis techniques presently being used reviewed. The results of the experiments are then summarised. The most important recent technique used in analysing PET and fMRI, that of structural modelling, is briefly described, results arising from it presented, and the problems this approach presents in bridging the gap to the underlying neural networks of the brain described. New neural networks approaches are summarised which are arising from these and related results, especially associated with internal models. The relevance of these for indicating future directions for the development of artificial neural networks concludes the article.

Keywords: Brain imaging, structural models, neural networks, working memory, internal models, intelligent systems

1 Introduction

There is increasing information becoming available from functional brain imaging on how the brain solves a range of tasks. The specific brain modules active while human subjects solve various cognitive tasks are now being uncovered [1],[2]. The data show that there are networks of modules active during task solution, with a certain amount of overlap between networks used to solve different problems. This use of non-invasive imaging to give a new 'window' on the brain has aroused enormous interest in the neuroscience community, and more generally begun to shed light on the way the brain solves hard tasks. The nature of the networks used to support information processing is now being explored by various means, especially in terms of the creation of internal models of the environment. This indicates a clear direction for the future development of artificial neural networks, a feature to be considered at the end of the paper.

There are many problems to be faced in interpreting functional brain images (those obtained whilst the subject is performing a particular task). Although the experimental paradigms used to obtain the functional brain imaging data already contain partial descriptions of the functions being performed by the areas detected, the picture is still clouded. The overall nature of a task being performed while subjects are being imaged can involve a number of more primitive sub-tasks which themselves have to be used in the determination of the underlying functions being performed by the separate modules. This means that there can be several interpretations of the roles for

S. Wermter et al. (Eds.): Emergent Neural Computational Architectures, LNAI 2036, pp. 20–38, 2001.

these modules, and only through the convergence of a number of experimental paradigms will it be possible to disentangle the separate primitive functions.

It is the hope that use of the latest techniques of analysis of resulting data, as well as the development of new techniques stretching the machines to their very limits, will allow solution to these problems of ambiguity, and a resulting truly global model of the brain will result. The paper starts with a review of brain imaging techniques and data analysis. A survey of results obtained by analysis of PET and fMRI data, termed structural modelling is then given. This approach has the potential to lead to a global processing model of the brain. As part of that, the connection between structural models and the underlying neural networks of the brain is then explored. Several recent brain imaging results are then considered which indicate new neural network architectures and processing styles. The manner in which the brain uses internal models of the environment, including its own active sensors and effectors as well as of its inputs, is then explored, leading to a general program of analysis of the global processing by the brain. This leads to a possible approach to building more advanced neural network architectures which is discussed briefly in the final section.

2 A Survey of Brain-Imaging Machines

2.1 PET & fMRI

These machines investigate the underlying neural activity in the brain indirectly, the first (the acronym PET= positron emission tomography) by measuring the 2 photons emitted in the positron annihilation process in the radio-active decay of a suitable radio-nuclide such as H_2 ^{15}O injected into a subject at the start of an experiment, the second (fMRI = functional magnetic resonance imaging) that of the uneven distribution of nuclear spins (effectively that of the proton) when a subject is in a strong magnetic field (usually of 1.5 Tessla, although a 12 T machine is being built especially for human brain imaging studies). The PET measurement allows determination of regions of largest blood flow, corresponding to the largest 2-photon count. The fMRI measurement is termed that of BOLD (blood oxygen-level dependent). This signal stems from the observation that during changes in neuronal activity there are local changes in the amount of oxygen in tissue, which alters the amount of oxygen carried by haemoglobin, thereby disturbing the local magnetic polarisability.

Spatially these two types of machines have a few millimetres accuracy across the whole brain. However temporally they are far less effective. PET measurements need to be summed over about 60-80 seconds, limiting the temporal accuracy considerably. fMRI is far more sensitive to time, with differences in the time of activation of various regions being measurable down to a second or so by the 'single event' measurement approach. This has already produced the discovery of dissociations in the time domain between posterior and anterior cortical sites in working memory tasks [3].

That regions of increased blood flow correspond exactly to those of increased neural activity, and these also identify with the source of the BOLD signal, is still the subject of considerable dispute. The sitting of the BOLD signal in the neurally most active region was demonstrated recently [4] by a beautiful study of the positioning of the rat whisker barrel cortex from both 7T fMRI measurement and by direct electrode penetration. The present situation was summarised effectively in [5], with a number of hypotheses discussed as to the sources of the blood flow and BOLD signals. I will assume that these signals are all giving the same information (at the scale of size we are considering).

Many cognitive studies have been performed using PET; there are now as many using fMRI, some duplicating the PET measurements. These results show very clear localisation of function and the involvement of networks of cortical and subcortical sites in normal functioning of the brain during the solution of tasks. At the same time there has been considerable improvement in our understanding of brain activity in various mental diseases, such as schizophrenia, Alzheimer's and Parkinson's diseases. There have also been studies of patients with brain damage, to discover how the perturbed brain can still solve tasks, albeit slowly and inefficiently in many cases.

2.2 MEG & EEG

The magnetic field around the head due to neural activity, although very low, is measurable by sensitive devices, such as SQUIDs (superconducting quantum interference devices). Starting from single coils to measure the magnetic field at a very coarse level, MEG (MEG = magneto-encephalography) measurements are now being made with sophisticated whole-head devices using 148 [6] or even 250 measuring coils [7]. Such systems lead to ever greater spatial sensitivity, although they have a number of problems before they can be fully exploited. In particular it is first necessary to solve the inverse problem, that of uncovering the underlying current sources producing the magnetic field. This is non-trivial, and has caused MEG not to be as far advanced in brain imaging as PET and fMRI systems. However that situation is now changing. There is good reason to bring MEG up to the same standard of data-read-out simplicity as PET and fMRI since, although it does not have the same spatial sensitivity as the other two it has far better temporal sensitivity - down to a millisecond. Thus MEG fills in the temporal gap on the knowledge gained by PET and fMRI. This is also done by EEG (EEG = electro-encephalography), which is being consistently used by numbers of groups in partnership with PET or fMRI so as to determine the detailed time course of activation of known sites already implicated in a task by the other machines [2].

3 Data Analysis

3.1 Statistical Parameter Maps (SPMs)

The data arising from a given experiment consists of a set of data points, composed of a time series of activations collected during an experiment, for a given position (termed a pixel) in the head. The is reasonably immediate for PET and fMRI, although there must be careful preprocessing performed in order to remove movement artefacts and to relate the site being measured to its co-ordinates in the head according to some standard atlas. All the machines now use a static MRI measurement of the brain of a subject as a template, to which the data being measured are referred. Some analyses use a more detailed warping of the brain of a given subject to that of a standard brain, as given especially by the standard brain atlas arrived at by the anatomical analysis of a number of human brains [8]. However this can introduce distortions so a time-consuming but more accurate method is to identify similar regions from the brains of different people by common landmarks, and then compare (or average) the activity at the similar regions decided on in this manner over the group of subjects to remove noise. The use of independent component analysis (ICA) has recently proved of value in data cleaning and especially in the removal of artefacts such as effects of breathing and heart beat.

The data at a given pixel, then, is a set of activation levels. These have been obtained from a number of measurements taken under a set of known conditions. For example, I am involved in studying the motion after-effect (MAE) by fMRI [9]. This occurs due to adaptation to movement in one direction, and arises, for example, if you look at a waterfall for a period of about 20 or so seconds and then turn your gaze to the side. The static rock face then seems to move upwards for about 10 seconds afterwards. Our measurements were taken using an 'on-off' paradigm, with a set of moving bars being observed by a subject during 10 measurements (each lasting 3 seconds), then 10 with static bars, then 10 with the bars moving up and down, then another 10 with static bars. The MAE occurs at the end of a period of movement in one direction only, so the purpose of the study was to determine the change of BOLD signal after the one-way movement, in comparison to the two-way movement. Regions were searched for with a suitably high correlation of their activation with the 'box-car' function, equal to +1 during movement and just afterwards and -1 during other static periods and the up-and-down movement. This correlation at each pixel in the head leads to the statistical parameter map of the heading of this sub-section. Significance levels can then be attached to the value at any one point in terms of the difference of that value as compared to that for a control condition in which there is no condition of interest being applied. This leads to the standard t-test and to maps of t- or z-parameters throughout the brain. More sophisticated analysis can then be performed to detect regions of interest (a number of adjacent pixels with significant z-scores) and their significance. There are now a number of software packages available to perform such analysis, and they have been recently been compared [10]. Fast data analysis techniques are now available so that on-line results can be obtained and thereby allow for optimisation of paradigms being employed [11].

3.2 Inversion Problem

There is a hard inverse problem - to determine the underlying current distribution causing the observed field strengths - for both EEG and MEG. That for the former is more difficult due to the conduction currents that 'wash out' the localisation of deep sources. This does not occur for MEG, but there is still the problem that certain currents, such as radial ones in a purely spherical head, are totally 'silent', leading to no external magnetic field. Modulo this problem, the standard approach to uncovering the underlying neural generators has been to assume that there is a limited set of current dipoles whose parameters (orientation, strength and position) can be varied to optimise the mean square reconstruction error (MSE) of the measured data. Such approaches are limited, especially when fast temporal effects are being investigated. More recently magnetic field tomography (MFT) has been developed to give a distributed source representation of the measurements [12].

The use of MFT and similar continuous source distribution systems is now becoming more commonplace in MEG, so that high-quality data are now becoming available for an increasing number of cognitive tasks similar to those from fMRI and PET but with far greater temporal sensitivity.

3.3 Structural Modelling

Initially the results of PET and fMRI studies have uncovered a set of active brain sites involved in a given task. This is usually termed a 'network', although there is no evidence from the given data that a network is involved but only an isolated set of regions. It is possible to evaluate the correlation coefficients between these areas, either across subjects (as in PET) or for a given subject (in fMRI). There is great interest in using these correlation coefficients to determine the strength of interactions between the different active areas, and so uncover the network involved. Such a method involves what is called 'structural modelling', in which a linear relation between active areas is assumed and the path strengths (the linear coefficients in the relation) are determined from the correlation matrix. In terms of the interacting regions of figure 1:

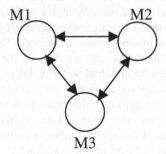

Fig. 1. A set of three interacting modules in the brain, as a representative of a simple structural model. The task of structural modelling is to determine the path strengths with which each module effects the others in terms of the cross-correlation matrix between the activities of the modules.

Corresponding activities z_i (i =1,2,3) of the modules satisfy a set of linear equations

$$z_i = \sum a_{ij} z_j + \eta_i \tag{1}$$

where the variables η_I are assumed to be independent random noise variables. It is then possible to calculate the coefficients a_{ij} from the correlation coefficients $C(i,j)$ between the variables z. This can be extended to comparing different models by a chi-squared test so as to allow for model significance testing to be achieved.

3.4 Bridge the Gap

An important question is as to how to bridge the gap between the brain imaging data and the underlying neural network activity. In particular the interpretation of the structural model parameters will then become clearer. One way to achieve this is by looking at simplified neural equations which underpin the brain activations. We can describe this activity by coupled neural equations, using simplified neurons:

$$\tau \, dU(i, t)/dt = - U(i,t) + \sum_{neurons, j} C(i,j) \, f[U(j, t\text{-}t(i,j))] \tag{2}$$

where $U(i, t)$ is the vector of membrane potential of the neurons in module i at time t, $C(i,j)$ is the matrix of connection strengths between the modules, and $t(i,j)$ the time delay (assumed common for all pairs of neurons in the two modules).

There are various ways we can reduce these coupled equations to structural model equations. One is by the mean field approximation $<U> = u1$ (where 1 is the vector with components equal to 1 everywhere) so that the particular label of a neuron in a given module is averaged over. The resulting averaged equations are:

$$\tau \, du(i,t)/dt = - u(i,t) + \sum_{j} c(i,j) \, f[u(j, t\text{-}t(i,j))] \tag{3}$$

where $c(i,j)$ is the mean connection strength between the modules i and j. In the limit of $\tau = 0$ and for linearly responding neurons with no time delay there results the earlier structural equations:

$$u(i,t) = \sum_{j} C(i,j) \, u(j, t) \tag{4}$$

These are the equations now being used in PET and fMRI. The path strengths are thus to be interpreted as the average connection weights between the modules. Moreover the path strengths can be carried across instruments, so as to be able to build back up to the original neural networks. This involves also inserting the relevant time delays as well as the connection strengths to relate to MEG data.

The situation cannot be as simple as the above picture assumes. Connection strengths depend on the paradigm being used, as well as the response modality (such as using a mouse versus verbal report versus internal memorisation). Thus it is necessary to be more precise about the interacting modules and their action on the inputs they receive from other modules; this will not be a trivial random linear map but involve projections onto certain subspaces determined by the inputs and outputs. The above reduction approach can be extended to such aspects, using more detailed descriptions of the modules [14].

4 Results of Static Activation Studies

There are many psychological paradigms used to investigate cognition, and numbers of these have been used in conjunction with PET or fMRI machines. These paradigms overlap in subtle ways so that it is difficult at times to 'see the wood for the trees'. To bring in some order we will simplify by considering a set of categories of cognitive tasks. We do that initially along lines suggested by Cabeza and Nyberg [15], who decomposed cognitive tasks into the categories of:

- attention (selective/sustained)
- perception (of object/face/space/top-down)
- language (word listening/word reading/word production)
- working memory (phonological loop/visuospatial sketchpad)
- memory (semantic memory encoding & retrieval/episodic memory encoding & retrieval)
- priming
- procedural memory (conditioning/skill learning)

So far the data indicate that there are sets of modules involved in the various cognitive tasks. Can we uncover from them any underlying functionality of each of the areas concerned? The answer is a partial 'yes'. The initial and final low-level stages of the processing appear more transparent to analysis than those involved with later and higher level processing. Also more study has been devoted to primary processing areas. Yet even at the lowest entry level the problem of detailed functionality of different regions is still complex, with about 30 areas involved in vision alone, and another 7 or 8 concerned with audition. Similar complexity is being discerned in motor response, with the primary motor area being found to divide into at least two separate subcomponents. However the main difficulty with the results of activated areas is that they are obtained by subtraction of a control condition, so that areas activated in common will disappear in the process. That can be avoided by using structural modelling introduced in the previous section. We will therefore turn to survey the still small but increasing sets of data now being analysed by that approach.

5 Structural Models of Particular Processes

5.1 Early Spatial and Object Visual Processing

This has been investigated in a series of papers concerned with the dorsal versus ventral routes of visual processing as related to spatial and object processing respectively. The object experiment used simultaneous matching of a target face to two simultaneously presented faces. The spatial task used a target dot located in a square with one side a double line to be compared to a simultaneously presented pair of similar squares containing dots; the matching test stimulus had a dot in the same position as the test stimulus in relation to the double line. The results of these researches [17], [18] showed that there are indeed two pathways for spatial and object processing respectively, the former following the ventral pathway from the occipital lobe down to the temporal lobe, the latter the dorsal route from occipital up to the parietal lobe, as shown in figure 2.

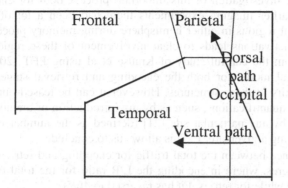

Fig. 2. A schematic of the cortex, showing its subdivision into the frontal, parietal, temporal and occipital lobes, and the vantral and dorsal routes for visual input to further cortex from early visual processing.

5.2 Face Matching

A careful study was performed as to how the paths involved change with increase of the time delay between the presentation of one face and the pair to which the original target face must be matched [19]. The delays ranged from 0 to 21 seconds, with intermediate values of 1, 6, 11 and 16 seconds. The lengthening of the time duration for holding in mind the target face caused interesting changes in the activated regions and their associated pathways. In particular the dominant interactions during perceptual matching (with no time-delay) are in the ventral visual pathway extending into the frontal cortex and involving the parahippocampal gyrus (GH) bilaterally, with cingulate (in the middle of the brain). The former of these regions is known to be crucial for laying down long-term memory, the latter for higher-order executive decision-making. This activity distribution changed somewhat as the delays increased, first involving more frontal interactions with GH in the right hemisphere and then becoming more bilaterally symmetric and including cingulate-GH interactions, until for the longest delays the left hemisphere was observed to have more interactions with GH and anterior occipital areas on the left, although still with considerable feedback from right prefrontal areas. These changes were interpreted [19] as arising from different coding strategies. At the shortest time delays the target face can be held by a subject in an 'iconic' or image-like representation, most likely supported by right prefrontal cortices. As the time delay becomes larger there is increasing difficulty of using such an encoding scheme, and the need for storage of more detailed facial features and their verbal encoding. That will need increasing use of the left hemisphere. However there is also a greater need for sustained attention as the duration gets longer, so requiring a greater use of right hemisphere mechanisms.

5.3 Memory Encoding and Retrieval

There is considerable investigation of this important process both for encoding and retrieval stages. In earlier imaging experiments there had been a lot of trouble in detecting hippocampal regions in either hemisphere during memory processing. The use of unsubtracted activations leads to clear involvement of these regions in both stages, as is clear from the careful study of Krause et al using PET [20]. This has resulted in a structural model for both the encoding and retrieval stages. There is considerable complexity in these structures. However it can be teased out somewhat by the use of various quantifications, such as by my introduction of the notion of the total 'traffic' carried by any particular site [21] (defined as the number of activated paths entering or leaving a given area). This allows us to conclude:

- there is a difference between the total traffic for encoding and retrieval between the two hemispheres, where in encoding the L/R ratio for the total traffic in all modules is 51/40 while for retrieval it has reversed to 46/50;
- the anterior to posterior cortex ratio is heavily weighted to the posterior in retrieval, the anterior/posterior traffic ratios for encoding being 32/38 and for retrieval 28/54;
- the sites of maximum traffic are in encoding left precuneus (in the middle of the brain at the top of the parietal lobes) with a traffic of 11, and right cingulate (with 8) and in retrieval the right precuneus (with 11) and its left companion (with 9). By comparison the hippocampal regions have traffic of (5 for R, 6 for L) in encoding and (8 for L, 6 for R) in retrieval.

We conclude that the first of these result is in agreement with the HERA model of Tulving and colleagues [22]: there is stronger use of the left hemisphere in encoding while this changes to the right hemisphere in retrieval. As noted by Tulving et al [22] there is considerable support for this asymmetry from a number of PET experiments. The structural models show that in both conditions there is more bilateral prefrontal involvement.

The anterior to posterior difference is not one which has been noted in the context of the paired-associate encoding and recall tasks. However there are results for posterior/anterior asymmetry associated with the complexity of the working memory tasks, as in the n-back task [23] (where a subject has to indicate when a particular stimulus of a sequence of them has occurred n times previously in the sequence). When n becomes greater than 2 (so for delay times in the task longer than 20 seconds) it has been reported that anterior sites become activated and there is an associated depression in posterior sites which were previously active for the case of n=0 and 1. The difference between the anterior and posterior traffic in our case indicates that the processing load is considerably lower in the retrieval part of the task than in the encoding component. This is consistent with expectations: the hard work in the task involves setting up the relations between the words as part of the encoding process. Once these relations are in place then there can be a reasonable level of automaticity.

5.4 Hearing Voices in Schizophrenia

A study has been reported by Bullmore and colleagues [24], in which subjects had to decide, for a set of 12 words presented sequentially, which of them referred to living entities, which to non-living ones.   Subjects were asked to rehearse their reply subvocally. This was compared to a baseline condition in which subjects looked at a featureless isoluminant screen.   A path analysis of the data showed a network involving extrastriate cortex ExCx, posterior superior temporal gyrus STG (Wernicke's area), dorsolateral prefrontal cortex DLPFC, inferior frontal gyrus IFG (Ba 44/45 L) and supplementary motor area SMA, where B, L and M denote bilateral, left-sided or mesial respectively.

The path analysis for normal subjects showed the flow pattern

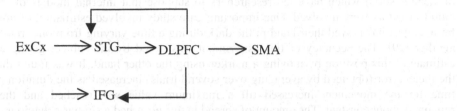

whereas for schizophrenic subjects there was no such clear path model in which there was SMA feedback to IFG and STG. This fits well with the model of Frith [25] of lack of control of feedback from the voice production region by SMA to STG in schizophrenics. They do not know they are producing internal speech and think they are hearing voices speaking to them. Awareness of this process could thus be in STG.

6 New Paradigms for Neural Networks?

The results now pouring in from brain imaging, as well as from single cell and deficit studies, lead to suggestions of new paradigms for neural networks. In brief, some of these are:

1) recurrent multi-modular networks for temporal sequence processing, based on cartoon versions of the frontal lobes, with populations of excitatory and inhibitory cells similar to those observed in cortex and basal ganglia. These are able to model various forms of temporal sequence learning [26] and delayed tasks and deficits observed in patients [27].

2) Attention is now recognised as sited in a small network of modules: parietal and prefrontal lobes and anterior cingulate. This supports the 'central representation' [28], with central control modules (with global competition) coupled to those with semantic content, and extends feature integration.

3) Hierarchically coded modules, with overall control being taken by a module with (oscillatory) coupling to the whole range of the hierarchy

However the above set of paradigms do not take account of the input→output nature of the brain which has developed across species so as to be increasingly effective in granting its possessor better responses for survival. One of the most important of these is, through increased encephalisation, the ability of the brain to construct internal models of its environment. This includes its own internal milieu, leading to ever better control systems for response in the face of threats. Such internal models are now being probed by a variety of techniques that are increasingly using brain imaging to site the internal models. It is these models which will be elaborated on next before being used to give a new functional framework for overall brain processing.

6.1 Spatial Arm Movement Estimation

There are many paradigms involving motor actions (posture control, eye movements of various sorts) which have led researchers to suppose that internal models of the motor effectors were involved. One important case study involved estimation of how far a subject has moved their hand in the dark during a time varying from one trial to another [29]. The accuracy of this assessment was tested by the subject giving an estimate of this position by moving a marker using the other hand. It was found that the mean error (obtained by averaging over several trials) increased as the duration of time for the movement increased till a maximum value was reached and then remained about constant. The amount of spread in this error had a similar behaviour.

To explain this result it was assumed that an internal model of the hand position had been already created in the brain, so that it could be updated by feedback from expected sensory signals arising in the arm from the change of its position as compared to further actual feedback from the proprioceptive sensors in the arm muscles activated during the movement. There would also be visual feedback, although in this paradigm (carried out in the dark) this would not be present (but will be used in the following paradigm). The total model of this system is shown in figure 3.

The over-all system of figure 3 uses a linear differential equation for obtaining an initial update of the position x(t) of the arm at time t:

$$d^2 x/dt^2 = -(b/m)dx/dt + F \tag{5}$$

where m is the weight of the arm, F the force being applied to it by the muscles and b a constant. The first term on the right hand side of the equation is a counter-force arising from the inertia of the arm muscles. The gain term is of form

$$K(t)[y - Cx(t)] \tag{6}$$

where C is the scaling factor modifying the position to give the expected position from proprioceptive feedback signals up the arm to the brain. The gain factor K(t) is time dependent, with time dependence obtainable from a well-defined set of equations (termed the Kalman filter equations) [30]. The results of using this model to determine the time development of the mean value of x(t) and of the variance of x(t) were in close agreement with the observed curves from these quantities for a set of subjects [29].

This result is strong, but indirect support, for the existence of internal models of motor action in the brain. In particular they indicate the presence of a monitor system, enclosed in the broken-line rectangle in fig 2, which assesses errors between predicted values of feedback and the actual feedback experienced. Both this and the internal arm model, however, have not been directly observed in the brain. To develop this 'internal model' or so-called 'observer' approach [30] further other evidence will now be considered which helps to site the internal model and monitor system more precisely.

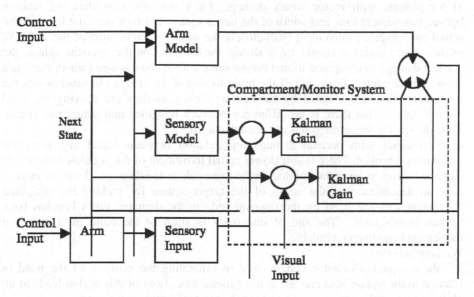

Fig. 3. Internal arm control model. The arm model in the upper box updates the position of the arm by a linear differential equation, using the external control signal (from elsewhere in the brain). Feedback from sensory and visual input is used to give correction terms (the 'Kalman gain' terms) which are added to the next state value arising from the linear arm model, to give an estimate of the next arm position. This is then used for further updating.

6.2 Accuracy of Arm Movement

The above model can be applied to another set of results, now with comparison having been made with subjects with certain deficits. The paradigm being considered is that of the movement of a pencil held in the hand from a given start position to a marked square on a piece of paper. If the side of the marked square to be moved to is denoted by W and A is the amplitude of the total arm movement from the start position to the centre of the marked square, then the time T taken to make the movement to the marked square increase as the width W of the square is reduced according to what is known as Fitt's law [31] :

$$T = a + b\ln[A/W] \tag{7}$$

where ln is the natural logarithm, a and b are constants with b being positive.

An extension of this paradigm has been used by asking subjects to imagine they were making the movement, not actually moving their arm at all, and stating when they have reached, in their mind's eye, the required position in the target square. The results for normals is that there imagined movement times still obey Fitt's law, with the same constants a and b above. However there are two sorts of deficits which cause drastic alterations to this result [31]:

a) for patients with motor cortex damage, Fitt's law still describes the relation between movement time and width of the target square, but now the time for both the actual and imagined movement (still closely the same for each value of the width W of the target square) is considerably slower for the hand on the opposite side to the cortical damage as compared to that on the same side as the damage (where there is a cross-over of control from side of the brain to side of the body). In other words the movement is slowed - noted by subjects as feeling as they are moving the hand through 'glue' - but there is no difference between imagined and actual movement. Thus the site for 'imagining' has not been damaged.

b) for patients with parietal damage this relation between actual and imagined movement is broken. There is still slower actual movement of the opposite hand to the damaged cortex as compared to that on the same side as the damage. However there is now no dependence on the width of the target square for making the imagined movement with the hand on the opposite side to the damage. Fitt's law has been broken in this case. The site of imaginary location of the hand as it makes it movement is no longer available.

We conclude that

1) the comparator/monitor system used in estimating the position of the hand in relation to the square is contained in the parietal lobe. Loss of this region leads to no ability to assess the error made in making the imagined movement.

2) the internal arm model itself (of equation 1 above) is sited in the motor cortex. Its degradation leads to an increase of the friction term b in the above model, as reported in the experience of patients with such damage.

This approach through observer-based control theory needs to be extended, or blended in, to other brain processes. In particular it is needed to relate the internal models described above to the results of coupled networks of active modules observed by structural models and recounted in section 5 above. This extension is difficult to achieve since there is still little deep understanding of the underlying functionality of the various regions being observed. However such a blending will be started here, commencing with a small extension to the case of fading of sensations in a patient with parietal damage.

6.3 The Fading Weight

The nature of continued reinforcement of activity allowing the continued experience of an input has been demonstrated in a remarkable fashion by a case of fading of the experience of a weight placed on the hand of a patient who had suffered severe loss of parietal cortex due to a tumour in that region [32]. When a weight was placed on her hand the sensation of it faded, in spite of her directed attention to it, in a time proportional to the mass of the object placed in her hand: a 50 gram weight faded in 5 seconds, a 100 gram one in 8 seconds, a 150 gram weight in 10 seconds, and so on.

It was suggested that such fading occurs due to the presence of a store for holding an estimate of the state of the hand (in this case the mass it is supporting) which has been lost by the patient with parietal lobe damage. In that case, with no such store available, the nature of the model of equation (1) indicates that the natural time constant is m/b. Thus it is to be expected, on that model, that such fading would occur (with no store) in a time proportional to the mass placed on the hand. In the presence of a store there will be active refreshment of such knowledge, so avoiding the fading of experience in such a fashion. Active refreshment is a process that again requires action, as did the arm motion in section 6.1. It is to be expected that such 'action' is in motor cortex or related frontal areas. We turn in the next sub-section to consider experimental evidence for the presence of processes in frontal lobes achieving actions on activity in posterior regions.

6.4 Rotating in the Mind

Working memory [33] is a term introduced in the 70's to describe short-term memory, as used, say, when you look up a phone number and hold it in mind until you have made the phone call. You need not make a permanent memory of the number; it fades from your mind afterwards unless you deliberately attempt to remember it for later use. Such working memory has two components: one involved a buffer store, in which items are activated by decay away in a few seconds if not refreshed. Such buffer working memory stores have been observed by brain imaging in a number of modalities, such as for words (the phonological store) for space (the visuo-spatial sketch-pad) for timing of signals (on the left hemisphere). In general these buffer sites are placed in the parietal lobes [34]. At the same time there is a more active system of refreshment in the frontal lobes, so that when you held in mind the phone number it could be rehearsed (subvocally) so as to hold it over a number of seconds if need be.

The manner in which working memory is involved, and its sites in the brain exposed, was made clear in a beautiful fMRI brain imaging experiment of rotating images in the mind to compare two three dimensional figures, seen at different angles, as the same or mirror images [35]. It was found that the time for response increased linearly with the amount one figure had been rotated from the other. On imaging the brain, a number of regions were observed to have activity which also increased linearly with the angle required to line up the two figures with each other. This was especially clear in both parietal lobes, as well as to a lesser extent in the inferior temporal and motor cortical areas.

These results can be understood in similar terms to the internal action model of figure 3 by the model of figure 4.

The internal model works in a similar manner to that of figure 3, in that it has a module, entitled the 'image model', which has a representation of the dynamical equation for updating the position of the input image, in this case just the rotation parameter θ. The equation of motion being implemented by this model is the linear equation

$$d\theta/dt = \omega \tag{8}$$

where ω is the angular velocity used in the internal rotation process. It is very likely that ω is determined by eye movement machinery, as part of preparing saccades. The increase of the rotation parameter θ by the dynamics of frontal lobe activity (to which we turn in a moment) will be sent back to the rotation parameter module, which then acts to rotate the spatial image being held on the location working memory, being guided by the object codes module to keep the shape of the object during rotation consistent with its original shape, although now with a rigid rotation added. During this process the comparator module assesses the effects of the rotation change and brings it to s halt when there is identification between the rotated and stationary image. All the time there must be rehearsal of the held images (or just the rotated on, the static one being referred to from the outside picture). This rehearsal must also involve an activity, so will be in a frontal site involving motor acts.

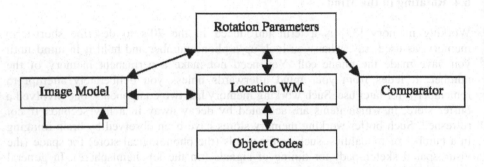

Fig. 4. Internal model of processing for the internal rotation of inputs associated with comparing pictures of two identical solid objects rotated/reflected with respect to each other. The image model is used both to refresh the image of the object held in the location module, in conjunction with the object code module and the rotation parameter module, and to achieve rotation of this image by augmenting the appropriate parameter. The comparator module is used to determine when the rotation has been completed and alignment between the two objects achieved (or not).

From the above account the identifications can be made:
1) the object code module is contained in the temporal lobe;
2) the location working memory module is in the parietal lobe (possibly in inferior parietal lobe);
3) the rotation parameter module is also in the parietal lobe (possibly in superior parietal lobe);
4) the comparator is also in the parietal lobe (as part of the monitoring system of sub-section 6.2);
5) the active rotation is performed in a frontal site, such as the supplementary eye field, where saccades are programmed.

 The above identifications are based on the identifications already made in the previous sub-section. However we can add further support for placing the active module of 5) above in the frontal lobes in terms of the general recurrent features of the frontal lobe connectivity. This is well known to be in terms of a set of recurrent neural loops involving not only the frontal cortices but also sub-cortical structures termed the basal ganglia and the thalamus. The latter is a way-station for all activity entering the cortex from outside, the former are involved in the recurrent pathway

cortex→basal ganglia→thalamus→cortex

This set of loops is in actuality far more complicated than the above simple flow diagram would indicate, and distinguishes the frontal lobes from posterior cortex, which lacks any similar recurrent system with basal ganglia as a controlling and guiding unit. Models of this frontal system indicate its power to enable the learning of temporal structure in response patterns to inputs, a quality lacking in the posterior cortex [26].

Such powers the frontal system posses indicate it is ideal to support an iterative updating of activity denoting a state of the body (or other parts of the brain). Thus the above equation of rotation parameter update can be written as the updating equation

$$\theta(t+\Delta t) = \theta(t) + \omega \Delta t \tag{9}$$

which can be seen to be achieved by updating the value of θ at any time by addition of the amount ω to it. Such a process can be achieved very straightforwardly by a linear neuron coding for θ being updated cyclically by a constant input ω input every time step. A similar interpretation of recurrent updating can be given for the mode of computation of the linear internal arm model considered in sub-section 6.1.

The various components of the model of figure 4 will have activity which in general will increase as the required amount of rotation is increased. The details of such increase still have to be worked out, and will add a valuable constraint to the modeling process, as well as amplify the sites involved and their functionality.

6.5 Conclusion on Processing

A number of experimental supports have been given for a style of processing in the brain based on internal 'active' models of either body parts or other brain activities, from buffer working memory sites. The actions involved have so been modeled above by simple linear differential equations; Th. true from of these models will in general be much more complex. However that is not a problem since neural networks are able to act as universal approximators of any dynamical system. These active models, based in the frontal lobes and depending on recurrent circuits for their powers. are coupled to posterior codes for inputs. Some of these codes have a categorical form, such as of object representations. Others are of a spatial form, giving the actual spatial parameters of the inputs to be available for active grasping, reaching or other actions to be made on the inputs. Finally there are also posterior monitoring systems for detecting mismatches and making corrections to frontal actions, which for fast response are based on assessments of inputs and their changes. The expected changes then have to be matched to actual changes as measured by inputs from outside (or from proprioceptive inputs from the body). The matching process takes place in posterior sites in the cortex.

In general, then, we have a division of labour in the brain as shown in the following table 1:

Table 1. Nature of the division between frontal and posterior sites from the control aspect.

SITE	POSTERIOR	FRONTAL
NATURE OF FUNCTION	Internal model state updating, rotation, rehearsal, maximum time scale: many seconds	Active, updating posterior representations & monitoring the relation between frontal & posterior sites
CONTROL ASPECT	Preprocessing, object representations, maximum time scale: 1-2 secs	Internal model state updating, rotation, rehearsal, maximum time scale: many seconds

It is possible to observe that the development of the action-frontally-based models of the above sub-sections attack the problem of understanding the networks of the brain in a different fashion from those described in section 4. There the structural models had a timeless character. Now the models from this section.

7 Conclusions

There are considerable advances being made in understanding the brain based on brain imaging data. This is made particularly attractive by the use of structural modelling to deduce the strengths of interconnected networks of active regions during task performance. Neural models are being developed to allow these results to be advanced, as well as indicate new paradigms for artificial neural networks. In particular the development of action-frontally-based models of the previous section attack the problem of understanding the networks of the brain in a different fashion from those described in section 4. The structural models derived there had a timeless character. The control models from the previous section involve a strict separation of duties: the posterior sites perform preprocessing, hold object and spatial representations, with short term memory components holding activity on-line for no more than 1-2 seconds. The anterior sites allow for voluntary actions to be taken on the posterior representations so as to achieve goals also held in frontal sites. Such a division of labour is an important result arising from fusing together the two ways of approaching the brain: structural models and analysis of active processing in terms of internal models.

Two approaches to understanding the brain have been described above: one through the use of CSEMs and the other by means of control models. The best way to reconcile them is to use the control metaphor as a guide to the search for functional assignments for the different modules and networks uncovered by the CSEM approach. Besides indicating more general functional connectivity, the various CSEMs give specific connection strengths between the modules in each case. These can then be used to determine, by detailed simulation, if a suggested control version of the CSEM network could and does function in the required control manner. The testing of control models in this manner will be improved by the use of temporal information obtained from MEG and EEG for the same paradigms. As data from completely multimodal experiments (PET/fMRI/MEG/EEG) become available the control models will be more effectively tested.

Finally what about artificial neural networks? These have already been used as important components of control models. The new aspect to be added to these models is inclusion of the cognitive aspect of brain processing, from a modular point of view. For example language and planning can be analysed using such a control view, as can attention and even consciousness. It is these human-like capabilities that are presently absent from artificial neural systems, in spite of the many attempts to build 'intelligent' systems. The above discussion hopefully gives an indication of an avenue of approach to these difficult but enormously important problems for artificial processing systems.

References

[1] Raichle ME, ed (1998)Proceedings of the colloquium "Neuroimaging of Human Brain Function", Proc Natl Acad Sci USA, vol 95, February, 1998.
[2] Posner M and Raichle ME (1994) Images of Mind, San Francisco: Freeman & Co
[3] Ungerleider LG, Courtney SM & Haxby JVA neural system for human visual working memory Proc Natl Acad Sci USA 95, 883-890
[4] Yang X, Hyder F & Shulman RG (1997) Functional MRI BOLD Signal Coincides with Electrical Activity in the Rat Whisker Barrels, NeuroReport 874-877
[5] Raichle M E (1998) Behind the scenes of functional brain imaging: A historical and physiological perspective. Proc Natl Acad Sci USA 95, 765-772
[6] Bti, Ltd, San Diego, CA
[7] NEC, private communication from AA Ioannides
[8] Talairach J & Tournoux P (1988) Co-planar steroetaxic atlas of the human brain. Stuttgart: G Thieme.
[9] Schmitz N., Taylor J.G., Shah N.J., Ziemons K., Gruber O., Grosse-Ruyken M.L. & Mueller-Gaertner H-W. (1998) The Search for Awareness by the Motion After-Effect, Human Brain Mapping Conference '98, Abstract; ibid (2000) NeuroImage 11:257-270
[10] Gold S, Christian B, Arndt S, Zeien G, Cizadio T, Johnoson DL, Flaum M & Andreason NC (1998) Functional MRI Statistical Software Packages: A Comparative Analysis. Human Brain Mapping 6, 73-84
[11] D Gembris, S Posse, JG Taylor, S Schor et al (1998) Methodology of fast Correlation Analysis for Real-Time fMRI Experiments, Magnetic Resonance in Medicine (in press).
[12] Ioannides AA (1995) in Quantitative & Topological EEG and MEG Analysis, Jena: Druckhaus-Mayer GmbH
[13] Taylor JG, Ioannides AA, Mueller-Gaertner H-W (1999) Mathematical Analysis of Lead Field Expansions, IEEE Trans on Medical Imaging. 18:151-163
[14] Taylor JG, Krause BJ, Shah NJ, Horwitz B & Mueller-Gaertner H-W (2000) On the Relation Between Brain Images and Brain Neural Networks, Human Brain Mapping 9:165-182.
[15] Cabeza R & Nyburg L (1997) Imaging Cognition. J Cog Neuroscience 9, 1-26; ibid (2000) J Cog Neuroscience 12:1-47.
[16] Gabrieli JDE, Poldrack RA & Desmond JE (1998)The role of left prefrontal cortex in memory Proc Natl Acad Sci USA, 95, 906-13
[17] McIntosh AR, Grady CL, Ungerleider LG, Haxby JV, Rapoport SI & Horwitz B (1994) Network analysis of cortical visual pathways mapped with PET, J Neurosci 14, 655-666
[18] Horwitz B, McIntosh AR, Haxby JV, Furey M, Salerno JA, Schapiro MB, Rapoport SI & Grdy CL (1995) Network analysis of PET-mapped visual pathways in Alzheimer type dementia, NeuroReport 6, 2287-2292

[19] McIntosh AR, Grady CL, Haxby J, Ungerleider LG & Horwitz B (1996) Changes in Limbic and Prefrontal Functional Interactions in a Working Memory Task for Faces. Cerebral Cortex 6, 571-584

[20] Krause BJ, Horwitz B, Taylor JG, Schmidt D, Mottaghy F, Halsband U, Herzog H, Tellman L & Mueller-Gaertner H-W (1999) Network Analysis in Episodic Encoding and Retrieval of Word Pair Associates: A PET Study. Eur J Neuroscience 11:3293-3301.

[21] Taylor JG, Krause BJ, Horwitz B, NJ Shah & Mueller-Gaertner H-W (2000) Modeling Memory-Based Tasks, submitted

[22] Tulving E, Kapur S, Craik FIM, Moscovitch M & Houles S (1995) Hemispheric encoding/ retrieval asymmetry in epsodic memory: Positron emission tomography findings. Proc Natl Acad Sci USA 91, 2016-20.

[23] Cohen JD, Perlstein WM, Braver TS, Nystrom LE, Noll DC, Jonides J& Smith EE (1997) Temporal dynamics of brain activation during a working memory task. Nature 386, 604-608

[24] Bullmore E, Horwitz B, Morris RG, Curtis VA, McGuire PK, Sharma T, Williams SCR, Murray RM & Brammer MJ (1998) Causally connected cortical networks for language in functional (MR) brain images of normal and schizophrenic subjects. submitted to Neuron

[25] Frith CD (1992) The Cognitive Neuropsychology of Schizophrenia. Hove UK: L Erlbaum Assoc.

[26] Taylor N & Taylor JG (2000) Hard wired models of working memory and temporal sequence storage and generation. Neural Networks 13:201-224; Taylor JG & Taylor N (2000) Analysis of recurrent cortico-basal ganglia-thalamic loops for working memory. 82:415-432.

[27] Monchi O & Taylor JG (1998) A hard-wired model of coupled frontal working memories for various tasks. Information Sciences 113:221-243.

[28] Taylor JG (1999) 'The Central Representation', Proc IJCNN99

[29] Wolpert DM, Ghahramani Z & Jordan M (1996) An Internal Model for Sensorimotor Integration. Science 269:1880-2

[30] Jacobs OLR (1993) Introduction to Control Theory (2nd ed). Oxford: Oxford University Press.

[31] Sirigu A, Duhamel J-R, Cohen L, Pillon B, DuBois B & Agid Y (1996) The Mental Representation of Hand Movements After Parietal Cortex Damage. Science 273:1564-8.

[32] Wolpert DM, Goodbody SJ & Husain M (1998) Maintaining internal representations: the role of the human superior parietal lobe. Nature neuroscience 1:529-534

[33] Baddeley A (1986) Working Memory. Oxford: Oxford University Press.

[34] Salmon E, Van der Linden P, Collette F, Delfiore G, Maquet P, Degueldre C, Luxen A & Franck G (1996) Regional brain activity during working memory tasks. Brain 119:1617-1625

[35] Carpenter PA, Just MA, Keller TA, Eddy W & Thulborn K (1999) Graded Functional Activation in the Visuospatial System with the Amount of Task Demand. J Cognitive Neuroscience 11:9-24.

[36] Taylor NT & Taylor JG (1999) Temproal Sequence Storage in a Model of the Frontal Lobes. Neural Networks (to appear)

Stimulus-Independent Data Analysis for fMRI

Silke Dodel, J. Michael Herrmann, and Theo Geisel

Max-Planck-Institut für Strömungsforschung, Bunsenstraße 10, 37073 Göttingen,
Germany

Abstract. We discuss methods for analyzing fMRI data, stimulus-based
such as baseline substraction and correlation analysis versus stimulus-
independent methods such as Principal Component Analysis (PCA) and
Independent Component Analysis (ICA) with respect to their capabilities of separating noise sources from functional activity. The methods
are applied to a finger tapping fMRI experiment and it is shown that
the stimulus-independent methods in addition to the extraction of the
stimulus can reveal several non-stimulus related influences such as head
movements or breathing.

1 Introduction

Functional Magnetic Resonance Imaging (fMRI) is a promising method to determine noninvasively the spatial distribution of brain activity under a given
paradigm, e. g. in response to certain stimuli. Beyond localization issues models
based on fMRI data could indicate functional connectivity among brain areas
such as pathways of feature processing, dynamical pathologies or even neural
correlates of consciousness, and may help to to determine restrictions of computational models for individual systems. Obviously, fMRI data are contaminated
by non-functional processes such as blood flow pulsation and metabolic variations, which have to be separated from the actual nervous activity. We will
emphasize the latter aspect and discuss here merely preconditions of such modeling which consists in the preprocessing of the raw data, but is crucial for further
exploitation of the data.

The functional signal in MRI depends on changes in the blood oxygen level
which are assumed to be causally related to variations in the neural activity at
the respective location. More specifically, the blood oxygen level is determined
by the amount of oxyhemoglobine. Oxygen consumption results in an increase of
desoxyhemoglobine. Oxy- and desoxyhemoglobine have different magnetic properties and thus have different influences onto the MR signal. In the presence of
oxyhemoglobine the signal is increased whereas it is decreased in the presence
of desoxyhemoglobine. This is called the BOLD (*b*lood *o*xygen *l*evel *d*ependent)
effect. However the hemodynamic response to neural activity is nonlinear and
not yet fully understood, although it can be roughly explained as follows: At
the onset of neural activity oxygen consumption surpasses the delivery of blood
oxygen. After a delay of 6-8 s the elevated consumption in the active region is
overcompensated by the delivery of oxyhemoglobine, such that the earlier activation becomes detectable by a relative increase of the MR signal of about

S. Wermter et al. (Eds.): Emergent Neural Computational Architectures, LNAI 2036, pp. 39–52, 2001.
© Springer-Verlag Berlin Heidelberg 2001

1-10% [3]. Apart from the neural activity also other influences occur to the signal and have to be taken into account when analyzing the data. Head movement and brain pulsation, which are hardly avoidable in experiments with behaving humans, can cause the signal to spill over to neighboring voxels. Further, in the vicinity of large vessels the signal is most likely driven by heart beat and breathing rather than by the neural activity. These components can be accounted for to some extent by the stimulus-independent analysis methods as will be shown in section 5 dealing with the results of applying stimulus-independent analysis methods to a finger tapping fMRI experiment. The experiment is described in section 4. Section 2 and 3 introduce stimulus-based and stimulus-independent data analysis methods, respectively.

2   Stimulus-Based Data Analysis Methods: Baseline-Difference and Correlation Analysis

In fMRI often paradigms are used where a task performed in an experiment is contrasted with a control condition in a second experiment. The data of the control condition experiment is taken as the so called baseline which is substracted from the data of the experiment under task condition. Taking the difference of data from the two experiments is rather critical because of variations in the hemodynamic response. Even in a single experiment, where a stimulus or behavior is alternately present or not present such variations can be observed. In single experiments the baseline-difference method is commonly replaced by correlating the time course of the stimulus with the activity. The time course is often modeled as a boxcar function, which is zero at the time of the control condition and unity at the task condition (cf. Fig. 1). Sometimes a modified or shifted version of the boxcar function is used to account for the hemodynamic response.

Recently measurements of the resting brain, i. e. with no predefined stimulus are of increasing interest. To analyze data of the resting brain stimulus independent methods are required. These methods can also be used to clean the data from artifacts as described above.

3   Stimulus-Independent Data Analysis: Principal Component Analysis (PCA) and Independent Component Analysis (ICA)

3.1   Data Space

fMRI data consists of time sequences of images reflecting the anatomy and areas of activity in one or more beforehand determined two-dimensional layers of the brain. The number m of pixels per image and the number k of images usually are of the same order of magnitude, but mostly the resolution exceeds the patience of the subjects, so $k < m$. The data can be considered as a set of m time sequences of length k each or as a set of k images of m pixels, emphasizing either temporal or spatial aspects. The pixel time sequences and the images are considered as vectors spanning the row space and the column space of the data

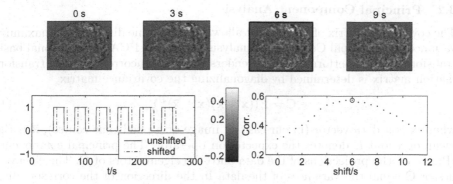

Fig. 1. Stimulus time course and correlation analysis.

Correlation analysis for the second part of the finger tapping experiment (cf. section 4). The stimulus time course is modeled as a boxcar function which is shifted to account for the hemodynamic delay. Every image pixel is correlated with the shifted stimulus time course. The resulting gray scale images of correlation values for different shifts are depicted on the top of the figure. On the left hand side of the bottom the stimulus time course modeled as a boxcar function is depicted. The right hand side shows the highest correlation value present in the respective correlation image versus different shifts of the stimulus time course. The highest correlation value is obtained for a shift of 6 s which is a reasonable value for the hemodynamic delay.

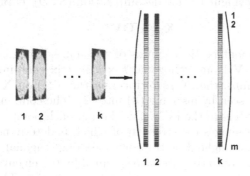

Fig. 2. Illustration of the data matrix $\mathbf{X}$ in which each column represents an image vector and each row a pixel time course. The left hand side shows the image data measured at k points in time. Rearranging the pixels of each image into a column vector gives the data matrix $\mathbf{X}$ shown on the right hand side.

matrix $\mathbf{X}$, respectively, shown in Fig. 2. Throughout this chapter we will apply the following notational convention: Matrices are indicated by upper case bold letters, vectors by lower case bold letters and scalars by normal lower case letters.

3.2   Principal Component Analysis

The covariance matrix of the data set allows to determine directions of maximum variance via Principal Component Analysis (PCA). In PCA an orthogonal basis transformation is performed which renders the data uncorrelated. The transformation matrix is determined by diagonalizing the covariance matrix

$$\mathbf{C} = E((\mathbf{x} - \bar{\mathbf{x}})(\mathbf{x} - \bar{\mathbf{x}})^T) \qquad (1)$$

where $\mathbf{x}$ is a data vector (in our case an image or a pixel time course), $\bar{\mathbf{x}}$ is the mean of $\mathbf{x}$ and E denotes the expectation operator. The principal components (PC's) are the projections of the data onto the eigenvectors of $\mathbf{C}$. The eigenvalues of $\mathbf{C}$ equal the variances of the data in the direction of the corresponding eigenvectors. An ordering of the PC's is given by the size of the eigenvalues, beginning with the largest eigenvalue, accounting for prevalent data features. In practical applications PCA is performed on the sample covariance matrix computed from the data matrix $\mathbf{X}$. In this contribution we are dealing with temporal PCA, which means that we are looking for uncorrelated time courses underlying the data. The sample covariance matrix then writes

$$\hat{\mathbf{C}} = \frac{1}{k}\bar{\mathbf{X}}\bar{\mathbf{X}}^T \qquad (2)$$

where $\bar{\mathbf{X}}$ is the temporally centered data matrix having the elements $\bar{x}_{ij} = x_{ij} - \frac{1}{k}\sum_{l=1}^{k} x_{il}$ $(1 \leq i \leq m, 1 \leq l \leq k)$. In the following we will identify $\mathbf{X}$ with $\bar{\mathbf{X}}$. In order to save computation time, instead of performing a diagonalization of $\hat{\mathbf{C}}$ we performed a singular value decomposition (SVD) [6] on $\mathbf{X}$. This yields

$$\mathbf{X} = \mathbf{U}\mathbf{D}\mathbf{V}^T \qquad (3)$$

where $\mathbf{U}$ is an $m \times k$ matrix, the columns of which are orthogonal, $\mathbf{D}$ is a $k \times k$ diagonal matrix and $\mathbf{V}$ is an orthogonal $k \times k$ matrix containing uncorrelated time courses as columns, the PC's. Performing the SVD on $\mathbf{X}$ is equivalent to diagonalizing $\hat{\mathbf{C}}$ as is seen by inserting (3) into (2). The columns of $\mathbf{U}$ represent the images corresponding to the PC's. The data at each point in time is a linear combination of these images the strength of which is determined by the values of the respective PC's weighted by the corresponding diagonal element of $\mathbf{D}$.

By projecting the data onto the subspace spanned by the eigenvectors with large eigenvalues the data dimension can be reduced considerably. The main PC's can be identified with main features of the data, if these factors are orthogonal, i. e. temporally uncorrelated. Usually the first few PC's account for most of the total variance (the latter equals the trace of the covariance matrix). Although the combined effect of the noisy factors which cannot be attributed to any feature still accounts for a considerable part of the total variance their individual variances are small and the factors do not contain any visible structure.

3.3   Independent Component Analysis (ICA)

To reconstruct factors underlying the data in a statistically independent fashion independent component analysis (ICA) can be applied. The assumption is here, that the data is a linear superposition (also called mixing in the following) of statistically independent processes.

$$\mathbf{y}(t) = \mathbf{A}\mathbf{s}(t) \tag{4}$$

Here $\mathbf{y}(t)$ is a vector of measurements, $\mathbf{A}$ is a (constant) mixing matrix, $\mathbf{s}(t)$ is a random vector the components of which are statistically independent and t is a temporal index. $\mathbf{y}(t)$ equals the t-th column of the data matrix $\mathbf{X}$. For simplicity we assume in this section the data matrix $\mathbf{X}$ as well as the mixing matrix $\mathbf{A}$ to be $n \times n$ square matrices.

As we are dealing with a finite number of pixels, we can write eq. (4) as a matrix equation

$$\mathbf{X}^T = \mathbf{A}\mathbf{S} \tag{5}$$

with a $n \times n$ matrix $\mathbf{S}$ the rows of which are statistically independent. Thus we consider each pixel time course in the data matrix $\mathbf{X}$ as the linear superposition of n statistically independent pixel time courses.

Eq. (5) has to be determined for both, $\mathbf{A}$ and $\mathbf{S}$ simultaneously. Since it is $\mathbf{X}$ that is given, in ICA the demixing matrix $\tilde{\mathbf{A}} = \mathbf{A}^{-1}$ is determined rather than the mixing matrix $\mathbf{A}$ directly. However, $\mathbf{A}$ and $\mathbf{S}$ are defined only up to permutation and scaling, since $\mathbf{A}' = \mathbf{A}\mathbf{P}\mathbf{\Lambda}^{-1}$ multiplied with $\mathbf{S}' = \mathbf{\Lambda}\mathbf{P}\mathbf{S}$, where $\mathbf{\Lambda}$ is a (diagonal) scaling matrix and $\mathbf{P}$ a permutation matrix, leaves the data matrix $\mathbf{X}$ invariant.

Statistical independence means that the joint probability density p of the random vector $\mathbf{s} = (s_1, \ldots, s_n)$ is a product of the marginal probability densities $p_i(s_i)$ ($1 \le i \le n$) of its components s_i.

$$p(\mathbf{s}) = \prod_{i=1}^{n} p_i(s_i), \tag{6}$$

ICA can be understood as minimizing a distance criterion $D(p(\mathbf{s}), \prod_{i=1}^{n} p_i(s_i))$ between the joint probability density of an estimation of the sources $\mathbf{s}$ and their factorized density with respect to the elements of the demixing matrix $\tilde{\mathbf{A}}$. Fig. 3 gives an illustration of the concept of ICA.

ICA versus PCA. PCA determines the directions of the maximum variances in the data distribution, thus it diagonalizes the covariance matrix, while ICA takes into account also statistics higher than second order and reveals the statistically independent processes underlying the data. In case of the data being multivariate Gaussian distributed ICA is equivalent to PCA, since in this case all higher than second order moments equal zero and thus uncorrelatedness is equivalent to statistical independence.

To our data we have applied the algorithm of Cardoso [1] which is based on the diagonalization of fourth order cumulant matrices and which will be described in the following.

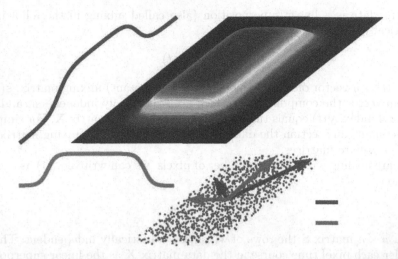

Fig. 3. Toy example to illustrate ICA. The surface depicts the joint probability density p of a 2-dimensional random vector $\mathbf{y}$. The point cloud represents the data (realizations of $\mathbf{y}$). Each coordinate y_i of a realization is a linear superposition of two statistically independent random variables s_1 and s_2, since the joint probability density p was created by multiplying the two marginal probability densities $p_1(s_1)$ and $p_2(s_2)$ shown along the two axes. PCA reveals the orthogonal directions with respect to which the data is uncorrelated. ICA reveals the directions with respect to which the data is statistically independent. Referring to our data the coordinates of each point would reflect an image (consisting of only two pixels) and the PCA vectors would indicate the images which have uncorrelated time courses whereas the ICA vectors indicate the images having statistically independent time courses. Thus the PCA vectors would correspond to the column vectors of the matrix $\mathbf{\Lambda}^{-\frac{1}{2}}\mathbf{R}^T$ where $\mathbf{\Lambda}$ and $\mathbf{R}$ are defined as in eq. (9) and the ICA vectors would correspond to the column vectors of the demixing matrix $\hat{\mathbf{A}}$.

Diagonalizing 4-th Order Cumulants. The approach of Cardoso [1] is based on the fact that for independent random variables the cross cumulants are zero, and at least for bivariate random variables the reverse is also true, provided that the joint distribution is determined by the moments ([5], p. 36). The n-th order cumulant $\kappa^{(n)}$ of a one dimensional random variable Y is given by the n-th derivative of the cumulant generating function $K_Y(\tau)$ at $\tau = 0$ [7]:

$$K_Y(\tau) = \ln \int e^{i\tau y} p(y) dy \tag{7}$$

$$\kappa^{(n)} = (-i)^n \frac{\partial^n K_Y}{\partial \tau^n}\bigg|_{\tau=0} \tag{8}$$

with p being the probability density of the random variable Y. The first and second order cumulants correspond to the first and second order moments, respectively. For higher dimensional random variables the cumulant $\kappa^{(n)}$ is a contravariant tensor of the order n [5]. The algorithm in [1] aims at diagonalizing the fourth order cumulant tensor. The third order cumulant tensor could have been used as well, although then the algorithm would fail in the important case of symmetrical source distributions.

Before diagonalizing the fourth order cumulant it is useful to whiten the data, i. e. to remove the cross correlations of second order and to rescale the data to unit variance. This can be achieved by PCA. If $\mathbf{R\Lambda R}^T$ denotes the eigendecomposition of the covariance matrix $\mathbf{C} = E(\mathbf{xx}^T)$ of the data then the covariance matrix of the whitened data $\mathbf{z} = \mathbf{\Lambda}^{-\frac{1}{2}}\mathbf{R}^T\mathbf{x}$ equals the identity matrix $\mathbf{I}$. It is assumed here that the covariance matrix has full rank, which is not the case for our data (see section 5.3 for an explanation how we used the ICA algorithm). If the independent components have unit variance (which can always be achieved by appropriate scaling) then the mixing matrix $\mathbf{W}^T$ of the whitened data is orthogonal:

$$\mathbf{z} = \mathbf{\Lambda}^{-\frac{1}{2}}\mathbf{R}^T\mathbf{x} = \mathbf{\Lambda}^{-\frac{1}{2}}\mathbf{R}^T\mathbf{As} = \mathbf{W}^T\mathbf{s} \tag{9}$$

$$E(\mathbf{zz}^T) = \mathbf{W}E(\mathbf{ss}^T)\mathbf{W^T} = \mathbf{WW^T} = \mathbf{I} \tag{10}$$

The goal is now to determine an orthogonal matrix $\mathbf{W}$ such that the fourth order cumulant tensor of $\mathbf{Wz}$ is diagonal.

In [1] an approach is taken which is based on the following two properties: First it can be shown that $\mathbf{W}$ jointly diagonalizes so called 'cumulant matrices', matrices resulting from a weighted sum over two dimensions of the fourth order cumulant tensor with arbitrary weights. Second, an orthogonal data transformation does not effect the total sum of the squared elements of the fourth order cumulant tensor, so the latter can be diagonalized by maximizing the sum of squares of its diagonal elements with respect to $\mathbf{W}$. The approach in [1] maximizes the sum of of squares of the cumulant tensor elements having identical first and second indices. This is equivalent to jointly diagonalizing a set of n^2 cumulant matrices. This set can be reduced to the n 'most significant' cumulant matrices by diagonalizing an $n^2 \times n^2$ matrix. In the following we will state the approach taken in [1] in more detail.

'Cumulant matrices' $\mathbf{N}$ are defined by a weighted sum of the elements of the fourth order cumulant tensor along two dimensions.

$$n_{ij} := \sum_{r=1}^{n}\sum_{s=1}^{n}\kappa^{(4)}(z_i, z_j, z_r, z_s)\, m_{rs}, \qquad 1 \le i, j \le n \tag{11}$$

where m_{rs} are the elements of an arbitrary matrix $\mathbf{M}$ and $\kappa^{(4)}(z_i, z_j, z_r, z_s)$ is the entry having the subscript $ijrs$ of the fourth order cumulant tensor $\kappa^{(4)}$ of the whitened data. Using the linearity of $\kappa^{(4)}$ and the diagonality of the cumulants for the statistically independent components s_i we can write

$$\kappa^{(4)}(z_i, z_j, z_r, z_s) = \sum_{a=1}^{n}\sum_{b=1}^{n}\sum_{c=1}^{n}\sum_{d=1}^{n} w_{ia}\, w_{jb}\, w_{rc}\, w_{sd}\, \kappa^{(4)}(s_a, s_b, s_c, s_d)$$

$$= \sum_{a=1}^{n} w_{ia}\, w_{ja}\, w_{ra}\, w_{sa}\, k_a \tag{12}$$

where k_a is the fourth order cumulant of the a-th independent component s_a (the a-th diagonal element of $\kappa^{(4)}(\mathbf{s},\mathbf{s},\mathbf{s},\mathbf{s})$. Inserting eq. (12) into (11) yields

$$n_{ij} = \sum_{a=1}^{n} w_{ia}\, w_{ja}\, k_a \sum_{r=1}^{n}\sum_{s=1}^{n} m_{rs}\, w_{ra}\, w_{sa}$$
$$= \sum_{a=1}^{n} w_{ia}\, w_{ja}\, \lambda_a \tag{13}$$

with $\lambda_a = \sum_{r=1}^{n}\sum_{s=1}^{n} m_{rs}\, w_{ra}\, w_{sa}$. Thus for every weight matrix $\mathbf{M}$ the demixing matrix $\mathbf{W}$ diagonalizes the corresponding cumulant matrix $\mathbf{N}$. While this is true in principle, for real world data where sample cumulants are used the resulting matrix $\mathbf{W}$ usually depends on the choice of the weight matrix $\mathbf{M}$. One could now proceed by jointly diagonalizing a set of arbitrary cumulant matrices but it is not clear a priori which is a good set of weight matrices. In [1] a link is established to the fact that the total sum of the squared elements of $\kappa^{(4)}$ is invariant under orthogonal data transformation $\mathbf{W}$:

$$\sum_{i,j,r,s=1}^{n} |\kappa^{(4)}(z_i,z_j,z_r,z_s)|^2 = \sum_{a,b=1}^{n}\sum_{i,j,r,s=1}^{m} w_{ia}\, w_{ib}\, w_{ja}\, w_{jb}\, w_{ra}\, w_{rb}\, w_{sa}\, w_{sb}\, k_a\, k_b$$
$$= \sum_{a=1}^{n} k_a^2 \tag{14}$$

This means that a transformation by the demixing matrix $\mathbf{W}$ maximizes the sum of the squared diagonal elements of $\kappa^{(4)}$. If so, $\mathbf{W}$ also maximizes the sum of squared elements with identical first and second indices.

$$\sum_{i=1}^{n}\sum_{r=1}^{n}\sum_{s=1}^{n} |\kappa^{(4)}(z_i,z_i,z_r,z_s)|^2 \tag{15}$$

This criterion is equivalent to joint diagonalizing the cumulant matrices with weight matrices $M^{(rs)}$ having one unit element at the subscript rs and the other elements being zero (cf. eq. (11)). Joint diagonalizing of several matrices $N^{(rs)}$ in practice means determining an orthogonal matrix $\mathbf{W}$ which maximizes the criterion

$$\sum_{r=1}^{n}\sum_{s=1}^{n} |diag(\mathbf{W}^T \mathbf{N}^{(rs)} \mathbf{W})|^2 \tag{16}$$

where $diag(.)$ is the sum of squares of the diagonal elements of the matrix argument. The sum in eq. (16) consists of n^2 addends and can be reduced to n addends by considering the following equality for $\mathbf{W}$ being a joint diagonalizer of the cumulant matrices $N^{(rs)}$. For $1 \le i \le n$

$$\sum_{a=1}^{n}\sum_{b=1}^{n} w_{ai}\, w_{bi}\, n_{ab}^{(rs)} = \sum_{a=1}^{n}\sum_{b=1}^{n} w_{ai}\, w_{bi}\, \kappa^{(4)}(z_a,z_b,z_r,z_s) = w_{ri}\, w_{si}\, k_i \tag{17}$$

This is an eigenvalue decomposition of an $n^2 \times n^2$ matrix $\mathbf{K}$ given by reordering the elements $\kappa^{(4)}(z_a, z_b, z_r, z_s)$ to the row indices $l = a + (n-1)b$ and column indices $c = s + (n-1)t$. The eigenvectors are $(w_{i1}w_{i1}, w_{i2}w_{i1}, \ldots, w_{ia}w_{ib}, \ldots, w_{in}w_{in})^T$ with $\sum_a \sum_b w_{ai}w_{bi}w_{aj}w_{bj} = \delta_{ij}$. Since there are at most n nonzero fourth order cumulants k_i of the independent components there are at most n significant cumulant matrices corresponding to the diagonal elements k_i of the cumulant tensor. These matrices are determined by the eigenvectors rearranged as weight matrices $\mathbf{w_i}\mathbf{w_i}^T$, where $\mathbf{w_i}$ denotes the i-th column of $\mathbf{W}$. This yields the sum in eq. (17) to be only over n elements

$$\sum_{i=1}^{n} |diag(\mathbf{W}^T \mathbf{N}^{(\mathbf{w_i})} \mathbf{W}|^2 \qquad (18)$$

$$\text{with} \quad \mathbf{N}^{(\mathbf{w_a})} = (\sum_{r=1}^{n} \sum_{s=1}^{n} \kappa^{(4)}(z_i, z_j, z_r, z_s) \, w_{ar} \, w_{as})_{ij}$$

In [1] the first n eigenvectors of the $n^2 \times n^2$ matrix $\mathbf{K}$ are determined and then the criterion in eq. (19) is maximized by Jacobi rotations.

4   The Experiment

In the following we show the results of analyzing fMRI data of a motor task experiment. The experiment consisted of three parts each taking 280 s. In the first part the subject should rest and refrain from moving. In the second part after a 60 s rest the subject was asked to tap fingers of both hands for 20 s and then rest again for 20 s. This finger tapping cycle was repeated six times, followed by a 20 s rest. The third part was identical to the second except that the subject was not supposed to actually do the finger tapping but only to imagine it according to the same time course as for the actual finger tapping. The images were taken with a frequency of 2 Hz, allowing a frequency resolution of up to 1 Hz. The finger tapping frequency of about 3 Hz was beyond this limit, but the finger tapping *cycle* frequency of $\nu = 1/40$ s $= 0.025$ Hz was resolved. Each image vector contained $m = 4015$ pixels arranged as a 55×73 gray scale image and a total of $k = 560$ images was taken. In the analysis, the first 20 images, corresponding to the first 10 s were omitted for magnetization steady state reasons. Simple correlation analysis, PCA and ICA were performed for each of the three parts of the experiment.

5   Results

5.1   Results Applying Correlation Analysis

The results of the correlation analysis of the actual finger tapping is shown in Fig. 1. An activated area is visible which can be identified as the motor cortex. The results of the correlation analysis of the other parts of the experiments is not shown here. For part one of the experiment no structure in the correlation images were visible, whereas for the imagined finger tapping an activated area identified as the supplementary motor cortex was determined.

5.2   Results Applying PCA

The results of applying PCA to the actual finger tapping are shown in Fig. 4. The PC's are depicted as time courses in the middle with the corresponding image on the left and the power spectrum on the right hand side, respectively. In Fig. 4 the components with the highest peaks at the finger tapping cycle frequency are shown. Computing the PC's no information other than the data matrix $\mathbf{X}$ was used and it can be seen that even without this information PCA reveals components which can be a posteriori related to the stimulus, e. g. for periodic stimuli by the power spectrum of the time courses.

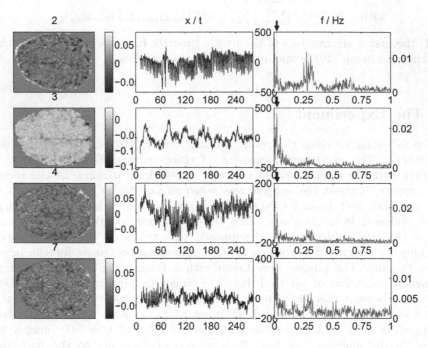

Fig. 4. PCA of actual finger tapping.
Left hand side: Images corresponding to the PC's 2, 3, 4 and 7. These PC's were the ones with the highest peaks at the finger tapping cycle frequency of 0.025 Hz. Middle: PC's (time courses) indicating the strength with which the PC is present in the measured image at the respective point in time.
Right hand side: Square root of the power spectrum of the PC's time course normalized to enable comparison. The finger tapping cycle frequency is indicated by arrows and dotted lines.
The "brain-like" pattern in the second PC most likely indicates head movements due to breathing an assumption which is also supported by the broad peak at the frequency around 0.3 Hz. The stimulus time course is best reproduced by the third PC, but with a weak the activation in the motor cortex. Note that another weak activation is visible in an area which can be identified as the supplementary motor cortex, which is active also in imagined finger tapping (cf. Fig. 5). The most prominent activation of the motor cortex is present in PC 7.

The results of PCA for the imagined finger tapping is shown in Fig. 5. An activated area similar to the one determined by correlation analysis is visible, which is identified as the supplementary motor cortex.

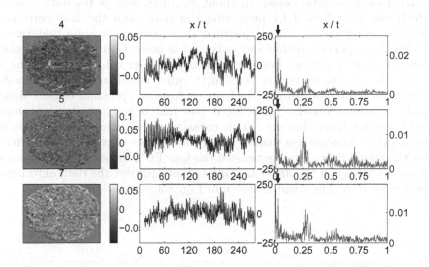

Fig. 5. PCA of imagined finger tapping.
PC's 4, 5 and 7 with the highest peaks at the tapping cycle frequency. In the fourth PC an activated area which can be identified as the supplementary motor cortex is visible. The other two PC's have prominent peaks at 0.025 Hz, however they do not show spatial structure.

Fig. 6 shows the sagittal sinus revealed by PCA of the first part of the experiment as an example of the ability of the stimulus-independent data analysis methods to reveal also non-stimulus related components.

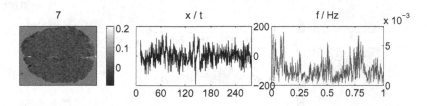

Fig. 6. PCA of resting brain.
Fourth Principal Component of the first part of the experiment where the subject should rest. The active areas represent the sagittal sinus.

However, the fact that several PC's show a strong relationship to the stimulus shows that the stimulus is not completely separated into one component by PCA. We thus proceed to the application of ICA hoping that the stimulus contribution is merged into fewer components by including higher order statistics.

5.3   Results Applying ICA

As the diagonalization of an $n^2 \times n^2$ matrix is prohibitive for high dimensional fMRI data, a reduction down to about 30 dimensions of the data is necessary which can be achieved by projecting the data onto the first corresponding principal components. This is justified since one can assume that the higher principal components mainly reflect Gaussian noise. Gaussian components are a nuisance in ICA anyway, because their higher order cumulants are zero, which makes the diagonalization of the fourth order cumulant tensor difficult. The dimension reduction furthermore frees us from the problem of rank deficiency of the covariance matrix. The reduced data matrix has more columns than rows and it can be shown that the sample covariance matrix for data matrices having a number of columns less than or equal the number of rows is rank deficient.

For the actual finger tapping we used the four PC's showing the strongest peaks at the frequency of the finger tapping cycle to project the data onto before we performed ICA. The result is shown in Fig. 5.3.

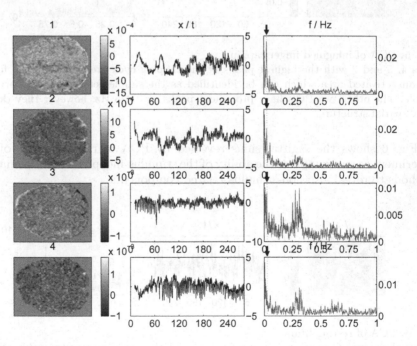

Fig. 7. ICA of actual finger tapping.
Independent Components obtained by ICA of the projection of the data onto the PC's numbered 2, 3, 4 and 7. For simplicity the IC's are numbered as well although no intrinsic order exists for Independent Components. Obviously the first IC contains most of the finger tapping cycle frequency and also an activated area in the motor cortex is visible. This indicates that indeed by ICA the stimulus related activity has been assigned to mainly one single component.

As a result of using the stimulus to chose the components for dimension reduction the stimulus related activity is mainly concentrated on the first independent component. A similar, slightly less concentrated result is obtained by projecting the data onto up to the first 30 principal components regardless of their relatedness to the stimulus.

For the imagined finger tapping we used the three PC's in Fig. 5 to project the data onto. The results are shown in Fig. 8.

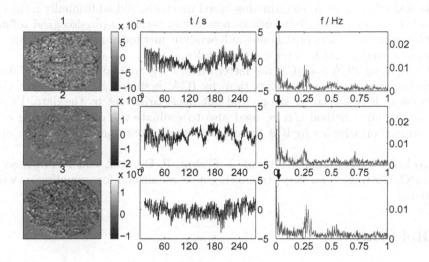

Fig. 8. PCA of imagined finger tapping.
Independent Components obtained by ICA of the projection of the data onto the PC's numbered 4, 5 and 7. The second IC contains most of the finger tapping cycle frequency, however the activated areas are not so clearly visible as in the fourth PC in Fig. 5. It seems that in this case ICA did not yield an improvement over PCA though the peak at 0.025 Hz is very pronounced in the second IC compared to the others.

In our analysis using PCA and ICA we have assumed that the interesting factors are characterized by a time course which is uncorrelated to or independent of the time courses of other influences of the data. However, one could also assume that the factors are characterized by a certain activity distribution. These factors could then be revealed by spatial PCA and ICA. This was done by McKeown et al. in [4]. For our data spatial ICA indeed seems to outperform temporal ICA in separating the stimulus-related activity [2].

6   Conclusion

We have shown that with a combination of PCA and ICA it is possible to extract stimulus related activity from the data without the need of prior information about the stimulus for the application of the methods. For the a-posteriori interpretation of the obtained results, however, information about the stimulus and non-stimulus-related influences is necessary. For periodical stimuli the stimulus related component can successfully be identified by the power spectrum of the respective time course. This is of course true also for periodically changing non-stimulus-related factors, e. g. many physiological influences. By using the power spectrum we identified stimulus-related as well as non-stimulus-related components such as blood vessels, head movements and breathing.

The comparison of stimulus-based and stimulus-independent methods reveals that the latter are more powerful since they extract the stimulus-related activity to a similar extent as the stimulus-based methods, and additionally return many other features of the data such as non-stimulus-related physiological influences. However the results of stimulus-independent methods are more complex and it can be challenging to interpret the results.

Comparing PCA and ICA the latter often leads to a concentration of different factors into fewer components than by ICA, however, the underlying assumptions in ICA are rather strong and not necessarily met by the data. PCA as a more robust method can be used also to evaluate the results of ICA, e. g. by comparison whether by ICA factors present in more than one PC have merged.

Acknowledgement. We thank J. Frahm, P. Dechent, and P. Fransson (MPI BPC, Göttingen) for very stimulating discussions and for providing us with the data.

References

1. J. F. Cardoso, A. Souloumiac, 1993, Blind Beamforming for Non Gaussian Signals, *IEE-Proceedings-F*, **140**, N° 6, 362-370
2. S. Dodel, J. M. Herrmann, T. Geisel, 2000, Comparison of Temporal and Spatial ICA in fMRI Data Analysis, *Proc. ICA 2000*, Helsinki, Finland, 543-547
3. J. Frahm, 1999, Magnetic Resonance Functional Neuroimaging: New Insights into the Human Brain, *Current Science* **76**, 735-743
4. McKeown MJ. Sejnowski TJ, 1998, Independent Component Analysis of fMRI Data - Examining the Assumptions, *Human Brain Mapping* 6(5-6), 368-372,
5. P. McCullagh, 1987, *Tensor Methods in Statistics*, Chapman and Hall, New York
6. W.H. Press, 1994, *Numerical recipes in C*, Cambridge Univ. Press
7. A. Stuart, K. Ord, 1994, *Kendall's Advanced Theory of Statistics, Vol. 1 Distribution theory*, Halsted Press

Emergence of Modularity within One Sheet of Neurons: A Model Comparison

Cornelius Weber and Klaus Obermayer

Dept. of Computer Science, FR2-1, Technische Universität Berlin
Franklinstr. 28/29, D-10587 Berlin, Germany. cweber@cs.tu-berlin.de

Abstract. We investigate how structured information processing within a neural net can emerge as a result of unsupervised learning from data. The model consists of input neurons and hidden neurons which are recurrently connected. On the basis of a maximum likelihood framework the task is to reconstruct given input data using the code of the hidden units. Hidden neurons are fully connected and they may code on different hierarchical levels. The hidden neurons are separated into two groups by their intrinsic parameters which control their firing properties. These differential properties encourage the two groups to code on two different hierarchical levels. We train the net using data which are either generated by two linear models acting in parallel or by a hierarchical process. As a result of training the net captures the structure of the data generation process. Simulations were performed with two different neural network models, both trained to be maximum likelihood predictors of the training data. A (non-linear) hierarchical Kalman filter model and a Helmholtz machine. Here we compare both models to the neural circuitry in the cortex. The results imply that the division of the cortex into laterally and hierarchically organized areas can evolve to a certain degree as an adaptation to the environment.

1 Introduction

The cortex is the largest organ of the human brain. However, a mammal can survive without a cortex and lower animals do not even have a cortex. Essential functions like controlling inner organs and basic instincts reside in other parts of the brain. So what does the cortex do? An other way to put this question is: what can the lower animals **not** do? If lower animals cannot learn a complex behavior we may infer that they cannot understand a complex environment. In other words: the cortex may provide a representation of a complex environment to mammals.

This suggests that the cortex is a highly organized structure. On a macroscopic scale, the cortex can be structured into dozens of areas, anatomically and physiologically. These are interconnected in a non-trivial manner, making neurons in the cortex receive signals primarily from other cortical neurons rather than directly from sensory inputs. In the absence of input the cortex can still generate dreams and imagery from intrinsic spontaneous activity. Recurrent connectivity between its areas may be the key to these capabilities.

S. Wermter et al. (Eds.): Emergent Neural Computational Architectures, LNAI 2036, pp. 53–67, 2001.

The cortex is, nevertheless, a sheet of neuronal tissue. Across its two dimensions, it hosts many functionally distinct areas (e.g. 65 areas in cat [12]) which process information in parallel as well as via hierarchically organized pathways [5]. The earliest manifestations of areas during corticogenesis are regionally restricted molecular patterns ("neurochemical fingerprints" [6]) which appear before the formation of thalamo-cortical connections [4].

Considering connectivity, there are up to ten times more area-to-area connections than areas. Thus, the description of cortico-cortical connectivity is more complex and requires modeling to be understood. Abstract geometrical models [12] suggest that topological neighborhood plays an important but not exclusive role in determining these connections.

Recently we presented computational models [14][13] in which the connections between areas are trained from neuronal activity. The maximum likelihood framework makes the network develop an internal representation of the environment, i.e. of the causes generating the training data.

Our recent models belong to two groups: one [14] we may call a non-linear Kalman filter model [11][9] in which neurons are deterministic and have a continuous transfer function. The other [13] is a Helmholtz machine [3] where hidden neurons are stochastic and binary. The dynamics of neuronal activations also differ: in the deterministic model, neurons integrate the inputs from several sources and over time. In the stochastic model, computations are separated and do not need to be integrated over time. The activation terms at different times are then summed up in the learning rules.

The model areas are determined a priori by the intrinsic functional properties of their neurons. More precisely, hidden neurons are divided into two groups which differed in their firing properties by corresponding parameter changes. Thus one group responds with stronger activity to a given input than the other group. In consequence the first group learns to process activity patterns which occur more frequently. The input space can thereby be divided into two groups. Using lateral weights among the hidden neurons, these two models allow hidden neurons to code using two hierarchical levels. When presented with hierarchically generated data, some of the hidden neurons establish a second hierarchical level by grouping together other neurons via their lateral weights while their weights to the input neurons decline.

In this contribution we want to inquire principles according to which the connectivity between cortical areas arises. We set up a model for the development of the connectivity between connectionist neurons and consider the macroscopic areas to be made up of a (small) group of microscopic neurons. A key idea is that such groups are distinguished by their neuronal properties prior to learning. We thereby do not address the possibility that intrinsic neuronal properties (other than those determined by their weights) can be dynamically changed during development.

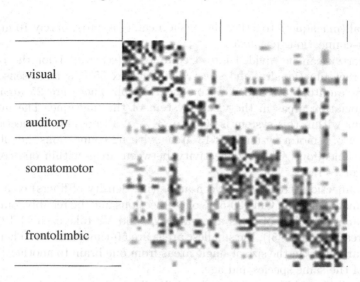

visual

auditory

somatomotor

frontolimbic

Fig. 1. The cortico-cortical connection matrix of the cat. On both axes, the 65 cortical areas are arranged into the four complexes, visual system, auditory system, somatosensory system and fronto-limbic system. Connections along the horizontal axis are afferent, and those along the vertical axis are efferent. Strong connections are depicted dark, weaker connections are depicted in a lighter grey, missing connections white. Self connections (diagonal) are omitted. Data was taken from [12].

1.1   Review of Biological Data

Cortical areas can be hard to distinguish. For this reason, different criteria may be combined:

- Chemical properties: recent findings suggest that a "neurochemical fingerprint" [6] determines the earliest compartmentalization in corticogenesis [4].
- Architecture: staining can reveal a different structure. This method is reliable only for a few areas [5].
- Physiological properties. As an example, topographic organization is measurable w.r.t. the visual field in half of all visual cortical areas [5].
- The connectivity "fingerprint", i.e. the connectivity pattern to other cortical areas. So, if two areas had similar connectivity patterns, then they would be the same.

An estimate of the number of cortical areas is 65 in the cat [12] and 73 in the macaque [15]. The number of connections reported by Young [12] between these areas is 1139 in the cat (Fig. 1) which represents 27.4% of all possible connections (only ipsilateral connections are considered). The number of connections reported in the macaque [15] is 758 which represents 15% of all possible connections. Most of the connections are bidirectional: they consist of axons going into both directions. There are only 136 reported one-way connections which is 18%

of the total (macaque). Together there is a mean of approximately 10 input- and 10 output-connections per area.

The cortex can be divided into 4 complexes (see Fig. 1 for the cat data) with different functionality and sizes: visual cortex 55%, somato-sensory 11%, motor 8%, auditory 3% and 23% for the rest [5]. There are 25 areas plus 7 visual-association areas in the visual system of the macaque. The number of connections within this system is 305, i.e. 31% of all possible connections. The somato-sensory/motor system has 13 areas with 62 connections, i.e. 40% of all possible connections. Thus, connectivity between areas within one complex is higher than average.

The connection strengths between areas (i.e. density of fibers) comprise two orders of magnitude [5]. Only 30-50% of connections may be reliably found across different animals. Sizes of areas also vary: V1 and V2 take each 11-12% of the surface area (macaque [5]), the smallest areas are 50-times smaller. There is even a 2-fold variability in the size of single areas from one brain to another [5] within animals of the same species and age.

Finally, it should be noted that every visual area is connected to non-cortical areas. The number of these connections may out-range the number of cortico-cortical connections.

Table 1. The roles of intrinsic and activity dependent developmental mechanisms. Neural connections as well as areas are the result of an interplay between both intrinsic and activity dependent mechanisms.

	intrinsic	activity dependent
what is meant	genetic description	learning
how does it work	chemical markers	Hebbian learning
when does it appear	early	late
its targets	cell movement, cell differentiation, connections	connections
its results	layers, areas	receptive field properties, barrels, areas

2 Methods

Commonly used model architectures have a pre-defined structure because the order in which neuronal activations are computed depends on the architecture. An architecture as in Fig. 2 **b)** for example does not allow a two-stage hierarchy to develop. An architecture as in Fig. 2 **c)** does not allow the hierarchically topmost units (the three dark units) to develop connections to the input.

A more general structure is a full connectivity between all neurons, as in Fig. 2 **a)**. The only restriction we have chosen here is that input units are not

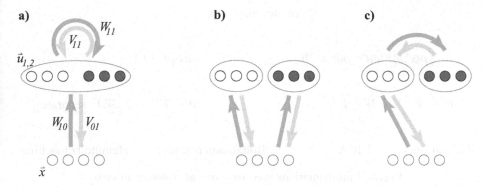

Fig. 2. Three different model architectures. In each of them the activations x on the input units are represented by hidden unit activations u. W are recognition weights, V are generative weights, indexed in the left figure with the number of the layer of termination and origin. **a)** architecture of our model. The lateral recognition weights W_{11} and generative weights V_{11} (top) allow each hidden neuron to take part in a representation u_1 on a lower and u_2 on a higher hierarchical level. Dark units differ from the white hidden units by a parameter of their transfer function only. Depending on the structure of the data training will result in one of the two other architectures shown: **b)** a parallel organization and **c)** a hierarchical organization of the two areas.

connected to other input units. Learning rules for such architectures exist, like the well-known Boltzmann machine learning rule. The purpose of our study is to let structures as in Fig. 2 **b)**,**c)** self-organize from this general, full connectivity.

2.1 Different Approaches to Maximum Likelihood

The goal of maximum likelihood is to find the model which can best explain, i.e. generate, the whole data set $\{x^\mu\}$. If the data are independent, we have:

$$p(\{x^\mu\}|V) \;=\; \prod_\mu p(x^\mu|V) \;=\; \int \prod_\mu p(x^\mu, u|V)\, du$$

where u is the hidden unit activation vector. The data x^μ and the hidden representation u are related to each other, so u is an internal representation of a data point. Through learning the model parameters V adjust to the whole data set, so the weights are a representation of the whole environment.

The integration across the hidden code u which is necessary to obtain the probability distribution of the model generating the data is computationally infeasible. Different approximations therefore must be done which lead to different models. Examples are shown in Fig. 3.

On the left branch of Fig. 3, the integral over the hidden code is replaced by the "optimal" hidden code vector u^{opt} which generates a certain data item with highest probability. This estimates a current system state u which relates to an observed process through $x = Vu + e$ with noise e. With a linear transform

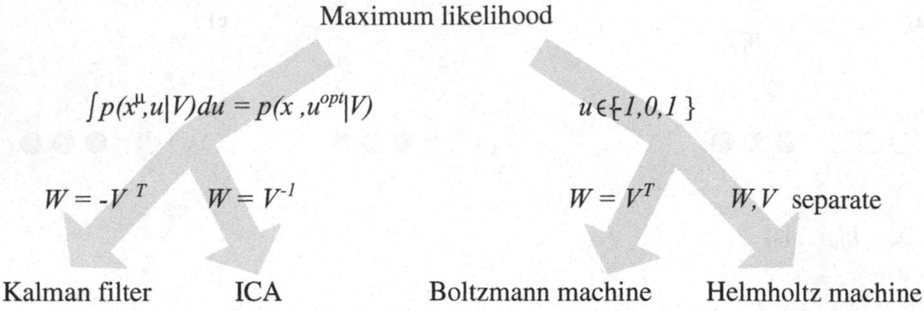

Maximum likelihood

$$\int p(x^{\mu},u|V)du = p(x,u^{opt}|V) \qquad\qquad u \in \{-1,0,1\}$$

$$W = -V^{T} \qquad W = V^{-1} \qquad\qquad\qquad W = V^{T} \qquad W,V \ \text{separate}$$

Kalman filter   ICA   Boltzmann machine   Helmholtz machine

Fig. 3. Approximations used to arrive at different models.

V and in the case of Gaussian noise e we would arrive at a Kalman filter [11]. However, we will assume a non-Gaussian "sparse" prior in our model which we then term a non-linear Kalman filter. In the special case of an invertible weight matrix, one arrives at an ICA algorithm [2] (see [10] for comparison).

On the right branch, the hidden code vector is discretized such that the integral may be computed. Having weight symmetry and a strict gradient ascent algorithm for training the weights which maximizes the likelihood we arrive at the Boltzmann machine (standard literature, e.g. [7]). Separate treatment of recognition weights and generative weights leads to the Helmholtz machine [3]. A practical but heuristic algorithm to train this network is the wake-sleep algorithm [8].

For gradient based training of the weights, the performance of the network in generating the data is permanently measured. The way in which the data set is used for the generation of the data throughout learning distinguishes the two models which we explore in this contribution:

- **Kalman filter model:** when the model generates data, a "target" data point x^{μ} is always selected which has to be reconstructed by the model.
- **Helmholtz machine:** without any data point selected, the model has to generate any of the given data points with the appropriate probability.

Thus, whether there is or there is no data point present influences the way in which the fast changing model parameters, the activations, are used:

- **Kalman filter model:** the goal is to find, given a data point, the posterior distribution of the fast changing model variables, i.e. the hidden unit activations. As an approximation, the optimal representation u^{opt} which maximizes the posterior probability to generate that data point is selected.
- **Helmholtz machine:** generation of the data is separated in time from the process of recognition: a "wake phase" characterized by presence of data is distinguished from a "sleep phase" during which the net generates its own "fantasy" data.

2.2 Architecture and Notation

Weights: The network architecture as well as some of the variables are depicted in Fig. 2, left. Input units (below) are linear and receive the data. Recognition weights W_{10} transfer information from the input units to all hidden units. Generative weights V_{01} transfer information from the hidden units to the input units. Lateral weights W_{11} and V_{11} transfer information from all hidden units to all hidden units. They are also distinguished into recognition weights, W_{11}, and generative weights, V_{11}, depending on whether they are used to transfer information upwards within the functional hierarchy or downwards towards the input units, respectively. In the Kalman filter model, $W_{10} = V_{01}^T$ and $W_{11} = V_{11}^T$.

Activations: The data cause an activation vector x on the input neurons. Hidden unit activation vectors have to be assigned to a virtual hierarchical level: in a hierarchical organization, the order of neuronal activation updates is a series where the number of update steps which are needed to propagate information from a data point to a neuron defines its hierarchical level. A hidden neuron is regarded to code on the first hierarchical level if it is activated by the input units. We denote this activation u_1. A hidden neuron is regarded to code on the second hierarchical level if it is activated by its input from other hidden neurons via lateral weights only but not from the input units directly. We denote this activation u_2. Note that both activations, u_1 and u_2, occur on all hidden units. Activations are distinguished by the way they arise on a neuron and may coexist in the Kalman filter model.

Parameters: Please see [14] and [13] for the values of all parameters in the non-linear Kalman filter model and the Helmholtz machine, respectively.

2.3 The Kalman-Filter Algorithm

The Kalman-filter algorithm can be separated into two steps: *(i)* calculation of the reconstruction errors $\tilde{x}_0(t)$ on the input units and $\tilde{x}_1(t)$ on the first hidden level and *(ii)* adjustment of activations in order to decrease these errors. Both steps are repeated until equilibrium is reached in which case the optimal hidden representation of the data has been found.

(i) compute reconstruction errors:
$$\tilde{x}_0(t) = x - V_{01}u_1(t)$$
$$\tilde{x}_1(t) = W_{10}\tilde{x}_0(t) - V_{11}u_2(t)$$

(ii) adjust hidden unit activations:
$$h_1(t) = u_1(t) + \varepsilon_u((2\beta - 1)W_{10}\tilde{x}_0(t) + (1 - \beta)V_{11}u_2(t))$$
$$h_2(t) = u_2(t) + \varepsilon_u W_{11}\tilde{x}_1(t)$$

where transfer functions $u_1(t+1) = f(h_1(t))$ and $u_2(t+1) = f(h_2(t))$ are used. ε_u denotes the activation update step size and β denotes the trade-off between bottom-up and top-down influence.

After repetition of these two steps until convergence the weights are updated (with step sizes ε^w) according to:

$$\Delta w_{10}^{ij} = \varepsilon_{10}^w u_1^i \tilde{x}_0^j \qquad\qquad \Delta w_{11}^{ik} = \varepsilon_{11}^w u_2^i \tilde{x}_1^k$$

2.4   The Wake-Sleep Algorithm (Helmholtz Machine)

The wake-sleep algorithm consists of two phases.

First, in the wake phase, data are presented. Then the net finds a hidden representation of the data and based on this representation, the net re-estimates the data.

Wake phase

infer hidden code: $u_1^w = f_m^w(W_{10}x)$ $u_2^w = f_m^w(W_{11}u_1^w)$

reconstruct input: $s_1^w = V_{11}u_2^w$ $s_0^w = V_{01}u_1^w$

After one-sweep computation of these equations the difference between the data and the re-estimation is used to train the generative weights (ε are the respective learning step sizes):

$$\Delta V_{11} = \varepsilon_{11}\left(u_1^w - s_1^w\right) \cdot \left(u_2^w\right)^T \qquad\qquad \Delta V_{01} = \varepsilon_{01}\left(x - s_0^w\right) \cdot \left(u_1^w\right)^T$$

Secondly, in the sleep phase, a random "fantasy" activation vector is produced in the highest level. The net then generates the corresponding "fantasy" data point and based on this, the net re-estimates the activation vector on the highest level.

Sleep phase

initiate hidden code at highest level: $s_2^s = f_b^s(0)$

generate input code: $s_1^s = f_b^s(V_{11}s_2^s)$ $s_0^s = V_{01}s_1^s$

reconstruct hidden code: $u_1^s = f_m^s(W_{10}s_0^s)$ $u_2^s = f_m^s(W_{11}s_1^s)$

After obtaining these variables the difference between the original activation vector and the re-estimation is used to train the recognition weights:

$$\Delta W_{10} = \varepsilon_{10}\left(s_1^s - u_1^s\right) \cdot \left(s_0^s\right)^T \qquad\qquad \Delta W_{11} = \varepsilon_{11}\left(s_2^s - u_2^s\right) \cdot \left(s_1^s\right)^T$$

2.5   Weight Constraints

In order to discourage a hidden neuron to be active at all processing steps, competition between all incoming weights of one neuron is introduced by an activity-dependent weight constraint for both models. This encourages a hidden neuron to receive input from the input neurons via W_{10} or from other lateral neurons via W_{11} but not both.

– In the Kalman filter model, recognition weights w_{10}^{ij} from input neuron j to hidden neuron i and lateral recognition weights w_{11}^{ik} from hidden neuron k to hidden neuron i receive the following activity dependent weight constraint which is added to the weight learning rules:

$$\Delta w_{10}^{ij^{constr}} = -\lambda^w \, |\bar{h}^i| \, w_{10}^{ij} \, \|\boldsymbol{w}^i\|^2$$
$$\Delta w_{11}^{ik^{constr}} = -\lambda^w \, |\bar{h}^i| \, w_{11}^{ik} \, \|\boldsymbol{w}^i\|^2$$

where λ^w is a scaling factor. $\|\boldsymbol{w}^i\|^2 = \sum_l^N (w_{10}^{il})^2 + \sum_l^H (w_{11}^{il})^2$ is the sum of the squared weights to all N input units and all H hidden units and $|\bar{h}^i| = |h_1^i| + |h_2^i|$ is the mean of absolute values of the inner activations of hidden neuron i at the final relaxation time step. Generative weights are made symmetric to the recognition weights, i.e. $V_{01} = W_{10}^T$ and $V_{11} = W_{11}^T$.

– For the Helmholtz machine, the weight constraint term which is added to the learning rule treats positive and negative weights separately. Using the Heaviside function $\Theta(x) = 1$, if $x > 0$, otherwise 0, we can write:

$$\Delta w^{ij^{constr}} = -\lambda^w \, \bar{h}^i \, \Theta(w^{ij}) \, w^{ij} \sum_{j'} \Theta(w^{ij'}) \, (w^{ij'})^2$$

where $\bar{h} = \boldsymbol{u}_1^w + \boldsymbol{u}_2^w + \boldsymbol{u}_1^s + \boldsymbol{u}_2^s$ is the sum of all activations which have been induced by the recognition weights. The indices j, j' extend over all input and hidden units. The wake-sleep algorithm is not a gradient descent in an energy space. It easily gets stuck in local minima. To improve the solutions found, generative weights V_{01} and V_{11} as well as lateral recognition weights W_{11} were rectified, i.e. negative weights were set to zero.

The weight constraints scale the length but do not change the direction of a hidden neuron weight vector. They are local in the sense that they do not depend on any weight of any other hidden neuron.

2.6 Distinguishing the Modules

In order to help two distinct areas to evolve, the hidden neurons are separated into two groups by assigning them different intrinsic parameters of their transfer functions. The key idea is that neurons from one area are more active than neurons from the other. The more active neurons will then respond to the more frequent features that can be extracted from the data; the less active neurons are expected to learn those features which do not occur as often. Note that these differences can distinguish hierarchical levels.

– In the Kalman filter model, a difference in one parameter among the hidden neurons will be sufficient to initiate the segregation process, either in parallel or hierarchically, depending on the data. The transfer function

$$f(h_i) = h_i - \lambda \cdot \frac{2h_i}{1 + h_i^2}$$

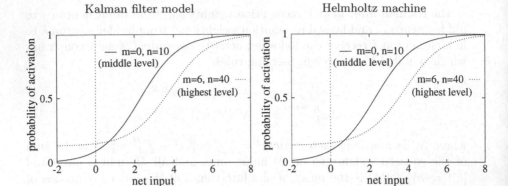

Fig. 4. Neuronal transfer functions.

is depicted in Fig. 4, left, with the two different values of the parameter λ which controls the sparseness of neuronal firing. Larger values of λ make a neuron respond with smaller activation to a given input.

- For the Helmholtz machine a stochastic transfer function is chosen for the hidden neurons (Fig. 4, right):

$$f_m(h_i) = \frac{e^{h_i} + m}{e^{h_i} + m + n}$$

By this function, the stochastic "ON"-state can be traced back to two distinguished sources. First, the activation from the neurons input, h_i and second, the parameter m. Both increase the probability for the neuron to be "ON". We refer to the latter as spontaneous or endogenous activity. The parameter n adds some probability for the neuron to be "OFF", thus encourages sparse coding.

We chose the parameters such that they matched the precise role the neurons should play in the wake-sleep algorithm. Especially the hierarchical setting has to be considered as it is more difficult to achieve this kind of structure. Two distinct physiological properties of the neurons are important for the role they play in the wake-sleep algorithm and thus two parameters are varied. (i) Neurons in the highest hierarchical level are responsible for initiation of the hidden code, i.e. they have to be spontaneously active without any primary input. For this reason we assigned high resting activity to the units which are designed for the higher level by setting the parameter m to a high value. (ii) Lower level neurons should respond to input. This applies both to recognition when there is input from the input neurons and to generation when there is input from the highest units. High responsivity is achieved by a strong gain – or a small sparseness parameter n.

3   Results

3.1   Generation of the Training Data

The data consist of discrete, sparsely generated elements. These are lines of 4 different orientations on a 5×5 grid of input neurons. In the parallel paradigm, horizontal and 45° lines are generated with probability 0.05 whereas the other group, vertical and 135° lines are generated twice as often, with probability 0.1 each (Fig. 5, **a)**). As a result of training, the group of more active neurons is expected to preferably code for the more frequent data elements, and vice versa (cf. [14][13]).

a)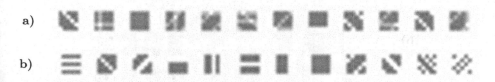
b)

Fig. 5. Examples of stimuli x used for training. **a)** Stimuli generated by two models in parallel. **b)** Stimuli generated by a hierarchical model. White means a positive signal on a grey zero value background.

In the hierarchical paradigm, one of 4 orientations are chosen, which represents a decision process within a higher hierarchical level. Then, on the lower level, lines from the formerly chosen orientation only are generated (with probability 0.3 each). See Fig. 5, **b)**.

3.2   Results of Training

Parallel Structure: Parts **a)** of Figs. 6 and 7 show the resulting weight matrices after both nets have been trained on the parallelly generated data. The neurons have extracted the independent lines from the data. In general, neurons in the upper half (Kalman filter) or upper third (Helmholtz machine) code for the more frequent lines (90° and 135°, in polar coordinates), whereas neurons in the lower parts code for the less frequent lines (0° and 45°).

The reason for this division of labor is that neurons in the upper part are more active. In case of the Helmholtz machine, the critical factor is the resting activity, more than the responsitivity to input. The reason for this may be that in the initial phase of training, input is generally low and the resting activity accounts for most of the learning.

Hierarchical Structure: Parts **b)** of Figs. 6 and 7 show the resulting weight matrices after training on the hierarchically generated data. Only in the case of the Helmholtz machine, the input is decomposed into the independent lines; the Kalman filter model generates the data using a superposition of several, more complex features each of which reflect one orientation but not an individual line.

W_{10} W_{11}

a)

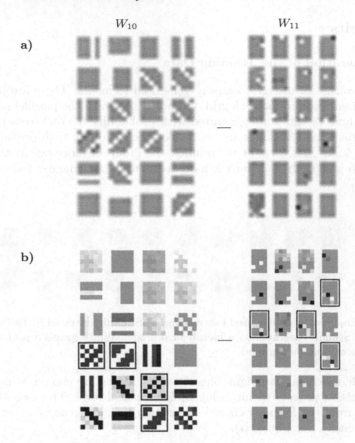

b)

Fig. 6. Kalman filter model results. **Left:** the recognition weight matrices W_{10} and **right:** the lateral recognition weight matrices W_{11} after training. Each square of the weight matrices shows a receptive field of one of the 4×6 hidden neurons. Each neuron has weights to one of the 5×5 input neurons (**left**) and lateral weights (**right**). Here, black indicates negative, white positive weights. Contrast is sharpened by a piecewise linear function such that weights weaker than 10 percent of the maximum weight value are not distinguished from zero and weights stronger than 60 percent of it appear like the maximum weight value. Short lines between left and right matrices indicate the area boundaries.
a) Parallel organization of areas: weights W_{10} to the inputs generally code for $0°$ and $45°$ lines in the lower half and on $90°$ and $135°$ lines in the upper half. **b)** Hierarchical organization of areas: neurons in the lower half code for the input via W_{10} while mainly neurons in the upper half group together some of the neurons in the lower half via W_{11}. Receptive fields which code for lines of $45°$ orientation are marked by a frame.

In the case of the Helmholtz machine, such a superposition cannot generate the data because the generative weights V are constrained to be positive.

Hierarchical structure has emerged such that, first, the less active neurons in the lower parts have pronounced weights to the input, and second, the more

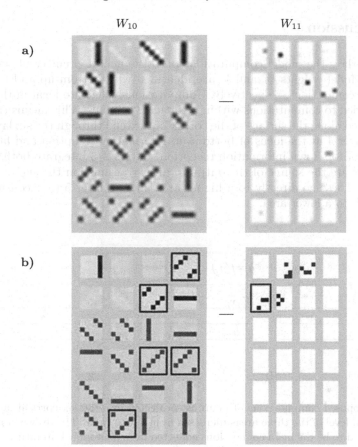

Fig. 7. Helmholtz machine model results. Architecture and display as in Fig. 6. Negative weights are brighter than the background (frame), positive weights are darker. For lateral weights (**right**), zero weights are depicted white (there are no negative weights). **a)** Areas have organized in parallel: weights W_{10} to the inputs code for 90° and 135° lines in the upper third and predominantly for 0° and 45° lines in the lower two thirds. **b)** Areas have organized hierarchically: neurons in the lower two thirds code for the input via W_{10} while four neurons in the upper third each integrate via W_{11} units from the lower two thirds which code stimuli of one direction. Neurons which code for lines of 45° orientation are marked by a frame.

active neurons in the upper parts have more lateral weights to the less active hidden neurons. These lateral connections group together neurons which code for the same orientation. In the Kalman filter model, results are fuzzy, because even neurons on the second level code in a distributed manner. Some neurons also code on both the first and second hierarchical level.

Outliers are found in both models. Note that it is only the data that changed between the parallel and vertical setting. All parameters and also initial conditions like randomized weights were the same in both settings for each model.

4   Discussion

Fig. 8 overlays the model computations onto the neural circuitry of the cortex (cf. [9]). Model weights W and V are identified with bottom-up and top-down inter area connections, respectively. Computations which are local in the model are identified to computations within a cortical "column". This means that they are localized along the surface of the cortex but extend through the six layers. We identify layer 4 as the locus of bottom-up information reception and layer 6 as the locus of top-down information reception. Layers 2/3 integrate both of these inputs and are the source of the outputs (after transmission through a transfer function). These go directly to a higher area and via layer 5 (not considered in the model) to a lower area.

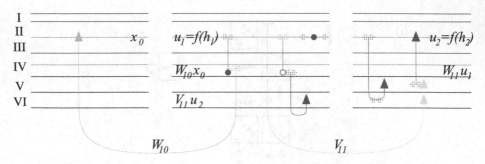

Fig. 8. Proposed computations of neurons, projected into the six cortical layers. The hierarchical level of the three areas increases from left to right. The drawn connections are proposed to be the main connections collected from anatomical literature. The area depicted to the left corresponds to the input layer of the models. There, data x_0 origin from layers 2/3. Alternatively, one can treat this as thalamic input.

Both models, the non-linear Kalman filter model and the Helmholtz machine, can be identified with this connection scheme and the notations of the model equations can be applied to Fig. 8. Both models use basic computations like the scalar product between the inter-area weights and the activations from the source area of an input. These computations follow directly from the anatomy.

The differences between the models lie in the way bottom-up and top-down input is integrated within one area. In the Kalman filter model, all activation values u_1, u_2 which are needed for training are computed incrementally, thus, they are maintained over a whole relaxation period. It is biologically implausible that each hidden neuron keeps track of both values. In the Helmholtz machine (wake-sleep algorithm), activations from the bottom-up and the top-down path are computed serially, they flow from one hierarchical level to the next. Even though a neuron belongs logically to two hierarchical levels, when it is activated on a certain level it can forget the previous activation on another level. The update dynamics is biologically plausible and reminiscent of a synfire chain model [1]. However, during recognition there is no feedback from higher cortical areas.

In the case of the Kalman filter model, when the learning rule is evaluated, all relevant terms are present. In the case of the Helmholtz machine, the full learning rule is split into two parts which are evaluated at separate times, a "wake" mode when data activate the input units and a "sleep" mode when hidden neurons become spontaneously active. If data is missing, we have shown that spontaneous activity on its own can in principle mould the internal structure of the network [13] in a restricted way.

More subtle differences between the two models certainly cannot show up in the anatomy: the time order of the neuronal computations, the initialization of activities and the learning rules. On the other hand, there are restrictions in the models: only the main streams are considered, lateral connections are omitted. Furthermore, in the models, no learning takes place within a cortical "column".

Acknowledgements. The writing-up was done in the lab of Alexandre Pouget and Sophie Deneve and Suliann Ben Hamed corrected this work.

References

1. M. Abeles. *Corticonics. Neural Cirquits of the Cerebral Cortex.* Cambridge University Press, 1991.
2. Anthony J. Bell and Terrence J. Sejnowski. An information-maximization approach to blind separation and blind deconvolution. *Neur. Comp.*, 7(6):1129–1159, 1995.
3. P. Dayan, G. E. Hinton, R. Neal, and R. S. Zemel. The Helmholtz machine. *Neur. Comp.*, 7:1022–1037, 1995.
4. M. J. Donoghue and P. Rakic. Molecular evidence for the early specification of presumptive functional domains in the embryonic primate cerebral cortex. *J. Neurosci.*, 19(14):5967–79, 1999.
5. D.J. Felleman and D.C. Van Essen. Distributed hierarchical processing in the primate cerebral cortex. *Cerebral Cortex*, 1:1–47, 1991.
6. S. Geyer, M. Matelli, G. Luppino, A. Schleicher, Y. Jansen, N. Palomero-Gallagher, and K. Zilles. Receptor autoradiographic mapping of the mesial motor and premotor cortex of the macaque monkey. *J. Comp. Neurol.*, 397:231–250, 1998.
7. S. Haykin. *Neural Networks. A Comprehensive Foundation.* Macmillan College Publishing Company, 1994.
8. G. E. Hinton, P. Dayan, B. J. Frey, and R. Neal. The wake-sleep algorithm for unsupervised neural networks. *Science*, 268:1158–1161, 1995.
9. M. Kawato, H. Hayakawa, and T. Inui. A forward-inverse optics model of reciprocal connections between visual cortical areas. *Network*, 4:415–422, 1993.
10. B.A. Olshausen. Learning linear, sparse, factorial codes. A.I. Memo 1580, Massachusetts Institute of Technology., 1996.
11. R.P.N. Rao and D.H. Ballard. Dynamic model of visual recognition predicts neural response properties of the visual cortex. *Neur. Comp.*, 9(4):721–763, 1997.
12. J.W. Scannell, C. Blakemore, and M.P. Young. Analysis of connectivity in the cat cerebral cortex. *J. Neurosci.*, 15(2):1463–1483, 1995.
13. C. Weber and K. Obermayer. Emergence of modularity within one sheet of intrinsically active stochastic neurons. In *Proceedings ICONIP*, 2000.
14. C. Weber and K. Obermayer. Structured models from structured data: emergence of modular information processing within one sheet of neurons. In *Proceedings IJCNN*, 2000.
15. M.P. Young. The organization of neural systems in the primate cerebral cortex. *Proc. R. Soc. Lond. B*, 252:13–18, 1993.

Computational Investigation of Hemispheric Specialization and Interactions

James A. Reggia[1,2], Yuri Shkuro[1], and Natalia Shevtsova[3]

[1] Dept. of Computer Science, A. V. Williams Bldg., University of Maryland, College Park, MD 20742 USA; reggia@cs.umd.edu, merlin@cs.umd.edu
[2] UMIACS and Department of Neurology, UMB
[3] A. B. Kogan Research Institute for Neurocybernetics, Rostov State University, Russia, nisms@krinc.rnd.runnet.ru

Abstract. Current understanding of the origins of cerebral specialization is fairly limited. This chapter summarizes some recent work developing and studying neural models that are intended to provide a better understanding of this issue. These computational models focus on emergent lateralization and also hemispheric interactions during recovery from simulated cortical lesions. The models, consisting of corresponding left and right cortical regions connected by the corpus callosum, handle tasks such as word reading and letter classification. The results demonstrate that it is relatively easy to simulate cerebral specialization and to show that the intact, non-lesioned hemisphere is often partially responsible for recovery. This work demonstrates that computational models can be a useful supplement to human and animal studies of hemispheric relations, and has implications for better understanding of modularity and robustness in neurocomputational systems in general.

1 Introduction

The modularity of the cerebral cortex can be viewed at many levels of abstraction: neurons, columns, areas, or lobes/hemispheres. Here we consider the large-scale interconnected network of cortical areas/regions that has been the subject of intense recent experimental investigation by neuroscientists. During the last several years, we and others have studied computational models of such areas and their interactions to better understand how these modules distribute and acquire functions. In this paper we focus on cerebral specialization, and describe results obtained with computational simulations of paired, self-organizing left and right cortical areas and their response to sudden localized damage. These studies have implications for the design and development of large-scale neurocomputational systems, particularly with respect to modularity and robustness.

In the following, we first review some basic facts about cerebral functional asymmetries, the issue of transcallosal hemispheric interactions, and interhemispheric effects of ischemic stroke, to provide the reader with relevant background information. Uncertainty about how to interpret some of this experimental data

S. Wermter et al. (Eds.): Emergent Neural Computational Architectures, LNAI 2036, pp. 68–82, 2001.

provides the motivation for turning to computational studies. Models of interacting left and right hemispheric regions are then summarized for two separate tasks: phoneme sequence generation and visual identification of letters. While these two models differ in the tasks they address, their architecture, and the learning rules they use, they lead to similar conclusions about hemispheric interactions and specializations, which are summarized in a final Discussion section.

2 Hemispheric Asymmetries and Interactions

A number of functional cerebral specializations exist in humans and these have been frequently reviewed (e.g., [12,22]). These cognitive/ behavioral lateralizations include language, handedness, visuospatial processing, emotional facial expression, and attention. There is great plasticity of the brain with respect to such functional asymmetries. For example, left hemispherectomy in infants can result in the right hemisphere becoming skilled in language functions [14], suggesting that the cerebral hemispheres each have the capacity at birth to acquire language.

The underlying causes of hemispheric functional lateralization are not well understood and have been the subject of investigation for many years. The many anatomic, cytoarchitectonic, biochemical and physiological asymmetries that exist have been reviewed elsewhere [18,22]. Asymmetries include a larger left temporal plane in many subjects, greater dendritic branching in left speech areas, more gray matter relative to white matter in the left hemisphere than in the right, different distributions of important neurotransmitters such as dopamine and norepinephrine between the hemispheres, and a lower left hemisphere threshold for motor evoked potentials.

Besides intrinsic hemispheric differences, another potential factor in function lateralization is hemispheric interactions via the corpus callosum. Callosal fibers are largely homotopic, i.e., each hemisphere projects in a topographic fashion so that mostly mirror-symmetric points are connected to each other [23,38]. It is currently controversial whether each cerebral hemisphere exerts primarily an excitatory or an inhibitory influence on the other hemisphere. Most neurons sending axons through the corpus callosum are pyramidal cells, and these end mainly on contralateral pyramidal and stellate cells [23]. Such apparently excitatory connections, as well as transcallosal diaschisis and split-brain experiments, suggest that transcallosal hemispheric interactions are mainly excitatory in nature [5], but this hypothesis has long been and remains controversial: several investigators have argued that transcallosal interactions are mainly inhibitory or competitive in nature [10,13,17,36]. Support for this latter view is provided by the fact that transcallosal monosynaptic excitatory postsynaptic potentials are subthreshold, of low amplitude, and followed by stronger, more prolonged inhibition [54]. Also, behavioral studies suggest that cerebral specialization for language and other phenomena may arise from interhemispheric competition or "rivalry" [10,17,27], while transcranial magnetic stimulation studies clearly indicate that activation of one motor cortex inhibits the contralateral one [16, 33].

Computational research on modeling hemispheric interactions and specialization is motivated by current uncertainties about how to interpret the enormous, complex body of existing experimental data. While there have been many previous neural models of cerebral cortex, very few have examined aspects of hemispheric interactions via callosal connections. One past model showed that oscillatory activity in one hemisphere can be transferred to the other via interhemispheric connections [1,2]. Another demonstrated that inhibitory callosal connections produced slower convergence and different activity patterns in two simulated hemispheres [11]. Additionally, a pair of error backpropagation networks were trained to learn a set of input-output associations simultaneously, and it was shown that slow interhemispheric connections were not critical for short output times [45]. None of these past neural models of hemispheric interactions examined the central issues of interest here: lateralization of functionality through synaptic weight changes (learning), and the interhemispheric effects of simulated focal lesions.

Other studies have related simulation results to their implications for lateralization, but did not actually model two hemispheric regions interacting via callosal connections [7,29]. A recent mixture-of-experts backpropagation model examined how unequal receptive field sizes in two networks could lead to their specialization for learning spatial relations, but did not incorporate callosal connections or represent cortical spatial relationships [25]. Another recent model focusing on receptive field asymmetries from a different perspective has also been used to explain several experimentally-observed visual processing asymmetries [24]. Also of relevance to what follows are a number of neural models that have been used to study acute focal cortical lesions [19,41,52]. This past research has generally examined only unilateral cortical regions and local adaptation, and most often has looked at post-lesion map reorganization. An exception is some recent work on visual information processing where both left and right hemispheric regions have been simulated [35,40], and then one hemispheric region removed/isolated to simulate unilateral neglect phenomena. However, these latter studies have not modeled hemispheric interactions via the corpus callosum, the effects of underlying hemispheric asymmetries, recovery following damage, or variable lesion sizes as are considered below.

Finally, in order to understand the motivations for and results of the modeling studies that follow, it is important to be aware of certain changes in the brain following damage. Acutely, following hemispheric infarction (focal brain damage due to localized loss of blood flow), there is an immediate depression of neural activity, metabolism and cerebral blood flow contralaterally in the intact hemisphere [15,34]. Such changes are referred to as *transcallosal diaschisis*. Their severity is proportional to the severity of the infarct and they persist for roughly 3 to 4 weeks after a stroke. It is often accepted that transcallosal diaschisis is responsible for part of the clinical deficit in stroke [34], including sensorimotor findings ipsilateral to the infarct [9], although this view has been challenged

[6]. Further, it is often accepted that an important mechanism responsible for transcallosal diaschisis is loss of excitatory callosal inputs to otherwise intact contralateral cortex, although there may be other contributing factors [34].

The importance of hemispheric interactions during recovery from deficits such as aphasia (impaired language abilities) following left hemisphere language area damage is underscored by evidence that the right hemisphere plays a crucial role in the language recovery process in adults. Many studies have indicated substantial right hemisphere responsibility for language recovery after left hemisphere strokes [26,28,30,39,51]. Several functional imaging studies, primarily positron emission tomography (PET) investigations, show that recovered aphasics have increased activation in the right hemisphere in areas largely homotopic to the left hemisphere's language zones [8,37,53,55]. However, there are also PET studies that, although sometimes finding increased right hemisphere activation in individuals with post-stroke aphasia, have questioned how well these changes correlate with the recovery process [4,20,21]. While much experimental data indicates some right hemisphere role in recovery, currently this appears to depend on lesion location and size, and perhaps other factors [47]. This issue remains controversial and an area of very active experimental research.

Within the context of the above information and uncertainties, our research group recently developed and studied a variety of neural models of corresponding left and right cortical regions interacting via a simulated corpus callosum. While these models are not intended as detailed, veridical representations of specific cortical areas, they do capture many aspects of cortical connectivity, activation dynamics, and locality of synaptic changes, and illustrate ways in which cortical modules can interact and became specialized. With each model we systematically determined conditions under which lateralization arose and the effects of simulating acute focal lesions. Multiple independent models were studied so that their collective results would be reasonably general, i.e., not tied to the specific choices of task, architecture, learning method, etc. that are necessarily made for any single neural model. The lateralization and lesioning results in two of these models using supervised learning are discussed in the following. Other models using unsupervised learning and self-organizing cortical maps have been described elsewhere [31,32].

3   Phoneme Sequence Generation

The phoneme sequence generation model was our first attempt to investigate computationally the factors influencing lateralization [42]. Its recurrently connected network is trained using supervised learning to take three-letter words as input and to produce the correct temporal sequence of phonemes for the pronunciation of each word as output. Fig. 1 schematically summarizes the network architecture, where input elements (I) are fully connected to two sets of neural elements representing corresponding regions of the left (LH) and right (RH) hemisphere cortex. These regions are connected to each other via a simulated corpus callosum (CC), and are also fully connected to output elements (O) re-

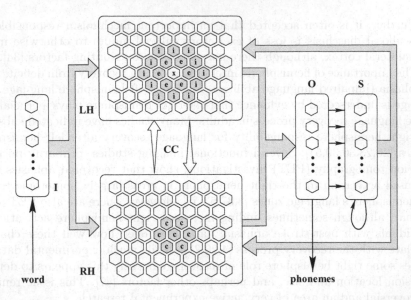

Fig. 1. Network architecture for phoneme sequence generation. Given an input word (bottom left), a temporal sequence of phonemes (bottom right) is generated. Individual neural elements are indicated by small hexagons, while sets of related elements are indicated by boxes. Labels: I = inputs, O = outputs, LH/RH = left/right hemispheric region, CC = corpus callosum, S = state elements, **e**/**i** = excitatory/inhibitory intra-hemispheric connections from element **x**, **c** = homotopic corpus callosum connections from element **x**.

presenting individual phonemes. State elements (S) provide delayed feedback to the hemispheric regions via recurrent connections. Learning occurs on all connections in Fig. 1 except callosal connections. The supervised learning rule used to train the model is a variant of recurrent error backpropagation.

Over a thousand simulations were done with different versions of this model [42,50]. The results obtained when intracortical connections were present and when they were absent were similar, so we just describe the results when such connections were present in the following [50]. The effects of different hemispheric asymmetries (relative size, maximum activation level, sensitivity to input stimuli, learning rate parameter, amount of feedback, etc.) were examined one at a time so that their individual effects could be assessed. For each hemispheric asymmetry and a symmetric control version of the model, the uniform value of callosal connection influences was varied from -3.0 to +3.0. This was done because, as noted above, it is currently controversial whether overall callosal influences of one hemisphere on the other are excitatory or inhibitory. Lateralization was measured as the difference between the output error when the left hemispheric region alone controlled the output versus when the right hemispheric region alone did.

Simulations with this model showed that, within the limitations of the model, it is easy to produce lateralization. For example, lateralization occurred toward the side with higher excitability (Fig. 2b), depending on callosal strength (contrast with the symmetric case, Fig. 2a). Lateralization tended to occur most readily and intensely when callosal connections exerted predominantly an inhibitory influence, i.e., Fig. 2b was a commonly occurring pattern of lateralization. This supports past arguments that hemispheric regions exert primarily a net inhibitory influence on each other. However, with some asymmetries significant lateralization occurred for all callosal strengths or for weak callosal strengths of any nature. In this specific model, the results could be interpreted as a "race-to-learn" involving the two hemispheric regions, with the "winner" (dominant side) determined when the model as a whole acquired the input-output mapping and learning largely ceased. Among other things, the results suggest that lateralization to a cerebral region can in some cases be associated with increased synaptic plasticity in that region relative to its mirror-image region in the opposite hemisphere, a testable prediction. This hypothesis seems particularly interesting in the context of experimental evidence that synaptogenesis peaks during the first three years of life and may be an important correlate of language acquisition [3].

For lesioning studies, multiple versions of the intact model were used (callosal influence excitatory in some, inhibitory in others), representing a wide range of prelesion lateralization [43,50]. Each lesion was introduced into an intact model by clamping a cortical area in one hemispheric region to be non-functional. Performance errors of the full model and each hemisphere alone were measured immediately after the lesion (acute error measures) and then after further training until the model's performance returned to normal (post-recovery error measures). Lesion sizes were varied systematically from 0% to 100%. The results show that acutely, the larger the lesion and the more dominant the lesioned hemispheric region, the greater the performance impairment of the full model. Fig. 3a and b show examples of this for a symmetric model version with inhibitory (a) and excitatory (b) callosal connections and left side lesions. The lesioned left hemispheric region exhibited impaired performance when assessed in isolation, as did the model as a whole. With excitatory callosal connections, large lesions also led to a small impairment of the intact right hemisphere when it was tested in isolation.

When the lesioned model was allowed to undergo continued training, recovery generally occurred with performance of the full model eventually returning to prelesion levels (Fig. 3c and d, horizontal dashed line with open triangles). The non-lesioned hemisphere very often participated in and contributed to post-lesion recovery, more so as lesion size increased. Recovery from the largest lesions was especially associated with marked performance improvement of the right un-lesioned hemispheric region (for example, compare Fig. 3c vs. Fig. 3a). This is consistent with a great deal of experimental evidence. Our results thus support the hypothesis of a right hemisphere role in recovery from aphasia due to some left hemisphere lesions. They also indicate that one possible cause for apparent

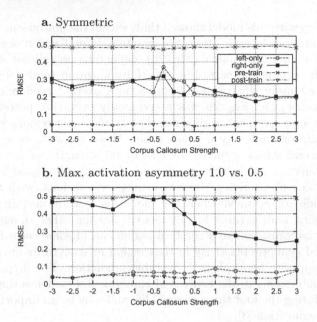

a. Symmetric

b. Max. activation asymmetry 1.0 vs. 0.5

Fig. 2. Root mean square error (RMSE) vs. callosal strength for different asymmetry conditions with the phoneme sequence generation model. In each graph the top (x's) and bottom (open triangles) lines indicate pre-training and post-training errors, respectively, for the full intact model where the two hemispheric regions jointly determine model output. The two middle lines are the post-training errors when the left (open circles) or right (filled squares) hemispheric region alone sends activation to output elements. Results are shown when (**a**) the hemispheric regions are symmetric, and (**b**) when the left hemisphere is more excitable.

discrepancies in past studies may be inadequate control for the effects of lesion size, a factor that should be carefully analyzed in future experimental studies.

When one measured the mean activation in the model cerebral hemispheres following a unilateral lesion, quite different results were obtained depending on whether the callosal connections were inhibitory or excitatory. With a left hemispheric lesion, mean activity fell in the lesioned left hemisphere, regardless of the nature of callosal connections. With inhibitory callosal connections, mean activity in the unlesioned right hemispheric region increased substantially following a left hemispheric lesion and remained high over time (Fig. 4a). In contrast, when excitatory callosal influences existed, the intact, non-lesioned hemispheric region also often had a drop in mean activation (Fig. 4b), representing the model's analog of transcallosal diaschisis. As noted earlier, regional cerebral blood flow and glucose metabolism are found experimentally to decrease bilaterally following a unilateral stroke. Thus, to the extent that coupling exists between neuronal activity and blood flow/oxidative metabolism, the mean activation shifts seen

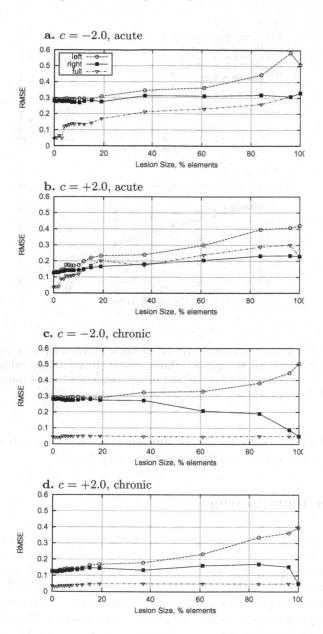

Fig. 3. Root mean square error ($RMSE$) vs. lesion size for (**a,c**) inhibitory and (**b,d**) excitatory callosal influences, both acutely and following further training in symmetric versions of the model. Lesions in the left hemisphere. Plots are of error for full model (dot-dash line with open triangles), and for when the left hemisphere alone (dashed line with open circles) and right hemisphere alone (solid line with closed boxes) control output.

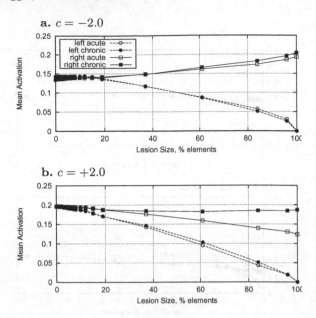

Fig. 4. Mean activation (measured over all elements in a hemispheric region and over all input stimuli) vs. lesion size for (**a**) inhibitory and (**b**) excitatory callosal influences, both acutely and following further training in symmetric versions of the model. Lesions are in the left hemispheric region.

with excitatory callosal influences in this model are most consistent with those observed experimentally following unilateral brain lesions.

4 Letter Identification

Our second model uses a combination of unsupervised and supervised training to learn to classify a small set of letters presented as pixel patterns in the left (LVF), midline, and right (RVF) visual fields [48]. The network again consists of interconnected two-dimensional arrays of cortical elements (Fig. 5). Each visual field projects to the contralateral primary visual cortex, which extracts only one important type of visual features: orientation of local edges. The primary cortex layers project onto ipsilateral visual association cortex layers. For simplicity and because biological primary visual cortex is largely lacking callosal connections, only the association layers in the model are homotopically connected via a simulated corpus callosum. Association layers project to an output layer where ideally only one output element is activated, representing the correct identity of the input element. Primary-to-association connection strengths are modified using a competitive unsupervised Hebbian rule, while associative-to-output weights are learned using a supervised error connection rule. Model performance is measured

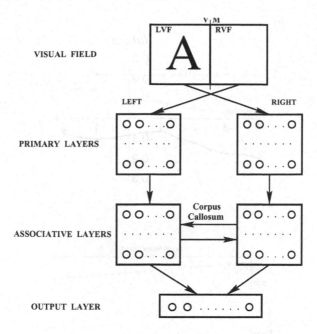

Fig. 5. Network for visual character recognition task. VM = vertical meridian, LVF = left visual field, RVF = right visual field.

using root mean square error of output activation patterns; lateralization is measured as the difference in contribution of the two hemispheres to performance.

In this model, persistent partial lateralization occurred toward the side having larger size, higher excitability, or higher unsupervised/supervised learning rate (e.g., Fig. 6a). Lateralization tended to be more pronounced when callosal connections were inhibitory and to decrease when they were excitatory, as was often the case with the phoneme sequence generation model. Lesioning to the associative layers also produced results similar to those seen with the former models: greater performance deficit with larger or dominant side lesions, acute post-lesion fall in activation in the intact hemispheric region as well as the lesioned hemisphere with excitatory but not inhibitory callosal influences (e.g., Fig. 6b), and frequent participation of the non-lesioned hemisphere in performance recovery regardless of callosal influences, especially with larger lesions [49]. While unilateral cortical lesions generally produced the expected contralateral visual field impairment, an unexpected finding was that mild impairment also occurred in the visual field on the *same* side as the lesion. This occurred with excitatory but not inhibitory callosal influences. It is of particular interest because such surprising ipsilateral visual field impairment has been found in recent clinical studies with unilateral visual cortex lesions [46].

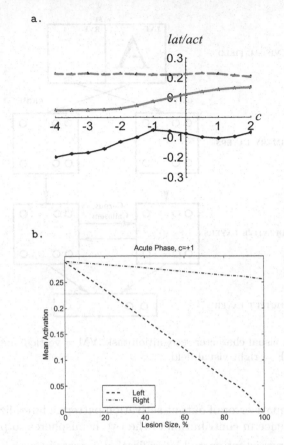

Fig. 6. a Post-training lateralization (thin, solid dark line; negative values indicate left lateralization) versus callosal strength c in a letter identification model version where the left hemispheric region is more excitable than the right. Solid (right hemisphere) and dashed (left hemisphere) thicker gray lines show individual hemisphere mean pre-training activation values. **b.** Mean activation levels in left (dashed line) and right (dot-dash line) associative cortex regions vs. size of a left associative cortex region lesion in a symmetric version of the model with excitatory callosal influences.

5 Discussion

Advances in neuroscience and in neural modeling methods during the last several decades have provided a number of suggestions for how more powerful computational techniques might be developed. These "hints" include: the use of massive, fine-grained parallel processing as a mechanism for speeding up computations; the emergence of complex global behavior from numerous relatively simple local interactions between processing units; and the importance of adaptability and learning as central features of any neural computation. Use of these principles has led to numerous impressive applications of neural networks during the last decade, several of which have been fielded commercially. Research on VLSI computer chips that are based on neural computation methods, and experiments where biological neurons are grown in culture on chips and interact with their electrical circuitry, suggest that we have just begun to realize the potential of this technology.

Understanding brain modularity is another area where advances may lead to new ideas in neural computation. Experimental research has established that the cerebral architecture is highly modular, and that cortical modules can be associated with specific behavioral functions. From an abstract perspective, computational studies like those described here are providing insights into how individual neural modules in multi-modular systems can become specialized for specific tasks. Conditions under which specialization can arise, what factors influence the amount of specialization, and how modules contribute to system robustness are all becoming more clearly determined.

In this context, the work described here has systematically examined whether underlying asymmetries can lead to hemispheric specialization, how assumptions about callosal influences affect lateralization, and the interhemispheric effects of focal cortical lesions. Use of multiple different models, including a purely unsupervised one not described here [31,32], insures that the overall results obtained are reasonably general, i.e., they are not tied to a specific task, architecture, or learning method. Our models have been limited in size and scope, have each involved lateralization of only a single task, have generally had only a single underlying asymmetry, and have been greatly simplified from neurobiological and behavioral reality. Nevertheless, in a simplified fashion they are all based on some basic principles of biological neural elements, circuits and synaptic plasticity.

Our models exhibit four results in common. First, each of a variety of underlying hemispheric asymmetries in isolation can lead to specialization, including asymmetric size, excitability, feedback intensity, and synaptic plasticity, lending support to past arguments that a single underlying hemispheric asymmetry is unlikely to account for language and other behavioral lateralizations [22]. Further, the extent of lateralization generally increases with increased underlying asymmetry. These results were not obvious in advance: some asymmetries might have led to lateralization and others not, and lateralization might have turned out to be a largely all-or-none effect. Second, in the intact models, lateralization generally appeared most intensely with inhibitory interhemispheric interactions (although in some cases it also appeared with excitatory or absent callosal connections), lending support to past arguments that whatever the actual neurophysiological nature of callosal synapses, callosal influences are effectively inhibitory/suppressive in nature. At present it appears that it is interhemispheric competition, and not necessarily the requirement that it be brought about by transcallosal inhibition, that is essential. Third, following focal lesions, an acute decrease was generally observed in activation and sometimes performance in the contralateral intact hemispheric region with excitatory but not inhibitory callosal interactions. These latter changes resemble those of diaschisis, which is seen experimentally in acute stroke patients, lending support to past arguments that callosal influences are predominantly excitatory. Resolving the apparent conflict between results 2 and 3 is a focus of ongoing research [44]. Of course, it is also possible that the corpus callosum may play an excitatory role under some conditions and an inhibitory role under other conditions. Fourth, after a post-lesion recovery period, the intact, non-lesioned model hemisphere was often partially responsible for recovery, supporting the controversial hypothesis that the right cerebral hemisphere plays a role in recovery from some types of aphasia due to left hemisphere lesions. This effect increased with increasing lesion size, suggesting that future experimental studies of this issue should carefully consider lesion size in interpreting data.

Our results support the idea that computational models have a substantial future role to play in gaining a deeper understanding of hemispheric specialization and neural modularity in general. While questions about the causes and implications of module specialization must ultimately be resolved by experiment, formal models provide a complementary approach in which non-trivial implications of hypotheses can be demonstrated. It is the complexity and non-linearity of brain dynamics that make computational models of the sort described here useful, just as they are in understanding other complex systems. One can anticipate an increasing use of such models of interacting hemispheric regions both in interpreting experimental data and in guiding future experiments.

Acknowledgement. Work supported by NIH award NS 35460.

References

1. Anninos P, et al. A Computer Model for Learning Processes and the Role of the Cerebral Commissures, *Biol. Cybern*, 50, 1984, 329-336.
2. Anninos P & Cook N. Neural Net Simulation of the Corpus Callosum, *Intl. J. Neurosci.*, 38, 1988, 381-391.
3. Bates E, Thal D & Janowsky J. Early Language Development and Its Neural Correlates, in S. Segalowitz & I. Rabin (eds.), *Handbook of Neuropsychology*, 7, 1992, 69-110.
4. Belin P, Van Eeckhout P, Zilbovicius M, et al. Recovery From Nonfluent Aphasia After Melodic Intonation Therapy: A PET Study, *Neurol.*, 47, 1996, 1504-1511.
5. Berlucchi G. Two Hemispheres But One Brain, *Behav Brain Sci*, 6, 1983, 171-3.
6. Bowler J, Wade J, Jones B, et al. Contribution of Diaschisis to the Clinical Deficit in Human Cerebral Infarction, *Stroke*, 26, 1995, 1000-1006.
7. Burgess C & Lund K. Modeling Cerebral Asymmetries in High-Dimensional Semantic Space, in *Right Hemisphere Language Comprehension*, M. Beeman & C. Chiarello (eds.), Erlbaum, 1998, 215-244.
8. Cappa S, Perani D, Grassi F, et al. A PET Follow-UP Study of Recovery After Stroke in Acute Aphasics, *Brain and Language*, 56, 1997, 55-67.
9. Caselli R. Bilateral Impairment of Somesthetically Mediated Object Recognition in Humans, *Mayo Clin. Proc.*, 66, 1991, 357-364.
10. Cook N. *The Brain Code*, Methuen, 1986.
11. Cook N & Beech A. The Cerebral Hemispheres and Bilateral Neural Nets, *Int. J. Neurosci.*, 52, 1990, 201-210.
12. Davidson R & Hugdahl K (eds.), *Brain Asymmetry*, MIT Press, 1995.

13. Denenberg V. Micro and Macro Theories of the Brain, *Behav. Brain Sci.*, 6, 1983, 174-178.
14. Dennis M. & Whitaker H. Language Acquisition Following Hemidecortication. Linguistic Superiority of Left Over Right Hemisphere, *Brain and Language*, 3, 1976, 404-433.
15. Feeney D & Baron J. Diaschisis, *Stroke*, 17, 1986, 817-830.
16. Ferbert A, et al. Interhemispheric Inhibition of the Human Motor Cortex, *J. Physiol.*, 453, 1992, 525-546.
17. Fink G, Driver J, Rorden C, et al. Neural Consequences of Competing Stimuli in Both Visual Hemifields, *Annals of Neurology*, 47, 2000, 440-446.
18. Geschwind N & Galaburda A. *Cerebral Lateralization*, MIT Press, 1987.
19. Goodall S, Reggia J, et al. A Computational Model of Acute Focal Cortical Lesions, *Stroke*, 28, 1997, 101-109.
20. Heiss W, Karbe H, et al. Speech-Induced Cerebral Metabolic Activation Reflects Recovery From Aphasia, *J. Neurol. Sci.*, 145, 1997, 213-217.
21. Heiss W, Kessler J, Thiel A et al. Differential Capacity of Left and Right Hemispheric Areas for Compensation of Poststroke Aphasia, *Annals of Neurology*, 45, 1999, 430-438.
22. Hellige J. *Hemispheric Asymmetry*, Harvard, 1993.
23. Innocenti G. General Organization of Callosal Connections in the Cerebral Cortex, *Cerebral Cortex*, 5, E. Jones & A. Peters (eds) Plenum, 1986, 291-353.
24. Ivry R & Robertson L. *The Two Sides of Perception*, MIT Press, 1998, 225-255.
25. Jacobs R & Kosslyn S. Encoding Shape and Spatial Relations, *Cognitive Science*, 18, 1994, 361-386.
26. Kinsbourne M. The Minor Cerebral Hemisphere as a Source of Aphasic Speech, *Arch. Neurol.*, 25, 1971, 302-306.
27. Kinsbourne M. (ed.) *Asymmetrical Function of the Brain*, Cambridge, 1978.
28. Knopman D, Rubens A, Selnes O, Klassen A, & Meyer M. Mechanisms of Recovery from Aphasia, *Annals of Neurology*, 15, 1984, 530-535.
29. Kosslyn S, Chabris C, et al. Categorical Versus Coordinate Spatial Relations: Computational Analyses and Computer Simulations, *J. Exper. Psych: Human Perception and Performance*, 18, 1992, 562-577.
30. Lee H, Nakada T, Deal J, et al. Transfer of Language Dominance, *Annals of Neurology*, 15, 1984, 304-307.
31. Levitan S & Reggia J. A Computational Model of Lateralization and Asymmetries in Cortical Maps, *Neural Computation*, 12, 2000, 2037-2062.
32. Levitan S & Reggia J. Interhemispheric Effects on Map Organization Following Simulated Cortical Lesions, *Artificial Intelligence in Medicine*, 17, 1999, 59-85.
33. Meyer B. et al. Inhibitory and Excitatory Interhemispheric Transfers Between Motor Cortical Areas in Normal Humans and Patients with Abnormalities of Corpus Callosum, *Brain*, 118, 1995, 429.
34. Meyer J, et al. Diaschisis, *Neurol. Res.*, 15, 1993, 362-366.
35. Monoghan P & Shillcock R. The Cross-Over Effect in Unilateral Neglect, *Brain*, 121, 1998, 907-921.
36. Netz J, Ziemann U & Homberg V. Hemispheric Asymmetry of Transcallosal Inhibition in Man, *Experimental Brain Research*, 104, 1995, 527-533.
37. Ohyama M, Senda M, Kitamura S, et al. Role of the Nondominant Hemisphere During Word Repetition in Poststroke Aphasics, *Stroke*, 27, 1996, 897-903.
38. Pandya D. & Seltzer B. The Topography of Commissural Fibers, in *Two Hemispheres - One Brain*, F. Lepore et al (eds.), Alan Liss, 1986, 47-73.

39. Papanicolaou A, Moore B, Deutsch G, et al. Evidence for Right-Hemisphere Involvement in Aphasia Recovery, *Arch. Neurol.*, 45, 1988, 1025-1029.
40. Pouget A. & Sejnowski T. Lesion in a Brain Function Model of Parietal Cortex, *Parietal Lobe Contribution in Orientation in 3D Space*, Thier, P. & Karnath, H. (eds.), 1997, 521-538.
41. Reggia J, Ruppin E, and Berndt R (eds.). *Neural Modeling of Brain and Cognitive Disorders*, World Scientific, 1996.
42. Reggia J, Goodall S, & Shkuro Y. Computational Studies of Lateralization of Phoneme Sequence Generation, *Neural Computation*, 10 1998, 1277-1297.
43. Reggia J, Gittens S. & Chhabra J. Post-Lesion Lateralization Shifts in a Computational Model of Single-Wood Reading, *Laterality*, 5, 2000, 133-154.
44. Reggia J, Goodall S, Shkuro Y & Glezer M. The Callosal Dilemma: Explaining Diaschisis in the Context of Hemispheric Rivalry, 2000, submitted.
45. Ringo J, Doty R, Demeter S & Simard P. Time Is of the Essence: A Conjecture that Hemispheric Specialization Arises from Interhemispheric Conduction Delay, *Cerebral Cortex*, 4, 1994, 331-343.
46. Rizzo M. & Robin D. Bilateral Effects of Unilateral Visual Cortex Lesions in Humans, *Brain*, 1996, 119, 951-963.
47. Selnes O. Recovery from Aphasia: Activating the "Right" Hemisphere, *Annals of Neurology*, 45, 1999, 419-420.
48. Shevtsova N & Reggia J. A Neural Network Model of Lateralization During Letter Identification, *J. Cognitive Neurosci.*, 11, 1999, 167-181.
49. Shevtsova N & Reggia J. Interhemispheric Effects of Simulated Lesions in a Neural Model of Letter Identification, *Brain and Cognition*, 2000, in press.
50. Shkuro Y, Glezer M & Reggia J. Interhemispheric Effects of Lesions in a Neural Model of Single-Word Reading, *Brain and Language*, 72, 2000, 343–374.
51. Silvestrini M, Troisi E, Matteis M, Cupini L & Caltagirone C. Involvement of the Healthy Hemisphere in Recovery From Aphasia and Motor Deficits in Patients with Cortical Ischemic Infarction, *Neurology*, 45, 1995, 1815-1820.
52. Sober S, Stark D, Yamasaki D & Lytton W. Receptive Field Changes After Stroke-like Cortical Ablation, *J. Neurophys.*, 78, 1997, 3438-3443.
53. Thulborn K, et al. Plasticity of Language-Related Brain Function During Recovery from Stroke, *Stroke*, 30, 1999, 749-754.
54. Toyama et al. Synaptic Action of Commissural Impulses upon Association Efferent Cells in Cat Visual Cortex, *Brain Res.*, 14, 1969, 518-520.
55. Weiller C, et al. Recovery from Wernicke's Aphasia: A PET Study, *Annals of Neurology*, 37, 1995, 723-732.

Explorations of the Interaction between Split Processing and Stimulus Types

John Hicks and Padraic Monaghan

University of Edinburgh, Edinburgh, UK
jwjhix@cogsci.ed.ac.uk, pmon@cogsci.ed.ac.uk
http://www.iccs.informatics.ed.ac.uk/~jwjhix

Abstract. This chapter concerns the influence of the bihemispheric structure of the brain on processing. The extent to which the hemispheres operate on different aspects of information, and the nature of integrating information between the hemispheres are vexed topics open to artificial neural network modelling. We report a series of studies of split-architecture neural networks performing visual word recognition tasks when the nature of the stimuli vary. When humans read words, the exterior letters of words have greater saliency than the interior letters. This "exterior letters effect" (ELE) is an emergent effect of our split model when processing asymmetrical (word-like) stimuli. However, we show that the ELE does not emerge if the stimuli are symmetrical, or are mixed (symmetrical and asymmetrical). The influence of split processing on task performance is inextricably linked to the nature of the stimuli, suggesting that the task determines the nature of the separable processing in the two hemispheres of the brain.

1 Introduction

We have been interested in exploring the nature of information processing mediated by the neuroanatomical structure of the brain. In particular, we have been concerned with the influence of separate processing due to the bihemispheric structure of the brain. The human fovea, for example, is precisely split about the midline of the visual field [25], such that the visual hemifields project contralaterally to the two hemispheres [4]. This means that when visual fixation on an object is central, identification of the stimulus, in many cases, requires the integration of initially divided information. This is crucial when the stimulus is asymmetrical, for instance in the case of word recognition, or face recognition [2].

Psychological studies of the divided processing between two halves of the brain were advanced by studying the cognitive deficits and advantages of patients with severed corpus callosum, the main connection between the two cerebral hemispheres. Commissurotomy patients showed that the two halves of the brain could function autonomously when disconnected [6], though there is some debate on the possible routing of information via subcortical structures [20]. The division of visual information between the hemispheres is also evident in cases

S. Wermter et al. (Eds.): Emergent Neural Computational Architectures, LNAI 2036, pp. 83–97, 2001.

of unilateral neglect [7], [19]. Neglect is a common deficit following right brain damage, particularly damage to the parietal cortex, where there is a deficit in attending to the contra-lesional side of the stimulus. This suggests that visual information is not completely integrated between the hemispheres in later stages of the visual pathway.

The influence of the hemispheric division in processing is apparent in such cases, but clarifying the nature of the effect on cognition of such a processing division is ongoing. In particular, we are interested in the following questions:

- To what extent do the two hemispheres of the brain process information independently?
- Do the two hemispheres of the brain pick up on different aspects of the same stimulus?
- If aspects of processing are separated, at what stage in processing does integration of information occur, and what are the mechanisms involved in integration?

We believe that neural network modelling of cognitive functions implementing split processors supplements and inspires neuropsychological research into these issues.

In this chapter we present neural network models of visual word recognition that have split inputs and split processing layers. This work replicates previous research showing that psychological effects emerge from the split architecture. This work is extended to show that such effects are due in part to the nature of the stimulus. When stimuli are more predictable, then the way information is integrated changes. In understanding cognitive functioning, then, we call for particular attention to be paid towards the nature of the stimuli interacting with the architecture of the processor.

2   Neuroanatomically Inspired Architectures

The questions raised in the introduction concerning separable and integrated processing in split architectures has been explored in several neural network research programs.

Kosslyn et al. [11] have explored the extent to which separate processing layers in neural networks may pick up on different aspects of the stimulus. They presented simple spatial stimuli to a neural network with separate hidden layers. The network had to solve both a "categorical" and a "coordinate" task. The categorical task required the network to judge relations between features in the stimulus, for example, whether a stimulus appears above or below a bar, or assess the prototypical version of varied instances of the stimulus. The coordinate task required judgements about exact distances between stimuli. They found that each hidden layer of the network became attuned to one or other task, demonstrating that these processes are effectively maintained as separate processes.

Jacobs and Kosslyn [8] investigated the effect of varying receptive field sizes of units in the hidden layer on task performance. Larger receptive fields correspond to coarser-coding, whereas smaller receptive fields relate to finer-coding. Brown and Kosslyn [1] hypothesize that the right hemisphere of the brain performs coarse-coding on visual stimuli, whereas the left hemisphere performs finer-coding. They found that networks with smaller receptive field units solved categorical tasks better than networks with larger receptive fields. Networks with larger receptive field units solved coordinate tasks more effectively.

These models address the question of what aspects of stimuli the two hemispheres of the brain pick up on, implementing general hemispheric asymmetries within the models. This perspective entails that processing of the hemispheres is complementary, modules processing different features of the stimulus lead to more effective task solution. Committees of networks can solve tasks better than networks performing singly, in particular they form more effective predictions on new data [15], [16]. Two alternative solutions to the same problem in the two hemispheres may contribute towards a better solution overall (or at least guard against forming the worst solution to a problem).

Shillcock, Ellison, and Monaghan [22] have explored the split nature of the visual field and shown that this split contributes towards more effective visual word recognition than in nonsplit architectures. Words tend to be fixated slightly to the left of centre, which means that during reading the two halves of a word are initially projected to different hemispheres of the brain. Word recognition therefore requires the integration of this information. A mathematical model of a large corpus of language indicated that this initial splitting of words, so that processing was initially separate, meant that word recognition was less effortful than if the whole word was available to both halves of the brain. If words are divided into two, then the clusters of letters in the two halves is sufficient to identify the word in most cases. Considering the split in information processing for reading also provides suggestions for the occurrence of the different subtypes of dyslexia, emerging developmentally and after brain damage [23].

Shevtsova and Reggia [21] examined various physiological mechanisms as contributors towards functional asymmetry in the two hemispheres. Their model had a split input visual field, projecting to separate hidden layers, which in turn projected to two "associative layers" connected by a "corpus collosum". The model was presented with representations of letters either on the left, central or right regions of the input field, the network having to identify the letter. Three different mechanisms were investigated in order to assess their contribution towards lateralization in the two hemispheres. Asymmetries in the excitability of neurons in the associative layers led to lateralization of function, which was more pronounced with inhibitory callosal connections. Similar lateralization of functioning emerged when the associative layers were different sizes, and when learning rates in the associative layers were different. In all cases, most lateralization of function occurs when the callosal connections are inhibitory.

Levitan and Reggia [13] conducted complementary investigations on an unsupervised neural network model with callosally-connected hemispheric regions.

When callosal connections were excitatory or weakly inhibitory, then complete and symmetric mirror-image topographic maps emerge in the two hemispheres. When callosal connections are inhibitory then partial to complete topographic maps develop which were complementary. When the size or excitability of the layers was varied then lateralization occurred. As mentioned above, such complementarity between hemispheres may be essential to optimal problem solving.

These studies indicate that asymmetries in split neural networks can model functional asymmetries in the two hemispheres of the brain. In addition, the role of the corpus callosum in contributing towards the lateralization of function is explored, lateralization occurring most readily when interhemispheric connections are inhibitory. The variety of approaches to modelling hemispheric functioning shows the effectiveness of using split neural network architectures for addressing questions of intra- and inter-hemispheric processing. We turn now to a model of visual word recognition that reflects the influence of split processing on task performance.

3   Modelling Visual Word Recognition

Shillcock and Monaghan [24] present a model of visual word recognition that imposes the split visual field in a neural network. Their aim was to explore the extent to which this split reflected psychological phenomena in reading words by humans, in particular the priority in processing given to exterior letters, these apparently having greater saliency than the interior letters. The exterior letters of words (e.g., $d\_\_k$) are reported more accurately than the interior letters (e.g., $\_ar\_$) when followed by a postmask that matched the boundary of the exterior letters [3]. Such an effect does not occur when the letters are not from a plausible word, so $d\_\_x$ showed no advantage [9], [10]. Furthermore, presenting the exterior letters of words primes the word but presenting the interior letters does not [5]. Collectively, these phenomena are referred to as the "exterior letters effect" (ELE).

Two input layers of four letter slots, representing the two halves of the visual field, projected to separate hidden layers. The 60 most frequent four letter words were presented to the model in all five positions across the two input layers. In three of these positions the word was split across the input layers. The model was required to reproduce the word at the output, abstracted from presentation position. A nonsplit control model was also trained, where the input layers projected to both hidden layers. The split architecture model spontaneously produced the ELE: identification of exterior letters was better than for interior letters, and exterior letters primed the whole word better than interior letters. The ELE resulted from the interaction of the split architecture with the shifted presentation of the words in the input, as the effect was not observed to the same extent in the nonsplit control model.

The ELE in the split model was taken to be due to a "hemispheric division of labour". For certain presentations of the input, the exterior letter of the word was presented to only one half of the model. This means that one half of the resources of the model was dedicated to reproducing this letter, with the other half of the model having to process the other three letters. Exterior letters benefit from this dedication of resources over interior letters, as each interior letter is always presented with at least one other letter. Division of labour between the hemispheres is beneficial for processing in this task. During training, presentations of words that fall across the two input fields were learned more quickly than presentations that fell entirely within one input field. The ELE emerges as a consequence of resource allocation due to the split structure of the model.

4   Divided Processing and Stimulus Types

Under what conditions is this division of labour beneficial for information processing? Are their particular aspects of the "word" stimuli that lead to the ELE via division of labour? If the stimuli are less complex, or more predictable, will division of labour be a good approach to solving the problem? We argue here that the effect of divided processors on task performance depends very much on the nature of the task. We will show that differences in performance between split models and nonsplit models depend on the characteristics of the stimuli that they are trained on.

The visual word recognition model described above coded letters as 8-bit features [17], each feature representing an aspect of letter orthography such as "contains closed area". Such simple feature-based stimuli can be varied in three ways. The stimuli can be asymmetrical, where redundancy in the stimulus is very low, but the relationship between input and output is systematic. We term these "word" stimuli. These stimuli were used in Shillcock and Monaghan's [24] model. Visual stimuli that are asymmetrical are perhaps the exception, words being unique in that one half of a word can seldomly be predicted from the other half. When such stimuli are split across input fields, the model requires information from both halves of the model in order to solve the task.

An alternative stimulus type is when the stimuli are symmetrical and the input-output relationship is again an identity mapping. In this case, the stimuli will contain a greater degree of redundancy. We term this stimulus type "palindromes". When such stimuli are split across input fields then each processor can predict the other half of the stimulus. Division of labour is therefore not so critical for this stimulus type. We predict that the ELE will not emerge in split models trained on such stimuli.

Mixing the two types of stimulus (words and palindromes) may also lead to different processes operating on the stimuli. The model would then have to recognize the cases where division of labour is beneficial, and when it is not necessary. We predict that the ELE will emerge in the split model, but not to the same extent as for word stimuli.

A third stimulus type is when the mapping between input and output is less systematic, or even arbitrary. A prime example is the case of orthography to phonology mappings. This type of stimulus presents difficulties for split networks without recurrent connections, as stimuli falling across the split input can in some cases be akin to a perceptron presented with the XOR problem. For example, the word *beat* split between the *e* and the *a* has a different vowel sound to that of *best* or *bent*, and yet the left input field cannot distinguish this without feedback from the other half of the model. The extent to which information has to be integrated and at what point in the process is a topic of investigation. We do not consider this further here, but take it as an indication of the fundamental importance of data types interacting with model structure.

The remainder of this chapter explores the performance of a split neural network model under different stimulus conditions.

5   Modelling Different Stimulus Types

Hicks, Oberlander and Shillcock [12] showed that the ELE in split models of visual word recognition extended to six-letter words. They found that when mean squared error (MSE) was measured for each letter position after training to a summed-error criterion, exterior letters had lower error than interior letters for the split model. The distinction did not emerge so clearly for the nonsplit model. This means that MSE can be used to reflect the ELE. This also means that masking of letters does not have to be performed for all the letter positions independently, an advantage when more than 4 letter stimuli are used. Using 6-letter stimuli meant that the influence of the greater complexity of the or-thotactic structure of the input could be assessed, and also that intermediate letter positions could also be assessed. These intermediate positions have not been assessed experimentally, human studies of the ELE have all used 4-letter words.

The "word" model is shown in Figure 1. Each input layer has six letter slots, and words of length six were presented across the input position one at a time. As mentioned above, the letters within each word were represented in terms of 8-bit features [17]. In the Figure, the word "better" is shown presented in its seven possible positions. For five of the seven positions, the word was split between the two halves of the model. Each input field is fully connected to a hidden layer with 20 units, both of which are fully connected to the output layer. The network was trained on the 60 most frequent 6-letter words, presented in all possible positions, and had to reproduce the word at the output layer.

The network was trained by backpropagation and training was halted when the overall MSE fell below 800. This corresponds to each unit being within approximately 0.2 of its target for each pattern. After training the MSE for each letter position was assessed for all words presented across all positions, and for all words presented in the central position. Ten simulations of each model were performed, and each model was treated as an individual subject in the statistical analyses.

The same model was also trained on "palindrome" stimuli, where 60 6-letter strings were generated randomly such that they were symmetrical (so *betteb* was a possible pattern for this model). The model was trained to the same criterion.

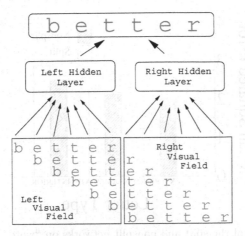

Fig. 1. The split network, each hidden layer seeing only half of the visual field, across which stimuli can be presented at different positions.

The "mixed" stimuli were 30 randomly generated "palindromes" and 30 randomly generated asymmetrical 6-letter strings. Again the model was trained to the same level of error.

For each stimulus type, a nonsplit control model was constructed with each input layer connecting to both hidden layers. The connectivity between the input and hidden layers underwent a random pruning of half of the connection. This was to ensure that the network's power was consistent with its split counterparts, network power being directly proportional to the number of weighted connections.

A Sun Ultra 5 workstation ran the simulations, using the PDP++ Neural Nets software [14].

6 Results

6.1 Learning

The models learned more quickly according to the different types of stimuli they were trained on. Table 1 shows the mean number of epochs of training for models to meet the training criterion. The time taken to reach criterion did

Table 1. Epochs of training required for models for each stimulus type.

Model	Stimulus		
	Words	Palindromes	Mixed
Split	146.7	48.1	124.6
Nonsplit	239.7	49.4	136.1

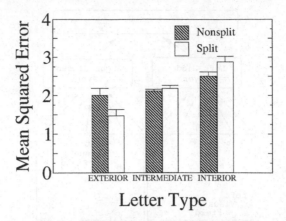

Fig. 2. Performance of the split and nonsplit networks on "word" stimuli for exterior, intermediate and interior letters. Errors are summed across all presentation positions.

not differ for the palindromes and the mixed stimuli (both t < 1, $p > 0.3$), but the split model learned significantly quicker for word stimuli ($t(18) = 6.37, p < 0.001$). As predicted, the palindrome stimuli were easiest to learn, containing less complexity, then the mixed stimuli, and finally the word stimuli. The nonsplit model had particular difficulty with the word stimuli, although this seemed to be an effect of a very small proportion of the simulations getting "lost" around local minima and skewing the mean epoch number.

6.2   The ELE

Does the ELE emerge as a result of any of the models? For split word models, the ELE clearly emerges, so this effectively reproduces the effects of Shillcock and Monaghan [24], and Hicks, Shillcock and Oberlander [12], as shown in Figure 2. Errors were summed for exterior letters (positions 1 and 6), intervening letters (positions 2 and 5), and interior letters (positions 3 and 4). The three positions were entered as within-subjects variable in a repeated measures ANOVA, with split and nonsplit as between-subjects variable. Each of the 10 runs of the models were counted as subjects. There was a main effect of position ($F(2, 36) = 161.93$, $p < 0.001$), there is also a main effect of split/nonsplit model ($F(1, 18) = 10.41$, $p < 0.01$), and an interaction between letter position and model ($F(2, 36) = 30.60, p < 0.001$). There is lower error for exterior letters and higher error for interior letters in the split model, as compared to the nonsplit model.

For the palindromes stimuli, however, the ELE does not seem to emerge in the split model (see Figure 3). However, the nonsplit model reflects the *opposite* effect: prioritising interior letters over the exterior letters. We refer to this effect as the "interior letters effect" (ILE). There is a main effect of letter position ($F(2, 36) = 115.91, p < 0.001$), no main effect of model type ($F(1, 18) = 0.00, p = 0.97$), but an interaction of letter position and model type was significant ($F(2, 36) = 13.60, p < 0.001$). For the split model, there is no significant difference in

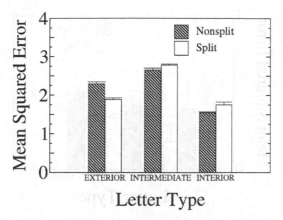

Fig. 3. Performance of the split and nonsplit networks on "palindrome" stimuli for exterior, intermediate and interior letters. Errors are summed across all presentation positions.

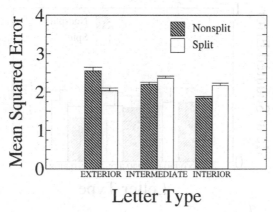

Fig. 4. Performance of the split and nonsplit networks on "mixed" stimuli for exterior, intermediate and interior letters. Errors are summed across all presentation positions.

the error for the different letter positions, but for the nonsplit model, the interior letters indicate less error than the exterior letter positions.

For the mixed stimuli, a similar pattern emerges to that of the palindromes stimuli (see Figure 4). The split model does not prioritize any letter position, whereas the nonsplit model performs better on the interior letters than the exterior letters. Again, there is a main effect of letter position (F(2, 36) = 16.86, $p < 0.001$)), there is no main effect of model type (F(1, 18) = 0.22, $p = 0.65$), and an interaction of letter position and model type was significant (F(2, 36) = 32.62, $p < 0.001$). There seems to be no effect of letter position for the split model, but an ILE for the nonsplit model.

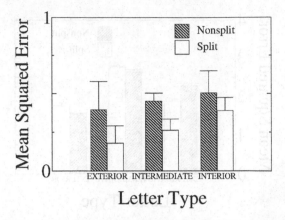

Fig. 5. The comparison of letter position (exterior, intermediate, interior) for the split and nonsplit networks when processing centrally presented words.

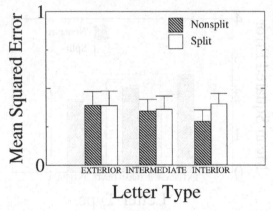

Fig. 6. The comparison of letter position (exterior, penultimate, interior) for the split and nonsplit networks when processing centrally presented palindromes.

When word positions are considered singly, similar effects emerge as for the analyses summed across position. The following analyses are for centrally presented stimuli, so three letters are projected to each half of the split model.

For word stimuli, as before, the ELE emerges clearly from the centrally presented condition (see Figure 5). There is a main effect of letter position ($F(2, 36)$ = 81.62, $p < 0.001$). There is a main effect of model type ($F(1, 18)$ = 155.89, $p <$ 0.001. There is also a significant interaction between letter position and model type ($F(2, 36)$ = 8.45, $p < 0.005$).

For palindromes, the ELE does not seem to emerge for the split model, but there is an ILE for the nonsplit model (see Figure 6). There is no significant main effect of letter position ($F(2, 36)$ = 1.79, p = 0.18), nor is there a significant main effect of model type ($F(1, 18)$ = 2.36, p = 0.14). There is a significant interaction between letter position and model type, however ($F(2, 36)$ = 3.28, $p < 0.05$).

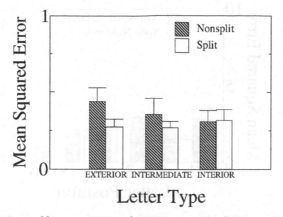

Letter Type

Fig. 7. The comparison of letter position (exterior, intermediate, interior) for the split and nonsplit networks when processing centrally presented mixed stimuli.

Finally, for mixed stimuli there is again no suggestion of an ELE for the split model, but there is an ILE for the nonsplit model (see Figure 7). There are main effects of letter position ($F(2, 36) = 6.04$, $p <¡ 0.01$), of model type ($F(1, 18) = 18.19$, $p < 0.001$), and a significant interaction between letter position and model type ($F(2, 36) = 19.01$, $p < 0.001$).

7 Discussion

The different architectures used in our neural network models pick up on the properties of the stimuli in different ways. For the word stimuli, the ELE emerges for the split model, replicating the studies of Shillcock and Monaghan [24] and Hicks, Oberlander and Shillcock [12]. Previous accounts of the ELE in the split model have been discussed in terms of the contributions of division of labour, as discussed above, and superpositional storage. The latter explanation accounts for the slight ELE found in the nonsplit model. The interior letters tend to occur in positions in the input where there is a greater density and variety of letter presentations during training. Prioritising exterior letters is an effective solution to this problem as the interior letters are presented at positions of greater density of information, and are therefore more difficult to process.

However, the nonsplit model picks up on other aspects of the word stimuli in its solution of the task. Figure 8 shows the MSE for each letter position within the word stimuli for split and nonsplit models. For the split model, MSE is lower for exterior letter positions, and higher for interior letter positions. In addition, the right exterior letter position has lower MSE than the left exterior letter position. For the nonsplit model, there is a general tailing off of MSE as letter position moves from left to right. The nonsplit model has a tendency to train in accordance with the by-letter entropy of the word stimuli (compare with Figure 9). The nonsplit model solves the task by reflecting the characteristics

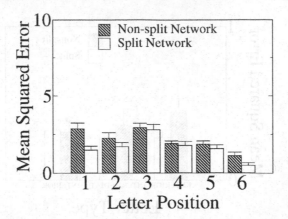

Fig. 8. Performance of the split and nonsplit networks on "word" stimuli for each letter position within the word. Errors are summed across all word positions.

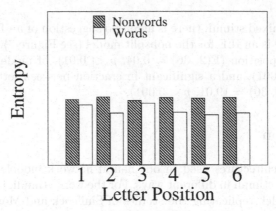

Fig. 9. By-letter entropy for word stimuli, compared with the flat entropy pattern for generated random strings. The nonsplit network in Figure 1 is picking up the structure of the word stimuli.

of the data set as a whole, which is a usual strategy for neural networks solving language-based problems (see [26]). The split model's architecture, however, overrides this approach to solving the problem, with entropy only seeming to influence exterior letters (entropy is lower for the rightmost exterior letter position). The division of labour approach interferes with solution to the task that is based primarily on the statistical profile of the stimuli. That is, the architecture acts as a filter on the influence of stimulus characteristics on task solution.

For the palindromes stimuli, the ELE was not seen to emerge for either the split or nonsplit model. Instead, the nonsplit model demonstrated an ILE. Our hypothesis was that the ELE would not be found in the split model as division of labour is not necessary for solving the task for this easier stimulus set. This is a plausible explanation for the lack of an ELE in the split model, but does not adequately explain the ILE in the nonsplit model. The nonsplit model always has two identical letters juxtaposed in its input. This means that processing can be done across these two positions, precise identification by each letter position is not so necessary. Therefore, the interior letters are going to be easier to process. In contrast, for the split model, when the palindrome is centrally presented, the geminate is divided between the two halves of the model. This makes the structure of the stimulus harder to pick up on, and means that the ILE does not emerge in the split model.

However, the ELE may not emerge for the split model for palindrome stimuli, not because division of labour is no longer necessary but because the geminates at the interior of the strings are easier to process when palindromes are not centrally presented. There would therefore be an ILE effect in the split model, but also an ELE due to the division of labour. The sum of these effects would give a flat portrait as shown in Figures 6. The summation of the ILE and ELE in the split model is plausible given Figure6, where performance is worst on intermediate positions. The presence of geminates lowers the error on interior letters, whereas the division of labour lowers error on exterior letters. Superpositional storage is certainly overruled by the combined factors of stimulus type and model architecture, as the intermediate letter positions do not demonstrate a MSE between that for exterior and interior positions.

For the mixed stimuli, too, the ELE seems to survive in the split model, albeit to a lesser degree than for word stimuli, but the nonsplit model continues to prioritize interior letters. The nonsplit model can pick up on the palindrome stimulus characteristics, even when these can't be relied upon for every stimulus in the environment. The split model maintains the prioritising of exterior letters for such a training set.

8   Conclusion

The simple neural network architectures that we have been using have explored the extent to which learning is determined by architecture and stimulus. The "distributed" approach of nonsplit neural network modelling has been contrasted with establishing modular structure on the networks by instantiating the split visual field input and split processing inspired by the two hemispheres of the brain.

Our models have shown that learning takes place within the context of the system, where this context is defined by the interaction between the model architecture and the nature of the stimulus. We have shown that different neural network architectures pick up on different aspects of stimuli, which influences the approach they have in solving the task. Nonsplit models can pick up on distributional features of the stimulus as a whole (Figures 8 and 9), and can also exploit the redundancy within a stimulus (for palindrome stimuli). However, the division of labour approach enforced by a split architecture reflects the psycholinguistic data of the ELE. The current study indicates that the ELE requires both a split model and asymmetric stimuli in order to emerge. This means that division of

labour and hemispheric interaction are different according to the structure of
the data being processed.

In the previous modelling of split functioning surveyed at the beginning of
this chapter, the characteristics of the stimuli are generally unexamined. We
suggest, however that such factors are of significant importance in exploring the
influence of network structure on task performance. Our starting point has been
to establish the influence that gross splits in processing have on solving simple
visual tasks. We have not as yet examined in detail the nature of hemispheric
interaction, or the different types of processing that might occur in the two
hemispheres. In the light of the studies presented here, we suggest that stimulus
type will be a factor in exploring these issues.

References

1. Brown, H. D., Kosslyn, S. M.: Cerebral Lateralization (Review). Current Opinions
   in Neurobiology **3** 183–186
2. Bruce, V., Cowey, A. , Ellis, A. W., Perrett, D. I. (Eds.): Processing the facial
   image (1992). Oxford: Clarendon Press/Oxford University Press.
3. Butler, B. E., Merikle, P. M.: Selective masking and processing strategy. Quarterly
   Journal of Experimental Psychology **25** (1972) 542–548
4. Fendrich, R., Gazzaniga, M.S.: Evidence of foveal splitting in a commissurotomy
   patient. Neuropsychologia **27** (1989) 273–281
5. Forster, K. I., Gartlan, G.: Hash coding and search processes in lexical access.
   Paper presented at the Second Experimental Psychology Conference, University of
   Sydney (1975).
6. Gazzaniga, M.S.: The bisected brain (1970). New York: Appleton-Century-Crofts.
7. Halligan, P. W., Marshall, J. C.: Visuo-Spatial neglect: The ultimate deconstruc-
   tion? *Brain & Cognition*, **37:3** (1998) 419–438
8. Jacobs, R. A., Kosslyn, S. M.: Encoding shape and spatial relations: The role
   of receptive field size in coordinating complementary representations. Cognitive
   Science **18** (1994) 361–386
9. Jordan, T. R.: Presenting words without interiro letters: Superiority over single
   letters and influence of postmask boundaries. Journal of Experimental Psychology:
   Human Perception and Performance **16** (1990) 893–909
10. Jordan, T. R.: Perceiving exterior letters of words: Differential influences of letter-
    fragment and non-letter-fragment masks. Journal of Experimental Psychology: Hu-
    man Perception and Performance **21** (1995) 512–530
11. Kosslyn, S. M., Chabris, C. F., Marsolek, C.J., Koenig, O.: Categorical versus
    coordinate spatial representations: Computational analyses and computer simula-
    tions. Journal of Experimental Psychology: Human Perception and Performance
    18 (1992) 562–577
12. Hicks, J., Oberlander, J. Shillcock, R.: Four letters good, six letters better: Ex-
    ploring the exterior letters effect with a split architecture. Proceedings of the 22nd
    Annual Conference of the Cognitive Science Society. 2000. Madison, WI: Lawrence
    Erlbaum Associates.
13. Levitan, S., Reggia, J. A.: A computational model of lateralization and asymme-
    tries in cortical maps. Neural Computation **12** (2000) 2037–2062
14. O'Reilly, R. C., Dawson, C. K., McClelland, J. L.: PDP++ neural network simu-
    lation software (1995). Pittsburgh, PA: Carnegie Mellon University.

15. Perrone, M. P.: General averaging results for convex optimization. In M. C. Mozer et al. (Eds.), Proceedings 1993 Connectionist Models Summer School (1994) 364–371. Hillsdale, NJ: Lawrence Erlbaum.
16. Perrone, M. P., Cooper, L. N.: When networks disagree: Ensemble methods for hybrid neural networks. In R. J. Mammone (Ed.), Artificial Neural Networks for Speech and Vision. (1993) 126–142. London: Chapman and Hall.
17. Plaut, D. C., Shallice, T.: Connectionist modelling in cognitive neuropsychology: A case study. (1994). Hillsdale, NJ: Lawrence Erlbaum.
18. Quartz, S. R., Sejnowski, T. J.: The neural basis of cognitive development: A constructivist manifesto. Behavioral and Brain Sciences **20** (1997) 537-596
19. Reuter-Lorenz, P. A., Posner, M. I.: Components of neglect from right-hemisphere damage: An analysis of line bisection. Neuropsychologia **28:4** (1990) 327–333
20. Sergent, J.: A new look at the human split brain. Brain **110** (1987) 1375–1392
21. Shevtsova, N., Reggia, J. A.: A neural network model of lateralization during letter identification. Journal of Cognitive Neuroscience **11** (1999) 167–181
22. Shillcock, R. Ellison, M.T., Monaghan, P.: Eye-fixation behaviour, lexical storage and visual word recognition in a split processing model. Psychological Review (in press)
23. Shillcock, R., Monaghan, P.: Using physiological information to enrich the connectionist modelling of normal and impaired visual word recognition. Proceedings of the 20th Annual Conference of the Cognitive Science Society. 1998. Madison, WI: Lawrence Erlbaum Associates
24. Shillcock, R.C.,Monaghan, P.: The computational exploration of visual word recognition in a split model. Neural Computation (in press)
25. Sugishita, M., Hamilton, C. R., Sakuma, I., Hemmi, I.: Hemispheric representation of the central retina of commisurotimized subjects. Neuropsychologia **32** (1994) 399–415
26. Yannakoudakis, E. J., Hutton, P. J.: An assessment of N-phoneme statistics in phoneme guessing algorithms which aim to incorporate phonotactic constraints. Speech Communication **11** (1992) 581–602

Modularity and Specialized Learning: Mapping between Agent Architectures and Brain Organization

Joanna Bryson and Lynn Andrea Stein[1]

Artificial Intelligence Laboratory, MIT
545 Technology Square, Cambridge MA 02139, USA
joanna@ai.mit.edu and las@ai.mit.edu

Abstract. This volume is intended to help advance the field of artificial neural networks along the lines of complexity present in animal brains. In particular, we are interested in examining the biological phenomena of *modularity* and *specialized learning*. These topics are already the subject of research in another area of artificial intelligence. The design of *complete autonomous agents* (CAA), such as mobile robots or virtual reality characters, has been dominated by modular architectures and context-driven action selection and learning. In this chapter, we help bridge the gap from neuroscience to artificial neural networks (ANN) by incorporating CAA. We do this both directly, by using CAA as a metaphor to consider requirements for ANN, and indirectly, by using CAA research to better understand and model neuroscience. We discuss the strengths and the limitations of these forms of modeling, and propose as future work extensions to CAA inspired by neuroscience.

Keywords: Spatial, Structural and Temporal Modularity; Complete Autonomous Agents; Behavior-Based AI; Brain Organization; Action Selection and Synchronization; Perceptual, Episodic, and Semantic Memory

1 Introduction

Although artificial neural networks (ANN) are vast simplifications of real neural systems, they have been a useful technology for helping us think about and model highly distributed systems of representation, control and learning. This work has proven useful both in science, by providing models, paradigms and hypotheses to neuroscientists; and to engineering, by providing adaptive control and classifier systems. In this chapter, we will propose that another area of AI, the agent literature, may also further both science and engineering. We also use agent software architectures to propose an ANN model that addresses issues of modularity, specialized learning, and to some extent synchronization.

---

[1] LAS: also Computers and Cognition Group, Franklin W. Olin College of Engineering, 1735 Great Plain Avenue, Needham, MA 02492. las@olin.edu

S. Wermter et al. (Eds.): Emergent Neural Computational Architectures, LNAI 2036, pp. 98–113, 2001.
© Springer-Verlag Berlin Heidelberg 2001

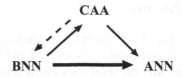

Fig. 1. This chapter addresses the goal of advancing artificial neural networks (ANN) through understanding neuroscience (BNN) via a third field, complete autonomous agents (CAA). This assistance comes both directly, by providing understanding and support for the requirements of such artificial systems, and indirectly, by providing another AI model of biological intelligence.

ANNs can not so far be used for control systems that attempt to replicate the behavioral complexity of *complete* animals. One reason is that the complexity of such systems effectively requires decomposition into modules and hierarchy (see [10] for discussion and references.) This requirement is not theoretical, but practical. In theory, monolithic systems may be Turing complete; but whether attacked by design or by learning, in practice complex control requires decomposition into solvable subproblems.

This principle of modularity is one of several features that characterize the control architectures that *have* been successfully used for complete autonomous agents (CAA). Such agents include autonomous mobile robots [35], virtual reality characters [55], and intelligent environments or monitoring systems [17]. CAA, like animals, must process complex, ambiguous perceptual information. They must also control many forms of action and expression, often with multiple means for achieving any particular behavior. They must also manage a large number of competing and possibly contradictory goals. For example, an animal or animal-like robot may need to balance the need to find food with the need to remain inconspicuous; a character-based tutoring system may need to be clear but not boring; a household robot may need to do the laundry and cook dinner. CAA are situated in the real world in real time. Consequently they must deal with timeliness and synchrony: the order and duration of their expressed behaviors matter.

We begin this chapter with a discussion of modularity as found in nature, then review the literature on CAA in this light. Next, we demonstrate the relevance of CAA to neuroscience by describing a mapping between the common features of CAA architectures and mammalian brain organization. We then examine the implications for learning in the CAA context, which is necessarily specialized due to the representational architecture. We also propose a neural architecture for an agent capable of full, animal-like learning, and future directions, inspired by neuroscience, for CAA itself.

2 Modularity in Nature

There are at least three types of modularity in mammalian brains. First, there is
architectural modularity. Neuroanatomy shows that the brain is composed of different organs with different architectural structures. The types and connectivity
of the nerve cells and the synapses between them characterize different brain
modules with different computational capabilities. Examples of architectural
modules include the neocortex, the cerebellum, the thalamus, the hippocampus, periaqueductal gray matter and so forth: the various organs of the fore, mid
and hindbrains.

Second, there is *functional modularity*. This modularity is characterized by
differences in utility which do not seem to be based on underlying differences in
structure or computational process. Rather, the modules seem to have specialized due to some combination of necessary connectivity and individual history.
Gross examples include the visual vs. the auditory cortices. Sur et al. [56] have
shown at least some level of structural interchangeability between these cortices
by using surgery on neonate ferrets. There is also other convincing and less inva-
sive evidence. For example, many functionally defined cortical regions such as V1
are in slightly different locations in different people [45]. Many people recover capacities from temporarily debilitating strokes that permanently disable sections
of their brains, while others experience cortical remaps after significant alterations of there body, such as the loss of a limb [50]. This evidence indicates that
one of the brain's innate capabilities is to adaptively form functionally modular
organizations of neural processing.

Thirdly, there is *temporal modularity*. This is when different computational
configurations cannot exist contemporaneously. There are at least two sorts of
evidence for temporal modularity. First, many regions of the brain appear to
have local "winner take all" connection wiring where a dominant "impulse" will
inhibit competing impulses [26, 28]. This neurological feature has been used to
explain the fact that humans can only perceive one interpretation of visually
ambiguous stimuli at a time [48]. Second, many cells in the brain are members
of more than one assembly, and can perform substantially different roles in only
subtly different contexts (e.g. in the hippocampus [34, 65].) This sort of temporal
modularity is not yet well understood, but it could have implications for individual differences in intellectual task performance such as insight and metaphoric
reasoning.

The presence of these forms of modularity in mammalian brains motivates
modular architectures in two ways. First, if we are interested in modeling the
brain as a matter of scientific interest, we will need to be able to replicate its
modularity. Second, the presence of modularity in the best examples of intelligent control available indicates that modularity is a useful means of organizing
behavior. Evolution is not a perfect designer — the mere presence of a solution
in nature does not prove it is optimal. However, given the extent and complexity to which the brain has evolved, it is at least worth treating the utility of
modularity as a hypothesis. We now consider further evidence for the utility of
modularity in another domain of intelligent control.

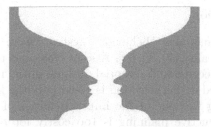

Fig. 2. An ambiguous image. This figure can be seen either as a vase or as two faces, but not both at the same time. From [25, p. 213].

3 Common Features of Complete Autonomous Agents

Modularity is a key feature of most successful CAA architectures. Other architectural features that have been found useful in CAA architectures are hierarchical and sequential structures for ordering the expression of behaviors, and dedicated, reactive alarm systems capable of switching the focus of attention to different points in a control hierarchy. In this section we review the evidence for the importance of these features. A more extensive review is available in [7].

3.1 Skill Modules

One of the fundamental incentives for AI is the idea of having intelligent technology as a companion, situated with us in our environment. However, this goal was long seen as not technically feasible. Consequently, most early AI systems focussed on only a restricted set of capabilities such as reasoning in a particular domain [6, 60]. Such systems were relatively unconstrained by issues such as timeliness or uncertainty about their perceptions or actions. As a result, early attempts at building CAA, which utilized these techniques, tended to be both slow and enormously costly [e.g. 31, 44, 46].

Around 1986 a new paradigm based on greater modularity became established in mobile robotics [40]. This paradigm, behavior-based AI (BBAI), favored niche-specific solutions for CAA. BBAI agents feature modules composed of sensing and control specialized to a particular task to be executed in a particular situation. Such skill modules are often referred to as "behaviors" in the CAA literature to this day, despite the fact that they actually *generate* behavior rather than represent it. Further, much of expressed behavior is supposed to emerge from the interaction of these 'behaviors' rather than being generated by any one-to-one mapping. The most difficult part of the BBAI discipline falls to the engineer: determining what behavioral modules the agent needs, and guaranteeing that these modules do not interfere with each other in the normal operation of the robot.

3.2   Action Selection

By the late 1990's, modular "behavior" systems were ubiquitous in mobile robotics and virtual reality [30, 35, 54]. Also by this time, the need for structured action selection had become well accepted, despite initial resistance in the BBAI community. Structured action selection is essentially the following of plans, and conventional planning was one of the impracticalities of old-style CAA. However, although constructive planning is too costly for real-time systems [16], action selection based on *established* plans is not. This technique, called *reactive planning*, has been facilitated technologically as well as culturally. Technologically, new plan representations were developed which were more flexible and adaptive to dynamic environmental considerations [e.g. 12, 22, 24, 47].

Some of the systems using these representations derive from traditional planning, with behavior modules relegated to being mere plan primitives, but others maintain the autonomy of the behavior modules, while allowing for an extra system to arbitrate the timing of expressed actions. This latter approach also characterizes the relatively new paradigm of multi-agent systems (MAS) which are now sometimes used to control intelligent systems and environments [61]. Their coordination issues are similar to the arbitration issues of BBAI, and will probably converge on similar solutions.

Structured action selection normally consists of both hierarchical and sequential components. Hierarchies are necessary to reduce the search space for the next action. They focus attention on a particular set of actions likely to be useful in the current context. Sequences are a special case of this process where each action can be followed very quickly and reliably by another, a necessity in certain kinds of fine control [32, 38]. Sequences are not adequately represented by simple chains of productions, partly because their elements may participate in more than one sequence, but also because of timing issues. This problem was discovered and addressed in the Soar architecture, the best established production-based AI architecture, when it was applied to the CAA problem [36].

3.3   Environment Monitoring

The difficulty with structured action selection in a dynamic environment is that it can leave the agent unprepared to cope with sudden, unexpected events. Consequently, the third necessary feature of CAA architectures is an independent, parallel environment monitoring or 'alarm' system for switching attention. Such a system must operate in parallel with the agent's main attention, and must be able to recognize critical situations very rapidly with minimal 'cognitive' overhead. That is, it cannot require the reactive planning system itself, but must rely on simple perception. All modular CAA systems have this feature. In the early BBAI system, *every* action was continuously monitoring the environment with its specialized perception, prepared to take control of the agent whenever necessary [5]. In systems closer to conventional planning, such as PRS [24], control cycles between the conventional action selection and the monitoring system. In our own system, context is quickly re-checked at the top, motivational level of the hierarchy between every arbitration step, and the entire hierarchy is revisited on the termination of any subtask [8].

4 Mapping CAA Features to Mammal Brain Structure

To summarize the last section, complete agent architectures have converged on three sorts of architectural modules in order to support complex, reactive behavior. First, skill modules — a functional decomposition of intelligent ability into skilled actions and their supporting specialized perception and memory. Second, hierarchical structures that support action selection, often known as reactive plans. These structures are used to focus attention on behaviors likely to be useful in a particular circumstance and provide temporal ordering for behavior. And third, an environment-monitoring or alarm system for switching the focus of action-selection attention in response to highly salient environmental events.

If this sort of organization is necessary or at least very useful for intelligent control, then it is also likely to be reflected in the organization of animal intelligence. As we explained in the introduction, animals have evolved to face similar problems of information management. In this section, we look at how these CAA principles relate to what is known of mammal brain architecture.

4.1 Skill Modules

In Section 2 we discussed modularity in mammalian brains. Using that terminology, we consider the skill modules of CAA to correspond roughly to functional modularity, particularly in the neocortex, and perhaps to some extent to temporal modularity. However, there is no direct correlation between brain modularity and CAA skill modules. For example, a skill module for grasping a visual target must incorporate the retinas, visual cortex, associative cortices, motor pre-planning and motor coordination. It would also need to exploit somatic and proprioceptive feedback from the grasping limb, though some of this complexity might be masked by interfacing to other specialist modules.

The reason there is not a direct correlation between CAA skill modules to mammalian functional modules is because CAA modules tend to be end-to-end. That is, they encapsulate both perception and action. Much of functional cortical modularity in mammals tends to be more general, for example the visual, auditory, somatic and motor cortices. Some of the temporal modularity in the parietal cortex and the hippocampal formation may correspond more directly to CAA modularity, but this lacks the parallelism typically recommended in CAA, and also the localized representation of perception and motor skills. Taking these sorts of context-specific parietal and hippocampal representations along with some supporting perception and action representations from other cortical areas is probably the best approximation of the typical CAA skill module.

Since a concern of this volume is modeling neural modularity in ANN, we should point out that even though CAA skill modules aren't typically analogous to cortical regions, this is for reasons of practicality. Within the agent discipline, modularity in CAA is primarily to support an orderly decomposition of intelligence into manageable, constructible units. But if one is more interested in modeling the brain directly, one could well use known or theorized cortical modularity as a blueprint for skill decomposition in a CAA agent [e.g. 63]. We would also like to note that more primitive neural systems such as those found in insects to some extent lack the generality of mammalian brains, and more closely match common CAA modularity. For example, spiders have multiple pairs of eyes, some of which seem to be dedicated to single skill modules, such as mating [37].

4.2   Action Selection

The basal ganglia has been proposed as the organ responsible for at least some aspects of action selection [27, 43, 49, 51]. In a distributed parallel model of intelligence, one of the main functions of action selection is to arbitrate between different competing behaviors. This process must take into account both the activation level of the various 'input' cortical channels and previous experience in the current or related action-selection contexts.

The basal ganglia is a group of functionally related structures in the forebrain, diencephalon and midbrain. Its main 'output' centers — parts of the substantia nigra, ventral tegmental area, and pallidum — send inhibitory signals to neural centers throughout the brain which either directly or indirectly control voluntary movement, as well as other cognitive and sensory systems [42]. Its 'input' comes through the striatum from relevant subsystems in both the brainstem and the forebrain. Prescott et al. [49] have proposed a model of this system whereby it performs action selection similar to that proven useful in CAA architectures.

Arbitrating between subsystems is only part of the problem of action selection. Action patterns must also be sequenced with appropriate durations to each step. The duration of many actions is too quick and intricate to be monitored via feedback, or left to the vagaries of spreading activation from competing but unrelated systems [32, 38]. Further, animals that have had their forebrains surgically removed have been shown capable of conducting complex species-typical behaviors — they are simply unable to apply these behaviors in appropriate contexts. In particular, the periaqueductal grey matter has been implicated in complex species-typical behaviors such as mating rituals and predatory, defensive and maternal maneuvers [39]. However, there appears to be little literature as to exactly how such skills are coordinated. There is also little evidence that learned skills would be stored in such areas. We do know that several cortical areas are involved in recognizing the appropriate context for stored motor skills [e.g. 1, 57]. Such cortical involvement could be part of the interface between skill modules and action selection (see further [8].)

4.3 Environment Monitoring

Our proposal for the mammalian equivalent to the environment monitoring and alarm systems is more straight-forward. It is well established that the limbic system, particularly the amygdala and associated nuclei, is responsible for triggering emotional responses to salient (particularly dangerous, but also reproductively significant) environmental stimuli. Emotional responses are ways of creating large-scale context shifts in the entire brain, including particularly shifts in attention and likely behavior [15, 18]. This can be in response either to basic perceptual stimuli, such as loud noises or rapidly looming objects in the visual field, or to complex cortical perceptions, such as recognizing particular people or situations [15]. Again, there can be no claim that this system is fully understood, but it does, appropriately, send information to both the striatum and the periaqueductal grey. Thus the amygdalic system meets our criteria for an alarm system being interconnected with action selection, as well as biasing cortical / skill-module activation.

4.4 Conclusion

In conclusion, it is difficult to produce an completely convincing mapping of CAA attributes to neural subsystems, primarily because the workings of neural subsystems are only beginning to be understood, but also because the levels of generalization are not entirely compatible. The primary function of a CAA architecture is to facilitate a programmer in developing an agent. Consequently, complexity is kept to a minimum, and encapsulation is maximized. Evolution, on the other hand, will eagerly overload an architectural module that has particular computational strengths with a large number of different functions. Nevertheless, we have identified several theories from neuroscience that are analogous to the features of CAA.

5 Adaptivity in Modular Systems

The modularity of an intelligence has obvious ramifications for learning. In particular, the generic CAA architecture described in Section 3, having three sorts of elements, has three sorts of adaptability.

First and foremost, there is adaptivity within the skill modules or behaviors. A behavior module must have the ability to represent perceptual experience in a way that provides for mapping motivation to appropriate actions. In other words, adaptive perceptual state *defines* what a behavior is and does. Specialized learning is ubiquitous in nature, to the point that leading researchers consider it by far the dominant form of natural adaptivity [23, 52]. Such attributes are naturally modeled in CAA under behavior oriented design [9].

Given that action selection requires structure, a natural extension of the CAA systems described above would allow the agent to learn new reactive plans. There are at least three means by which this could be done. The most commonly attempted in AI is by *constructive planning*. This is the process whereby plans are created by searching for sets of primitives which, when applied in a particular order to the current situation, would result in a particular goal situation [e.g. 21, 62]. Another kind of search that has been proposed but not seriously demonstrated is using a genetic algorithm (GA) or GA-like approach to combine or mutate existing plans [e.g. 14]. Another means of learning plans is to acquire them socially, from other, more knowledgeable agents.

Constructive planning is the most intuitively obvious source of a plan, at least in our culture. However, this intuition probably tells us more about what our consciousness spends time doing than about how we actually acquire most of our behavior patterns. The capacity for constructive planning is an essential feature of Soar and of most three-layer-architectures; however it is one that is still underutilized in practice. We suspect this will always be the case, as it will be for GA type models of "thinking", because of the combinatoric difficulties of planning and search [16]. Winston [66] states that learning can only take place when one nearly knows the answer already: this is certainly true of learning plans. Search-like algorithms for planning in real-time agents can only work in highly constrained situations, among a set of likely solutions.

Social or mimetic learning addresses this problem of constraining possible solutions. Observing the actions of another intelligent agent provides the necessary bias. This may be as simple as a mother animal leading her children to a location where they are likely to find food, or as complex as the imitation of complex, hierarchical behavioral patterns (in our terminology, plans) [13, 64]. This may not seem a particularly promising way to increase intelligence, since the agent can only learn what is present in its society, but this is not the case. First, since an agent uses its own intelligence to find the solution within some particular confines, it may enhance the solution it is being presented with. This is famously the case when young language learners regularize constructed languages [3, 33]. Secondly, a communicating culture may well contain more intelligence than any individual member of it, leading to the notion of cultural evolution and mimetics [20]. Thus although the use of social learning in AI is only beginning to be explored [e.g. 53], we believe it will be an important capacity of future artificial agents.

Finally, we have the problem of learning *new* functional and/or skill modules. Although there are many PhD theses on this topic [a good recent example is 19], in the taxonomy presented in this paper, most such efforts would fall under the parameter learning for a single skill module or behavior. Learning full new representations and algorithms for actions is beyond the current state of the art for machine learning. Such a system would almost certainly have to be built on top of a fine-grain distributed representation — essentially it should be an ANN. However, again, the state of the art in ANN does not allow for the learning and representation of such complex and diverse modules.

6   Requirements for a Behavior Learning System

If the current state of the art were not an obstacle, what would a system capable of *all three* forms of adaptivity described in the previous section look like? We think it would require at minimum the elements shown in Figure 3(a). In this section, we will explain this model.

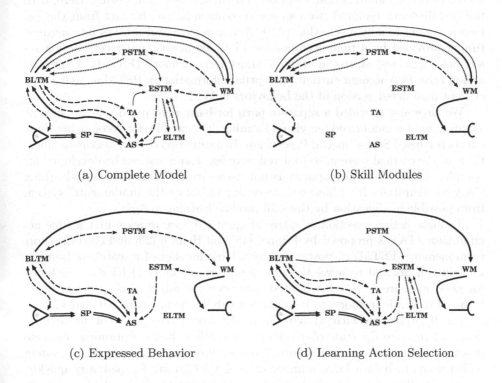

(a) Complete Model

(b) Skill Modules

(c) Expressed Behavior

(d) Learning Action Selection

Fig. 3. An architecture for allowing adaptation within skill modules, of new plans, and of new skill modules. Icons for sensing and action are on the lower left and right respectively. Dashed lines show the flow of information during an active, attending system. Dotted lines are pathways for consolidation and learning. The heavy solid line is the path of expressed behavior; the double line represents the constant perceptual pathway for environmental alerts. The fine lines indicate references: the system pointed to references representations in the system pointed from.

Consider first the behavior or skill module system (Figure 3(b)). The representation of the CAA skill modules has been split into two functional modules: the Behavior Long Term Memory (BLTM) and the Perceptual Short Term Memory (PSTM). The persistent representation of the skill modules' representations and algorithms belong in the former, the current perceptual memory in the latter. There is further a Working Memory (WM) where the representation of the behaviors from the BLTM can be modified to current conditions, for example compensating for tiredness or high wind. In a neurological model, some of these representations might overlap each other in the same organs, for example in different networks within the neocortex or the cerebellum. As in BBAI, the skill modules contain both perception and action, though notice the bidirectional arrows indicating expectation setting for perception.

The full path for expressed action is shown in Figure 3(c). This takes into account both standard action selection and environment monitoring. Here, with the learning arcs removed, we can see recommendations flowing from the behavior system to action selection (AS). Action selection also takes into account timing provided by a time accumulator (TA, see below) and recent action selections (decisions) stored in episodic short term memory (ESTM). Expressed action takes into account current perceptual information in PSTM as well as the current modulated version of the behaviors in WM.

We have also provided a separate path for basic perceptual reflexes such as alarm at loud noises or sudden visual looming. The module for recognizing these effects is labeled SP for Special Perception. In nature this system also has connections to the cortical system, so that reflexive fear responses can be developed for complex stimuli, but this capacity is not necessary for a minimal fully-learning CAA configuration. It is, however, necessary to isolate the fundamental system from possible modification by the skill module learning system.

To make action selection adaptive (Figure 3(d)) we provide first a time accumulator (TA) as proposed by Pöppel [48] and Henson [29] and episodic short term memory (ESTM) as postulated by a large number of researchers (see [41] for experiments and review.) Episodic long term memory (ELTM) is included for good measure — as consolidated experience, it might also represent other forms of semantic memory, or it might actually be homologous with BLTM.

Finally, in keeping with [41, 58], this model assumes that many of the modules make reference to the state of other modules rather than maintaining complete descriptions themselves. This is considered an important attribute of any system which needs to hold a large number of things which are learned very quickly, because it allows for a relatively small amount of state. Such reference is considered important in computer science as a means to reduce the probability of conflicting data sets, and is also a likely feature of evolved systems, where existing organization is often exploited by a variety of means.

7   Future Directions: From Neuroscience to CAA

Fortunately, we do not expect that implementing such a complex system is necessary for most CAA applications. In general, the adaptive needs of the agent can be anticipated in advance by the designer, or discovered and implemented during the process of developing the agent. We do, however, suspect that some of the systems being discovered and explored in neuroscience may soon become standard functional modules in CAA, in the same way that action selection and alarm systems are now. We finish our chapter with some lessons from neuroscience which might further CAA, and hopefully therefore indirectly benefit the ANN community.

We expect that one of the capacities often ascribed to the hindbrain, that of smoothing behavior, should be broken into a separate module. This allows modules that create motor plans to operate at a relatively coarse granularity. It also allows for the combination of influences from multiple modules and the current situation of the agent without complicating those skill modules. The only CAA architecture we know that explicitly has such a unit is Ymir [59], where the Action Scheduler selects the most efficacious way to express messages given the agent's current occupations. For example, if the agent has decided "cognitively" to agree with something, but it is currently engaged in listening rather than speaking, it may nod its head rather than say "yes" and interrupt the other speaker. This sort of capacity is also present in a number of new AI graphics packages which allow for the generation of smooth images from a script of discrete events [2, 4]. The fact that such work is not yet seen in robotics may be partially due to the fact that a physical agent can take advantage of physics and mechanics to do much of its smoothing [11], but as robots attempt more complex feats such as balancing on two legs, providing for smoothed or balanced motions may well deserve dedicated modules or models similar to those cited above.

We also expect that having comprehensive but sparsely represented records of episodic events will become a standard mechanism. Episodic records are useful for complimenting and simplifying reactive plans by recording state about previous attempts and actions, thus reducing the chance that an agent may show inappropriate perseveration or redundancy when trying to solve a problem. Further, as mentioned previously, episodic memory can be a good source for consolidating semantic information, such as noticing regularities in the environment or the agent's own performance. These records can in turn be used by specialized learning systems for particular problems, even if a full-blown skill learning system has not been implemented.

Many researchers are currently working on emotion modules for CAA. We remain skeptical of the need for an independent emotion module for two reasons. First, there is a great deal of evidence that the basic emotions evolved independently at different times in our history. This suggests that a single emotion module might not be appropriate. Second, emotions are intimately involved in action selection. In fact, Damasio [18] implies that any species-typical behavior pattern is effectively an emotional response. This suggests that it is impossible to separate emotions from motivation in action selection.

8 Conclusions

In this chapter we have attempted to further the advance of ANN by unifying it and neuroscience with another branch of artificial intelligence: complete autonomous agents. We have described both the utility and costs of modularity, and explained the sorts of mechanisms the CAA field has found necessary to work with these systems. We have discussed how these mechanisms relate to neuroscience, and in turn have suggested how current advances in neuroscience might further the field of CAA. We have also proposed a complex model for a modular agent intelligence capable of true mammal-like learning, which would rely heavily on ANN representations. ANN, with its distributed representations for machine learning, is the most promising field for developing a system capable of developing its own representations and algorithms. Unfortunately, to date such learning is, as far as we know, beyond the state of the art.

Although our future work section focussed on our own area, CAA, we hope that the advances in this area of AI can also contribute to the advance of neural networks research. We expect that a modular ANN would require both similar types of modules and similar interfaces between them as we have described here. Further, we hope that as the field of CAA advances, it can become as useful a tool for helping brain scientists think about and model the sorts of representations and interactions they are attempting to understand.

Acknowledgements. Thanks to Will Lowe for his comments and suggestions.

References

[1] W. F. Asaad, G. Rainer, and E. K. Miller. Task-specific neural activity in the primate prefrontal cortex. *Journal of Neurophysiology*, 84:451–459, 2000.

[2] A. Baumberg and D. C. Hogg. Generating spatio-temporal models from examples. *Image and Vision Computing*, 14(8):525–532, 1996.

[3] Derek Bickerton. *Language & Species*. The University of Chicago Press, Chicago, Illinois, 1987.

[4] Matthew Brand, Nuria Oliver, and Alex Pentland. Coupled hidden markov models for complex action recognition. In *Proceedings of IEEE Conference on Computer Vision and Pattern Recognition (CVPR97)*, 1997.

[5] Rodney A. Brooks. A robust layered control system for a mobile robot. *IEEE Journal of Robotics and Automation*, RA-2:14–23, April 1986.

[6] Rodney A. Brooks. Intelligence without reason. In *Proceedings of the 1991 International Joint Conference on Artificial Intelligence*, pages 569–595, Sydney, August 1991.

[7] Joanna Bryson. Cross-paradigm analysis of autonomous agent architecture. *Journal of Experimental and Theoretical Artificial Intelligence*, 12(2):165–190, 2000.

[8] Joanna Bryson. Hierarchy and sequence vs. full parallelism in reactive action selection architectures. In *From Animals to Animats 6 (SAB00)*, pages 147–156, Cambridge, MA, 2000. MIT Press.

[9] Joanna Bryson. Making modularity work: Combining memory systems and intelligent processes in a dialog agent. In Aaron Sloman, editor, *AISB'00 Symposium on Designing a Functioning Mind*, pages 21–30, 2000.

[10] Joanna Bryson. The study of sequential and hierarchical organisation of behaviour via artificial mechanisms of action selection, 2000. MPhil Thesis, University of Edinburgh.

[11] Joanna Bryson and Brendan McGonigle. Agent architecture as object oriented design. In Munindar P. Singh, Anand S. Rao, and Michael J. Wooldridge, editors, *The Fourth International Workshop on Agent Theories, Architectures, and Languages (ATAL97)*, pages 15–30, Providence, RI, 1998. Springer.

[12] Joanna Bryson and Lynn Andrea Stein. Architectures and idioms: Making progress in agent design. In C. Castelfranchi and Y. Lespérance, editors, *The Seventh International Workshop on Agent Theories, Architectures, and Languages (ATAL2000)*. Springer, 2000. *in press*.

[13] Richard W. Byrne and Anne E. Russon. Learning by imitation: a hierarchical approach. *Brain and Behavioral Sciences*, 21(5):667–721, 1998.

[14] William H. Calvin. *The Cerebral Code*. MIT Press, 1996.

[15] Niel R. Carlson. *Physiology of Behavior*. Allyn and Bacon, Boston, 2000.

[16] David Chapman. Planning for conjunctive goals. *Artificial Intelligence*, 32:333–378, 1987.

[17] Michael H. Coen. Building brains for rooms: Designing distributed software agents. In *Proceedings of the Ninth Conference on Innovative Applications of Artificial Intelligence (IAAI97)*, Providence, RI, 1997.

[18] Antonio R. Damasio. *The Feeling of What Happens: Body and Emotion in the Making of Consciousness*. Harcourt, 1999.

[19] John Demiris. *Movement Imitation Mechanisms in Robots and Humans*. PhD thesis, University of Edinburgh, May 1999. Department of Artificial Intelligence.

[20] Daniel Dennett. *Darwin's Dangerous Idea*. Penguin, 1995.

[21] Richard E. Fikes, Peter E. Hart, and Nils J. Nilsson. Learning and executing generalized robot plans. *Artificial Intelligence*, 3:251–288, 1972.

[22] James Firby. An investigation into reactive planning in complex domains. In *Proceedings of the National Conference on Artificial Intelligence (AAAI)*, pages 202–207, 1987.

[23] C.R. Gallistel, Ann L. Brown, Susan Carey, Rochel Gelman, and Frank C. Keil. Lessons from animal learning for the study of cognitive development. In Susan Carey and Rochel Gelman, editors, *The Epigenesis of Mind*, pages 3–36. Lawrence Erlbaum, Hillsdale, NJ, 1991.

[24] M. P. Georgeff and A. L. Lansky. Reactive reasoning and planning. In *Proceedings of the Sixth National Conference on Artificial Intelligence (AAAI-87)*, pages 677–682, Seattle, WA, 1987.

[25] Henry Gleitman. *Psychology*. Norton, 4 edition, 1995.

[26] Stephen Grossberg. How does the cerebral cortex work? learning, attention and grouping by the laminar circuits of visual cortex. *Spatial Vision*, 12:163–186, 1999.

[27] Kevin Gurney, Tony J. Prescott, and Peter Redgrave. The basal ganglia viewed as an action selection device. In *The Proceedings of the International Conference on Artificial Neural Networks*, Skovde, Sweden, September 1998.

[28] D.O. Hebb. *The Organization of Behavior*. John Wiley and Sons, New York, New York, 1949.

[29] Richard N. A. Henson. *Short-term Memory for Serial Order*. PhD thesis, University of Cambridge, November 1996. St. John's College.

[30] Henry Hexmoor, Ian Horswill, and David Kortenkamp. Special issue: Software architectures for hardware agents. *Journal of Experimental & Theoretical Artificial Intelligence*, 9(2/3):147–156, 1997.

[31] Berthald K. P. Horn and Patrick H. Winston. A laboratory environment for applications oriented vision and manipulation. Technical Report 365, MIT AI Laboratory, 1976.

[32] George Houghton and Tom Hartley. Parallel models of serial behavior: Lashley revisited. *PSYCHE*, 2(25), February 1995.

[33] Simon Kirby. *Function, Selection and Innateness: the Emergence of Language Universals*. Oxford University Press, 1999.

[34] T. Kobayashi, H. Nishijo, M. Fukuda, J. Bures, and T. Ono. Task-dependent representations in rat hippocampal place neurons. *JOURNAL OF NEUROPHYSIOLOGY*, 78(2):597–613, 1997.

[35] David Kortenkamp, R. Peter Bonasso, and Robin Murphy, editors. *Artificial Intelligence and Mobile Robots: Case Studies of Successful Robot Systems*. MIT Press, Cambridge, MA, 1998.

[36] John E. Laird and Paul S. Rosenbloom. The evolution of the Soar cognitive architecture. Technical Report CSE-TR-219-94, Department of EE & CS, University of Michigan, Ann Arbor, September 1994. also in *Mind Matters*, Steier and Mitchell, eds.

[37] M.F. Land. Mechanisms of orientation and pattern recognition by jumping spiders (salticidae). In R. Wehner, editor, *Information Processing in the Visual Systems of Arthropods*, pages 231–247. Springer-Verlag, 1972.

[38] K. S. Lashley. The problem of serial order in behavior. In L. A. Jeffress, editor, *Cerebral mechanisms in behavior*. John Wiley & Sons, New York, 1951.

[39] Joseph S. Lonstein and Judith M. Stern. Role of the midbrain periaqueductal gray in maternal nurturance and aggression: *c-fos* and electrolytic lesion studies in lactating rats. *Journal of Neuroscience*, 17(9):3364–78, May 1 1997.

[40] Chris Malcolm, Tim Smithers, and John Hallam. An emerging paradigm in robot architecture. In *Proceedings of IAS*, volume 2, Amsterdam, Netherlands, 1989.

[41] James L. McClelland, Bruce L. McNaughton, and Randall C. O'Reilly. Why there are complementary learning systems in the hippocampus and neocortex: Insights from the successes and failures of connectionist models of learning and memory. *Psychological Review*, 102(3):419–457, 1995.

[42] Frank A. Middleton and Peter L. Strick. Basal ganglia output and cognition: evidence from anatomical, behavioral, and clinical studies. *Brain and Cognition*, 42(2):183–200, 2000.

[43] Jonathon W. Mink. The basal ganglia: focused selection and inhibition of competing motor programs. *Progress In Neurobiology*, 50(4):381–425, 1996.

[44] Hans P. Moravec. The stanford cart and the CMU rover. In I. J. Cox and G. T. Wilfong, editors, *Autonomous Robot Vehicles*, pages 407–419. Springer, 1990.

[45] O. Nestares and David J. Heeger. Robust multiresolution alignment of MRI brain volumes. *Magnetic Resonance in Medicine*, 43:705–715, 2000.

[46] Nils Nilsson. Shakey the robot. Technical note 323, SRI International, Menlo Park, California, April 1984.

[47] Nils Nilsson. Teleo-reactive programs for agent control. *Journal of Artificial Intelligence Research*, 1:139–158, 1994.

[48] E. Pöppel. Temporal mechanisms in perception. *International Review of Neurobiology*, 37:185–202, 1994.

[49] Tony J. Prescott, Kevin Gurney, F. Montes Gonzalez, and Peter Redgrave. The evolution of action selection. In David McFarland and O. Holland, editors, *Towards the Whole Iguana*. MIT Press, Cambridge, MA, to appear.

[50] V. S. Ramachandran and S. Blakeslee. *Phantoms in the brain: Human nature and the architecture of the mind*. Fourth Estate, London, 1998.

[51] Peter Redgrave, Tony J. Prescott, and Kevin Gurney. The basal ganglia: a vertebrate solution to the selection problem? *Neuroscience*, 89:1009–1023, 1999.

[52] T. J. Roper. Learning as a biological phenomena. In T. R. Halliday and P. J. B. Slater, editors, *Genes, Development and Learning*, volume 3 of *Animal Behaviour*, chapter 6, pages 178–212. Blackwell Scientific Publications, Oxford, 1983.

[53] Stefan Schaal. Is imitation learning the route to humanoid robots? *Trends in Cognitive Sciences*, 3(6):233–242, 1999.

[54] Phoebe Sengers. Do the thing right: An architecture for action expression. In Katia P Sycara and Michael Wooldridge, editors, *Proceedings of the Second International Conference on Autonomous Agents*, pages 24–31. ACM Press, 1998.

[55] Phoebe Sengers. *Anti-Boxology: Agent Design in Cultural Context*. PhD thesis, School of Computer Science, Carnegie Mellon University, 1999.

[56] M. Sur, A. Angelucci, and J. Sharma. Rewiring cortex: The role of patterned activity in development and plasticity of neocortical circuits. *Journal of Neurobiology*, 41:33–43, 1999.

[57] J. Tanji and K. Shima. Role for supplementary motor area cells in planning several movements ahead. *Nature*, 371:413–416, 1994.

[58] T. J. Teyler and P. Discenna. The hippocampal memory indexing theory. *Behavioral Neuroscience*, 100:147–154, 1986.

[59] Kristinn R. Thórisson. A mind model for multimodal communicative creatures & humanoids. *International Journal of Applied Artificial Intelligence*, 13(4/5): 519–538, 1999.

[60] Alan M. Turing. Computing machinery and intelligence. *Mind*, 59:433–460, October 1950.

[61] Gerhard Weiss, editor. *Multiagent Systems: A Modern Approach to Distributed Artificial Intelligence*. MIT Press, Cambridge, MA, 1999.

[62] Daniel S. Weld. Recent advances in AI planning. *AI Magazine*, 20(2):93–123, 1999.

[63] Michael Wessler. A modular visual tracking system. Master's thesis, MIT, 1995. Artificial Intelligence Laboratory.

[64] Andrew Whiten. Primate culture and social learning. *Cognitive Science*, 24, 2000. in press.

[65] S. I. Wiener. Spatial, behavioral and sensory correlates of hippocampal CA1 complex spike cell activity: Implications for information processing functions. *Progress in Neurobiology*, 49(4):335, 1996.

[66] Patrick Winston. Learning structural descriptions from examples. In Patrick Winston, editor, *The Psychology of Computer Vision*. McGraw-Hill Book Company, New York, 1975.

Biased Competition Mechanisms for Visual Attention in a Multimodular Neurodynamical System

Gustavo Deco

Siemens AG, Corporate Technology, ZT IK 4
Otto-Hahn-Ring 6, D- 81739 Munich, Germany
Gustavo.Deco@mchp.siemens.de

Abstract. Visual attention poses a mechanism for the selection of behaviourally relevant information from natural scenes which usually contain multiple objects. The aim of the present work is to formulate a
neurodynamical model of selective visual attention based on the "biased competition hypothesis" and structured in several network modules
which can be related with the different areas of the dorsal and ventral
path of the visual cortex. Spatial and object attention are accomplished
by a multiplicative gain control that emerges dynamically through inter-
cortical mutual biased coupling. We also include in our computational
model the "resolution hypothesis" in order to explain the role of the
neurodynamics control of spatial resolution evidenced in psychophysical
experiments. We propose that V1 neurons have different latencies depending on the spatial frequency to which they respond more sensitively.
In concrete, we pose that V1 neurons sensitive to low spatial frequency
are faster than V1 neurons sensitive to high spatial frequency. In this
sense, a scene is first predominantly analysed at a coarse resolution level
and the dynamics enhances subsequently the resolution at the location
of an object until the object is identified.

1 Introduction

Visual attention can function in two distinct modes: spatial focal attention that
can be visualized as a spotlight that 'illuminates' a certain location of visual
space for focused visual analysis; dispersed object attention with which a target
object can be searched in parallel over a large visual space. Duncan [1] proposed
that the two modes of operation are both manifestation of a top-down selection
process. In spatial attention, the selection is focused in the spatial dimension and
spread in feature dimension; while in object attention, the selection is focused in
the feature dimension and spread in the spatial dimension. Experimental observations in functional imaging [2,3] and single-cell recording [4,5] provide strong
evidences for the biased competition hypothesis whichsuggests that attention
modulates visual processing by enhancing the responses of the neurons representing the features or location of the attended stimulus and reducing the suppressive interactions of neurons representing nearby distractors. In this work, we

S. Wermter et al. (Eds.): Emergent Neural Computational Architectures, LNAI 2036, pp. 114–126, 2001.

formulate a neurodynamical system that addresses the issues on how spatial and object attention mechanisms, presumably mediated by the dorsal visual stream and the ventral visual stream respectively, can be integrated and function as a unitary system in visual search and visual recognition tasks. The system is built upon the biased-competition hypothesis, attentional gain field modulation, and on interactive processes in visual processing. An important novel idea in this model is that the dorsal stream and ventral stream interact at multiple points and levels. The locus of intersection is a function of the scale of analysis. For example, detailed spatial and feature integration and attentive analysis would utilize area V1, whereas analysis of larger scale would involve V4. These early visual computational buffers interact with the dorsal stream and the ventral stream via the multiple feedforward/feedback inter-cortical loops. These loops, together with the competitive mechanism within each cortical area, enable the system to switch from one mode of operation (e.g. spatial attention) to another (e.g. object attention) depending on the biased input from the prefrontal cortex.

We will consider two important functions of the visual system, namely: object recognition and visual search. A phenomenological description of these two perceptual functions is schematically presented in Figure 1 in the context of a natural landscape. In the case of object recognition, a particular location in the natural scene is a priori specified with the aim of identification of the object which lies at that position. Therefore, object recognition asks for "what" is at a predefined particular spatial location. In the naive framework of the spotlight metaphor, one can describe the role of attention in object recognition by imagining that the prespecification of the particular spatial location is realized by fixing an attentional window or spotlight at that position. The features inside the fixed attentional window should be now bounded and recognized. On the contrary, by visual search, one specifies a priori a given target object (i.e. features) with the goal of finding if the target object is present in the scene and at which location. Consequently, visual search asks for "where" is a predefined bunch of given features. A naive description of visual search in the framework of the spotlight paradigm considers that during visual search attentional mechanisms shift a window along the entire scene in order to serially search at different positions for the target object.

Our computational neuroscience approach shows how spatial and object attention mechanisms can be integrated and function as a unitary system in visual search and visual recognition tasks. The dynamical intra- and intermodular interactions in our cortical model implement attentional top-down feedback mechanisms that embody a physiologically plausible system of active vision which unifies different perceptual functions. In our system, attention is a dynamical emergent property, rather than a separate mechanism operating independently of other perceptual and cognitive processes. The dynamical mechanisms that define our system work across the visual field in parallel but due to the different latencies show serial or parallel behaviour over the space mode, like it is observed in visual search. Neither explicit serial scanning with an attentional spotlight nor saliency maps need to be assumed.

Additionally, psychophysical evidence strongly suggests that selective attention can enhance the spatial resolution in the input region corresponding to the

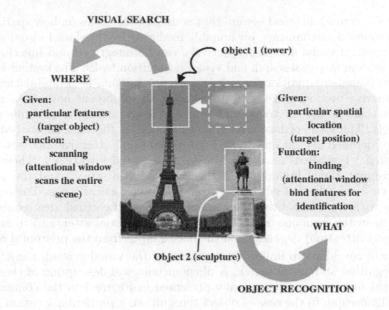

VISUAL SEARCH

Object 1 (tower)

WHERE

Given:
particular features
(target object)
Function:
scanning
(attentional window
scans the entire
scene)

Given:
particular spatial
location
(target position)
Function:
binding
(attentional window
bind features for
identification

WHAT

Object 2 (sculpture)

OBJECT RECOGNITION

Fig. 1. The role of attentional mechanisms in object recognition and visual search in a natural scene.

focus of attention. Hence, we include in our computational model a neurodynamical control of spatial resolution. We propose that V1 neurons have different latencies depending on the spatial frequency to which they respond more sensitively. In concrete, we pose that V1 neurons sensitive to low spatial frequency are faster than V1 neurons sensitive to high spatial frequency. In this sense, a scene is first predominantly analysed at a coarse resolution level and the dynamic enhances subsequently the resolution at the location of an object until the object is identified. We propose and simulate new psychophysical experiments where the effect of the attentional enhancement of spatial resolution can be demonstrated by predicting different reaction time profiles in visual search experiments where the target and distractors are defined at different levels of resolution.

2   Computational Model

We now briefly describe the cortical model of visual attention for object recognition and visual search based on the biased competitive hypothesis and the corresponding neurodynamical mechanisms [6]. The system is absolutely autonomous and each of its functionalities is explicitly described in a complete mathematical framework. The overall systemic representation of the model is shown in the Figure 2.

The system is essentially composed of three modules structured such that they resemble the two known main visual path of the mammalian visual cortex. Information from the retino-geniculo-striate pathway enters the visual cortex

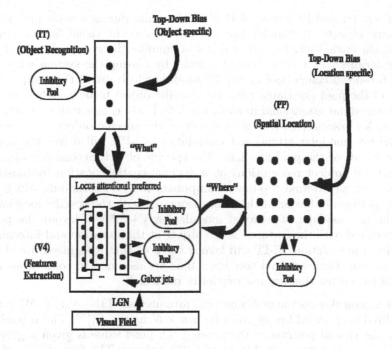

Fig. 2. A cortical architecture for visual attention.

through area V1 in the occipital lobe and proceeds into two processing streams. The occipital-temporal stream (the so-called "what" pathway) leads ventrally through V2, V4 and IT (inferotemporal cortex) and is mainly concerned with object recognition, independently of position and scaling. The occipito-parietal stream (the so-called "where" pathway) leads dorsally into PP (posterior parietal complex) and is concerned with the location and spatial relationships between objects. The first module (V4) of our system is engaged with the extraction of features and consists of pools of neurons with Gabor receptive fields tuned to different positions in the visual field, orientations and spatial frequency resolutions. The "where" pathway is given through the mutual connection with the second (PP) module that consists of pools codifying the position of the stimuli. The connections with the first module origin a top-down biasing attentional modulation associated with the location of the stimuli. At last, the third module (IT) of our system is engaged with the recognition of objects and consists of pools of neurons which are sensitive to the presence of a specific object in the visual field. The pools in (IT) are synaptically connected with translationally invariant receptive fields with pools of the first module (V4) such that based on the Gabor features specific objects are invariantly recognized. The mutual connections between IT and V4 modules represent a top-down biasing attentional modulation associated with specific objects.

The system operates in two different modes: the learning mode and the recognition mode. During the learning mode the synaptic connection between V4 and IT are trained by means of Hebbian learning during several presentations of specific objects at changing random position in the visual field. During the recognition mode there two possibilities of running the system. First, an object can be localized in a scene (visual search) by biasing the system with an external top-down component at the IT module which drives the competition in favour of the pool associated with the specific object to be searched such that the intermodular attentional modulation V4-IT will enhance the activity of the pools in V4 associated with the features of the specific object to be searched and the intermodular attentional modulation V4-PP will drives the competition in favour of the pool localizing the specific object. Second, an object can be identified (object recognition) at a specific spatial location by biasing the system with an external top-down component at the PP module which drives the competition in favour of the pool associated with the specific location such that the intermodular attentional modulation V4-PP will favour the pools in V4 associated with the features of the object at that location and intermodular attentional modulation V4-IT will favour the pool that recognized the object at that location. Both external top-down bias are assumed to come from frontal areas of the cortex that are not explicitly modelled.

Let us now discuss the mathematical formulation of the system. We consider a pixelized grey-scaled image given by a $n \times N$ matrix Γ_{ij}^{orig}. The subindices ij denote the spatial position of the pixel. Each pixel value is given a grey value coded in a scale between 0 (black) and 255 (white). The first step in the preprocessing consists in removing the DC component of the image (i.e. the mean value of the grey-scaled intensity of the pixels) which is probably done in the lateral geniculate nucleus (LGN) of the thalamus. The visual representation in LGN is essentially a contrast invariant pixel representation of the image, i.e. each neuron encodes the relative brightness value at one location in visual space referred to the mean value of the image brightness. Feedforward connections to a layer of V1 neurons perform the extraction of simple features. Theoretical investigations have suggested that simple cells in the primary visual cortex can be modelled by 2D-Gabor functions. The 2D-Gabor functions are local spatial bandpass filters that achieve the theoretical limit for conjoint resolution of spatial and frequency information, i.e. in the 2D-spatial and 2D-Fourier domains. The Gabor receptive fields have five degrees of freedom given essentially by the product of an elliptical Gaussian and a complex plane wave. The first two degrees of freedom are the 2D-location of the receptive field's centre, the third is the size of the receptive field, the fourth is the orientation of the boundaries separating excitatory and inhibitory regions, and the fifth is the symmetry. This fifth degree of freedom is given in the standard Gabor transformation by the real and imaginary part, i.e by the phase of the complex function representing it, whereas biologically this can be done by combining pairs of neurons with even or odd receptive fields. Let us consider the experimental neurophysiological constraints. There are three constraints fixing the relation between width, height, orientation and spatial frequency. The first constraint postulates that the aspect ratio of the elliptical Gaussian envelope is 2:1. The second constraint postulates that the orientation is aligned with the long axis of the elliptical Gaussian. The third constraint assumes that the half-amplitude bandwidth of the frequency response is about 1 to 1.5 octaves along the optimal orientation. Further, we assume that the mean is zero in order to have an admissible wavelet basis. A family of discretized 2D-Gabor wavelets that satisfy the wavelet theory and the neurophysiological constraints for simple cells as given by [7]

$$G_{kpql}(x,y) = a^{-k}\Psi_{\theta_l}\left(a^{-k}(x-2p) - a^{-k}(y-2q)\right)$$

$$\Psi_{\theta_l} = \Psi\left(x\cos(l\theta_0) + y\sin(l\theta_o), -x\sin(l\theta_o) + y\cos(l\theta_0)\right) \tag{1}$$

and the mother wavelet is given by

$$\Psi(x,y) = \frac{1}{\sqrt{2\pi}}e^{-\frac{1}{8}(4x^2+y^2)} \cdot \left[e^{i\kappa x} - e^{-\frac{\kappa^2}{2}}\right]. \tag{2}$$

In the above equations θ_0 denotes the step size of each angular rotation, l the index of rotation corresponding to the preferred orientation θ_l, k denotes the octave, and pq the position of the receptive field centre. In this form, the receptive fields at all levels cover the spatial domain in the same way, i.e. by overlapping always the receptive field in the same fashion. In this work we choose $\kappa = \pi$ that correspond to a spatial frequency bandwidth of one octave. The neurons in the pools in V1 have receptive fields performing a Gabor wavelet transform. Let us denote by l_{kpql}^{V1} the sensorial input activity to a pool in V1 which is sensitive to a determined spatial frequency given at octave k, to a preferred orientation defined by the rotation index l and to stimuli at the centre location specified by the indices pq. The sensorial input activity to a pool in V1 is therefore defined by the module of the convolution between the corresponding receptive fields and the image. Since in our numerical simulations the system needs only to learn a small number of objects (usually 2-4), in our current implementation, for simplicity, we temporarily eliminate the V4 module and fully connect V1 and IT cell assemblies directly together. We implement translation invariance by the attentional intermodular biasing interaction between pools in the modules V1 and PP. For each V1 neuron, the gain modulation observed decreases as the actual point where attention is being focused moves away from the centre of the receptive field in a Gaussian-like form. Consequently, the connections with the pools in the PP module are specified such that the modulation is Gaussian. Let us define in the PP module a pool A_{ij}^{PP} for each location ij in the visual field. The mutual connection (i.e. bilateral) between a pool A_{kpql}^{V1} in V1 (or V4) and a pool A_{ij}^{PP} in PP are therefore defined by

$$W_{pqij} = Ae^{-\frac{(i-p)^2+(j-q)^2}{2s^2}} - B \tag{3}$$

Let us now define the neurodynamical equations that regulates the evolution of the whole system. The activity level of the input current in the V1 module is given by

$$\tau\frac{\partial}{\partial t}A_{kpql}^{V1}(t) = -A_{kpql}^{V1}(t) + aF(A_{kpql}^{V1}(t)) - bF(A_k^{I,V1}(t)) + I_{kpql}^{V1}(t)$$
$$I_{pq}^{V1-PP}(t) + I_{kpql}^{V1-IT}(t) + I_0 + \nu \tag{4}$$

where the attentional biasing due to the intermodular "where" connections with the pools in the parietal module PP I_{ab}^{V1-PP} is given by

$$I_{pq}^{V1-PP}(t) = \sum_{i,j} W_{pqij} F\left(A_{ij}^{PP}(t)\right) \qquad (5)$$

and the attentional biasing due to the intermodular "what" connections with the pools in the temporal module IT I_{kpql}^{V1-IT} is defined by

$$I_{kpql}^{V1-IT}(t) = \sum_{c=1} w_{ckpql} F\left(A_c^{IT}(t)\right) \qquad (6)$$

where w_{ckpql} is the connection strength between the V1 pool A_{kpql}^{V1} and the IT pool A_c^{IT} corresponding to the coding of an specific object category c. We assume that the IT module has C pools corresponding to different object categories. We implement the resolution hypothesis by assuming different latencies τ in eq. (4) for pools corresponding to different octaves. Higher resolution octaves have a smaller τ than the ones corresponding lower resolutions. For each spatial frequency level, a common inhibitory pool is defined. The current activity of the inhibitory pools obey the following equations:

$$\tau_P \frac{\partial}{\partial t} A_k^{I,V1}(t) = -A_k^{I,V1}(t) + c \sum_{p,q,l} F\left(A_{kpql}^{V1}(t)\right) - dF\left(A_k^{I,V1}(t)\right) \qquad (7)$$

The current activity of the excitatory pools in the posterior parietal module PP are given by

$$\tau \frac{\partial}{\partial t} A_{ij}^{PP}(t) = -A_{ij}^{PP}(t) + aF\left(A_{ij}^{PP}(t)\right) - bF\left(A^{I,PP}(t)\right) + I_{ij}^{PP-V1}(t) + I_{ij}^{PP,A} + I_0 + \nu \qquad (8)$$

where $I_{ij}^{PP,A}$ denotes an external attentional object specific top-down bias and the intermodular attentional biasing I_{ij}^{PP-V1} through the connections with the pools in the module V4 is

$$I_{ij}^{PP-V1}(t) = \sum_{k,p,q,l} w_{pqij} F\left(A_{kpql}^{V1}(t)\right) \qquad (9)$$

and the activity current of the common PP inhibitory pool evolves according to

$$\tau_P \frac{\partial}{\partial t} A^{I,PP}(t) = -A^{I,PP}(t) + c \sum_{i,j} F\left(A_{ij}^{PP}(t)\right) - dF\left(A^{I,PP}(t)\right) \qquad (10)$$

The dynamics of the inferotemporal module IT is given by

$$\tau \frac{\partial}{\partial t} A_c^{IT}(t) = -A_c^{IT}(t) + aF\left(A_c^{IT}(t)\right) - bF\left(A^{I,IT}(t)\right) + I_c^{IT-V1}(t) + I_c^{IT,A} + I_0 + \nu \qquad (11)$$

where $I - c^{IT,A}$ denotes an external attentional spatial specific top-down bias and the intermodular attentional biasing I_c^{IT-V4} between IT and V1 pools is

$$I_c^{IT-V1}(t) = \sum_{k,p,q,l} w_{ckpql} F\left(A_{kpql}^{V1}(t)\right) \tag{12}$$

and the activity current of the common PP inhibitory pool evolves according to

$$\tau_P \frac{\partial}{\partial t} A^{I,IT}(t) = -A^{I,IT}(t) + c \sum_c F\left(A_c^{IT}(t)\right) - dF\left(A^{I,IT}(t)\right) \tag{13}$$

During a learning phase each object is learned. The external attentional location specific bias in PP $I_{ij}^{PP,a}$ is set so that only the pool ij corresponding to the spatial location where the object to be learned is, receives a positive bias. In this way the spatial attention defines the localization of the object to be learnt. The external attentional object specific bias in IT $I_c^{IT,A}$ is similarly set so that only the pool that will identify the object receives a positive bias. Therefore, we define in a supervised way the identity of the object. After presentation of a given stimulus, i.e. an specific object at an specific location, and the corresponding external bias, the system evolves until convergence. After convergence, the V1-IT connections w_{ckpql} are trained through the following Hebbian rule

$$w_{ckpql} = w_{ckpql} + \eta F\left(A_c^{IT}(t)\right) F\left(A_{kpql}^{V1}(t)\right) \tag{14}$$

being t large enough (i.e. after convergence) and η the learning coefficient. This procedure is repeated for all objects at all possible locations until the weights converge. During the recognition phase there are two possibilities, namely to search for an specific object (visual search) or to identify an object at a given spatial location (object recognition). In the case of visual search the stimuli are presented and the external attentional object specific bias in IT $I_c^{IT,A}$ is set so that only the pool c corresponding to the category of the object to be searched receives a positive bias while the external attentional location specific bias in PP $I_{ij}^{PP,A}$ is set zero everywhere. The external attentional bias $I_c^{IT,A}$ drives the competition in the IT module so that the pool corresponding to the searched object wins. The intermodular attentional modulation between IT and V1 bias the competition in V1 so that all features detected from the retinal inputs at different positions that are compatible with the specific object to be searched will now win. At last, the intermodular attentional bias between V1 and PP drives the competition in V1 and in PP so that only the spatial location in PP and the associated V1 pools compatible with the presented stimulus and with the top-down specific category of the object to be searched will remain active after convergence, i.e. reading the final state in PP the object is found. In this form, the final activation state is neurodynamically driven by stimulus, external top-down bias and intermodular bias, in a fully parallel way. The attention is not a mechanism involved in the competition but just an emergent effect that support the dynamical evolution to a state where all constraints are satisfied. In the case of object recognition, the external attentional bias in PP $I_{ij}^{PP,A}$ is set so that only the pool associated with the spatial location where the object to be identified is, receives a positive bias, i.e. an spatial region will be "illuminated". The other external bias $I_c^{IT,A}$ is zero everywhere. In this case, the dynamic evolve such that in PP only the pool associated with the top-down biased spatial location will win, and this drives the competition in V1 such that only the pools corresponding to features of the stimulus at that location will win biasing the dynamics in IT such that only the pool identifying the class of the features at that position will remain active indicating the category of the object at that predefined spatial location.

3   The Neurodynamics of Object Recognition

In this section, we present the simulation results corresponding to the external biasing of a specific spatial location and the resulting recognition of the object category at that position. We use for this example the natural scene shown in Figure 1.The input system processed a pixelized 66×66 image ($N = 66$). The V1 hypercolumns covered the entire image uniformly. They was distributed in 33×33 locations ($P = 33$) and each hypercolumn were sensitive to three spatial frequencies and to eight different orientations (i.e., $K = 3$ and $L = 8$). Consequently, the V1 module has pools and three inhibitory pools. The IT module utilized has two pools and one common inhibitory pool. Finally, the PP module contains pools corresponding to each possible spatial location, i.e. to each of the 66×66 pixels, and a common inhibitory pool. Two objects (also shown in Figure 1) are isolated in order to define two categories to be associated with two different pools in the IT module. During the learning phase, these two objects are presented randomly and at random positions in order to achieve translation invariance responses. The system required 1000000 different presentations for training the IT pools.

For object recognition, we utilize the system in the recognition mode by biasing externally only the pools in the PP module that correspond to a specific position. We choose as specific position the centre of the object 1 (at pixel coordinates $x = 22$, $Y = 8$). To visualize the temporal evolution of the units in different modules functioning in the two modes, we display the maximum activities of the cell assemblies in three different cortical areas, i.e. V1, IT and PP, associated with the target and the distractors respectively. Figure 3 shows that the maximum activities of target assemblies and the distractor assemblies in the different modules. The maximal activity associated with the target is denoted with the subindex T and the maximal activity of the distractor pools with the subindex D.

The system correctly identifies the object class (i.e. tower or sculpture) at the respective locations. In an object recognition task, the system functions in a spatial attention mode. A particular cell assembly at d46 is activated manually to provide a top-down bias to excite a particular cell assembly responsible for a spatial location in the PP module. When an image is presented, the spatial attention will highlight the visual information within its spotlight, enhancing the signals so that it will pass through the bottle-neck of competitive normalization to be recognized by the IT module. If it belongs to one of the learnt object classes, a cell assembly will be activated in IT, corresponding to recognition. This is precisely what happened in our system.

In Figure 3, we observe that the pools in the PP module polarizes in this case much rapid, due to the external spatial bias, and posteriorly drives the competition in the other modules. After enough time, the IT module competition signalizes the win of the pool corresponding to the category identifying the object, i.e. the tower.

The neural pools corresponding to spatial positions in the neighbourhood of the particular location preselected by the prefrontal area d46 are active whereby the pools corresponding to other locations are inactive. In area V1, only the neural pools corresponding to features of the retinal stimulus at the preselected location remain active after convergence. The intramodular competition in V1 is biased during the dynamical evolution by the top-down feedback connections coming from PP and IT (and reversely the evolution in IT and PP is influenced steadily by V1). In the case of object recognition, the rapid polarization of pools in PP to a particular location serves to modulate the competition in V1 in such a way that only the "illuminated" features at a particular location will survive and further influence the processing of object identification in IT. The interareal interactions which bias the intramodular competition at each different module are precisely the attentional mechanisms that underlie the process of object recognition. It is important to remark, that there is no explicit spotlight of attention in our model, but the parallel and global dynamical evolution of the entire system converges to a state where a spotlight of attention in PP appears and is explicitly used, even during its formation, for modulate the information processing channel for object recognition. The most interesting aspect of our theory is that the process of object recognition can not be separated in different processes acting in the ventral or dorsal stream, but is the result of a global dynamical steadily interaction between all brain areas. Of course each brain area has a defined functional role, but the global behaviour emerges from the constant cross-talk between the different modules in the ventral and dorsal visual paths.

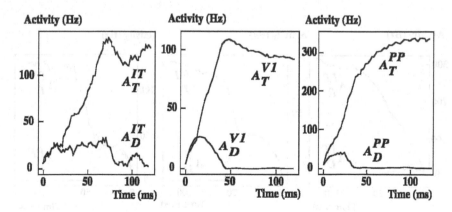

Fig. 3. Dynamical evolution of the system in an object recognition task at a specific location (the location of the tower).

4　The Neurodynamics of Visual Search

In this section we analyse the dynamical evolution of the different modules of our system in the case where a real object has to be found in a natural scene. During the recognition phase, the natural landscape of Figure 1 is presented as stimulus and an specific object ((1) tower or (2) sculpture) is selected by setting the external top-down bias to the IT module so that the corresponding pool associated with the selected object is positively biased. The results for the search of object 1 is shown in Figure 4. In the visual search task, the system functions in a object attention mode. A particular object class's cell assembly at v46 is activated manually to provide a top-down bias to the IT module. When an image containing the target object is presented to the 'retina', the system evolves automatically to localize the object in the image. As the dynamical system settles, only a very localized region in PP is activated. The dynamical evolution observed in Figures 4 shows a more rapid polarization of the competition in the IT module that after a certain time drives the competition in the V1 and PP module in favour of the neural pools corresponding to the target object, meaning that the spatial localization of the searched object was found.

5　The Resolution Hypothesis

We implement the resolution hypothesis by assuming different latencies τ in eq. (4) for pools corresponding to different octaves. Higher resolution octaves have a smaller τ than the one corresponding to lower resolution. We simulate visual search tasks in the context of the global-local paradigm, i.e. by using hierarchical patterns as target and distractors. As hierarchical patterns we use global letter shapes constructed with local letter elements. Let us denote as "Xy" an item corresponding to a global letter "X" composed as local letters "y". We define two different types of visual search experiments. The first experiment (VS1) defines the distractors and the target such that at the global (coarse) level the

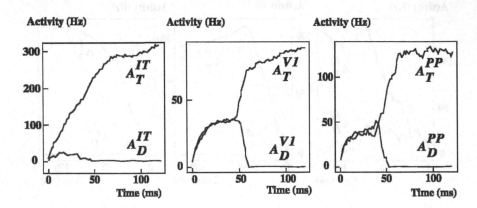

Fig. 4. Dynamical evolution of the system in a visual search task (search of the tower).

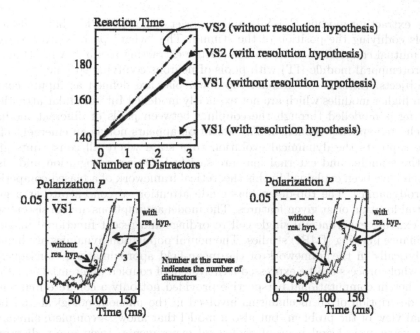

Fig. 5. Neurodynamics of visual search of hierarchical patterns.

target can pop-up from the distractors. In this case the target is defined for example as the pattern Hh and the distractors are patterns Tt or Th. The second experiment (VS2) defines the distractors and the target such that their global characteristics are indistinguishable and they only differ at the local level. The target is defined for example as the pattern Hh and the distractors are patterns Ht. The reaction time is defined by determining when the level of polarization of the pools in the PP module reach a certain threshold $\theta = 0.05$. The level of polarization P is defined as the difference of the maximal activity of pools corresponding to target and distractors. Figure 5 shows the dynamical evolution of P for experiments VS1 and VS2 and different number of distractors. When the resolution hypothesis is implemented, VS1 shows a parallel search instead of the serial search observed when the different octave latencies in V1 are equalized. On the contrary, in the case of VS2 always a serial search is observed. This result explain the neurodynamics underlying the global-local precedence effect in the active context of a visual search experiment.

6 Conclusions

We formulated a neurodynamical system based on the "biased competition" hypothesis that consists of interconnected populations of cortical neurons distributed in different brain modules which can be related with the different areas of the dorsal or "where" and ventral or "what" path of the primate cortex. The "where" pathway is reflected through the mutual connection between a feature extracting module (V1-V4) and a parietal module (PP) that consists of pools codifying the position of the stimuli. The "what" path is given through the mutual connections between the feature extracting module (V1-V4) and an inferotemporal module (IT) with pools of neurons codifying specific

objects. External attentional top-down bias are defined as inputs coming from higher modules which are not explicitly modelled Intermodular attentional biasing is modelled through the coupling between pools of different modules, which are explicitly modelled. The attention appears not as an emergent effect that supports the dynamical evolution to a state where all constraints given by the stimulus and external bias are satisfied. Object recognition and visual search have been explained in this theoretical framework of a biased competitive neurodynamics. The top-down bias guide attention to concentrate at a given spatial location or at given features. The model assumptions are consistent with the existing experimental single cell recordings and recent functional magnetic resonance imaging (fMRI) studies. The neural population dynamics are handled analytically in the framework of the mean-field approximation. Consequently, the whole process can be expressed as a system of coupled differential equations. The herein computational perspective provided not only a concrete mathematical description of all mechanisms involved in the phenomenological and functional view of the problem, but also a model that allows a complete simulation of psychophysical and neurophysiological experiments. Even more, disruption of computational blocks corresponding to submechanisms in the model can be used for simulations that predict impairment in visual information selection in patients suffering from brain injury.

References

[1] Duncan, J. (1980). The locus of interference in the perception of simultaneous stimuli. *Psychological Review*, **87**, 272–300.
[2] Kastner, S., De Weerd, P., Desimone, R., and Ungerleider, L. (1998). Mechanisms of directed attention in the human extrastriate cortex as revealed by functional MRI. *Science*, **282**, 108–111.
[3] Kastner, S., Pinsk, M., De Weerd, P., Desimone, R. and Ungerleider, L. (1999). Increased activity in human visual cortex during directed attention in the absence of visual stimulation. *Neuron*, **22**, 751–761.
[4] Reynolds, J., and Desimone, R. (1999). The role of neural mechanisms of attention in solving the binding problem. *Neuron*, **24**, 19–29.
[5] Reynolds, J., Chelazzi, L., and Desimone, R. (1999). Competitive mechanisms subserve attention in macaque areas V2 and V4. *Journal of Neuroscience*, **19**, 1736–1753.
[6] Deco, G. and Lee, T.S. (2000). A Multimodular Neurodynamical Model of Biased Competition Mechanisms for Visual Attention. *Vision Research*, submitted.
[7] Lee, T. S. (1996). Image representation using 2D Gabor wavelets. *IEEE Transactions on Pattern Analysis and Machine Intelligence*, **18**, 10, 959–971.

Recurrent Long-Range Interactions in Early Vision

Thorsten Hansen, Wolfgang Sepp, and Heiko Neumann

Universität Ulm, Abt. Neuroinformatik, D-89069 Ulm, Germany
(hansen,wsepp,hneumann)@neuro.informatik.uni-ulm.de

Abstract. A general principle of cortical architecture is the bidirectional flow of information along feedforward and feedback connections. In the feedforward path, converging connections mainly define the feature detection characteristics of cells. The computational role of feedback connections, on the contrary, is largely unknown. Based on empirical findings we suggest that top-down feedback projections modulate activity of target cells in a context dependent manner. The context is represented by the spatial extension and direction of long-range connections. In this scheme, bottom-up activity which is consistent in a more global context is enhanced, inconsistent activity is suppressed. We present two instantiations of this general scheme having complementary functionality, namely a model of *cortico-cortical V1–V2* interactions and a model of recurrent *intracortical V1* interactions. The models both have long-range interactions for the representation of contour shapes and modulating feedback in common. They differ in their response properties to illusory contours and corners, and in the details of computing the bipole filter which models the long-range connections. We demonstrate that the models are capable of basic processing tasks in vision, such as, e.g., contour enhancement, noise suppression and corner detection. Also, a variety of perceptual phenomena such as grouping of fragmented shape outline and interpolation of illusory contours can be explained.

1 Motivation: Functionality and Architecture

How does the brain manage to form invariant representations of the environment that are relevant for the current behavioral task? The sensory system is steadily confronted with a massive information flow that arrives via different channels. In vision, spatio-temporal pattern arrangements that signal coherent surface arrangements must be somehow reliably detected and grouped into elementary items even in changing situations and under variable environmental conditions. Such a grouping enables the segregation of figural components from cluttered background as well as the adaptive focusing of processing capacities, while suppressing parts of the input activity pattern that are less relevant to support the behavioral goal or task [14,28]. Grouping and segregation requires the interaction of several representations and activity distributions generated by different processing streams. Here we focus on the detection of contour features

S. Wermter et al. (Eds.): Emergent Neural Computational Architectures, LNAI 2036, pp. 127–138, 2001.
© Springer-Verlag Berlin Heidelberg 2001

such as smooth boundary patterns as well as corner and junction configurations by adaptive neural mechanisms.

A characteristic feature of cortical architecture is that the majority of visual cortical areas are linked bidirectionally by feedforward and feedback fiber projections to form cortico-cortical loops. So far, the precise computational role of the descending feedback pathways at different processing stages remains largely unknown. Empirical evidence suggests that top-down projections primarily serve to *modulate* the responsiveness of cells at previous stages of the processing hierarchy (e.g., [20]). We particularly investigated the recurrent interaction of areas V1 and V2. The results of this investigation suggest a novel interpretation of the role of contour grouping and subjective contour interpolation at V2 such that observable effects relate to the task of surface segmentation. This information is used to evaluate and selectively enhance initial measurements at the earlier stage of V1 processing of oriented contrasts.

Other architectural principles encountered in cortical architecture are long-range horizontal connections and intracortical feedback loops [9], among others. Via horizontal connections cells of like-orientation couple and thus cell responses are selectively influenced by stimuli outside their classical receptive field (RF). We propose a simplified model architecture of V1 that incorporates a sequence of preprocessing stages and a recurrent loop based on long-range interaction. The results demonstrate that noisy low contrast arrangements can be significantly enhanced to form elementary items of smooth contour segments which are precursory for subsequent integration and organization into salient structure. Beyond the formation of salient contour fragments this scheme of processing is able to enhance contour responses at corner and junction configurations. These higher order features have been identified to play a significant role in object recognition and depth segregation (e.g., [1]).

2   Empirical Findings

The computational models have the following key components:

- feedforward and feedback processing between two areas or layers
- localized receptive fields for oriented contrast processing
- lateral competitive interaction
- lateral horizontal integration

In order to motivate the model design, we summarize recent anatomical and physiological data on recurrent processing and horizontal long-range interaction in early visual areas. The summary is accompanied by a review of recent psychophysical data on visual grouping and context effects. A more detailed review is given in [29].

2.1   Anatomy and Physiology

Wiring schemes of projections. Feedback is a general principle of cortical architecture and arises at different levels. A coarse distinction can be made between cortico-cortical loops (e.g., V1–V2) and intracortical loops (e.g., V1 layer 4→2/3→5→6→4 [2,10]).

The pattern of *feedforward* projections preferably link patches of similar feature preference, as shown for orientation selective cells in V1 and V2 [12]. The pattern of *feedback* projections show a retinotopic correspondence [3], as suggested for the linking of cells in cytochrome oxidase blobs and bands [27]. However, the feedback connections diverge from V2 to multiple clusters in V1, which may reflect the convergence of information flow within V2 [34]. In V1, the intracortical feedback loop connects cells within the same column. Cells within one column have common receptive field properties, e.g., ocular dominance and orientation preference [2].—We conclude that the wiring scheme is specific for contrast orientation and curved shape outline.

Modulatory feedback. Several physiological studies indicate that feedback projections have a gating or modulating rather than generating effect on cell activities [19,35,20]. Feedback alone is not sufficient to drive cell responses [36, 19].

Context influences. The response of a target cell to an individual stimulus element is also modulated by the visual context. V1 cell responses to isolated optimally oriented bars are reduced if the bar is placed within a field of randomly oriented bars, but enhanced if the bar is accompanied by several coaligned bars [23]. A texture of bars of the same type has a suppressive effect, which is maximal for bars of the same orientation and weakest for orthogonal orientation [24].

Wiring of horizontal long-range connections. The grouping of aligned contours require a mechanism that links cells of proper orientation over larger distances. Horizontal long-range connections found in the superficial layers of V1 and V2 may provide such a mechanism: They span large distances [11] and selectively link cells with similar feature preference [12,37]. Receptive field sizes in V2 are substantially larger than in V1 [41].

Response to illusory contours. Contour cells in V2 respond both to oriented contrasts and to illusory contours [43]. Response is maximal for physical contrast [42], but there is also a response to coherent arrangements of two inducers of an illusory contour. If one inducer is missing, response drops to spontaneous activity [30]. Unlike V2, cells in macaque V1 do not respond to illusory contours induced by two flanking bars placed outside their classical receptive field, but show a response increase to the same configuration if the classical receptive field is also stimulated [23].

2.2   Psychophysics

Perceptual grouping is a key mechanism to bind coherent items and to form chunks of surface and object outline. Several studies investigated the dependence of target detection on visual context. Spatial arrangements of Gabor patches within a field of distractors are facilitated by other patches coaligned with the target patch [6,32]. In another study the context effect of flanking bars on contrast threshold for a target bar is investigated [23]. The distance along the axis of colinearity, orthogonal displacement and deviation in orientation are critical parameters for the optimal placement of flanking bars.

Grouping mechanisms help to form object boundaries which are precursory for surface segmentation and figure-ground segregation. Such processes necessitate contour completion over gaps where luminance differences are missing [31]. This completion can be initiated by inducers which are oriented in the direction of the interpolated contour. Completion occurs in the same direction as the inducing contrasts as well as orthogonal to line ends [22,33,38].

3   Computational Models

In this section we present the two models of recurrent processing, a model of cortico-cortical V1–V2 interaction [29] and a model of intracortical V1 interaction [17]. The two models are intended to selectively study different properties of intracortical and cortico-cortical processing. Both models have distinct and partly complementary features, and are designed to be integrated eventually within a single more complex model.

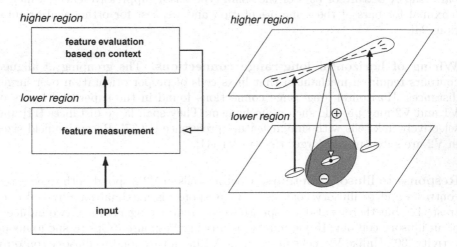

Fig. 1. Sketch of the general scheme of recurrent interaction (left) and of long-range interaction (right). Filled arrowheads indicate driving feedforward connections, unfilled arrowheads indicate modulating feedback connections. In the right sketch, input from the lower region provided by two cells with similar orientation preference is integrated by the long-range filter at the higher region. Integrated activity is fed back to modulate the response of the target cell. Inhibitory influence is generated on neighboring cells (gray circle). Together with the excitatory mechanism this defines a scheme of recurrent on-center/off-surround interaction

Common to both models is the response to oriented contrast and the basic interaction scheme of two bidirectionally linked regions. The term "region" refers to cortical areas (cortico-cortical V1–V2 model) or layers (intracortical V1 model). We propose that for a pair of bidirectionally connected cortical regions the "lower" region serves as a stage of feature measurement and signal detection. The "higher" region represents expectations about visual structural entities and context information to be matched against the incoming data carried by the feedforward pathway (see Fig. q1, left). The matching process generates a pattern of activation which is propagated backwards via the feedback pathway. This activation pattern serves as a signature for the degree of match between the data and possible boundary outlines. The activation is used to selectively enhance those signal patterns that are consistent with the model expectations. A gain control mechanism, that is accompanied by competitive interactions in an on-center/off-surround scheme, realizes a "soft gating" mechanism that selectively filters salient input activations while suppressing spurious and inconsistent signals. As a result the primary functional role of the feedback pathway realizes a gain control mechanism driven by top-down model information, or expectation [14,28,40]. The gain control mechanism enhances only cells which are already active [20]. In other words, feedback is *modulatory*, i.e., feedback alone is not sufficient to drive cell responses. The proposed scheme of driving feedforward and modulating feedback connections is consistent with the no-strong-loops hypothesis by Crick and Koch [4], which only forbids loops of *driving* connections.

The differences between both models are summarized in Table 1. These differences are motivated by the different properties of V1 and V2 as reviewed above. In the remainder of this section we briefly describe the two models, focusing on the basic computational principles employed. A detailed mathematical description, including parameter settings, can be found in the respective references.

Table 1. Different properties of V1–V2 and V1 model

	V1–V2 model	V1 model
location of model long-range connections	V2	V1
response to		
• illusory contours	yes	no
• corners	no	yes
bipole properties		
• RF size	≈ 8	≈ 3
	(multiple of resp. feedforward RF size)	
• combination of lobes	nonlinear "AND"-gate	linear
• feature compatibility	circular boundary segments	boundary segments of same orientation only
• subfields	on- and off-subfield	on-subfield only

3.1   Cortico-Cortical V1–V2 Interaction

We suggest that a variety of empirical findings about the physiology of cell responses in different contextual situations and about the psychophysics of contour grouping and illusory contour perception can be explained within a framework of basic computational mechanisms. We have realized an instantiation of this general interaction scheme described above to model the interaction between primary visual cortical areas V1 and V2. The model information is stored in "curvature templates" which represent shape segments of varying curvature. These templates are matched against the measurements of local oriented contrast. The matching is realized in a correlation process that utilizes oriented weighting functions which sample a particular segment of the spatial neighborhood. In order to combine significantly matching input from spatial locations from either side along the preferred orientation, a subsequent nonlinear accumulation stage integrates the activities from a colinear pair of lobes (compare [15]). An arrangement of consistent local contrast measurements activates a corresponding shape model which is represented in the spatial weights of double-lobed kernels and stands for model curve segments. This activation in turn enhances the activities of initial measurements by way of sending excitatory activation via the descending pathway. The net effect of bidirectional interaction generates a stabilized representation of shape in both model areas. A more detailed description of the mathematical definition of the model can be found in [29].

3.2   Intracortical V1 Interaction

In the model described above we have utilized center-surround feedback interaction for localized cells at the stage of model V1. We kept the model as simple as possible in order to study context effects that are exclusively generated at the higher cortical stages of model V2 to modulate the localized initial measurements. The effect of lateral oriented long-range interaction even at the stage of V1 is investigated in a second model described below.

In cortical area V1 layer 2/3, complex cells of like orientation are coupled via horizontal long-range connections which span two up to three hypercolumns on each side. We propose a model of V1 processing that incorporates both lateral horizontal interactions and recurrent intracortical processing. Oriented long-range interactions are utilized to enhance the significance of responses of coherent structure. We suggest that at a target cell location contrast activities from similar oriented cells are integrated via long-range excitatory connections. Unlike previous approaches, the integrated activation acts as a gain enhancer of activity that is already present by localized measurement of oriented contrast. In comparison to the long-range mechanism of the recurrent V1–V2 model, the bipole filter of long-range interaction is i) linear, adding the inputs of its two lobes, ii) connects cells of same orientation preference only and iii) is smaller in size compared to the size of the complex cell RFs in the feedforward stream (see Tab. 1).

The recurrent interaction at V1 enhances local coherent arrangements while incoherent noisy measurements is suppressed. In view of the cortico-cortical processing scheme described above, we claim that the localized interaction in V1 alleviates the detection of more global shape outline pattern in V2. By selectively enhancing coherent activity, this process maximizes the orientation significance at edges, compared to noisy arrangements of initial complex cell responses. At corner and junction configurations, significant responses for more than one single orientation emerge from the recurrent interaction, forming salient responses in independent orientation channels. In other words, contours and junctions are signaled by high orientation variance and high magnitude in individual orientation channels. This is consistent with recent studies [5], showing that correlated activities of V1 cells can signal the presence of smooth outline patterns as well as patterns of orientation discontinuity as they occur at corners and junctions. Our scheme generates such representations even without the requirement of specialized connectivity schemes between cells of different orientation preference. A more detailed description of the mathematical definition of the model can be found in [17].

4   Results

Simulation results demonstrate that the model predictions are consistent with a broad range of experimental data. The results further suggest that the different mechanisms of these models realize several key principles of feature extraction that are useful in surface segmentation and depth segregation.

The first two figures show results generated by the model of recurrent V1–V2 interaction. Figure 2 demonstrates the capability of grouping individual bar items of a fragmented shape into a representation of coherent activity in model V2. This activation is fed back to further enhance and stabilize those V1 activities that match the global structure. Figure 3 shows the correct prediction of illusory contour strength as a function of the ratio between inducer length and total contour length (Kanizsa figures) and as a function of line density (Varin figures).

Fig. 2. V1–V2 model: Grouping by cortico-cortical feedback processing: Input pattern of fragmented shape (left), model V1 cell responses after center-surround feedback processing (middle), model V2 contour cell responses (right). Reprinted with permission from [29]

Figure 4 and 5 show results generated by the model of recurrent V1 long-range interaction. Figure 4 demonstrates the functionality of lateral long-range interaction for the enhancement of coherent structure. Outline contrasts are detected and subsequently enhanced such that the activities of salient contrast as well as orientation significance is optimized. Figure 5 shows the results of processing an image of a laboratory scene. Initial complex cell activations generated for localized high contrast contours are further stabilized. Initially weak activations in coherent spatial arrangements are enhanced. Spatial locations where high amplitude contrast responses exist in multiple orientation channels are marked by circles. They indicate the presence of higher order structure such as corners and junctions.

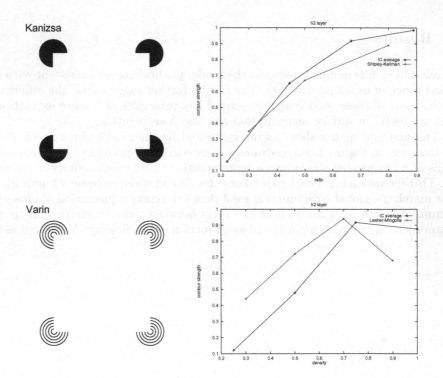

Fig. 3. V1–V2 model: Predictions for illusory contour strength after grouping (model V2 cell responses) for Kanizsa figure (top row) and Varin figure (bottom row). In the Kanizsa figure contour strength is displayed as a function of the ratio between increasing inducer radius and total length of the illusory contour for four different inducers sizes (top right). In the Varin figure contour strength is displayed as a function of line density. For a given radius the number of evenly spaced circular arcs determines the density of the inducers. Model predictions are shown for four different ratios (bottom right). Both graphs show model predictions (continuous lines) and psychophysical results (dashed lines) [39,25]

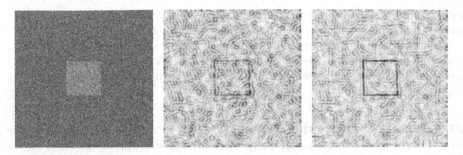

Fig. 4. V1 model: Processing of a square pattern with additive high amplitude noise: Input image (left), initial complex cell responses (middle), result of recurrent processing utilizing long-range interaction (right)

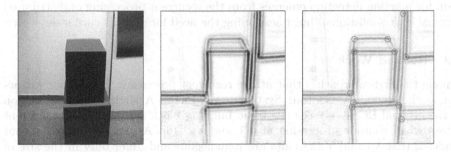

Fig. 5. V1 model: Enhancement of activity distribution in model V1 and detection of corner and junction features in a laboratory scene: Luminance distribution of input image (left), initial complex cell responses (middle), V1 cell responses generated by long-range interaction and recurrent processing (right). Locations of corners and junctions are marked and indicate positions with significant responses in more that one orientation channel

5   Summary and Discussion

5.1   Results

We propose a computational scheme for the recurrent interaction between two cortical regions. In this basic scheme of two interacting regions, the "lower region" serves as a stage of signal measurement and feature detection, while the "higher region" evaluates the local features within a broader context and selectively enhances those features of the "lower region" which are consistent within the context arrangement.

V1–V2 cortico-cortical interaction. The model of V1–V2 cortico-cortical interaction links physiological and psychophysical findings. The model predicts the generation of illusory contours both along (Kanizsa figures) and perpendicular to line ends (Varin figures) in accordance with psychophysical results [25] (see Fig. 3). Further successful predictions (see [29]) include responses to bar texture patterns [24,23], to abuting gratings, and to the suppression of figure contour when placed in a dense texture of similar lines [21].

For the processing of noisy and fragmented shape outline, the model groups coherent activity and completes contour gaps at the V2 stage (see Fig. 2, right) and shapes both spatial and orientational tuning of initial responses at the V1 stage.

V1 intracortical interaction. The model of V1 intracortical interaction, like the V1–V2 model, enhances consistent contours while suppressing noisy, inconsistent activity, both in space and orientation domain. Further, at locations of inherent orientation variability, such as corners or junctions, the relevant orientations remain. Such points of high orientational variance which "survive" the recurrent consistency evaluation reliably mark corners or junctions. This mechanism for junction detection emerges from the recurrent processing of distributed contrast representations, thus questioning the need for explicit corner detectors.

5.2   Related Work

Among the first approaches that utilize recurrent processing for contour extraction is the Boundary Contour System, e.g., [15,13]. A slightly revised version of the original BCS serves as the basic building block for a model of recurrent *intracortical* contour processing at V1 and V2 [16]. A main difference to our model is that V1 and V2 circuits are homologous and differ only in the size of the receptive fields, proposing that V2 is basically V1 at larger scale. In contrast, we propose that V1 and V2 have different and functional roles, such that, e.g., cells responding to illusory contours occur in V2 and corner selective cells occur in V1.

Other models selectively integrate activity from end-stop responses [42,18,7, 8], while we use activity from initial contrast measurement which is sharpened by feedback modulation.

A model architecture similar as our intracortical V1 model has been proposed by Li [26] that focuses on the detection of texture boundaries. The models differ in the feature compatibility used for contour integration: while we integrate activity between edges of same orientation only, Li uses a contour template of many orientations forming a smooth contour. Unlike in our model, feedback is not modulatory in Li's model.

5.3   Conclusion

We propose a computational framework, suggesting how feedback pathways are used to modulate responses of earlier stages. We particularly focus on the recurrent contour processing in V1 and V2. The models are not intended to generate biologically realistic responses, rather to elucidate the underlying computational principles. For the future, we are planning to integrate the two models within one framework. We claim that the proposed principles are not restricted to V1 and V2 but may be extended to recurrent interactions between other cortical areas, like V4 or MT.

References

1. I. Biederman. Human image understanding: Recent research and a theory. *CVGIP*, 32(1):29–73, 1985.
2. J. Bolz, C. D. Gilbert, and T. Wiesel. Pharmacological analysis of cortical circuitry. *TINS*, 12(8):292–296, 1989.
3. J. Bullier, M. E. McCourt, and G. H. Henry. Physiological studies on the feedback connection to the striate cortex from cortical areas 18 and 19 of the cat. *Exp. Brain Res.*, 70:90–98, 1988.
4. F. Crick and C. Koch. Why is the a hierarchy of visual cortical and thalamic areas: The no-strong loops hypothesis. *Nature*, 391:245–250, 1998.
5. A. Das and C. Gilbert. Topography of contextual modulations mediated by short-range interactions in primary visual cortex. *Nature*, 399:655–661, 1999.
6. D. J. Field, A. Hayes, and R. F. Hess. Contour integration by the human visual system: Evidence for local "association field". *Vision Res.*, 33(2):173 – 193, 1993.
7. L. H. Finkel and G. M. Edelman. Integration of distributed cortical systems by reentry: A computer simulation of interactive functionally segregated visual areas. *J. Neurosci.*, 9(9):3188–3208, 1989.
8. L. H. Finkel and P. Sajda. Object discrimination based on depth-from-occlusion. *Neural Comput.*, 4:901–921, 1992.
9. C. Gilbert. Horizontal integration and cortical dynamics. *Neuron*, 9:1–13, 1992.
10. C. D. Gilbert. Circuitry, architecture, and functional dynamics of visual cortex. *Cereb. Cortex*, 3(5):373–386, Sep/Oct 1993.
11. C. D. Gilbert and T. N. Wiesel. Clustered intrinsic connections in cat visual cortex. *J. Neurosci.*, 3:1116–1133, 1983.
12. C. D. Gilbert and T. N. Wiesel. Columnar specificity of intrinsic horizontal and corticocortical connections in cat visual cortex. *J. Neurosci.*, 9(7):2432–2442, 1989.
13. A. Gove, S. Grossberg, and E. Mingolla. Brightness perception, illusory contours and corticogeniculate feedback. In *Proc. World Conference on Neural Networks (WCNN-93), Vol. I-IV*, pages (I) 25–28, Portland (Oreg./USA), July 11–15 1993.
14. S. Grossberg. How does a brain build a cognitive code? *Psych. Rev.*, 87:1–51, 1980.
15. S. Grossberg and E. Mingolla. Neural dynamics of perceptual grouping: Textures, boundaries, and emergent segmentation. *Percept. Psychophys.*, 38:141–171, 1985.
16. S. Grossberg, E. Mingolla, and W. D. Ross. Visual brain and visual perception: how does the cortex do perceptual grouping? *TINS*, 20(3):106–111, 1997.
17. T. Hansen and H. Neumann. A model of V1 visual contrast processing utilizing long-range connections and recurrent interactions. In *Proc. ICANN*, pages 61–66, Edinburgh, UK, Sept. 7–10 1999.
18. F. Heitger, R. von der Heydt, E. Peterhans, L. Rosenthaler, and O. Kübler. Simulation of neural contour mechanisms: Representing anomalous contours. *Image Vis. Comp.*, 16:407–421, 1998.
19. J. A. Hirsch and C. D. Gilbert. Synaptic physiology of horizontal connections in the cat's visual cortex. *J. Neurosci.*, 11(6):1800–1809, June 1991.
20. J. M. Hupé, A. C. James, B. R. Payne, S. G. Lomber, P. Girard, and J. Bullier. Cortical feedback improves discrimination between figure and background by V1, V2 and V3 neurons. *Nature*, 394:784–787, Aug. 1998.
21. G. Kanizsa. Percezione attuale, esperienza passata 1′ "esperimento impossibile". In G. Kanizsa and G. Vicario, editors, *Ricerche sperimentali sulla percezione.*, pages 9–47. Università degli studi, Triente, 1968.
22. G. Kanizsa. Subjective contours. *Sci. Am.*, 234(4):48–52, 1976.

23. M. K. Kapadia, M. Ito, C. D. Gilbert, and G. Westheimer. Improvement in visual sensitivity by changes in local context: Parallel studies in human observers and in V1 of alert monkeys. *Neuron*, 15:843–856, Oct. 1995.
24. J. J. Knierim and D. C. Van Essen. Neuronal responses to static texture patterns in area V1 of the alert macaque monkey. *J. Neurophys.*, 67(4):961–980, 1992.
25. G. W. Lesher and E. Mingolla. The role of edges and line-ends in illusory contour formation. *Vision Res.*, 33(16):2253–2270, 1993.
26. Z. Li. Pre-attentive segmentation in the primary visual cortex. M.I.T. A.I. Lab., Memo No. 1640, 1998.
27. M. Livingstone and D. Hubel. Anatomy and physiology of a color system in the primate visual cortex. *J. Neurosci.*, 4(1):309–356, 1984.
28. D. Mumford. On the computational architecture of the neocortex II: The role of cortico-cortical loops. *Biol. Cybern.*, 65:241–251, 1991.
29. H. Neumann and W. Sepp. Recurrent V1–V2 interaction in early visual boundary processing. *Biol. Cybern.*, 81:425–444, 1999.
30. E. Peterhans and R. von der Heydt. Mechanisms of contour perception in monkey visual cortex. II. Contours bridging gaps. *J. Neurosci.*, 9(5):1749–1763, 1989.
31. E. Peterhans and R. von der Heydt. Subjective contours—bridging the gap between psychophysics and physiology. *TINS*, 14(3):112–119, 1991.
32. U. Polat and D. Sagi. The arcitecture of perceptual spatial interactions. *Vision Res.*, 34:73–78, 1994.
33. K. Prazdny. Illusory contours are not caused by simultaneous brightness contrast. *Percept. Psychophys.*, 34(4):403–404, 1983.
34. K. Rockland and A. Virga. Terminal arbors of individual "feedback" axons projecting from area V2 to V1 in the macaque monkey: A study using immunohistochemistry of anterogradely transported Phaseolus vulgaris-lencoagglunitin. *J. Comp. Neurol.*, 285:54–72, 1989.
35. P.-A. Salin and J. Bullier. Corticocortical connections in the visual system: Structure and function. *Physiol. Rev.*, 75(1):107–154, 1995.
36. J. Sandell and P. Schiller. Effect of cooling area 18 on striate cortex cells in the squirrel monkey. *J. Neurophys.*, 48(1):38 – 48, 1982.
37. K. Schmidt, R. Goebel, S. Löwel, and W. Singer. The perceptual grouping criterion of colinearity is reflected by anisotropies of connections in the primary visual cortex. *Europ. J. Neurosci.*, 9:1083–1089, 1997.
38. T. F. Shipley and P. J. Kellman. The role of discontinuities in the perception of subjective figures. *Percept. Psychophys.*, 48(3):259–270, 1990.
39. T. F. Shipley and P. J. Kellman. Strength of visual interpolation depends on the ratio of physically specified to total edge length. *Percept. Psychophys.*, 52(1):97–106, 1992.
40. S. Ullman. Sequence of seeking and counter streams: A computational model for bidirectional information flow in the visual cortex. *Cereb. Cortex*, 2:310–335, 1995.
41. R. von der Heydt, F. Heitger, and E. Peterhans. Perception of occluding contours: Neural mechanisms and a computational model. *Biomed. Res.*, 14:1–6, 1993.
42. R. von der Heydt and E. Peterhans. Mechanisms of contour perception in monkey visual cortex. I. Lines of pattern discontinuity. *J. Neurosci.*, 9(5):1731–1748, 1989.
43. R. von der Heydt, E. Peterhans, and G. Baumgartner. Illusory contours and cortical neuron responses. *Science*, 224:1260—1262, 1984.

Neural Mechanisms for Representing Surface and Contour Features

Thorsten Hansen and Heiko Neumann

Universität Ulm, Abt. Neuroinformatik, D-89069 Ulm, Germany
(hansen,hneumann)@neuro.informatik.uni-ulm.de

Abstract. Contours and surfaces are basic qualities which are processed by the visual system to aid the successful behavior of autonomous beings within the environment. There is increasing evidence that the two modalities of contours and surfaces are processed in separate, but interacting visual streams or sub-systems. Neurons at early stages in the visual system show strong responses only at locations of high contrast, such as edges, but only weak responses within homogeneous regions. Thus, for the processing and representation of surfaces, the visual system has to integrate sparse local measurements into a dense, coherent representation. We suggest a mechanism of confidence-based filling-in, where a confidence measure ensures a robust selection of sparse contrast signals. The new mechanism supports the generation of surface representations which are invariant against size and shape transformation. The filling-in process is controlled by contour or boundary signals which stop the filling-in of contrast signals at region boundaries. Localized responses to contours are most often noisy and fragmented. We suggest a recurrent processing scheme for the extraction of contours that incorporates long-range connections. The recurrent long-range processing enhances coaligned activity which is consistent within a more global context, while inconsistent noisy activity is suppressed. The capability of the model is shown for noisy synthesized and natural stimuli.

1 Introduction

Experimental studies indicate the existence of distinct perceptual subsystems in human vision, one that is concerned with *contour extraction* and another that assigns *surface properties* to bounded regions. The emerging picture from the experimental investigations is one in which shape outlines are initially extracted, followed by the assignment of attributes such as texture, color, lightness, brightness or transparency to regions [10,38,6,17]. Several perceptual completion phenomena [35] suggest that, on a functional level, regions inherit local border contrast information by means of "spreading mechanisms" or "filling-in" [32,7]. The assignment of surface properties would then be dependent on the determination of stimulus contrast in various feature dimensions, such as luminance, motion direction and velocity, and depth, that would be used to fill-in bounded regions.

S. Wermter et al. (Eds.): Emergent Neural Computational Architectures, LNAI 2036, pp. 139–153, 2001.
© Springer-Verlag Berlin Heidelberg 2001

2   Neural Mechanisms for Representing Surface Features

The problem of deriving a dense representation of surface quality, such as brightness or color, from local estimates, such as luminance or chromatic border contrast, is inherently ill-posed: there exists no unique solution nor is the solution guaranteed to be stable. Such an inverse problem needs to be regularized in the sense that certain constraints have to be imposed on the space of possible solutions. The constraint of generating a smooth surface, as formalized by minimizing the first order derivatives, leads to a linear diffusion process with a simple reaction term [30].

In filling-in theory, feature signals which provide the source term of the filling-in process are modeled as cells with circular receptive fields (RFs) such as retinal ganglion cells or LGN cells. In previous filling-in models, such cells are modeled to exhibit strong responses even to homogeneous regions [8,18]. Physiological studies however show that retinal ganglion cells respond strongly only at positions of luminance differences or contrasts [11]. Motivated by these results we use sparse contrast signals with no response to homogeneous regions. The sparse nature of signals necessitates additional confidence signals for the filling-in process. Confidence signals indicate the positions of valid contrast response to be taken as source for the filling-in process. Having established the link between models of perceptual data for biological vision and the mathematical frameworks of regularization theory this lead to the proposal of *confidence-based filling-in* [30].

2.1   Evidence for Neural Filling-in

Filling-in models are based on the assumption of distributed, topographically organized maps of boundaries and regions [25]. This assumption has been questioned in favor of a non-topographic or sparse coding of contrasts and boundaries using a compact symbolic code or a sparse localized code [37]. Regarding filling-in, both visual scientists and philosophers have argued against the logical need for a neural spreading of activity and against a continuous representation of the brightness profiles (for review, see [35]). It has been suggested that instead of filling-in, the brain could simply assign a symbolic brightness label to a bounded region or could ignore the absence of direct neural support. In support of the filling-in hypothesis, there is ample empirical evidence, mostly from psychophysics, that brightness perception indeed depends on a neural activity spreading.

Evidence comes from a study where a visual masking paradigm is used to investigate two issues [32]. First, the role of edge information in determining the brightness of homogeneous regions, and second the temporal dynamics of brightness perception. If brightness perception relies on some form of activity spreading, it should be possible to interrupt this spreading process. In the experiment, a target of a bright disk is followed by a mask (e.g., a smaller circle or a C-shape), which is presented at variable time intervals. For an interstimulus interval of about 50–100 ms, the brightness of the central area is highly dependent on the shape of the mask. For example, for a C-shaped mask, a darkening of the middle region is observed, with the bright region "protruding" inside the C. For a circular-shaped mask, an inner dark disk is perceived. Both these results are consistent with the hypothesis that brightness signals are generated at the borders of their target stimuli and propagate inward. Furthermore, it has been demonstrated that for larger stimuli maximal suppression occurs later. This finding supports the view that filling-in is an active spreading of neural activity, i.e., a process which takes time.

Recently, a similar masking paradigm has been used to investigate brightness filling-in within texture patterns [7]. Again the spreading could be blocked by the mask if the interstimulus interval is in accordance with the propagation rate required to travel the distance between boundary and mask position. Results of a study employing Craik-O'Brien-Cornsweet (COC) gratings point in the same direction: For higher spatial frequencies of the grating (i.e., for smaller distances) the effect was stronger and persisted to higher temporal frequencies of COC contrast reversal [9,34].

In summary, these studies are suggestive of active neural filling-in processes that are initiated at region edges. Using brightness filling-in, the brain generates a spatially organized representation through a continuous propagation of signals, a process that takes time [35,30].

2.2   Mathematical Models for Filling-in

To introduce concepts, we consider the task of generating a continuous representation of surface layout as one of painting or coloring an empty region [26]. The task thus consists of generating an internal representation of surface properties from given data. Individual surfaces occur at different sizes and with various shapes. Therefore, any such mechanism has to be insensitive to such size and shape variations.

Models of brightness perception were among the first to explore the dichotomy of boundary and surface subsystems. Based on stabilized image studies, it has been proposed that the perception of brightness can be modeled by filling-in processes. Filling-in models suggest that feature measures are used in the determination of surface appearance through a process of lateral spreading, or diffusion [12]. The basic ideas are formalized in a model of complementary boundary and surface systems (Boundary Contour System/Feature Contour System, BCS/FCS) [8,17]. In a nutshell, BCS/FCS processing occurs as follows: The BCS extracts boundaries via a hierarchy of processing levels, defining a segmentation of the initial input image into compartmental boundaries. Within the FCS, these boundaries control the lateral spreading or diffusion of contrast-sensitive input signals. This proposal qualitatively accounts for a white variety of brightness phenomena, including, e.g., simultaneous contrast, brightness assimilation and the COC effect [18]. An extension of the model accounts for trapezoidal and triangular Mach bands, low- and high-contrast missing fundamental stimuli and sinusoidal waves, among others [33].

Standard Filling-in Equation. The filling-in equation that is used in several models of early vision [8,18,14,33,16] is equivalent to a linear inhomogeneous diffusion with reaction term [30]. The reaction term consists of contrast-sensitive input signals K and a passive decay of activity (with rate α). The diffusion term describes the nearest-neighbor coupling $\mathcal{N}_i$ of filling-in activities which is locally controlled by permeability signals P (inhomogeneous diffusion). Permeability signals are a monotonically decreasing function of boundary or contour signals, i.e., high contour signals imply low permeability and vice versa. In all, the discretized equation for filling-in activity U reads

$$\partial_t U_i = \underbrace{- \alpha U_i + K_i}_{\text{reaction term}} + \underbrace{\sum_{j \in \mathcal{N}_i} (U_j - U_i) P_{ij}}_{\text{diffusion term}} , \tag{1}$$

where ∂_t denotes partial differentiation with respect to t. Discrete spatial locations are denoted by i and j. The nearest neighbor coupling is given by $\mathcal{N}_i = \{i-1, i+1\}$ for the 1D case and $\mathcal{N}_{ij} = \{(i-1,j),(i+1,j),(j-1,i),(j+1,i)\}$ for the 2D case.

Confidence-based Filling-in Equation. Previous models of filling-in use a dense representation of contrast-sensitive feature signals as source for the filling-in process. Cells at early stages of the visual system, such as retinal ganglion cells, show strong responses only at luminance discontinuities. Given the sparseness of contrast signals which are zero within homogeneous regions, the visual system has to compute a dense brightness surface from local contrast estimates. Such inverse problems are generally ill-posed in the sense of Hadamard [41,36, 2]. This means that the existence and uniqueness of a solution and its continuous dependence on the data cannot be guaranteed since the measurements are sparse and may be noisy. The solution to the problem has to be regularized such that proper constraints are imposed on the function space of solutions. Such a constraint is the *smoothness* of the solution, for example. Smoothness can be characterized by minimizing the first order derivatives of the desired solution. The goal is to minimize both the local differences between the measured data and the reconstructed function values (data term) and the stabilizing functional imposed on the function (smoothness term). Minimizing a quadratic functional finally leads to the discretized version of a new filling-in equation, where an additional *confidence signal* Z steers the contribution of the data term [30]:

$$\partial_t U_i = (-\alpha U_i + K_i) Z_i + \sum_{j \in \mathcal{N}_i} (U_j - U_i) P_{ij} . \tag{2}$$

Note that for constant unit-valued confidence signals $Z = 1$ confidence-based filling-in (Eq. 2) is equivalent to standard filling-in (Eq. 1).

Confidence signals are in the range $[0; 1]$. Zero confidence signals indicate positions where no data are available, while unit-valued confidence signals occur at region boundaries and signal positions of reliable contrast measurements.

Consequently, we suggest that an intermediate representation in the processing of contour signals, namely complex cell responses C, are involved in the computation of confidence signals. A candidate mechanism is

$$Z_i = \beta C_i + \varepsilon \ , \tag{3}$$

where β is a scaling parameter, and ε is a small tonic input to achieve well-posedness of the filling-in process. It is suggested that the complex cell interaction incorporates the self-normalizing properties of a shunting interaction [22], to generate signals in a bounded range such as $[0; 1]$.

For a detailed description of the model equations and the parameters used the reader is referred to [30].

2.3   Simulation Results

In this section we present simulation result which show that the proposed scheme of confidence-based filling-in exhibits basic properties which makes it suitable for the computation of surface properties in early vision. Results are compared for confidence-based filling-in and the corresponding standard filling-in by setting $Z = 1$.

First, we demonstrate the independence of the brightness predictions of confidence-based filling-in on the shape and size of the regions (Fig. 1 and 2). The mechanism of confidence-based filling-in is then applied to psychophysical stimuli (Fig. 3). In order to demonstrate the model's capacity to deal with real world data, we finally show results of processing real camera images (Fig. 4).

Invariance Properties. The first investigation focuses on the properties of the filling-in mechanisms and their dependency on the parameter settings and the size of the region to be filled-in. We start with a simple luminance pattern that shows a light square on a dark background (Fig. 1). The brightness signals generated by the standard filling-in mechanism tend to bow depending on the strength of the permeability coefficient. An increase in the permeability helps

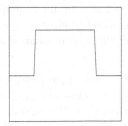 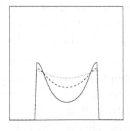 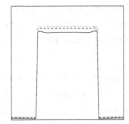

Fig. 1. Generation of brightness appearance for a rectangular test pattern utilizing mechanisms of standard and confidence-based filling-in. *Left to right:* Luminance profile, simulation results for standard filling-in and confidence-based filling-in under variations of the permeability parameter

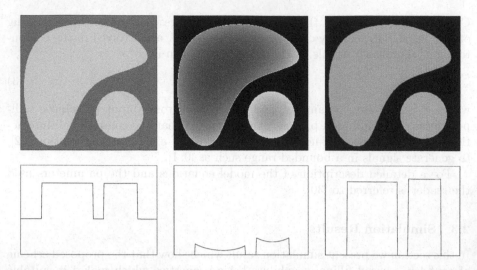

Fig. 2. Filled-in brightness signals for shapes of different size but same luminance level. Signal representations are generated by the filling-in mechanisms using the parameter settings that achieved proper results for the square test pattern in Fig. 1. *Top row, left to right:* Input luminance pattern, brightness signal generated by standard filling-in and by confidence-based filling-in. *Bottom row:* Corresponding profiles of the luminance function and the brightness patterns taken along the 2D picture diagonals (from upper left to lower right corner)

generating flat signals (Fig. 1, middle). The corresponding brightness patterns generated by confidence-based filling-in remain invariant against these parameter changes and are always flat (Fig. 1, right). Next, the same mechanisms have been applied to another test image that contains shapes of different form and size but the same luminance level. The results reveal the potential weaknesses of standard filling-in: Depending on the size or diameter of a pattern, which is unknown, the brightness signals appear at a different amplitude and show different amounts of bowing (Fig. 2, middle). With the confidence-based filling-in mechanism the brightness patterns appear homogeneous and of almost the same brightness (Fig. 2, right). We conclude that confidence-based filling-in helps to generate a brightness representation that is largely invariant against shape and size transformations, thus improving the robustness of filling-in mechanisms.

Psychophysical Data on Brightness Perception. In this section we demonstrate the ability of confidence-based filling-in to process classical luminance patterns that have been investigated in brightness perception. We particularly focus on remote border contrast effects and their creation of brightness differences. These cases provide examples of the crucial role of edges in determining the brightness appearance. For example, two regions of equal uniform luminance separated by a "cusp edge" appear differently bright, the so-called Craik-O'Brien-Cornsweet (COC) effect. These types of stimuli have been identified

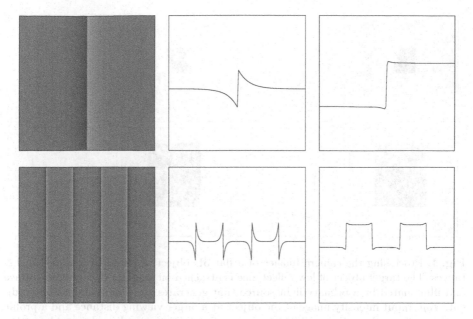

Fig. 3. Filled-in brightness signals for a standard COC stimulus and a COC grating *(bottom row)* made of cusps of opposite contrast polarity. *Left to right:* Input luminance pattern with corresponding profile, and the profile of the brightness pattern generated by confidence-based filling-in

as the most challenging ones for alternative theories of brightness perception, such as, for example, filter theories. In fact, yet, only filling-in models appear to properly predict the brightness appearance for COC stimuli and their variants (compare [4]). The processing of a standard COC stimulus is shown in Fig. 3 (top row). A COC stimulus consist of a cusp edge, separating two regions of equal luminance. Both regions seem to be of different brightness, where the region which is associated with the negative lobe of the cusp is perceived as a uniformly darker region compared to the right region. Confidence-based filling-in correctly predicts this effect (Fig. 3, top right), as do previous filling-in models.

A COC grating (Fig. 3, bottom row) consists of a sequence of cusp edges having pairwise opposite contrast polarities. This stimulus is perceived as a series of alternating dark and bright stripes similar to a square wave. The temporal dynamics of brightness perception in such COC arrangement is consistent with a filling-in mechanism. Confidence-based filling-in, at equilibrium, correctly predicts the appearance of the final brightness square wave pattern (Fig. 3, bottom right).

Real World Application. In order to demonstrate the functional significance of the proposed mechanism, we show the processing results for a camera image of a real object. In order to exclude any possible influences from 3D effects, e.g.,

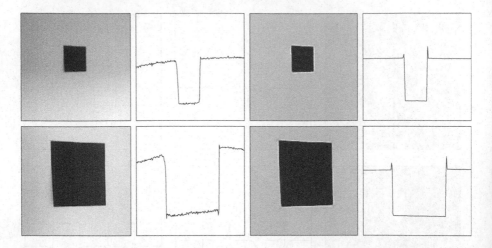

Fig. 4. Processing the camera images of a flat 3D object acquired from different distances. The target object of low reflectance is attached on a lighter background surface and illuminated by a primary light source that generates a visible illumination gradient. *Top:* Input intensity image of the object at a larger viewing distance and a profile section *(left pair)* together with the corresponding filling-in result and a profile section *(right pair)*. *Bottom:* Corresponding input representations and processing results for the object at a closer viewing distance

by shadowing or variations in surface orientations, we used a card-board that has been attached to a flat background surface. This intrinsically flat scene was directly illuminated by a point-like light source at a distance of approximately 2 m. This generates a significant intensity gradient in the original intensity image. The target surface has been imaged from two different distances at about 2 and 1 m, respectively.

Simulation results show that the mechanism of confidence-based filling-in is capable of generating a representation of homogeneous surface properties (Fig. 4). The result is independent of the projected region size, thus showing the property of size invariance. Also the illumination gradient is discounted and the noise is successfully suppressed.

2.4   Outlook

The proper restoration of reference levels remains a deficit of filling-in functionality. The use of DC-free contrast signals discounts the illuminant, but at the same time destroys all information about the reference levels of contrast signals. Several approaches have been advocated to solve this problem, such as directional filling-in [1] or an extra luminance channel [18,28], but fail to discriminate, e.g., COC stairs from luminance stairs, or are flawed by missing physiological evidence. We suggest that a multi-scale approach [40] together with the localized coding of luminance information at contrast positions may solve the problem.

3   Contour Processing

The generation of brightness representations by means of filling-in relies on the proper computation of contour signals. For the filling-in process, robust, reliable contour extraction is important, since contour signals are used to determine permeability signals which control the lateral spreading of activities (cf. Eqns. 1 and 2). Contour signals must not suffer from high amplitude variations to allow for a stable representation of brightness surfaces. Initial contrast measurements, however, which define the first processing stage in the computation of contour signals, are often noisy and fragmented. Therefore, an important task in early visual processing is to determine the salient or prominent contours out of an array of noisy, cluttered contrast responses.

How can this task be accomplished? We suggest a computational framework involving *long-range connections, feedback,* and *recurrent interactions.* The task of contour extraction cannot be solved solely on the basis on the incoming data alone, but requests for additional constraints and assumptions on the shape of frequently occurring contours. An important principle of salient contours is that they obey the Gestalt law of good continuation. It has been suggested that horizontal *long-range connections* found in the superficial layers of early visual areas like V1 and V2 provide a neural implementation of the law of good continuation [39]. The assumptions or a priori information such as expressed in the law of good continuation have to be carefully matched against the incoming data. We suggest that *feedback* plays a central role in this matching process by selectively enhancing those feedforward input signal which are consistent with the assumptions. The interaction between feedforward data and feedback assumptions requires certain time steps. In each step the result of the interactions is recursively fed into the same matching process. Such a process of *recurrent interaction* might be used by the brain to determine the most stable and consistent representation depending on both the assumptions and the given input data.

Motivated by empirical findings we present a model of *recurrent long-range interaction in the primary visual cortex* for contour processing.

3.1   Computational Model

The computational model incorporates feedforward and feedback processing, lateral competitive interaction and horizontal long-range integration, and localized receptive fields for oriented contrast processing. The model architecture is defined by a sequence of preprocessing stages and a recurrent loop based on long-range interaction. The model realizes a simplified architecture of V1 [13] and is outlined in Fig. 5. The computational role of the proposed circuit is to enhance the salient contours and to suppress noisy activities. The circuit compensates for variations of amplitude strength and orientation selectivity in the initial contrast measurements along the contour. This property allows for the robust computation of closed contours to be used in the filling-in process.

The model uses *modulating* feedback, i.e., initial bottom-up activity is necessary to generate activity. The model of V1 thus does not allow for the creation

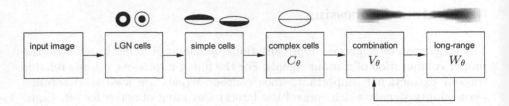

Fig. 5. Overview of the model stages together with a sketch of the sample receptive fields of cells at each stage for 0° orientation. For the long-range stage, the spatial weighting function of the bipole filter is shown

of illusory contours. Illusory contours evoke cell responses in V2 [42] and have been investigated in a model of V1–V2 interactions [31].

We propose a functional architecture for recurrent processing. In this architecture of two interacting regions, let them be cortical layers or areas, each region has a distinctive purpose. The lower region serves as a stage of feature measurement and signal detection. The higher region represents expectations about visual structural entities and context information to be matched against the incoming data carried by the feedforward pathway [31,21].

In the *feedforward path*, the initial luminance distribution is processed by isotropic LGN-cells, orientation-selective simple and complex cells. The interactions in the feedforward path are governed by basic linear equations to keep the processing in the feedforward path relatively simple and to focus on the contribution of the recurrent interaction. A more elaborated processing in the feedforward path would make use of, e.g., nonlinear processing at the level of LGN cells and simple cells [19,29]. The computation in the feedforward path is detailed in [20]. In our model, complex cell responses C_θ as output of the feedforward path (cf. Fig. 5) provide an initial local estimate of contour strength, position and orientation which is used as bottom-up input for the recurrent loop. The *recurrent loop* has two stages, a combination stage where bottom-up and top-down inputs are integrated, and a stage of long-range interaction. At the *combination* stage, feedforward inputs C_θ and feedback inputs W_θ are added and subject to shunting interaction

$$\text{net}_\theta = C_\theta + \delta_V W_\theta \ , \tag{4}$$

$$\partial_t V_\theta = -\alpha_V V_\theta - \beta_V V_\theta \, \text{net}_\theta + \text{net}_\theta \ . \tag{5}$$

The equation is solved at equilibrium, resulting in a normalization of activity

$$V_\theta = \beta_V \frac{\text{net}_\theta}{\alpha_V + \text{net}_\theta} \ . \tag{6}$$

The weighting parameter $\delta_V = 2$ is chosen so that dimensions of C_θ and W_θ are approximately equal, the decay parameter $\alpha_V = 0.2$ is chosen small compared to net$_\theta$ and $\beta_V = 10$ scales the activity to be sufficiently large for the subsequent long-range interaction. For the first iteration step, feedback responses W_θ are set to C_θ.

At the *long-range* stage, the contextual influences on cell responses are modeled. Directional sensitive long-range connections provide the excitatory input. The inhibitory input is given by undirected interactions in both the spatial and orientational domain. Long-range connections are modeled by a bipole filter [17]. The spatial weighting function of the bipole filter is narrowly tuned to the preferred orientation, reflecting the highly significant anisotropies of long-range fibers in visual cortex [39,5] (see Fig. 5, top right). The size of the bipole is about twice the size of the RF of a complex cell.

Essentially, excitatory input is provided by correlation of the feedforward input with the bipole filter B_θ. A cross-orientation inhibition prevents the integration of cells responses at positions where orthogonal responses also exists. The excitatory input is governed by

$$\text{net}_\theta^+ = [V_\theta - V_{\theta\perp}]^+ \star B_\theta \,, \tag{7}$$

where $\star$ denotes spatial correlation and $[x]^+ = \max\{x, 0\}$ denotes half-wave rectification.

The profile of the bipole filter is defined by a directional term D_φ and a proximity term generated by an isotropically blurred circle $C_r \star G_\sigma$ where $r = 25$, $\sigma = 3$. The detailed equations read

$$B_{\theta,\alpha,r,\sigma}(x, y) = D_\varphi \cdot C_r \star G_\sigma \tag{8}$$

$$D_\varphi = \begin{cases} \cos(\frac{\pi/2}{\alpha}\varphi) & \text{if } \varphi < \alpha \\ 0 & \text{otherwise} \,, \end{cases} \tag{9}$$

where φ is defined as $\text{atan2}\,(|y_\theta|, |x_\theta|)$ and $(x_\theta, y_\theta)^\mathsf{T}$ denotes the vector $(x, y)^\mathsf{T}$ rotated by θ. The parameter $\alpha = 10°$ defines the opening angle of 2α of the bipole. The factor $\frac{\pi/2}{\alpha}$ maps the angle φ in the range $[-\alpha; \alpha]$ to the domain $[-\pi/2; \pi/2]$ of the cosine function with positive range.

Responses which are not salient in the sense that nearby cells of similar orientation preference also show strong activity should be diminished. Thus an inhibitory term is introduced which samples activity from both orientational $\tilde{g}_{\sigma_o,\theta}$, $\sigma_o = 0.5$, and spatial neighborhood $G_{\sigma_{\text{sur}}}$, $\sigma_{\text{sur}} = 8$,

$$\text{net}_\theta^- = \text{net}_\theta^+ \circledast \tilde{g}_{\sigma_o,\theta} \star G_{\sigma_{\text{sur}}} \,, \tag{10}$$

where $\circledast$ denotes correlation in the orientation domain. The orientational weighting function $\tilde{g}_{\sigma_o,\theta}$ is implemented by a 1D Gaussian g_{σ_o}, discretized on a zero-centered grid of size $o_{\max}$, normalized, and circularly shifted so that the maximum value is at the position corresponding to θ. The parameterization of the spatial inhibitory neighborhood results in an effective spatial extension of about half the size of the bipole filter.

Excitatory and inhibitory term combine through shunting interaction

$$\partial_t W_\theta = -\alpha_W W_\theta - \eta^- W_\theta \,\text{net}_\theta^- + \beta_W V_\theta \left(1 + \eta^+ \,\text{net}_\theta^+\right) \,. \tag{11}$$

The equation is solved at equilibrium, resulting in a divisive interaction

$$W_\theta = \beta_W \frac{V_\theta \left(1 + \eta^+ \,\text{net}_\theta^+\right)}{\alpha_W + \eta^- \,\text{net}_\theta^-} \,. \tag{12}$$

where $\eta^+ = 5$, $\eta^- = 2$ and $\beta_W = 0.001$ are scale factors and $\alpha_W = 0.2$ is a decay parameter. The multiplicative contribution of V_θ ensures that long-range connections have a modulating rather than generating effect on cell activities [23, 24]. The result of the long-range stage is fed back and combined with the feedforward complex cell responses, thus closing the recurrent loop. The shunting interactions ensure a saturation of activities after a few recurrent cycles.

3.2 Simulation Results

In a first simulation a synthetic stimulus of a noisy square is employed. Figure 6 demonstrates the functionality of lateral long-range interaction for the enhancement of coherent structure. Outline contrasts are detected and subsequently enhanced such that the activities of salient contrast as well as orientation significance is optimized. Figure 7 shows the results of processing an image of a laboratory scene. Initial complex cell activations generated for localized high contrast contours are further stabilized. Initially weak activations in coherent spatial arrangements are enhanced. Spatial locations where high amplitude contrast responses exist in multiple orientation channels indicate the presence of corners and junctions. The results demonstrate that noisy low contrast arrangements can be significantly enhanced to form elementary items of smooth contour segments. Beyond the enhancement of coherent contours, the proposed scheme is able to enhance contour responses at corner and junction configurations. These higher order features play a significant role in object recognition and depth segregation (e.g., [3]).

4 Summary

We have presented a computational framework for the processing of discontinuities and homogeneous surface properties.

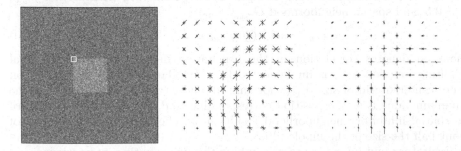

Fig. 6. Processing of a square pattern with additive high amplitude noise. *Left to right:* Input image and close-up of the upper left corner (white square inset in the input image) for complex cell responses and long-range responses. In the close-ups, three important properties of the long-range interaction can be seen: i) enhancement of the orientation coaligned to the contour, ii) suppression of noisy activity in the background, and iii) preservation of the significant orientations at corners

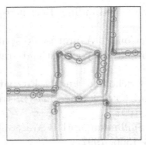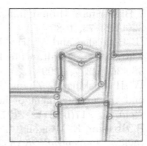

Fig. 7. Enhancement of activity distribution and detection of corner and junction features in a laboratory scene. *Left to right:* Input image, complex cell responses, and long-range responses. Locations of corners and junctions are marked with circles and indicate positions with significant responses in more than one orientation channel. At the complex cell stage, many false responses are detected due to noisy variations of the initial orientation measurement. Such variations have been reduced at the long-range stage, and only the positions of significant variations at corners are signaled

Surface properties such as brightness can be computed from sparse contrast data. Confidence signals are used to discriminate positions of reliable measurements of contrast data from positions where no data is available. The suggested mechanism of confidence-based filling-in allows to generate a size invariant brightness representation even on the basis of sparse input data. Furthermore, perceptual phenomena and real world applications are successfully processed.

For the filling-in process, the proper extraction of *contours* is important. For the processing of *discontinuities* such as contours and junctions, we have suggested a framework of recurrent interaction, using feature integration by long-range connections to evaluate feedforward signals within a broader context. Modulating feedback then selectively enhances those features which fit into the context. The suggested circuit of long-range interactions is an instantiation in the domain of early vision of this general scheme. We show that a single circuit is sufficient to solve basic tasks in early vision, such as contour enhancement, noise suppression and corner enhancement.

While the importance of contour signals for various tasks, such as object recognition, is generally acknowledged, the need for an explicit and intrinsically redundant representation of extended brightness regions is subject to intense debate. Whether such a representation is crucially involved in conscious human brightness perception or is helpful for behavioral tasks such as grasping or object recognition of occluded objects [15,27] is a challenging question to be answered by future research. Models of surface completion are helpful by integrating empirical results into a precise computational and algorithmic description.

References

1. K. F. Arrington. Directional filling-in. *Neural Comput.*, 8:300–318, 1996.
2. M. Bertero, T. Poggio, and V. Torre. Ill-posed problems in early vision. *Proc. IEEE*, 76(8):869–889, 1988.

3. I. Biederman. Human image understanding: Recent research and a theory. *CVGIP*, 32(1):29–73, 1985.
4. B. Blakeslee and M. E. McCourt. A multiscale spatial filtering account of the white effect, simultaneous brightness contrast and grating induction. *Vision Res.*, 38:4361–4377, 1999.
5. W. H. Bosking, Y. Zhang, B. Schofield, and D. Fitzpatrick. Orientation selectivity and the arrangement of horizontal connections in tree shrew striate cortex. *J. Neurosci.*, 17(6):2112–2127, 1997.
6. P. Bressan, E. Mingolla, L. Spillmann, and T. Watanabe. Neon color spreading: A review. *Perception*, 26:1353–1366, 1997.
7. G. Caputo. Texture brightness filling-in. *Vision Res.*, 38(6):841–851, 1998.
8. M. Cohen and S. Grossberg. Neural dynamics of brightness perception: Features, boundaries, diffusion, and resonance. *Percept. Psychophys.*, 36:428–456, 1984.
9. M. P. Davey, T. Maddess, and M. V. Srinivasan. The spatiotemporal properties of the Craik-O'Brien-Cornsweet effect are consistent with 'filling-in'. *Vision Res.*, 38:2037–2046, 1998.
10. R. L. Elder and S. Zucker. Evidence for boundary specific grouping. *Vision Res.*, 38:143–152, 1998.
11. C. Enroth-Cugell and J. G. Robson. Functional characteristics and diversity of cat retinal ganglion cells. *Invest. Ophthalmol. Visual Sci.*, 25:250–267, 1984.
12. H. J. M. Gerrits and A. J. H. Vendrik. Simultaneous contrast, filling-in process and information processing in man's visual system. *Exp. Brain Res.*, 11:411–430, 1970.
13. C. D. Gilbert. Circuitry, architecture, and functional dynamics of visual cortex. *Cereb. Cortex*, 3(5):373–386, 1993.
14. A. Gove, S. Grossberg, and E. Mingolla. Brightness perception, illusory contours and corticogeniculate feedback. *Visual Neurosci.*, 12:1027–1052, 1995.
15. S. Grossberg. 3-D vision and figure-ground separation by visual cortex. *Percept. Psychophys.*, 55(1):48–121, 1994.
16. S. Grossberg and N. McLoughlin. Cortical dynamics of three-dimensional surface perception: Binocular and half-occluded scenic images. *Neural Networks*, 10:1583–1605, 1997.
17. S. Grossberg and E. Mingolla. Neural dynamics of perceptual grouping: Textures, boundaries, and emergent segmentation. *Percept. Psychophys.*, 38:141–171, 1985.
18. S. Grossberg and D. Todorović. Neural dynamics of 1-D and 2-D brightness perception: A unified model of classical and recent phenomena. *Percept. Psychophys.*, 43:241–277, 1988.
19. T. Hansen, G. Baratoff, and H. Neumann. A simple cell model with dominating opponent inhibition for robust contrast detection. *Kognitionswissenschaft*, 9(2):93–100, 2000.
20. T. Hansen and H. Neumann. A model of V1 visual contrast processing utilizing long-range connections and recurrent interactions. In *Proc. ICANN*, pages 61–66, Edinburgh, UK, Sept. 7–10 1999.
21. T. Hansen, W. Sepp, and H. Neumann. Recurrent long-range interactions in early vision. In S. Wermter, J. Austin, and D. Willshaw, editors, *Emergent Neural Computational Architectures based on Neuroscience*, LNCS/LNAI. Springer, Heidelberg, 2000. In press.
22. D. Heeger. Normalization of cell responses in cat striate cortex. *Visual Neurosci.*, 9:181–197, 1992.
23. J. A. Hirsch and C. D. Gilbert. Synaptic physiology of horizontal connections in the cat's visual cortex. *J. Neurosci.*, 11(6):1800–1809, 1991.

24. J. M. Hupé, A. C. James, B. R. Payne, S. G. Lomber, P. Girard, and J. Bullier. Cortical feedback improves discrimination between figure and background by V1, V2 and V3 neurons. *Nature*, 394:784–787, 1998.

25. H. Komatsu, I. Murakami, and M. Kinoshita. Surface representation in the visual system. *Brain. Res. Cogn. Brain. Res.*, 5(1):97–104, 1996.

26. D. Mumford. Neural architectures for pattern-theoretic problems. In C. Koch and J. L. Davis, editors, *Large-scale neuronal theories of the brain*. MIT Press, Cambridge, MA,, 1994.

27. H. Neumann. Completion phenomena in vision: A computational approach. In L. Pessoa and P. de Weerd, editors, *Filling-in: From perceptual completion to skill learning*. Oxford Univ. Press. In preparation.

28. H. Neumann. Mechanisms of neural architecture for visual contrast and brightness perception. *Neural Networks*, 9(6):921–936, 1996.

29. H. Neumann, L. Pessoa, and T. Hansen. Interaction of ON and OFF pathways for visual contrast measurement. *Biol. Cybern.*, 81:515–532, 1999.

30. H. Neumann, L. Pessoa, and T. Hansen. Visual filling-in for computing perceptual surface properties. 2000. Submitted.

31. H. Neumann and W. Sepp. Recurrent V1–V2 interaction in early visual boundary processing. *Biol. Cybern.*, 81:425–444, 1999.

32. M. A. Paradiso and K. Nakayama. Brightness perception and filling-in. *Vision Res.*, 31:1221–1236, 1991.

33. L. Pessoa, E. Mingolla, and H. Neumann. A contrast- and luminance-driven multiscale network model of brightness perception. *Vision Res.*, 35(15):2201–2223, 1995.

34. L. Pessoa and H. Neumann. Why does the brain fill in? *Trends Cogn. Sci.*, 2(11):422–424, 1998.

35. L. Pessoa, E. Thompson, and A. Noë. Finding out about filling-in: A guide to perceptual completion for visual science and the philosophy of perception. *Behav. Brain. Sci.*, 21(6):723–802, 1998.

36. T. Poggio, V. Torre, and C. Koch. Computational vision and regularization theory. *Nature*, 317(26):314–319, 1985.

37. D. A. Pollen and S. F. Ronner. Visual cortical neurons as localized spatial frequency filters. *IEEE Transactions on Systems, Man, and Cybernetics*, SMC-13(5):907–916, 1983.

38. D. C. Rogers-Ramachandran and V. S. Ramachandran. Psychophysical evidence for boundary and surface systems in human vision. *Vision Res.*, 38:71–77, 1998.

39. K. Schmidt, R. Goebel, S. Löwel, and W. Singer. The perceptual grouping criterion of colinearity is reflected by anisotropies of connections in the primary visual cortex. *Europ. J. Neurosci.*, 9:1083–1089, 1997.

40. W. Sepp and H. Neumann. A multi-resolution filling-in model for brightness perception. In *Proc. ICANN*, Edinburgh, UK, Sept. 7–10 1999.

41. A. N. Tikhonov and V. Y. Arsenin. *Solutions of ill-posed problems*. V. H. Winston & Sons, Washington D. C., 1977.

42. R. von der Heydt, E. Peterhans, and G. Baumgartner. Illusory contours and cortical neuron responses. *Science*, 224:1260—1262, 1984.

Representations of Neuronal Models Using Minimal and Bilinear Realisations

Gary G.R. Green, Will Woods, and S. Manchanda

Department of Physiological Sciences, The Medical School, Newcastle upon Tyne,
NE2 4HH, UK
Gary.Green@ncl.ac.uk
http://www.staff.ncl.ac.uk/gary.green

Abstract. Construction of large scale simulations of neuronal circuits is often limited by the intractability of implementing numerical solutions of large numbers of differential equations describing the neuronal elements of the circuit. To make modelling more tractable, simplified models are often used. The relationship between these simplified models and real neuronal circuits is often only qualitative. We demonstrate differential geometric techniques that allow the formal construction of neuronal models in terms of their minimal realisation. A minimal model can be described in terms of a rational series with an associated formal language. These techniques preserve the fundamental behavior of the system. A Lie algebra approach is used to produce approximations of arbitrary order and of minimal dimension. It is shown that the dimension of the minimal representation of a neuronal model is determined by the order of approximation and not the number of states in the original description. A bilinear realisation of Hodgkin Huxley models shows that in the critical region of behaviour below the threshold for firing an action potential, the system should not be described as a leaky, linear, integrator, but as a non-linear integrator.

1 Introduction

Following an elegant series of observations, Hodgkin and Huxley described the flow of current across the squid axon membrane in terms of an equivalent electrical circuit. This circuit was also describable as a set of four non-linear differential equations [6]. These equations described the net current flow across the membrane and the rate of change of state variables that governed the flow of current through two ion selective channels. Since then many ion selective channels have been identified and characterised. When these channels are included in realistic neuronal models with many regions of active membrane then a large number of non-linear differential equations are required for the model formulation. The only practical way to explore the behaviour of such models is through extensive numerical simulation. A further difficulty in examining the input-output characteristics of a neurone is that many different combinations of the parameters of the Hodgkin-Huxley equation produce similar qualitative behaviour[1][4].

S. Wermter et al. (Eds.): Emergent Neural Computational Architectures, LNAI 2036, pp. 154–160, 2001.
© Springer-Verlag Berlin Heidelberg 2001

A tractable way of producing large scale models which are amenable to formal analysis as well as easing the needs on computer resources is to use simplifications of the Hodgkin-Huxley model. The simplest models treat the neurone as a linear, leaky integrator that fires an action potential when the membrane voltage passes a threshold. More sophisticated models use a wide range of polynomial differential representations which have similar qualitative behaviour to the Hodgkin-Huxley model (e.g. the Fitzhugh-Nagumo based models). However comparison with real neuronal behaviour is difficult and requires the examination of the invariant defining features of a system. These features are the coefficients or kernels that describe the local analytic solutions of the system. As with many systems of non-linear differential equations the analytic solutions of Hodgkin-Huxley equations are not found to have a simple closed form.

This paper is concerned with formal methods of finding the minimal realisation of Hodgkin-Huxley type models. It is shown that by examining the Lie series solutions of the neuronal model, a basis set can be identified, with an associated Lie algebra, which can be used to produce a minimal dimension realisation. A bilinear approximation of the Hodgkin-Huxley model is also realised. Although this representation is not necessarily minimal, it can be used to visualise the input-output defining Volterra kernels. It is shown that the Hodgkin-Huxley system of equations contain a large second order nonlinearity which means that this neuronal model is not simply a linear, leaky integrator.

2   An Alternative Form of the Hodgkin-Huxley Equations

The standard form of the Hodgkin-Huxley equations is

$$-C\dot{v} = \bar{g}_L(v - v_L) + \bar{g}_K n^4(v - v_K) + \bar{g}_{Na}m^3 h(v - v_{Na}) + u \tag{1}$$

$$\dot{n} = \alpha_n(1 - n) - \beta_n \tag{2}$$

$$\dot{m} = \alpha_m(1 - m) - \beta_m m \tag{3}$$

$$\dot{h} = \alpha_h(1 - h) - \beta_h h \tag{4}$$

$$y = v \tag{5}$$

where v is the voltage across the membrane of capacity C, n, m, h are state variable which determine the probability of ionic channels being open or closed. v_L, v_K, v_{Na} are the reversal potentials of the leakage, potassium and sodium channels respectively. The maximal conductances of these channels are g_L, g_K, g_{Na}. α_i and β_i are the forward and backward rate constants of the first order kinetics of the state variables n, m, h, In the original description it was empirically assumed that these rate constants were non-linear functions of the voltage v. An input current to the equivalent circuit enters linearly as u.

In most physiological situations there are many more currents than just the leakage, potassium and sodium channels proposed by Hodgkin and Huxley. In these cases the equivalent circuits require many more differential equation to model the system if it is assumed that the current flow has to satisfy the conditions of independence.

However many channels are included, the general form of the Hodgkin-Huxley equations are linear analytic such that the individual state equations can be represented as

$$\dot{x}_i = f_i(X) + g_i u_i$$

and the output

$$y(t) = h(X) \tag{6}$$

where x is a state variable of the original system (either v, n, m, h in the original Hodgkin-Huxley equations), f and h are analytic functions of the vector X of the states x_i and g_i are constants such that in the original equations only g_1 was non-zero.

The advantage of making the equations more abstract is that they can also be described in a coordinate free manner using a Lie derivative operator such that

$$\dot{X} = A_0(X) + A_1(X)u \tag{7}$$

where A_0 and A_1 are the Lie derivatives

$$A_0 = \sum_{i=1}^{m} f_i(X) \frac{\partial}{\partial x_i}$$

$$A_1 = \sum_{i=1}^{m} g_i(X) \frac{\partial}{\partial x_i}$$

where m is the number of states.

If the Lie operators are applied to the original vector of states X then the original Hodgkin-Huxley equations are retrieved. These operators can be used to produce a local functional expansion of the relationship between the output and the past history of the input, where the invariant features of the model are clearly shown. This functional form can then be used to find the minimal representation.

3   Generating Series Solutions

The Generating series solution of equations 6 and 7 is formed by Peano-Baker iteration. It has been shown to be [3]

$$y(t) = S = h_{|X_0} + \sum_{v \geq 0} A_{j_1} A_{j_i} \ldots A_{j_v}(h)_{|X_0} z_{j_v} \ldots z_{j_i} z_{j_1}$$

where S is the mapping between the free monoid Z^* constructed from the alphabet z_0, z_1 into $\Re$ and the subscript $|X_0$ means evaluated at the initial conditions of the state vector x. Each word $z_1 z_i \ldots z_v$ corresponds to the iterated integral

$$\int_0^t u_1(\tau_1) \int_0^{\tau_1} u_i(\tau_i) \ldots \int_0^{\tau_{v-1}} u_v(\tau_v) d\tau_v \ldots d\tau_i d\tau_1$$

defined recursively on its length, (u_0 is defined to be unity). The coefficient of the word $z_1 z_i \ldots z_v$ is the iterated Lie derivative $A_v \ldots A_i A_1$ operating on the output function h.

Two systems are locally equivalent, in state space and in time, *if and only if* their Generating series match. The Fliess generating series has a closed relationship with the popular Volterra series form of describing the input-output behaviour of a non-linear dynamic system. For the Hodgkin-Huxley equations the local analytic solutions, could, in principle, be found by using the Fliess or Volterra series forms of the input-output behaviour. Inspection rapidly shows that this is, in general, not a straightforward practical possibility. Except in very specific situations the solutions, i.e. the series, are infinite and no closed form can be easily found. This is because the successive use of the Lie derivative operators for either the coefficients of the Fliess series or the Volterra kernels results in a combinatorial explosion of terms. As the original formulation of the Hodgkin-Huxley equations was empirical and designed to approximate real behaviour, an argument can be made that approximate solutions may be used to examine the invariant features at different degrees of approximation from linear up to an arbitrary choice of degree of approximation.

4   Minimal Realisations

Truncation of the Generating series produces an approximation of the local input-output behaviour of the neuronal model. It is formally equivalent to the specification of the Runge-Kutta method of integration of the model if the input is piecewise constant. The truncated Generating series of arbitrary order k can be expressed in terms of its Lie-Hankel matrix [7]. A Lie-Hankel matrix of a Generating series is an array whose rows are indexed by a basis of the Lie algebra of Z and the columns are indexed by the components of the monoid Z^*. If this matrix if of finite rank then a minimal realisation exists and its rank then governs the dimension of the realisation[2]. The Lyndon basis of the Lie algebra of Z is chosen[9]. The Lie Hankel matrix of the Hodgkin-Huxley Generating series truncated at order two is

	ϵ	z_0	z_1
z_0	$A_0(h)$	$A_0^2(h)$	$A_1 A_0(h)$
z_1	$A_1(h)$	0	0
$[z_0, z_1]$	$A_1 A_0(h)$	0	0

and its rank is maximally three. Similar analyses can be made for high order truncations. Table 1 shows the rank of the Lie-Hankel matrix for the Hodgkin-Huxley equations at different orders of truncation.

Table 1.

Order of truncation	Rank
1	1
2	3
3	5
4	8

The rank and therefore the dimension of the minimal realisation is *not* determined by the number of states (ion channels and compartments) or the Hodgkin-Huxley model but just by the order of truncation.

5   Construction of the Minimal Realisation

The co-ordinates of the minimal realisation can be chosen as the Lyndon words that make up the Lie polynomials that indexed the rows of the Lie-Hankel matrix. This candidate basis set can then be used to reconstruct the Generating series using an algorithm proposed by Jacob & Oussous [7]. The algorithm iteratively searches for and identifies proper left factors of the truncated series. The minimal realisation of the 2_{nd} order truncation of the Hodgkin-Huxley equation is

$$\dot{q}_1 = 1$$
$$\dot{q}_2 = u$$
$$\dot{q}_3 = q_1 u$$
$$y = -A_1 A_0(h)q_3 + \frac{1}{2}A_0^2(h)q_2^2 + A_1 A_0(h)q_1 q_2 + A_1(h)q_2 + A_0(h)q_1$$

Higher order realisations have polynomial state equations.

6   Volterra Series Solutions

An alternative method for finding an approximate Generating series for the Hodgkin-Huxley equations is to approximate the system by a bilinear representation[8]. This form also has a finite rank Lie algebra which can be used to reconstruct the approximating *infinite* series upto Volterra order k.This approximation then consists of words from the free monoid that contain up to k instances of z_1 whose coefficients *match* those of the original Hodgkin-Huxley Generating series. The Generating series of such approximations differ from those studied in section 4. The minimal forms studied there match the Hodgkin-Huxley Generating series upto a given length of the series but do not necessarily have identical Volterra kernels up to a particular order. Bilinear approximations can be constructed to produce the correct Volterra kernel up to order k. Several algorithms exist for finding this representation but they are not necessarily minimal. Figure1 shows the second order Volterra kernel for the Hodgkin-Huxley equations just below the threshold for an action potential. It should be noted that there *is* a second order kernel and therefore the Hodgkin-Huxley equations cannot be replaced, below threshold, by a simple leaky, linear, integrator.

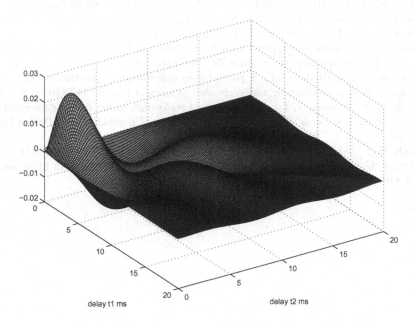

Fig. 1. 2nd Order kernel at 0mV - just before an Action Potential

7   Discussion

The formal links between minimal realisations and the Hodgkin-Huxley model of neurones have been introduced. The major conclusion is that minimal approximations can be constructed whose dimension is independent of the original model formulation. These approximations can be used to find the analytic solutions of the Hodgkin-Huxley equations up to an arbitrary order. Bilinear representations can be exploited to easily find the Volterra series solutions and demonstrate that a large second order non-linearity is present just below the threshold for firing an action potential. The two representations, the minimal realisations or bilinear approximants, are closely related and may allow the construction of modular forms of neuronal model and allow more extensive investigation of the behaviour of real neural circuits.

References

1. Bhalla, U.S. & Bower,J.M. *Exploring parameter space in detailed single neuron models: Simulations of the mitral and granule cells of the olfactory bulb* J. Neurophysiol. 69 (1992) p1948-1965
2. Fliess, M. *Realisation locale des systemes non lineaires, algebres de Lie filtrees transitives et series generatrices non commutatives* Invent. Math. 71 (1983) p521-537
3. Fliess, M.,Lamnabhi, M. & Lamnabhi-Lagarrigue, F. *An algebraic approach to nonlinear functional expansions* IEEE Trans. Circuits and Systems 30 (1983) p554-570
4. Foster, W.R., Ungar, L.H. & Schwaber, J.S. *Significance of conductances in Hodgkin-Huxley models* J. Neurophysiol. 70 (1993) p 2502-2517
5. Hespel, C. & Jacob, G. *Approximation of nonlinear dynamic systems by rational series* Theoretical Computer Science 79 (1991) p151-162
6. Hodgkin, A.L. & Huxley, A.F. *A quantitative description of membrane current and its application to conduction and excitation in nerve* J.Physiol.117 (1952) p500-544
7. Jacob, G. & Oussous, N. *Local and minimal realisation of nonlinear dynamical systems and Lyndon words* IFAC symposium: Nonlinear Control Systems Design (1989) p155-160
8. Rugh, W.J. *Nonlinear system theory* John Hopkins University Press (1981)
9. Viennot, G. *Algebre de Lie libres et monoides libres* Lecture Notes in Mathematics Vol. 691. Springer-Verlag (1978)

Collaborative Cell Assemblies: Building Blocks of Cortical Computation

Ronan G. Reilly

Department of Computer Science
National University of Ireland, Dublin
Belfield, Dublin 4
Ireland
Ronan.Reilly@ucd.ie

Abstract. The goal of this chapter is to propose a candidate for consideration as a building block of cortical computation. The candidate in question is the collaborative cell assembly (CCA), and I will make the case in the rest of this chapter that the concept provides a unifying framework for accounting for the mechanisms of perceptual "binding" in the sensory areas of the brain, as well as providing a mechanism for the incremental construction of cognitive and language systems from simpler components.

1 Computational Building Blocks

If one looks at how the brain processes sensory information or controls motor activity, one is struck by a surprising design feature. Rather than there being a single brain region involved in, say, object perception, a number of regions collaborate, contributing a specialization to the achievement of the overall perceptual goal. The cortex appears to function as a collection of collaborating specialists, none of which can solve the problem alone, but can only do so when each works together. Another feature of this pattern of specialization and modularity is a phenomenon of complementarity apparent in the organization of, for example, sensory processing streams. Grossberg (in press) has argued that pairs of complementary processing streams should be considered the basic cortical functional unit. A frequently cited example of complementarity in vision is the division of processing between the so-called "what" and "where" pathways. Ungerleider and Mishkin (1982) have proposed that the two major output fiber tracts from the occipital lobe, the inferior and superior longitudinal fasciculi, are responsible for carrying information about the object identity and object location, respectively. The inferior fasciculus travels into the temporal lobe, the superior fasciculus into the posterior parietal region. Notwithstanding their separate trajectories, these pathways communicate extensively. Consequently it is a good example of what Grossberg (in press) has identified as a complementary pair of collaborative processing streams, with information from both streams required to provide a coherent percept.

S. Wermter et al. (Eds.): Emergent Neural Computational Architectures, LNAI 2036, pp. 161–173, 2001.
© Springer-Verlag Berlin Heidelberg 2001

At first glance, it may seem an inefficient way of constructing a perceiving–acting system by fragmenting its operations along lines that often seem counterintuitive, and more importantly which apparently give rise to the computational cost of integration further downstream. Of course, part of the counterintuitive quality arises from the fact that we are conscious of the unity of our perceptual experience of the world. It therefore seems anomalous that different brain regions should be responsible for what appear to be intimately integrated aspects of a percept (e.g., its color and form, in the case of vision). There is then the further apparent challenge of integrating and coordinating the disparate outputs from each of the specialist modules in order for the system to mount a coherent response to a specific perceptual event. As I will argue later, the challenge may be more apparent than real.

1.1   Cortical and Software Engineering

Some informative analogies can be made between cortical "design" and approaches to the construction of large–scale and complex software systems. In the last decade, with the demand for increasingly complex operating system and application software, a monolithic approach to software development has no longer proved feasible. The dominant approach to software engineering is now object–oriented (OO) design and implementation. The advantage of object–oriented design is that it encourages the modularization and encapsulation of functionality, with communication between modules being limited to messages in a mutually agreed format. Flow of control in an object–oriented program is not centralized, but distributed throughout the objects or modules comprising the system. A key attraction of an object–oriented approach is that it facilitates, more readily than earlier approaches, the addition of new functionality, without the need to re–engineer the entire system. It also facilitates the re–use of already existing functionality for new purposes. It should be emphasized that re–use does not inevitably follow from an OO design, rather re–usability requires as a foundation the modularity that an OO approach provides. The price paid for the benefits of OO is the need for an agreed interface between objects or modules. Furthermore, if re–use is to be facilitated, this interface needs to be kept as simple as possible.

How does all of this relate to cortical design? In the domain of complex system design, be it cortical engineering or software engineering, several common constraints are at work: the need to manage the complexity of the problem, coordination of the elements of the solution, the need for incremental modification. In both the software and cortical cases, I will argue, complexity has been managed by decomposition of the task into smaller components. Coordination of these components has been achieved by means of a common interface, and incremental modification by keeping the interface as simple as possible. In the biological context, the issue of decomposition is uncontroversial. However, what I mean by a "common interface" will be the focus of a substantial part of this chapter.

1.2 Evolutionary Constraints

A number of additional factors constrain the space of possible solutions to designing an organism with a large peceptuo–motor repertoire. The first is that it must be constructed by the incremental tinkering of an evolutionary process driven by natural selection. An incremental approach to system construction will inevitably tend to favor the addition of small specialists modules rather than a re–engineering of the entire system. As Simon (1969) has pointed out, complex wholes will tend to be developed incrementally over evolutionary time. This is because, within an evolutionary context, the progression from simple to complex must entail viable intermediate stages.

Another, and not unrelated factor is that the hardware upon which any additional functionality is built also acts as a source of constraint. Kaas (1987) has shown that as the neocortex enlarges, there is a natural tendency for regions to form, with a high–degree of local inter–connectivity within a region, and limited inter–regional connectivity via reciprocal cortico–cortical projections. This phenomenon of regionalization arises from structural constraints imposed by the two–dimensional nature of the cortical sheet, and the physical limits on cortical connectivity. Although there is still considerable debate about the role played by neocortical size in human cognitive capacity (Passingham, 1982), some neuroscientists now argue that human brain size is only marginally larger than that for a primate scaled–up to human proportions (Roth, 2000). The uniqueness of the human brain does not rely so much on quantities of neurons, but rather on the increased structural complexity based on the formation and acquisition of additional cortical modules, specialized for a variety of evolutionarily advantageous task.

There are, however, dangers in seeing the human brain as simply a scaled–up primate brain, which in turn is a scaled–up mammalian. The view of evolution as a ladder leading inexorably to a human pinnacle has long been abandoned by evolutionary biologists, but seems still to have an inescapable hold on neuroscientists (Preuss, 1993). Contemporary evolutionary biologists prefer to use a tree–metaphor when describing the progress of evolution. Branches of the tree represent divergences from a common branch, leading to species–specific specializations. Therefore, while acknowledging that increased cortical complexity is an important part of the story, a more interesting and more challenging question is the nature of the species–specific specializations that took advantage of neocortical enlargement and that led us down the hominid branch of the tree.

An additional aspect of evolution which should not be ignored, and to which Clark (1997) very eloquently alerts us, is that it will often blur the boundary between the individual and the environment. The solutions it arrives at are very often distributed, not only across individuals, but also seamlessly across the individual and the environment. While this may seem self–evident to evolutionary biologists, it is still seen as somewhat radical in the cognitive sciences. For example, recent research in the area referred to as "interactive" vision emphasizes the need to take account of the informational content intrinsic in he variation and stability of the environment, and the fact that vision is just one element in the perception–action behavioral loop (Churchland, Sejnowski, & Ramachandran, 1997; O'Regan & Noe, in press; Ballard, Hayhoe, & Pook, 1998). Much of this work harks back to Gibson's pioneering theory of ecological optics (Gibson, 1966), but has been given a fresh impetus by attempts to build artificial vision systems. What it tells us is that adaptation is pervasive and unpredictable in the configuration of environment and organism that gets exploited to gain leverage in the competition for survival.

1.3   Towards a Computational Building Block

So where does that leave us in our quest for a cortical computational building block? Well, one element of the story must be the easy ability to put together and interface different specialized modules – the common interface mentioned earlier.

When I first read about the architecture of the human visual system as an undergraduate, I was impressed by the computational complexity that I thought must be involved to bind together the different sets of feature detectors active in different regions of the visual area. Only latterly, did it occur to me that perhaps, the so–called "binding" problem, is not really a problem at all, but something that is supported by an intrinsic property of the way the cortex computes. It is something so intrinsic, that evolution has exploited it on many occasions. Take for example the evolution of color vision in primates. It turns out that primates are in a minority among mammals in having color vision. The rodent–like ancestor of mammals, which emerged during dinosaur period, was of necessity a nocturnal creature. It took advantage of the nocturnal niche left by the dinosaurs who had poorer control of their body temperatures and were consequently restricted in their nighttime activities. As a nocturnal creature, there was a greater requirement for rods than cones in their visual system, consequently monochromatic vision was favored by evolution in this context. When the dominance of the dinosaurs waned, the daytime niches became occupied by these early mammals. Nonetheless, it appeared that only in certain of these new environments was color vision an advantage; one of these being he arboreal environment of early primates.

What this story of the re–emergence of color vision in mammals tells us is that the facility for adding new modules to the sensory system, and integrating their outputs to complement the output from other modules, is one that is supported by some intrinsic features of cortical computation. In the next section I will discuss what this computational support might be.

2   Collaborative Cell Assemblies

There is general agreement that some form of population coding involving an ensemble of neurons, along the lines first proposed by Hebb (1948), is implicated in cortical representation of sensory and motor patterns.

2.1   Representation

Before proceeding, however, it is important to clarify what I mean by the vexed term "representation." According to Haugeland (1991), a system that uses internal (in our case, cortical) representations must: (a) be able to map unreliable environmental inputs onto relevant representations; (b) use these representations to "stand in" for entities in the environment; and (c) manipulate the representations within a larger, systematic, framework. The main source of disagreement among cognitive scientists centers on criterion (c), though there have been one or two well–articulated proposals arguing against the need for representations of the "standing–in" variety altogether (e.g., Brooks, 1991).

For those committed to the concept of some form of representation, it is possible to identify two extremes: at one end, there are those who argue for representations with properties of context independence and compositionality of the type associated with formal linguistic and logical representations (Chomsky, 1986; Fodor & Pylyshyn, 1988; Hadley, 1999). The newer and more radical view is that the brain is ultimately a dynamical system, and that "classical" representations of the type just described are an entirely inadequate framework within which to view its operation (Kelso, 1997; Port & vanGelder, 1995). According to these dynamicists, the differential rather than the predicate calculus is a more appropriate tool for describing brain function. One finds, however, that the two opposing camps often draw on different psychological and developmental phenomena to support their respective cases. The classicists tend to focus on language and language–related reasoning tasks, whereas the dynamicists tend to emphasize action–related activities in ecologically realistic contexts. The view taken here will follow that of Clark (1997) in arguing that there is a role for representation, not necessarily of the punctate compositional variety favored by the classicists. These representations can be function specific, and particularly come into play for what Clark has called "representation–hungry" problems. In other cases, more dynamical approaches to solving the problem can be used. To paraphrase Clark (1997), we should (a) beware of putting too much into the head – the brain, body and the environment are all candidates for providing problem–solving support, and (b) beware of rigid assumptions about the nature of internal representations – they can be, as the need arises, special purpose, action related, or even classical.

There are a number of potential pitfalls in discussing representation in the brain. Implicit in many theories of neural representation is the notion of a given representation, or type of representation, being located or contained in some specific brain region. This is partly motivated by evidence from lesions, which suggest that localized damage to areas of the cortex (e.g., Broca's) has clearly defined behavioral consequences. However, there is a danger in going from these lesion studies to the position that the relevant brain region "contains" the representations relevant to a particular behavior. A more plausible account, and one in keeping with a more detailed analysis of the lesion evidence, has been proposed by Damasio and Damasio (1994) in their theory of convergence zones. They argue that complementing the observed parcellation of function is an interdependency of representation. Within their view, representations in the brain are dispositional; that is, regions of the brain implicated in a concept comprise its representation. Retrieval of information involves a process of reconstruction through the activation of these interconnected and distributed representations. Convergence zones serve to tie many of these distributed components together. As Damasio et al. put it:

The essence of this framework, then, comprises reconstruction of entities and scenes from component parts and integration of component parts by time correlations. The requisite reactivation is mediated by excitatory projections. (Damasio & Damasio, 1994; p.73)

Damasio et al. emphasize the dynamical nature of the retrieval process, and the fact that a lesion serves not just to remove one or two components from the distributed representation, but also to force a reconfiguration of the whole retrieval process. Consequently, a lesion to Broca's region can be viewed as not destroying syntactic knowledge stored at the locus of the lesion, but disrupting the coordination of widely distributed knowledge used in sentence construction.

It would appear from the foregoing that the brain employs highly distributed form of representation. Pulvermüller (1999) has persuasively argued this case for word representations. He has marshaled evidence to demonstrate that words relating to a particular sensory modality or motor activity depend for their representation on the brain regions responsible for that modality or activity. Related to this view, O'Regan and Noe (in press), argue that representations of the visual world are fundamentally sensorimotor. They reject the idea that seeing derives from the activation of an internal replica of the outside world. Instead, they propose that seeing is a way of acting, a particular way of exploring the environment.

There is no internal representation whose activity generates the experience of seeing. The outside world serves as its own, external, representation. The experience of seeing occurs when the brain has mastery of the currently occurring laws of sensorimotor contingency (O'Regan & Noe, 1999, p.1).

Their anti–representationalist argument has some echoes of Brooks' (1991) position. Nonetheless, I believe that their central message of action being an integral part of visual perception is very much in keeping with the argument being developed here.

To summarize, representation in the brain appears to involve the collaboration of many different brain regions, where the collaborating regions are involved both in the acquisition of the signal giving rise to the representation, and in actions resulting from the perception of the signal.

2.2   Collaboration

It is necessary to unpack what I mean by the term "collaborative cell assembly." As has already been discussed, there is some agreement among cognitive neuroscientists that cortical representations are realized as distributed patterns of activity over an assembly of neurons. The concept of a cell assembly is due originally to Hebb (1948) who defined an assembly as a set of neurons, either adjacent to, or distant from, each other that have become associated through the strengthening of their connections as a result of simultaneous activation. Pulvermüller (1999), among others, have proposed an additional characteristic, namely that uncorrelated firing of connected neurons will lead to the weakening of the connections between them. Abeles (1991) has also argued that the spatio–temporal pattern of firing in an assembly is a significant aspect of its functionality.

What size can a cell assembly be? Calvin (1996) defines a minimal cell assembly as a two–dimensional hexagonal arrangement of minicolumns about 0.5mm across, where a minicolumn is about 0.03 mm in diameter comprising about 100 neurons. The cell assembly involves a spatio–temporal pattern of activation within a hexagon. The motivation for the hexagon centers on the need for a reliable representation of a word or concept in the cortex. Calvin invokes a cortical selectional mechanism, which he refers to as a Darwin machine, for cloning and selecting concept representations. Within his framework, multiple copies ensure accurate transmission over noisy cortico–cortical connections. There is an implicit commitment within Calvin's work to a classical mode of representation, with the concomitant requirement that concept representations be discrete and composable. Although, the individual concepts themselves have a spatio–temporally distributed character, they are bounded and punctate when incorporated into larger sentence–like representations.

The cell assemblies envisaged by Pulvermüller (1999) involve both local clusters of neurons, along the lines proposed by Calvin, as well as neuronal groups at much larger distances, even involving traversal of the corpus callosum. The thrust of Pulvermüller's proposal is that cortical representations of words involve neuronal groups in regions that relate to their meanings. So, for example, cell assemblies representing words dealing with vision, will preferentially involve neuronal groups in the visual area of the cortex.

The collaborative cell assemblies proposed here are envisaged to be along the lines suggested by Pulvermüller; that is, involving neuronal groups encompassing all areas of the cortex. What is new about the proposal to be outlined below is that it focuses on the interaction between these types of cell assembly, and the possible role that development may play in this interaction. Of specific interest, are interacting cell assemblies in which there is an asymmetry in development, and the implication that this might have for building a cognitive system. Furthermore, we will not confine ourselves to assemblies representing words, but will extend the concept to deal with more general aspects of cortical functioning.

An important distinction among cell assembly theorists is what type of activity constitutes an active cell assembly. The simplest approach is to say that an assembly is a set of reciprocally connected neuronal groups that are, within some time window, above an excitation baseline. This is what is referred to as a "rate" code representational scheme. Problems arise when there is more than one object being attended to, and consequently, more than one cell assembly is active. There is a need to be able to distinguish among multiple objects. A number of researchers have proposed that a temporal code is required in this case (Gray, König, Engel, & Singer,1989; Crick & Koch, 1992; Singer, 1994). Therefore a set of neurons can be considered to be participating in a given cell assembly if its neurons are firing at a rate above some baseline, and have their activities synchronized.

Hebb's original proposal concentrated on the conditions for establishing and maintaining a single active cell assembly, which could in principle extend over a number of spatially separated cortical regions (Hebb, 1948). Things get a little complicated when we consider the collaboration of multiple cell assemblies. Clearly, given the scale of the human cortex, at any given time there are many millions of CAs active, some of which will be related to the same perceptual/cognitive event. How these CAs interact and possibly combine is an issue that needs to be addressed if we are to understand their functionality in more complex situations.

I propose that the best place to start looking for the simplest forms of collaborative cell assembly is in the sensory areas of the brain. Furthermore, by taking a "collaboration" perspective, I hope to show that some current problems in understanding the underlying neural mechanisms of perception may be cast in a more informative light.

2.3   Binding as Collaboration

Following stimulation of the relevant sensory cortex, activation spreads collatererally and "forward" to secondary and association areas of the cortex. In addition reciprocal activation from these forward cortical areas feeds backwards to the primary sensory areas. In the case of the primate visual system (Van Essen, et al., 1990), there are converging and diverging feedforward and feedback routes through a variety of specialized modules. This picture is also probably true of other sensory modalities.

Singer (1994) describes the hypothetical interplay of activity between two visual areas in this complex chain of processing, namely areas V1 and V5 of the visual cortex. If we assume that a subject is viewing a moving geometric shape, various features of the shape will cause neurons that are optimally responsive to these features to start responding. Tangential connections in V1 will cause this activity to spread to neighboring neurons responsive to the same features. This spreading will be enhanced by continuous or contiguously positioned features in the input. Meanwhile, in V5, where neurons are optimally responsive to direction of motion, groups of neurons that are responding to the motion of coherent contours will tend to synchronize their activity. By means of backprojections, the synchronized activity in V5 will support and selectively reinforce the feature–based activity in V1. As Singer (1994) puts it:

Responses to contour elements that are far apart and have different orientations have a low probability of being synchronized by local interaction within V1. However, if these contour elements move coherently, their coherence would be detected by neurons in V5. Responses to these contours would synchronize in V5, and through the backprojections increase synchronization probability for the respective set of neurons in V1. (Singer, 1994; p235)

The scenario that Singer describes above is paradigmatic of interacting cell assemblies. I will consider it to be the most fundamental form of interacting CA, since it is evolutionarily ancient, having been around as long has there has been sensory systems responsive to more than one stimulus dimension..

Let us look a little more closely at the possible dynamical properties of this basic form of CA interaction. The first question is how the separate assemblies start off synchronized in the first place. The simplest explanation is that it arises from the fact that the object that is the source of stimulation impinges on the different parts of the sensory system at the same time. So, for example, V1 and V5 receive input from the retina at the same time, and the cycle of synchronous activity is simultaneously triggered in both regions. Once a disparate set of sensory regions have become active, then through a network of reciprocal projections these regions influence each other. This interactive influence can serve to sharpen a noisy signal or augment an incomplete one. So, for example, the combination of information relating to motion, contour, color, can help uniquely to specify an object in the visual field. Additional top–down influences, such as activation of prior memories of the object will further reinforce and stabilize its associated cell assembly.

When the process of perception is viewed as a collaborative one, both within low–level areas of the visual cortex, and between lower and higher levels, it becomes clear that to characterize the task of integration of disparate features of the input as a binding "problem" (e.g., Crick & Koch, 1990) is perhaps to view it from the wrong angle. What evolution has done in constructing our sensory systems is to provide additional sources of constraint on the possible identity of an object in the visual world by tapping into the information immanent in the sensory signal. Relying on motion information, or contour information alone, will get you only so far (if you are an insect, however, this may be far enough). Each additional source of information combines multiplicatively to constrain the identity of an object. The more sources of constraint in the sensory signal that you can exploit to guide your actions, the greater your survival advantage. One can think of evolution in this context as striving to maximize information transmission, in the information theoretic sense, by enlarging an organism's available bandwidth.

2.4 From Symmetric to Asymmetric Collaboration

We can readily extend the mechanism described by Singer above from the domain of perceptual information integration to more system–wide collaboration. For example, Iverson and Thelen (1999) argue for the interdependence of cognition and motor control systems in the brain. They view brain computation as a dynamical as opposed to logical process. In their view, language, cognition, perception, and action are cut from the same computational cloth. They summarize their position as follows:

In sum, our argument for embodiment rests on the necessity for compatible dynamics so that perception, action, and cognition can be mutually and flexibly coupled. Such dynamic mutuality means that activity in any component of the system can potentially entrain activity in any other... (Iverson & Thelen, 1999, p. 37)

In my view, what they propose is functionally equivalent to the collaborative cell assembly described by Singer, but this time potentially involving cortical regions at a remove from the sensory–motor "surface", and traversing sensory modality boundaries.

In this broader view of collaboration, it is useful to make a distinction between collaborations involving cell assemblies that are equally well developed, and those in which one partner in the collaboration is more developed than the other. I will refer to the former as symmetric collaboration and the latter as asymmetric. In the case of asymmetric collaboration, there is the possibility for the less well–developed cell assembly to exploit the functionality of the more developed one. Elsewhere (Reilly, 1995), I have referred to this as cortical software re–use. The term is inspired by the software engineering concept of re–use (Bigerstaff & Perlis, 1989), and motivates the proposal that the developing partner in a collaboration is able to re–use the repertoire of cell assemblies already established in the more developed cortical region. Within this view, the sensory–motor systems, can be thought of as providing the basic computational building blocks for higher–level language and cognitive functions, which are then adapted and modified to fulfill their new role. The cortical connectivity provided by reciprocal cortico–cortical projections between sensory, motor, and association areas, and more frontal cortical areas is a key mechanism for providing access to a repertoire of re–usable functions. The style of computation that this connectivity supports is best viewed as a process of dynamical entrainment along the lines proposed by Iverson and Thelen (1999) involving the synchronization of firing patterns in reciprocally connected cortical areas. The selection of re–usable modules from one domain for re–use in another is based on a structural isomorphism, possibly supported by resonance, between the dynamical firing patterns of the source and target domains. This might mean, for example, that a motor plan for complex object–manipulation task gets re–used in the service of syntax in language production (e.g., Greenfield, 1991; Reilly, in press). The selection of relevant re–usable functions is domain or content independent, and relies on relatedness at an abstract structural level. In the case of the previous example, it might rely on the inherent hierarchical and recursive nature of both motor plans and utterance plans. The key point here is that there may be, but not necessarily, a "semantic" connection between the re–used component and its new application.

There is a wide variety of circumstantial evidence to support the re–use hypothesis. For example, Greenfield (1991) observed parallels in the developmental complexity of speech and object manipulation. In studying the object manipulation of children aged 11–36 months, she noted that the increase in complexity of their object combination abilities mirrored the phonological and syllabic complexity of their speech production. There are two possible explanations for this phenomenon: (1) It represents analogous and parallel development mediated by separate neurological bases; or (2) the two processes are founded on a common neurological substrate. Greenfield (1991) used evidence from neurology, neuropsychology, and animal studies to support her view that the two processes are indeed built upon an initially common neurological foundation, which then divides into separate specialized areas as development progresses. If Greenfield is correct in her analysis, this is good evidence of developmental re–use by the language domain of functionality initially established in the motor domain.

Another source of circumstantial evidence is Gerstmann's Syndrome (Restum & Sobota, 1983). This syndrome comprises the following cluster of co–occurring symptoms arising from left parietal lesions: (a) finger agnosia, (b) acalculia, (c) left–right disorientation, and (d) agraphia. What is particularly interesting from the re–use point of view is the association between finger agnosia and acalculia. The re–use interpretation of this association is that numerical abilities exploit the functionality already established for dealing with spatial arrangements of fingers, particularly the ability to individuate or recognize as separate and distinct each finger. One of the features of finger agnosia is not being able to tell without looking whether one or more fingers is being touched. Patients with this condition have sense of their fingers being an undifferentiated mass.

While the evidence just outlined is suggestive, more compelling evidence will come from direct testing of hypotheses derived from the CCA proposal, in conjunction with computer simulation studies. Some preliminary hypotheses are discussed in the next section.

3   Implications of the CCA Perspective

I will divide the implications into two broad categories: (a) those relating to CCA as a cognitive science theory, and (b) those relating to the interaction between artificial neural network research and neuroscience. From a cognitive science perspective there are a number of testable hypotheses that can be derived from the CCA proposal, particularly relating to its re–use aspect. A number of these center on the issue of task interference. If language is constructed on a sensory–motor foundation, and continues to draw upon that underlying functionality, one should expect to find competing non–linguistic demands on this resource having an impact on language performance. So, for example, we should find subjects' language performance impaired when asked to produce syntactically complex utterances while at the same time performing a motor task with, say, some form of recursive structure. Contrariwise, we would not expect to find motor performance impaired by the complexity of a language task. In a similar vein, subjects' performance on a mental arithmetic task should show some degradation if they are asked simultaneously to carry out a motor task involving their fingers. Obviously, considerable effort would have to be invested in designing appropriate control conditions for these type of experiments.

The CCA perspective also has a number of implications for the construction of computational artifacts to tackle difficult AI problems. One such implication is that an approach involving the incremental construction of solutions start-ing with simpler ones and using these as a foundation for more complex ones might be a fruitful one to explore. To some degree the work on neural network ensembles can be considered to be taking this approach, though motivated by reasons of reliability and redundancy rather than the learning of complex tasks (see Sharkey, 1999, for a comprehensive survey of this field). A marrying of CCA and ensemble approaches would, I believe, provide a powerful tool for the construction of large–scale neural network applications.

4   Conclusions

In this chapter I have proposed a candidate building block for cortical computation, which I refer to as a collaborative cell assembly (CCA). I have argued that CCAs arose originally from an evolutionary requirement for incremental increases in the functionality and bandwidth of sensory systems. However, in conjunction with the extended developmental period of the human cortex, they have provided a framework for re–using perceptuomotor "software" to construct the cognitive and linguistic systems. The CCA proposal can be used to generate a number of testable hypotheses relating primarily to resource competition in the execution of cognitive tasks that utilize underlying sensory–motor functions. Furthermore, implications are drawn regarding how best to construct artificial neural networks to tackle large scale problems.

References

Abeles, M. (1991) Corticotonics. Cambridge, UK: Cambridge University Press.

Ballard, D.H., Hayhoe, M.M., Pook, P.K. & Rao, R.P.N. (1997) Deictic codes for the embodiment of cognition. Behavioral and Brain Sciences, **20**, 723–767.

Biggerstaff, T.J. & Perlis, A.J. (1989). Software Reusability: Concepts & Models. New York: ACM (Addison–Wesley).

Brooks, R. A. (1991). Intelligence without representation. Artificial Intelligence, **47**, 139–59

Calvin, W.H. (1996). The cerebral code. Cambridge, MA: MIT Press.

Chomsky, N. (1986). Knowledge of language. New York: Praeger.

Churchland, P., Sejnowski, T.,& Ramachandran, V.S. (1995). A critique of pure vision. In C. Koch & J.L. Davis (eds.), Large–scale neuronal theories of the brain. Cambridge, MA: MIT Press.

Clark, A. (1997). Being there: Putting brain, body, and the world together again. Cambridge, MA: MIT Press.

Crick, F., & Koch, C. (1990) Toward a neurobiological theory of consciousness. Sem. Neurosci., **2**, 263–275.

Damasio, A.R., & Damasio, H. (1994). Cortical systems for retrieval of contents knowledge: The convergence zones framework. In C. Koch & J.L. Davis (eds.), Large–scale neuronal theories of the brain. Cambridge, MA: MIT Press, pp 201–238.

Fodor J. A., & Pylyshyn, Z.W. (1988). Connectionism and cognitive architecture: A critical analysis. Cognition, **28**, 3-71.

Gibson, J.J. (1966). The senses considered as perceptual systems. Boston, MA: Houghton Mifflin.

Gray, C. M., König, A., Engel, A. K., & Singer, W. (1989). Oscillatory responses in cat visual cortex exhibit inter–columnar'synchronization which reflects global stimulus properties. Nature, **338**, 334–337.

Greenfield, P. (1991). Language, tool and brain: The ontogeny and phylogeny of hierarchically organized sequential behavior. Behavioral and Brain Sciences, **14**, 531–595.

Grossberg, S. (in press). The complementary brain: A unifying view of brain specialization and modularity. Trends in Cognitive Sciences.

Hadley, R.F. (1994). Systematicity in connectionist language learning. Mind and Language, **9**, 247-271.

Haugeland, J. (1991). Representational genera. In W. Ramsey et al. (Eds.), Philosophy and connectionist theory. Hillsdale, NJ: Erlbaum.

Hebb, D.O. (1948) The organization of behavior. New York: Basic Books.

Iverson, J.M., & Thelen, E. (1999). Hand, mouth, and brain: The dynamic emergence of speech and gesture. Journal of Consciousness Studies, 6, 19–40.

Kaas, J.H. (1987). The organization of neocortex in mammals: Implications for a theory of brain function. Annual Review of Psychology, 38, 124–151.

Kelso, J.A. S. (1997). Dynamic patterns. Cambridge, MA: MIT Press.

O'Regan, J.K., Noe, A. (in press). A sensorimotor account of vision and vision consciousness. Behavioral and Brain Sciences.

Passingham, R.E.(1982). The human primate. San Francisco: W.H. Freeman.

Port, R. & vanGelder, T. (Eds.) (1995). Mind as motion. Cambridge, MA: MIT Press.

Preuss, T. M. (1993). The role of the neurosciences in primate evolutionary biology. Historical commentary and prospectus. In R.D.E. McPhee (Ed.), Primates and their relatives in phylogenetic perspective. New York: Plenum Press, pp. 333–362.

Pulvermüller, F. (1999). Words in the brains language. Behavioral and Brain Sciences, 22, 253–336.

Reilly, R.G. (in press). On the relationship between object assembly and language production: A connectionist simulation of Greenfield's hypothesis. Behavioral and Brain Sciences.

Reilly, R.G.(1995). Sandy ideas and coloured days: The computational implications of embodiment. Artificial Intelligence Review, 9, 1–18.

Restum, W. & Sobota, W.L. (1983). Gerstmann's syndrome: A new perspective to some controversial issues. Archives of Physical Medicine and Rehabilitation, 64, 499.

Roth, G. (2000). Is the human brain unique. Paper presented to Mirror Neurons and the Evolution of Brain and Language, Hanse Institute for Advanced Study, Delmenhorst, Germany.

Sharkey, A. (1999). Combining artificial neural networks: Ensemble and modular multi-net systems. London: Springer–Verlag.

Simon, H. (1969). The architecture of complexity. In H. Simon (Ed.), The sciences of the artificial. Cambridge, UK: Cambridge University Press.

Singer, W. (1994). Synchronization of cortical activity and its putative role in information processing and learning. In C. Koch & J.L. Davis (eds.), Large–scale neuronal theories of the brain. Cambridge, MA: MIT Press, pp 201–238.

Thatcher, R.W. (1992). Cyclic cortical reorganization during early childhood. Brain and Cognition, 20, 24-50.

Ungerleider, L.G., & Mishkin, M. (1982). Two cortical visual system. In J. Ingle, M.A. Goodale, & Mansfield, R.J.W. (Eds.), Analysis of visual behavior. Cambridge, MA: MIT Press, pp. 549–586.

Van Essen, D.C., & Anderson, C.H. (1990). Information processing strategies and pathways in the primate retina and visual cortex. In S.F. Zornetzer, J.L. Davis, & C. Lau (Eds.), An introduction to neural and electronic networks. New York: Academic Press, pp. 43–72.

On the Influence of Threshold Variability in a Mean-Field Model of the Visual Cortex

Hauke Bartsch, Martin Stetter, and Klaus Obermayer

Technische Universität Berlin, Germany,
hauke@cs.tu-berlin.de,
http://ni.cs.tu-berlin.de/
Technische Universität Berlin, Fachbereich 13 - Informatik Sekretariat FR 2-1,
Franklinstr. 28/29, D-10587 Berlin, Germany

Abstract. Orientation–selective neurons in monkeys and cats show contrast saturation and contrast–invariant orientation tuning (Albrecht and Hamilton, 1982). Recently proposed models for orientation selectivity predict contrast invariant orientation tuning but no contrast saturation at high strength of recurrent intracortical coupling, whereas at lower coupling strengths the contrast response saturates but the tuning widths are contrast dependent (Hansel and Sompolinsky, 1997; Bartsch, Stetter and Obermayer, 1997). In the present work we address the question, if and under which conditions the incorporation of a stochastic distribution of activation thresholds of cortical neurons leads to the saturation of the contrast response curve as a network effect. We find that contrast saturation occurs naturally if two different classes of inhibitory inter-neurons are combined. Low threshold inhibition keeps the gain of the cortical amplification finite, whereas high threshold inhibition causes contrast saturation.

1 Introduction

One of the most prominent features of neurons in the primary visual cortex of cats and monkeys is their preference for oriented stimuli within their classical receptive field (Hubel and Wiesel, 1962). Their tuning widths are weakly dependent or even independent of the contrast of the applied stimulus (Sclar and Freeman, 1982). Also, for many but not all cells the contrast–response function saturates well below 100% contrast (Albrecht and Hamilton, 1982). Saturation of these neurons is stronger than for their geniculate input neurons. Different models have been carried out to explain these results by intracortical circuitry (Todorov, Siapas, Somers and Nelson., 1997), but it often remains unclear which property of the underlying model assumptions is responsible for a given effect. In this work we address the question, if and under which conditions the inclusion of a variability of the activation thresholds can stabilize contrast saturation also for high intracortical excitation. For this we consider a mean–field model which describes activations of neurons with different activation threshold distributions. We start with the derivation of a mean-field description for the dynamics of a neuron population that, in the limit of low mean activation, describes a population of spiking neurons.

S. Wermter et al. (Eds.): Emergent Neural Computational Architectures, LNAI 2036, pp. 174–187, 2001.

2 Model

2.1 Modelling the Dynamics of a Neuron Population

In this section we write down a simplified description of neuronal population dynamics. We assume individual neurons as binary stochastic units, which can flip forth and back between an active (spiking) and inactive (non-spiking) state. The probability per unit time for an inactive neuron to be activated is denoted by the activation rate γ and its inactivation rate by δ (1a). Below, we will relate γ

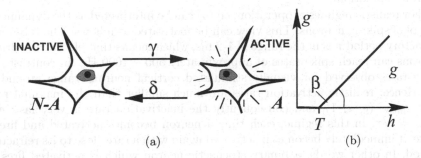

$$(a) \qquad\qquad (b)$$

Fig. 1. (a) Principle of binary stochastic neurons: The neuron randomly toggles forth and back between an active (right) and inactive (left) state. If the neuron is driven by excitatory input, the probability for activation increases and inactivation probability decreases resulting in a higher mean activity. **(b)** Semilinear activation function $g(h)$.

and δ to the mean synaptic input of the population: A high excitatory input will lead to a high activation rate and a low inactivation rate, which causes a higher fraction of neurons to be active. Low input or inhibitory input, in contrast will reduce the activation rate and the neurons settle down to the inactive state. If we consider a population of N binary stochastic neurons with identical activation and inactivation rates, the number of active neurons A changes within the small time interval Δt according to

$$\Delta A = \gamma \Delta t (N - A) - \delta \Delta t A \;=\; (\gamma N - (\gamma + \delta)A)\, \Delta t. \qquad (1)$$

In the limit $\Delta t \to 0$, this relation transforms to a rate equation for the fraction of neurons $m = A/N$ that are active at time t:

$$\frac{d}{dt} m = \gamma - (\gamma + \delta)m. \qquad (2)$$

Now we assume for simplicity, that the inactivation rate behaves inversely to the activation rate. This is reasonable because neurons which are strongly driven by input are less likely to stop firing spontaneously. If $\gamma_{\max}$ denotes the maximum possible activation rate, we arrive at $\delta = \gamma_{\max} - \gamma$ and

$$\frac{d}{dt} m = \gamma - \gamma_{\max} m. \qquad (3)$$

One possible interpretation of the maximum activation is related to the refractory period of biological neurons. For a neuron to undergo two subsequent activations, it must at least fire a spike and wait for the refractory period τ until it can be activated to fire the next spike. Hence, we can identify the maximum activation rate with $\gamma_{max} = 1/\tau$. The rate equation for the population activity becomes

$$\tau \frac{d}{dt}m = -m + \tau\gamma =: -m + g,$$ (4)

where $0 \leq g = \tau\gamma \leq 1$ denotes the activation probability (relative to τ) for the neuron population.

For realistic regimes of operation, eq (4) can be interpreted as the dynamics of a pool of spiking neurons. This view can be motivated as follows: The (absolute) refractory period τ is in the range of 1-2 ms, which means that electrically driven neurons can reach spike rates of approximately $500 - 1000$ Hz. In contrast, the spike rates observed for visually stimulated cortical neurons range around 50 Hz. Hence, realistic activation rates are much smaller than the maximal rate, $\gamma \ll \gamma_{max}$, ($g \ll 1$), and consequently the inactivation rate is very fast: $\delta \approx \gamma_{max} = 1/\tau$. In this regime, each time a neuron becomes activated and fires a spike, it immediately becomes inactivated again with a rate close to its refractory period. In other words, a binary stochastic neuron which is activated fires an individual spike and automatically inactivates again: In the limit of low mean activation, our mean-field model describes a population of spiking (instead of binary) neurons.

2.2   Mean–Field Model of an Orientation Hypercolumn

We take into account the fact that cortical tissue contains many different cell types and combine each of these cell types to a separate population. In general, an orientation column, indexed by its preferred orientation θ, contains N_e populations with different excitatory neuron types and N_i different populations of inhibitory neurons (figure 2b). The n-th excitatory population is indexed by (e, n) and the n-th inhibitory population by (e, n). We henceforth refer to the subpopulations as model neurons or simply "neurons".

The strength of recurrent intracortical couplings is assumed to depend only on the source and target orientation columns but not on the particular target neuron. The mean connection strength from neuron (α, n) within column θ' to neuron β, m in column θ ($\alpha, \beta = $ e,i) is given by

$$S^{m,n}_{\beta,\alpha}(\theta, \theta') \equiv S_\alpha(\theta - \theta').$$ (5)

We keep all properties of the neurons up to their activation functions identical for the present considerations and assume that the neurons differ only in their mean activation thresholds. The activity of neuron (α, n), $\alpha = e, i$ in response to synaptic input h is given by a semi-linear activation function

$$g_{\alpha,n}(h) = \max(\beta_\alpha(h - T_{\alpha,n}), 0),$$ (6)

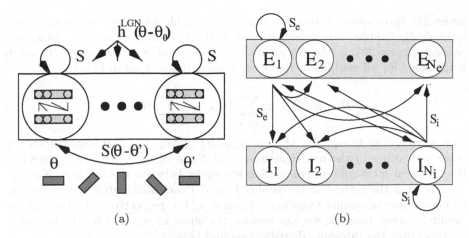

Fig. 2. (a) Setup of a mean-field hypercolumn with many different cell types. Recurrent couplings depend on the source and target orientations only. **(b)** Structure of a single orientation column. It consists of populations of N_e excitatory and N_i inhibitory cell types with in general different properties.

where β_α denotes its slope and $T_{\alpha,n}$ its activation threshold (c.f. figure 1b). The activities of neurons (e, n) and (i, n) in column θ, $m_{e,n}(\theta, t)$ and $m_{i,n}(\theta, t)$, evolve according to

$$\frac{d}{dt} m_{\alpha,n}(\theta, t) = -m_{\alpha,n}(\theta, t) + g_{\alpha,n} \left(h^{\text{lat}}(\theta, t) + h^{\text{LGN}}(\theta, t) \right) \tag{7}$$

$$h^{\text{lat}}(\theta, t) = \sum_{\beta=e,i} \sum_n \int_{-\pi/2}^{\pi/2} d\theta' S_\alpha(\theta - \theta') m_{\beta,n}(\theta', t) \tag{8}$$

$$h^{\text{LGN}}(\theta - \theta_0) = c(1 - \varepsilon + \varepsilon \cos(2(\theta - \theta_0))). \tag{9}$$

Note that h^{LGN} and h^{lat} are identical for all subpopulations.

Because intracortical couplings often connect cell populations with similar orientation preference (Kisvarday, Kim, Eysel and Bonhoeffer, 1994) we describe the connection patterns by Gaussian functions in orientation space,

$$S_\alpha(\theta - \theta') = \frac{S_\alpha}{C_\alpha} \exp \frac{(\theta - \theta')^2}{2\sigma_{S\alpha}^2} \tag{10}$$

where C_α is a normalization constant, S_α is the integral over the coupling strengths, and $\sigma_{S\alpha}$ is the variance of the coupling. Until further notice we assume $\varepsilon = 0.5$ and a balanced excitation and inhibition, i.e., $S_\alpha = S$, $\alpha = e, i$.

3   Analytical Treatment of Contrast Saturation

We wish to understand how the contrast-response curve of the orientation column – or a representative subpopulation therein – depends on the distribution of activation thresholds. To keep the analysis tractable we analyze an isolated but intrinsically coupled orientation column with N_e excitatory and N_i inhibitory neurons (figure 2b). In the stationary state, the total synaptic input, H, which is the same for all neurons in the orientation column, is given by

$$H = h^{\text{LGN}} + S_e \sum_{n=1}^{N_e} M_{e,n} - S_i \sum_{n=1}^{N_i} M_{i,n}, \tag{11}$$

where, according to eq. (6), $M_{\alpha,n} = g_{\alpha,n}(H) \equiv M_{\alpha,n}(T_{\alpha,n}, H)$ are the steady state activations of the model neurons and $S_\alpha \equiv S_\alpha((\theta - \theta') = 0)$ abbreviate the identical intra-column connection strengths between the neurons. Now we assume, that the activation thresholds T_e and T_i are distributed over the orientation column according to pdfs $p_e(T_e)$ and $p_i(T_i)$, respectively. In the limit of infinitely many neurons, we can replace the sums in eq. (11) by the ensemble averages over the threshold distributions and obtain

$$H = h^{\text{LGN}} + S_e \int_{-\infty}^{\infty} M_e(T_e, H)p_e(T_e)\, dT_e - S_i \int_{-\infty}^{\infty} M_i(T_i, H)p_i(T_i)\, dT_i. \tag{12}$$

Because of the definition of the semi-linear transfer function eq. (6), we know that neurons with $T_\alpha \geq H$ are silent and therefore do not contribute to the sums or integrals in eqs. (11) and (12). Conversely, for $T_\alpha < H$, the activation function can be replaced by its linear part, $M_\alpha(T_\alpha, H) = \beta_\alpha(H - T_\alpha)$. Therefore we can replace the upper limits of the integrals in equation 12 by H:

$$H = h^{\text{LGN}} + \beta_e S_e \int_{-\infty}^{H} (H - T_e)p_e(T_e)\, dT_e - \beta_i S_i \int_{-\infty}^{H} (H - T_i)p_i(T_i)\, dT_i. \tag{13}$$

Equation (13) represents a self-consistent relation between the total synaptic input H and the afferent input h^{LGN}. By solving this equation, we can write down an analytical solution for the stationary activation

$$M_e(T_e, H) \equiv M_e(T_e, H(h^{\text{LGN}})) = M_e(T_e, h^{\text{LGN}}) \tag{14}$$

as a function of the *external* instead of the total synaptic input, which is the contrast-response function of the neurons. Carrying out the integrals in eq. (13) yields

$$H = h^{\text{LGN}} + S_e\beta_e G_e(H) - S_i\beta_i G_i(H) \tag{15}$$

$$G_\alpha(H) = \int_{-\infty}^{H} dH' \int_{-\infty}^{H'} dT p_\alpha(T), \quad \frac{d^2}{dT^2}G_\alpha(T) = p_\alpha(T), \quad \alpha = e, i. \tag{16}$$

By defining the function

$$F(H) = H - \beta_e S_e G_e(H) + \beta_i S_i G_i(H) \tag{17}$$

equation 15 reduces to $F(H) = h^{\text{LGN}}$ and we can express the steady state activations M_α by

$$M_\alpha(T_\alpha, h^{\text{LGN}}) = \beta_\alpha (H - T_\alpha) = \beta_\alpha \left(F^{-1}\left(h^{\text{LGN}}\right) - T_\alpha\right) \tag{18}$$

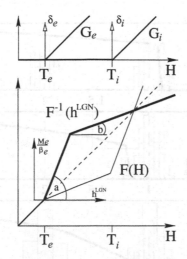

Fig. 3. Analytical solution eq. (18) for one isolated orientation column and δ-peaked threshold distributions. **Top**: The distributions and the resulting second integrals $G_\alpha(H)$. **Bottom**: The function $F(H)$ eq. (17) (thin line) and its inverse (thick line) as resulting from the scenario in the top part. The thick line relative to the small coordinate system schematically illustrates the behavior of the contrast response function.

Equation (18) provides an analytical relationship between geniculate input and the response of the recurrent cortical circuit. Note that it only holds for one isolated orientation column and if F is invertible. The latter condition corresponds to the boundary condition for the linear phase.

Figure 3 illustrates the meaning of eq. (18) for the special case of only one excitatory and one inhibitory neuron type and only two threshold values T_e and T_i. In this case, the two threshold distributions reduce to Kronecker delta functions around the two thresholds, $p_e(T) = \delta(T - T_e)$ and $p_i(T) = \delta(T - T_i)$ and their second integrals become semilinear functions. $G_\alpha(H) = \max(H - T_\alpha, 0)$ (figure 3 top). The function $F(H)$ (Eq. 17) becomes

$$F(H) = \begin{cases} H & H \le T_e \\ H - \beta_e S_e(H - T_e) & T_e < H \le T_i \\ H - \beta_e S_e(H - T_e) + \beta_i S_i(H - T_i) & H > T_i \end{cases} \qquad (19)$$

In this scenario, the resulting contrast response function eq. (18) shows saturating behavior. The gradients of $F^{-1}(h^{\text{LGN}})$ are $a = (1 - \beta_e S_e)^{-1}$, where $a\beta_e$ is the initial contrast gain of the contrast-response function, and $b = (1 - \beta_e S_e + \beta_i S_i)^{-1}$ for higher contrast levels.

3.1   Numerical Simulations of Contrast Response

For the following simulations we assumed threshold distributions, which for excitatory neurons are Gaussian, $p_e(T_e) = \mathcal{N}(\mu_e, \delta_e)$, and for inhibitory neurons

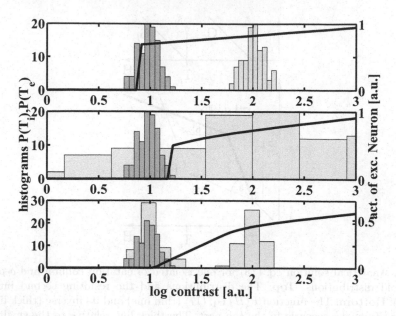

Fig. 4. Simulation for a set of 400 coupled model neurons (200 exc., 200 inh.) in the marginal phase ($S_e = S_i = 6$). **Solid lines:** Contrast-response curves; **dark gray:** Histograms of excitatory thresholds $p_e(T_e) = \mathcal{N}(1, 0.1)$; **light gray:** Histograms of inhibitory thresholds $p_i(T_i)$. **Top:** $p_i(T_i)$ unimodal, small variance ($\mathcal{N}(2, 0.1)$); **Middle:** $p_i(T_i) = \mathcal{N}(2, 1)$ unimodal, large variance. **Bottom:** A bimodal distribution of $p_i(T_i)$ is used ($\mu_{i,1} = 1, \mu_{i,2} = 2, \delta_{i,1} = \delta_{i,2} = 0.1$). A bimodal distribution p_i is necessary and sufficient for graded contrast-response and contrast saturation also in the marginal phase.

are either also Gaussian, $p_i(T_i) = \mathcal{N}(\mu_i, \delta_i)$) or bimodal according to two superimposed Gaussian functions $p_i = 0.5(\mathcal{N}(\mu_{i,1}, \delta_{i,1}) + \mathcal{N}(\mu_{i,2}, \delta_{i,2}))$. Inhibitory mean activation thresholds are set to be higher ($\mu_i = 2$) than excitatory mean activation thresholds ($\mu_e = 1$). Also, simulations will use $\beta_e = 0.5$ and $\beta_i = 1$, but the special choice of parameters does not strongly influence the results.

Figure 5 compares the numerical solution of the differential equation eq. (7) (solid line) with the analytical expression eq. (18) (circles) for two unimodal and fairly narrow threshold distributions (histograms) in the linear phase. It demonstrates that the analytical solution approximates the solution of the differential equation very well. The dashed and dash-dotted lines plot G_e and G_i for the distributions used. The behavior of this system can be understood as follows: First, only excitatory neurons are active and, because we operate in the linear phase, act as linear amplifiers. For higher contrast levels, more and more inhibitors become active and reduce the contrast gain. Different from the case of only two thresholds, the contrast-response curve gradually changes its gain over contrast. A gradual contrast saturation can be qualitatively understood as follows: With increasing afferent input h^{LGN}, more and more inhibitory neuron subpopulations are recruited (become active): The increase in number is pro-

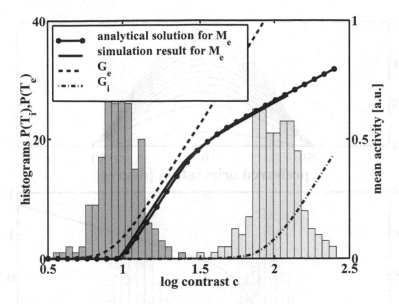

Fig. 5. Simulation of a contrast-response curve for a set of 400 coupled model neurons (200 exc., 200 inh.) in the linear phase ($S_e = 1$, $S_i = 1$) and for unimodal Gaussian threshold distributions $p_e(T_e) = \mathcal{N}(1, 0.1)$, $p_i(T_i) = \mathcal{N}(2, 0.1)$ (cf. threshold histograms in the diagram). **Solid line:** Numerical solution of the differential equation eq. (7). **Circles:** Evaluation of the analytical expression (18). Both curves agree very well. Dashed and dash-dotted lines show G_e and G_i, respectively.

portional to $p_i(F^{-1}(h^{\mathrm{LGN}}))$. The more neurons are recruited, the stronger the decrease in contrast gain. In other words, we expect a relationship between the second derivative of the contrast-response function at h^{LGN} and the density of neurons with activation thresholds $T_i = F^{-1}(h^{\mathrm{LGN}})$.

This relationship can be quantified by by forming the 2nd derivative of the steady state activation eq. (18) with respect to the LGN input. We arrive at the following relationship between the curvature of the contrast-response function and the distributions of activation thresholds $p_\alpha(T_\alpha)$:

$$\frac{d^2}{d(h^{\mathrm{LGN}})^2} M_e = \beta_e \frac{S_e\beta_e p_e(H) - S_i\beta_i p_i(H)}{(-1 + S_e\beta_e G'_e(H) - S_i\beta_i G'_i(H))^3}, \quad (20)$$
$$H = F^{-1}(h^{\mathrm{LGN}})$$

The denominator of eq. (20) is positive in the linear phase, because the gain of F has to be finite (invertibility of F). The contrast-response curve shows a negative curvature or saturation, if more inhibitory than excitatory neurons are recruited by a small increase in the input, i.e. if $S_e\beta_e p_e(H) < S_i\beta_i p_i(H)$ holds. Otherwise, the contrast-response function increases its gain.

Besides a quantitative understanding of the structural origin of contrast gain in the linear phase, it might be even more important to see, whether a gradually increasing and finally saturating contrast-response can also be stabilized in the

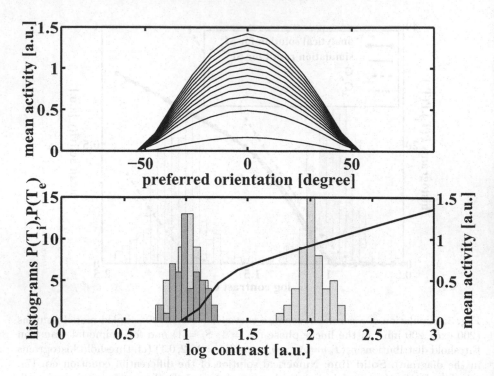

Fig. 6. Top Orientation tuning curve and **Bottom** contrast-response curve of an excitatory neuron with preferred orientation $0°$ for a hypercolumn with 21 orientation columns (50 excitatory neurons and 100 inhibitory neurons each) in the marginal phase. The system shows a graded and saturating contrast response, which is combined with a contrast-invariant orientation tuning width. Parameters: $S_e = 6, S_i = -6, \sigma_e = 34$ deg, $\sigma_i = \infty$, $\delta_e = \delta_{i,1} = \delta_{i,2} = 0.1, \mu_e = \mu_{i,1} = 1, \mu_{i,2} = 2$.

marginal phase by some threshold distribution. If this could be achieved, we would succeed in formulating necessary conditions for cortical circuitry to show a constant orientation tuning and contrast saturation for a single parameter setting.

Figure 4 shows the contrast-response curve of an excitatory neuron with threshold $T_e = 1$ for different cases of the inhibitory threshold distribution in the marginal phase. If the threshold-distribution is small and unimodal (top), the contrast-response shows a pseudo-binary switch-on behavior as observed in the marginal phase with two neuron types. This behavior remains stable, as long as the distribution is unimodal, even if it is very wide (figure 4, middle). However, as soon as the threshold distribution becomes bimodal (figure 4, bottom), the contrast-response first increases from zero and later saturates, as observed in biology. This demonstrates that two inhibitory neuron populations, one with low and the other with higher activation threshold, are necessary and sufficient to stabilize contrast saturation in the marginal phase.

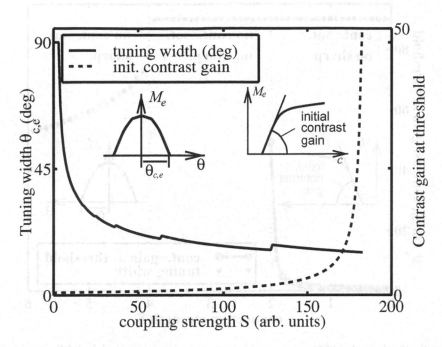

Fig. 7. The behavior of the contrast gain at activation threshold (solid line) and the orientation tuning width (crosses) as a function of the connection strength $S \equiv E_0 = E_2 = I_0$ $(I_2 = 0)$ for a hypercolumn with one excitatory and two inhibitory neuron populations. Other parameters were: $\beta_e = 0.5, \beta_i = 1, T_e = 1, T_{i1} = 1, T_{i2} = 2, \varepsilon = 0.01$, contrast for the orientation tuning width: $c = 2.0$. Insets illustrate criteria used for calculation of the curves. There is a wide range $(4 \leq S \leq 180)$, over which the linear and marginal phase coincide. Steps in the solid line are finite-size effect.

4   Orientation and Contrast Response with Three Neuron Types

Now we are ready to combine many structured orientation columns as analyzed previously to a full hypercolumn. The orientation columns are mutually coupled by lateral connections with Gaussian profiles as specified in eq. (10), and are driven by weakly orientation biased input ($\varepsilon = 0.1$).

Figure 6 shows orientation tuning curves of the $m_{e,1}(\theta)$ excitatory populations of 21 orientation columns (top) and the contrast-response curves of a subset of 5 excitatory subpopulations of the $\theta = 0°$ column (bottom) for unimodal $p_e(T_e)$ and bimodal $p_i(T_i)$. Even though the system operates in the marginal phase, where orientation tuning is independent of contrast, the contrast-response curve shows expressed saturation at the same time. This behavior is independent of the detailed shape of the threshold distributions, as long as it is bimodal.

A phase-diagram determined by the initial contrast gain and the orientation sharpening (cf. figure 8) for a hypercolumn with one excitatory and two

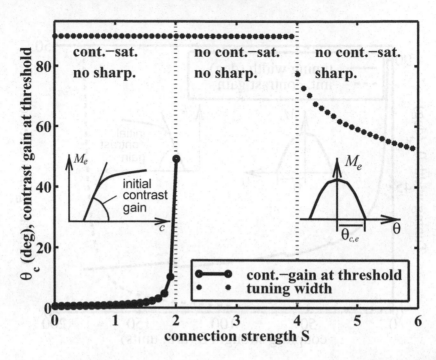

Fig. 8. The behavior of the contrast gain at activation threshold (solid line) and the orientation tuning width (crosses) as a function of the connection strength $S \equiv E_0 = E_2 = I_0$ ($I_2 = 0$) for a hypercolumn with cosine/flat connectivity. Vertical dotted lines mark the analytical results. Other parameters were: $\beta_e = 0.5, \beta_i = 1, T_e = 1, T_i = 2, \varepsilon = 0.01$, contrast for the orientation tuning width: $c = 2.0$. Insets illustrate criteria used for calculation of the curves. Both analytical and numerical results predict, that there is no overlapping regime of co-occurrence of contrast saturation and contrast-invariant orientation tuning.

inhibitory (low- and high-threshold) neuron types is plotted in Figure 7. Due to low-threshold inhibition, the linear phase with finite initial contrast gain is stabilized up to very strong recurrent excitation strengths ($S \approx 180$ compared to $S = 2$ for two-neuron hypercolumns), and there is a wide range of coupling strengths, in which orientation tuning is invariant and the contrast-response saturates. In summary, this finding predicts that the experimentally observed cortical response properties require essentially two functionally distinct inhibitory neuron types to be present: Inhibitors with a low activation threshold (or tonically active inhibitors) stabilize the contrast gain at near the contrast threshold to finite values, whereas inhibitors with high activation thresholds cause the saturation of the contrast-response curves at higher contrast levels.

In figure 6, both types of inhibitors were assumed to distribute lateral inhibition between different orientations. Possible candidates for such inhibitors are basket cells with axonal arborizations up to 1200 μm (Lund, 1987). However, it seems more reasonable to identify the two functionally different inhibitors

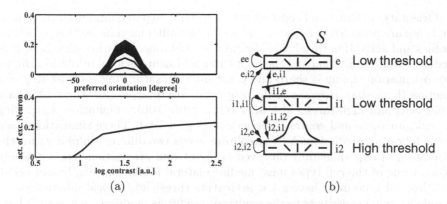

(a)   (b)

Fig. 9. (a) Orientation tuning curves (top) and the contrast-response function of the
zero-deg. orientation column (bottom). (b) Schematic illustration of the corresponding
wiring scheme: low-threshold lateral inhibitors (e.g. basket neurons) and high-threshold
local inhibitors (e.g. chandelier cells). Parameters: $S_e = S_{i1} = S_{i2} = 50$; (marginal
phase) (b) $\sigma_e = 34$ deg; $(\sigma_{i1}, \sigma_{i2}) = (\infty, 34)$ deg. $T_e = T_{i1} = 1; T_{i2} = 1.5$. The
hypercolumn properly operates as in figure 6.

with two anatomically distinguishable biological neuron types. However, many
inhibitors apart from basket cells are local companions, which contact only postsynaptic neurons within the same or closely adjacent orientation columns. One
important local inhibitor is the chandelier cell. Therefore, we may ask, under
which conditions a hypercolumn with pyramidal neurons as the excitatory population, basket cells as lateral inhibitors and chandelier cells as local inhibitors,
still show the behavior seen in figure 6.

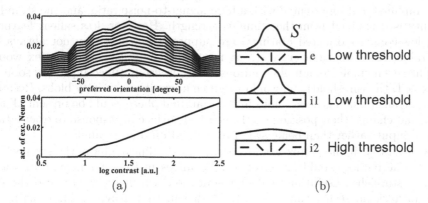

(a)   (b)

Fig. 10. (a) Orientation tuning curves (top) and the contrast-response function of the
zero-deg. orientation column (bottom) for reverse properties of the inhibitory neurons:
(b) low-threshold local inhibitors and high-threshold lateral inhibitors. Parameters:
$S_e = S_{i1} = S_{i2} = 50$; (marginal phase) (b) $\sigma_e = 34$ deg; $(\sigma_{i1}, \sigma_{i2}) = (34, \infty)$ deg.
$T_e = T_{i1} = 1; T_{i2} = 1.5$.

Orientation tuning and contrast response for a hypercolumn with three neuron types are provided in figures 9 and 10 for two different combinations of wiring profiles and activation thresholds of inhibitory neurons. If the low-threshold cells mediate lateral inhibition and the high-threshold neurons local inhibition (figure 9b), orientation tuning is sharp and constant with saturating contrast response function (figure 9a). However, if the properties are reversed (low-threshold chandelier cells and high-threshold basket cells, figure 10b), orientation sharpening is weak, unstable and contrast-dependent (figure 10a). These simulations lead to the following prediction: A hypercolumn needs two different inhibitors for the generation of experimentally observed contrast and orientation representation. At least one of the cell types must mediate lateral inhibition (e.g. basket cells), and this cell type must have a low activation threshold. If local inhibitors (e.g. chandelier cells) contribute to the recurrent circuit as modeled, they should have a high activation threshold.

5   Conclusions

Because individual hypercolumns within the model show a very different behavior depending on the parameter regime (either linear or marginal), it is important to determine, which phase, if any, might be implemented in the primary visual cortex. Unfortunately, this cannot be tested directly, because the phase boundaries provided above cannot be translated directly into biologically accessible quantities, and also because the total strength between biological neuron populations cannot be easily measured. However, in both the linear and the marginal phase there are model predictions which can help testing whether biological brain states can be properly described by a mean-field model in one or the other regime. Both regimes have some advantages and drawbacks.

In the marginal phase, salient features (e.g. high contrast oriented gratings) are amplified and represented with a high signal-to-noise ratio, and the stimulus quality is decoupled from the stimulus strength. However, less salient features (e.g low-contrast oriented gratings) are suppressed, which may not always be desirable. Also, for sufficiently strong recurrent connections, the cortex would amplify even small random fluctuations in otherwise untuned input (as soon as it exceeds the threshold) and would produce nonuniform activity blobs. Because neural transmission is noisy, blobs in the marginal phase would be present all the time and change their position on the cortical surface in response to orientation biased input rather than being switched on and off due to input.

Conversely, if the cortex would operate in the linear phase, the strength of cortical activation would be sensitive to changes in the afferent activities and would establish a useful internal representation, which represents also the details of a visual scene. Simulations have shown, that a hypercolumn acting in the linear phase (as opposed to the marginal phase) seems better capable of representing more than one stimulus-orientation simultaneously and showing cross-stimulation effects (Stetter, Bartsch and Obermayer, 2000). Recently, fast synaptic depression of intracortical connections has been proposed as a mechanism to combine the advantages of both phases. Immediately after a saccade, the

- Development produces a huge variety of neural systems re-using a hierarchy of compactly encoded programmes. Variety is also generated based on self-organising interactions between neural components.
- Developmental programmes are robust to genetic errors and variations in the developmental environment, allowing non-trivial ANNs to be produced without the need for an exact codification.
- Complex developmental programmes involving multiple variables and relations can be controlled by only a few genes (or parameters).

The long-term aim of this work is to develop computational neural systems that are more dynamic and adaptive. We would like a generic set of developmental mechanisms capable of creating neural systems with a wide variety of different architectures and functionalities, not limited to stereotypical features.

This chapter briefly reviews our work to date. It begins by introducing the developmental environment which we have implemented, giving examples of the variety of potential neuron morphologies that it can produce. By combining the developmental simulator with an evolutionary algorithm, experiments to evolve network connectivity and individual neuron morphologies have also been investigated. These experiments are summarised before future work is discussed.

2   Developmental Modelling

The modelling of neural development can be approached in many different ways. One approach, where the aim is to understand the principles of development, requires highly detailed modelling of molecular and bio-physical processes [4, 5]. Abstract simulations on the other hand seek to capture mechanisms of neural development using compact mathematical descriptions [6, 7, 8]. The goal of the simulation environment reported here is a combination of these 2 approaches, namely to extend the minimal mathematical approaches with further developmental mechanisms whilst retaining a model which is adaptable and robust.

We have been implementing a 3 dimensional simulation of biological development, in which neuron-to-neuron connectivity is created through interactive self-organisation [9, 10]. Development occurs in a number of overlapping stages, which govern how neurons extend axons and dendrites (collectively termed neurites). Neurons grow within an artificial, embryonic environment, into which they and their neurites emit local chemical gradients. The growth of neurites is influenced by these local gradients and the following sets of interacting, developmental rules [11]:

- **Interactive growth rules** enable growing neurites to branch in response to gradient conditions within their local developmental environment.
- Interspersed with growth, **spontaneous neural activity** dynamically regulates the growth rate of neurites.
- The cumulative effects of activity are used by **pruning rules** to remove ineffective neurons and synapses.

Mathematical descriptions of the interactive growth rules can be found in the Appendix. These selected descriptions illustrate how the developmental rules are conceived and designed. The complete set of mathematical descriptions, including the activity-driven and pruning rules, can be found in [11].

2.1   Examples of Neuron Morphologies

The developmental rules are controlled by parameters, in the same way as gene expression levels can be thought of as parameters for biological development. By varying these parameters, a variety of neuron and network morphologies can be achieved. Example morphologies are illustrated in Figure 1.

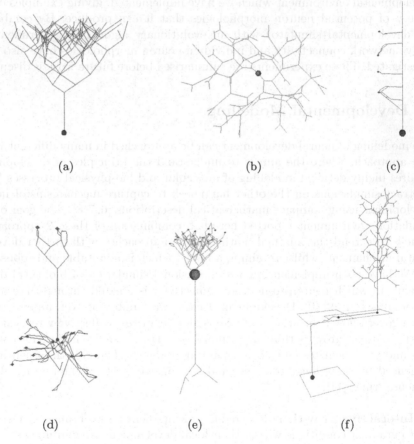

(a)   (b)   (c)

(d)   (e)   (f)

Fig. 1. Examples of 3 dimensional neurons grown using the developmental simulator. Somas are represented as spheres and synapses as smaller polygons. (a) to (c) are examples of neurons grown in one single phase. The morphologies of (d) to (f) result from combining multiple phases of growth.

Figures 1(a) and (b) are examples of morphologies where the frequencies of branching are different (rule 13). Neurite branching in (b) occurs with regular frequency whilst in (a) the initial branch occurs approximately midway through growth followed by increasingly frequent branching. An instance of multiple initial dendrites (rule 7) is further illustrated in Figure 1(b). Branching based on selected directions (rule 14d) is demonstrated in Figure 1(c), which seeks to mimic axon branching as observed in biological systems [12].

The creation of complex neural morphology does not, however, have to be the result of a single phase of growth. Figures 1(d) to (f) demonstrate how different phases of growth, subject to different developmental parameters, can be combined to create neural morphologies. For example, (e) is the combination of a lower phase of long-range neurite navigation followed by low frequency branching, together with an upper phase of short-range navigation with high frequency branching. The combination of 4 growth phases was used to create the morphology in Figure 1(f), which mimics the process of navigation via guidepost neurons [13].

2.2 Benefits of Developmental Self-Organisation

Developmental self-organisation provides the simulator with a number of attributes beneficial for neural system design. These include:

- **Self-Regulation.** Neurons are able to self-regulate their levels of innervation by interacting with in-growing neurites. In this way neurons can adaptively determine both the number of neurons to which they connect and the number of connections made. Self-regulating growth also enables neurons to modify their growth rates without an external stopping criteria having to be imposed.
- **Robustness.** Neurons and networks are robust to noise during development, e.g. variations in the positions of neurons and parameter values. Networks with the same connectivity patterns can be achieved using sets of similar parameters. Network design therefore does not rely upon identifying a single, specific set of parameters which works under only one set of initial conditions.
- **Error Correction.** Developing neurons can overgrow and make connections with neurons that are ultimately of little significance. It is not always possible to compensate for such eventualities in advance and thus pruning is able to correct for these wiring errors. (Overgrowth can be viewed as a positive feature since it enables systems to explore alternative configurations before pruning determines more optimal arrangements.)

3 Developmental Evolution

Evolving the developmental simulator for a specific network then becomes equivalent to the search for optimal sets of developmental parameters. A genetic algorithm (GA) is used as the search tool in our approach.

3.1   Evolution of Connectivity

Previously we have used a GA to evolve developmental programmes which lead to the creation of edge-detecting retinas. Various approaches to guide the evolutionary process have been used, such as specific target retina architectures [14,10] and a desired functionality [15]. An example of this work is shown in Figure 2.

In the absence of any initial neuron placement noise, constrained or intrinsic developmental rules alone can result in retina structures whose functional response (Figure 2(e)) matches the desired response (Figure 2(b)). However, if the positions of neurons are perturbed, representing real-world noise, the intrinsic rules fail to produce adequate functionality as illustrated in Figure 2(f).

The developmental programme was made more robust to noise by incorporating rules that permit growing neurons to produce extra branches via interactions with the local developmental environment. The addition of interactive overgrowth rules, results in a significant improvement in structure and hence functionality over intrinsic rules only (Figures 2(g) to (i)). The target response is far more distinguishable in Figure 2(i) compared to Figure 2(f). However, black and white pixels indicate that those particular bipolar neurons are saturated due to having too many connections.

A further improvement in performance is achieved by incorporating pruning rules with the intrinsic and overgrowth rules, to control the extent of overgrowth. With the addition of the pruning parameters, the effects of noise on the functionality of the retina are again reduced, as seen in Figure 2(l).

In these experiments evolution was carried out in stages, incorporating parameter values from previous stages and evolving them alongside new parameters. For example, the intrinsic rule parameters are evolved in the first stage and the values obtained are used to seed the subsequent stages of development including the interactive branching and pruning rules. In this way the evolutionary search space is increased in an orderly manner where evolution is channelled through developmental constraints.

This contrasts with other models where all the developmental rules are coevolved. Presenting such large, global search spaces can cause evolution to stall in the early generations [16]. If the search space is large then all the networks in a population may have the same fitness value. Evolution is therefore given no clear trajectory along which to progress. Other approaches to this problem have incrementally increased the complexity of the fitness function [17] or subdivided a fitness function into a series of sub-tasks [16]. However, we feel that this approach of staged developmental evolution is more easily generalised and more biologically plausible.

3.2   Evolution of Multi-compartmental Neurons

Single biological neurons, let alone networks and systems, exhibit highly dynamic, adaptive computational capabilities [18]. One of the key determinants of these capabilities is the relationship between a neuron's morphology and its function [19]. The form or geometry of a neuron directly influences how action

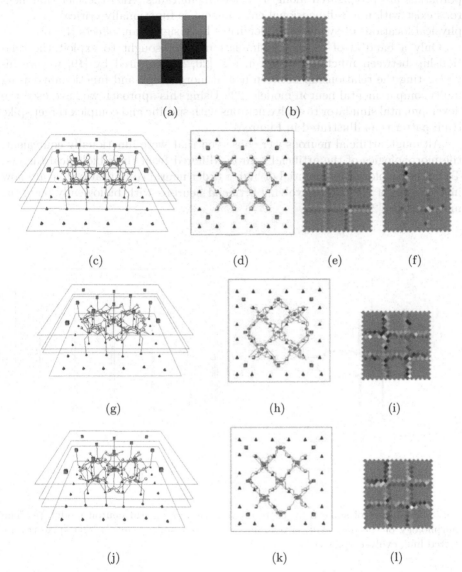

Fig. 2. Architectures and functionality of evolved retinas. (a) Input image. (b) Target output image. (c) and (d) are side and top views respectively of a portion of the best evolved retina using unperturbed, symmetric neuron placement and intrinsic rules only. (e) and (f) are outputs from unperturbed and perturbed retinas respectively. (g) and (h) are the side and top views respectively of a portion of the best evolved retina using perturbed neuron placement, evolving intrinsic and interactive branching rules. (i) Output image from the best evolved retina. (j) and (k) are the side and top views respectively of a portion of the best evolved retina using perturbed neuron placement, evolving intrinsic, interactive branching and pruning rules. (l) Output image from the best evolved retina.

potentials are propagated along its active membranes. Also the fact that neurons exist within a 3-dimensional environment is functionally critical, since the physical locations of synapses on dendrites have significant effects [13, 18].

Only a handful of computer simulations have sought to exploit the relationship between function and form, e.g. [20, 21]. Inspired by [19] we are investigating the relationship between neural morphology and functionality using multi-compartmental neuron models [22]. Using this approach we have used the developmental simulator to evolve neurons with specific and complex target spike train patterns as illustrated in Figure 3.

Although artificial neurons were observed that were functionally equivalent, the morphologies of the artificial neurons differed from their biological targets. To further explore the potential of higher order neural computation, we are now investigating a method of evolving artificial neurons with biologically-plausible morphologies [23].

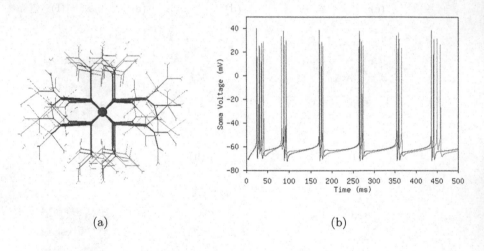

(a)   (b)

Fig. 3. Evolution of spike train patterns in a compartmental neuron model. (a) The morphology of the best evolved neuron. (b) Output spike trains: target spike train - dotted line, evolved spike train - solid line.

4   Discussion and Future Work

The focus of current work, the creation of neurons with biologically-defensible morphologies, is of interest from 2 perspectives. Firstly, can abstract rules develop neuron morphologies that can mimic the richness of neural form? Secondly, plausible neuron morphologies can be utilised on 2 levels of modelling abstraction. One is at the artificial neural network level to investigate more biologically-inspired processing in neurons. The other lies in the domain of computational neuroscience. Computational neuroscience uses highly-detailed models of neuron morphology together with fully, dynamic membrane models (such as GENESIS [24]) to simulate neural behaviour. These models are however computationally-intensive such that a methodology of providing synthetic neurons with reduced computational requirements, would be of benefit.

Future work will explore the following topics:

- **Neural Functionality.** Combining 3 dimensional morphology from the simulator and multi-compartmental models will enable more neural models to be investigated, beyond simple 'integrate-and-fire' models. Areas of potential interest include active dendrite models and the effects of the spatial arrangement of synapses on the firing patterns of neurons [20].
- **Libraries of Neurons.** Evolving classes of neurons with specific geometries and functionality will allow such classes to be simply called into the design process of neural systems. The classes can provide basic neuron primitives with which to instigate designs. Developmental self-organisation will enable the basic classes of neurons to adapt and fine-tune themselves to their environment. This provides a stepping-off point to investigate the evolution of modular neural systems where the morphology and connectivity of neurons within separate modules can be different.
- **Dynamic Architectures.** The developmental rules may be arbitrarily combined to build programmes of development with differing complexities. Rules may be continuously active or be activated in particular sequences. This permits the neural architectures on which the rules act to be dynamic and adaptive. For example, architectures could self-repair through neurons resprouting axons and dendrites to form new connections.

References

1. Fahlman S.E., Lebiere C. The cascade-correlation learning architecture. In Touretzky D.S., editor, *Advances in Neural Information Processing Systems 2*, pages 524–532, San Mateo, California, 1990. Morgan Kaufman.
2. Reed R. Pruning algorithms - a survey. *IEEE Transactions on Neural Networks*, 4(5):740–747, 1993.
3. Yao X. Evolving artificial neural networks. *Proceedings of the IEEE*, 87(9):1423–1447, 1999.
4. Fleischer K. *A Multiple-Mechanism Developmental Model for Defining Self-Organizing Geometric Structures*. PhD dissertation, California Institute of Technology, May 1995.
5. Hely T.A., Willshaw D.J. Short-term interactions between microtubules and actin filaments underlie long-term behaviour in neuronal growth cones. *Proceedings of the Royal Society of London: Series B*, 265:1801–1807, 1998.
6. Nolfi S., Parisi D. Growing neural networks. Technical Report PCIA-91-15, Institiute of Pyschology, Rome, December 1991.
7. Gruau F. *Neural Network Synthesis Using Cellular Encoding and the Genetic Algorithm*. Thesis, Ecole Normale Superieure de Lyon, January 1994.

196   A.G. Rust et al.

8. Burton B.P., Chow T.S., Duchowski A.T, Koh W., McCormick B.H. Exploring the brain forest. *Neurocomputing*, 26-27:971–980, 1999.
9. Rust A.G., Adams R., George S., Bolouri H. Activity-based pruning in developmental artificial neural networks. In Husbands P., Harvey I., editors, *Procs. of the 4th European Conference on Artificial Life (ECAL'97)*, pages 224–233, Cambrdige, MA, 1997. MIT Press.
10. Rust A.G. *Developmental Self-Organisation in Artificial Neural Networks*. PhD thesis, Dept. of Computer Science, University of Hertfordshire, July 1998.
11. Rust A.G., Adams R., Bolouri H. Developmental rules for the evolution of artificial neural systems. Technical Report 347, Department of Computer Science, University of Hertfordshire, May 2000.
12. Bastmeyer M., O'Leary D.D.M. Dynamics of target recognition by interstitial axon branching along developing cortical axons. *Journal Science*, 16(4):1450–1459, 1996.
13. Hall Z.W. *An Introduction to Molecular Neurobiology*. Sinauer Associates, Sunderland, MA, 1st edition, 1992.
14. Rust A.G., Adams R., George S., Bolouri H. Developmental evolution of an edge detecting retina. In Niklasson L., Boden M., Ziemke T., editors, *Procs. of the 8th International Conference on Artificial Neural Networks (ICANN'98)*, pages 561–566, London, 1998. Springer-Verlag.
15. Bolouri H., Adams R., George S., Rust A.G. Molecular self-organisation in a developmental model for the evolution of large-scale artificial neural networks. In Usui S., Omori T., editors, *Procs. of the International Conference on Neural Information Processing and Intelligent Information Systems (ICONIP'98)*, pages II 797–800, 1998.
16. Nolfi S. Evolving non-trivial behaviors on real robots: A garbage collecting robot. *Robotics and Autonomous Systems*, 22:187–198, 1997.
17. Harvey I., Husbands P., Cliff D. Seeing the light: Artificial evolution, real vision. In Cliff D., Husbands P., Meyer J-A., Wilson S., editors, *From Animals to Animats 3: Procs. of the 3rd International Conference on Simulation of Adaptive Behavior (SAB94)*, Cambridge, MA, 1994. MIT Press.
18. Koch C., Idan Segev I. *Methods in Neuronal Modeling: From Ions to Networks*. MIT Press, Cambridge, MA, 2nd edition, 1998.
19. Mainen Z.F., Sejnowski T.J. Influence of dendritic structure on firing pattern in model neocortical neurons. *Science*, 382:363–366, 1996.
20. Mel B.W. Information-processing in dendritic trees. *Neural Computation*, 6(6):1031–1085, 1994.
21. Graham B.P. The effects of intrinsic noise on pattern recognition in a model pyramidal cell. In *Procs. of the 9th International Conference on Artificial Neural Networks*, pages 1006–1011, London, 1999. IEE.
22. Rust A.G., Adams R. Developmental evolution of dendritic morphology in a multicompartmental neuron model. In *Procs. of the 9th International Conference on Artificial Neural Networks (ICANN'99)*, pages 383–388. London, 1999. IEE.
23. Rust A.G., Adams R., Bolouri H. Evolutionary neural topiary: Growing and sculpting artificial neurons to order. In Bedau M.A., McCaskill J.S., Packard N.H., Rasmussen S., editors, *Artificial Life VII: Procs. of the 7th Interntional Conference*, Cambridge, MA, 2000. MIT Press.
24. Bower J., Beeman D. *The Book of GENESIS: Exploring Realistic Neural Models with the GEnesis NEural SImulation System*. Springer-Verlag, New York, 2nd edition, 1998.

Appendix: Developmental Growth Rules

Descriptions of the complete set of developmental rules - growth, activity-driven events and pruning mechanisms - can be found in [11] [1].

Table A1. Initial developmental configuration rules.

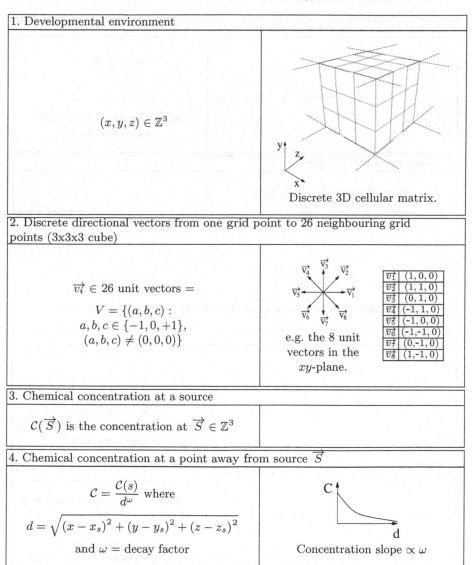

1. Developmental environment

$(x, y, z) \in \mathbb{Z}^3$

Discrete 3D cellular matrix.

2. Discrete directional vectors from one grid point to 26 neighbouring grid points (3x3x3 cube)

$\vec{v_i} \in 26$ unit vectors =

$V = \{(a, b, c) :$
$a, b, c \in \{-1, 0, +1\},$
$(a, b, c) \neq (0, 0, 0)\}$

e.g. the 8 unit vectors in the xy-plane.

$\vec{v_1}$	$(1, 0, 0)$
$\vec{v_2}$	$(1, 1, 0)$
$\vec{v_3}$	$(0, 1, 0)$
$\vec{v_4}$	$(-1, 1, 0)$
$\vec{v_5}$	$(-1, 0, 0)$
$\vec{v_6}$	$(-1, -1, 0)$
$\vec{v_7}$	$(0, -1, 0)$
$\vec{v_8}$	$(1, -1, 0)$

3. Chemical concentration at a source

$\mathcal{C}(\vec{S})$ is the concentration at $\vec{S} \in \mathbb{Z}^3$

4. Chemical concentration at a point away from source $\vec{S}$

$C = \dfrac{\mathcal{C}(s)}{d^\omega}$ where

$d = \sqrt{(x - x_s)^2 + (y - y_s)^2 + (z - z_s)^2}$

and ω = decay factor

Concentration slope $\propto \omega$

[1] The authors thank Neil Davey of UH for improving the readability of the mathematical descriptions of the growth rules.

Table A2. Initial developmenal configuration rules (continued).

5. Grid resolved directional concentration gradients	
Gradients in directions $\vec{v_i} = \nabla \mathcal{C} \cdot \vec{v_i}$	

6. Background concentration gradient	
$\vec{g_b} = (0, g_b, 0)$	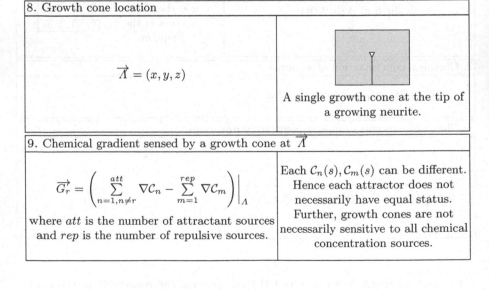 $\vec{g_b} = 0$ along x and z, and constant along y.

7. Number of initial dendrites	
$N_d = 2^\beta$ where $\beta \in \{0, 1, 2, 3\}$	

Table A3. Developmental rules that determine neurite navigation.

8. Growth cone location	
$\vec{\Lambda} = (x, y, z)$	A single growth cone at the tip of a growing neurite.

9. Chemical gradient sensed by a growth cone at $\vec{\Lambda}$	
$\vec{G_r} = \left(\overset{att}{\underset{n=1, n \neq r}{\sum}} \nabla \mathcal{C}_n - \overset{rep}{\underset{m=1}{\sum}} \nabla \mathcal{C}_m \right)\Bigg|_\Lambda$ where att is the number of attractant sources and rep is the number of repulsive sources.	Each $\mathcal{C}_n(s), \mathcal{C}_m(s)$ can be different. Hence each attractor does not necessarily have equal status. Further, growth cones are not necessarily sensitive to all chemical concentration sources.

Table A4. Developmental rules that determine neurite navigation (continued).

10. Growth cone gradient resolution	
$$g_m = \max_i \{\overrightarrow{G_r} \cdot \overrightarrow{v_i}\}$$ $$\text{for } i = 1 \text{ to } 26$$	

11. Growth cone movement	
	Growth cones hill-climb within the gradient landscape.
$$\overrightarrow{A}(t+1) = \overrightarrow{A}(t) + \overrightarrow{v_j}\,\Delta l$$ $$\text{where } \overrightarrow{v_j} = \begin{cases} \overrightarrow{v_m} & : \ (g_m - g_c) > \Gamma \\ \overrightarrow{v_c} & : \ (g_m - g_c) \le \Gamma \end{cases}$$ $\overrightarrow{v_m} = \overrightarrow{v_i} \in V$ s.t. $\overrightarrow{G_r} \cdot \overrightarrow{v_i} = g_m$, g_c is the concentration gradient in the current direction, $\overrightarrow{v_c}$ is the unit vector in the current direction, Γ is a turning threshold and $\Delta l \in \{1, \sqrt{2}, \sqrt{3}\}$	Γ low: growth cone instantly responds to gradient changes. Γ high: growth cone responds only when close to the concentration source. t $\qquad\qquad$ $t + \Delta t$ Neurites can extend at different rates, implemented using time delays.

Table A5. Developmental rules that control neurite branching.

12. Interactive branching	
$$\text{Prob branch} = P(\vec{\mathcal{A}}(t)) = \frac{g_m \times g_{m'}}{\tau + g_m{}^2}$$ where $g_{m'} = \max\limits_i \{\vec{G_r} \cdot \vec{v_i}, \vec{G_r} \cdot \vec{v_i} \neq g_m\}$ where g_m and $g_{m'}$ are the largest and second largest gradients respectively, and τ is a concentration gradient threshold.	t $t+1$

13. Intrinsic branching	
Branch times $\subset \mathbb{Z}^+$	Growth cones branch at specific times.

14. Branch direction selection:	
$\vec{v_j}$ and $\vec{v_k}$ are potential branch directions	
14a. Intrinsic Any $\vec{v_j}, \vec{v_k} \in V$	Two arbitrary directions
14b. Randomly selected orthognal pairs $\vec{v_j}, \vec{v_k} \in V$ s.t. $\vec{v_j} \cdot \vec{v_k} = 0$	
14c. Interactive $\vec{v_j} = \vec{v_m}$ and $\vec{v_k} = $ direction of $g_{m'}$ if $g_m \times g_{m'} > 0$	
14d. Along current direction of elongation and other selected direction $$\vec{v_j} = \vec{v_c}$$ $$\vec{v_k} \in V$$	t $t+1$

Table A6. Developmental rules that control neurite branching (continued).

15. Branch inhibition:	

15a Branch delay
 	
    if $(t - t_l) \leq Branchdelay$
 	
  where t_l = time of last branch
  then $branch\_allow := false$
  else $branch\_allow := true$	

15b Environmental
 	
    if $g_m > Saturation$
 	
  then $branch\_allow := false$
  else $branch\_allow := true$
	Inhibits branching when a neurite is close to a concentration source.

15c Branching angle
 	
    if $\vec{v_j} \cdot \vec{v_k} \geq cos\theta$
 	
  then $branch\_allow := false$
  else $branch\_allow := true$	

16. Concentration reduction upon elongation	
$$\mathcal{C}(\Lambda(t+1)) = \mathcal{R}_e\mathcal{C}(\Lambda(t))$$ where $0 \leq \mathcal{R}_e \leq 1$	 The growth cone of a neurite may produce a larger chemical gradient than its trailing neurite.
17. Concentration reduction upon branching	
$$\mathcal{C}(\Lambda_i(t+1)) = \mathcal{R}_b\mathcal{C}(\Lambda(t))$$ where $0 \leq \mathcal{R}_b \leq 1$ and $i = 1, 2$	 t $t + 1$

Table A7. Developmental rules that control neurite branching physiology and synaptogenesis.

18. Diameter change upon branching	
$$\Omega(t+1) = \mathcal{R}_d\Omega(t)$$ where $0 \leq \mathcal{R}_d \leq 1$ and Ω is the current neurite diameter	
19. Growth termination and synapse formation	
if $\vec{\Lambda} = \vec{\mathcal{S}_i}$ where i is an attractant then $growth := false$ $synapse := true$	When axons and dendrites encounter each other, or the somas of neurons, their growth is terminated.
20. New synapse value	
$\epsilon = 1$ where ϵ is synapse efficacy	
21. Concentration reduction at source upon synaptogenesis	
$$\mathcal{C}(s') := \mathcal{R}_s\mathcal{C}(s)$$ where $0 \leq \mathcal{R}_s \leq 1$	

The Complexity of the Brain: Structural, Functional, and Dynamic Modules

Péter Érdi and Tamás Kiss

Department of Biophysics KFKI Research Institute for Particle and Nuclear Physics
of the Hungarian Academy of Sciences P.O.Box 49, H-1525 Budapest, Hungary;
erdi@rmki.kfki.hu

Abstract. The 'complexity of the' brain is reviewed in terms of structural, functional and dynamic modules. Specifically, the modular concept is supplemented by using statistical approach for the dynamic behavior of large population of neurons. The relationship between anatomical cortical connectivity and (over)synchronized spatiotemporal activity propagation was studied by using large-scale brain simulation technique.

1 Introduction

Structural modular architectonics is an important concept of neural organization. Neuronal connectivity in a typical neural center is sufficiently specific to permit disassembly of the whole network into neuronal modules of characteristic internal connectivity; and the larger structure can be reconstituted by repetition of these modules.

Functional modules or schemas are behavioral units. Arbib [2] provided a neural theory of functional modules (schemas). It is typically a top-down theory, but it is time to talk about a partially new topic, what we call structure-based modeling of schemas [7].

Can we talk also about dynamic modules? Networks of neurons organized by excitatory and inhibitory synapses are structural units subject to time-dependent inputs and they also emit output signals. From a dynamical point of view, a single neural network may be considered sometimes as a pattern generating and/or pattern recognizing device. More often it is not a single network but a set of cooperating neural networks that forms the structural basis of pattern generation and recognition. (Pattern generating devices are functional elements of schemas.)

2 Complexity of the Brain: The Need of Integrative Approach

It is often said in colloquial sense that the brain is a prototype of complex systems. A few different notions of complexity may have more formally related to neural systems. First, structural complexity appears in the arborization of the

S. Wermter et al. (Eds.): Emergent Neural Computational Architectures, LNAI 2036, pp. 203–211, 2001.
© Springer-Verlag Berlin Heidelberg 2001

nerve terminals at single neuron level, and in the complexity of the graph structure at network level. Second, functional complexity is associated to the set of tasks performing by the neural system. Third, dynamic complexity can be identified with the different attractors of dynamic processes, such as point attractors, closed curves related to periodic orbits, and strange attractors expressing the presence of chaotic behaviour.

The understanding of neural organization requires the integration of structural, functional and dynamic approaches [3], [4]. Structural studies investigate both the precise details of axonal and dendritic branching patterns of single neurons, and also global neural circuits of large brain regions. Functional approaches start from behavioural data and provide a (functional) decomposition of the system. Neurodynamic system theory [6] offers a conceptual and mathematical framework for formulating both structure-driven bottom-up and function-driven top-down models.

3   Structural Modules

Experimental facts from anatomy, physiology, embryology, and psychophysics give evidence of highly ordered structure composed of 'building blocks' of repetitive structures in the vertebrate nervous system. The building block according to the modular architectonic principle is rather common in the nervous system. Modular architecture is a basic feature of the spinal cord, the brain stem reticular formation, the hypothalamus, the subcortical relay nuclei, the cerebellar and especially the cerebral cortex. After the anatomical demonstration of the so-called cortico-cortical columns it was suggested by Szentágothai [15] that the cerebral cortex might be considered on a large scale as a mosaic of vertical columns interconnected according to a pattern strictly specific to the species.

Szentágothai applied the modular architectonics principle to the cerebral cortex, linking observations of anatomical regularities to the observations of Mountcastle on physiological "columns" in somatosensory cortex and of Hubel and Wiesel on visual cortex. Other important anatomical regularities are the quasi-crystalline structure of the cerebellar cortex and the basic lamellar structure of the hippocampus.

Vernon Mountcastle, one of the pioneers of the columnar organization of the neocortex, states [13] that modules may vary in cell type and number, in internal and external connectivity, and in mode of neuronal processing between different large entities, but within any single large entity they have a basic similarity of internal design and operation. A cortical area defined by the rules of classical cytoarchitectonics may belong to different systems. Therefore distributed structures may serve as the anatomical bases of distributed function.

4   Functional Modules

Michael Arbib's schema theory is a framework for the rigorous analysis of behavior which requires no prior commitment to hypotheses on the localization of each schema (unit of functional analysis), but which can be linked to a structural analysis as and when this becomes appropriate.

Complex functions such as the control of eye movements, reaching and grasping, the use of a cognitive map for navigation, and the roles of vision in these behaviors, by the use of schemas in the sense of units which provide a functional decomposition of the overall skill or behavior.

5 Dynamic Modules

Neural systems can be studied at different levels, such as the molecular, membrane, cellular, synaptic, network, and system levels. Two classes of the main neurodynamical problems exist: (i) study of the dynamics of activity spreading through a network with fixed wiring; (ii) the study of the dynamics of the connectivity of networks with modifiable synapses - both in normal ontogenetic development, and in learning as a network is tuned by experience. The key dynamical concept is the attractor, a pattern of activity which "captures" nearby states of an autonomous system. An attractor may be an equilibrium point, a limit cycle (oscillation), or a strange attractor (chaotic behavior). We have the structure-function problem: for what overall patterns of connectivity will a network exhibit a particular temporal pattern of dynamic behavior?

5.1 Neural Networks and Dynamics

Depending on its structure, an autonomous NN may or may not exhibit different qualitative dynamic behavior (convergence to equilibrium, oscillation, chaos). Some architectures shows unconditional behavior, which means that the qualitative dynamics does not depend on the numerical values of synaptic strengths. The behavior of other networks can be switched from one dynamic regime to another by tuning the parameters of the network. One mechanism of tuning is synaptic plasticity, which may help to switch the dynamics between the regimes (e.g. between different oscillatory modes, or oscillation and chaos, etc. ...)

5.2 Computation with Attractors: Scope and Limits

"Computation with attractors" became a paradigm which suggests that dynamic system theory is a proper conceptual framework for understanding computational mechanisms in self-organizing systems such as certain complex physical structures, computing devices, and neural networks. Its standard form is characterized by a few properties. Some of them are listed here: (i) the attractors are fixed points; (ii) a separate learning stage precedes the recall process whereby the former is described by a static 'one-shot' rule; (ii) the time-dependent inputs are neglected; (iv) the mathematical objects to be classified are the static initial values: those of them which are allocated in the same basin evolve towards the same attractor, and can recall the memory trace stored there. In its extended form, not only fixed points, but also limit cycles and strange attractors can be involved; a continuous learning rule may be adopted but, in this case, the basins of the attractors can be distorted which may even lead to qualitative changes in the nature of the attractors. Realistic models, which take explicitly into account the continuous interaction with the environment, however, are nonautonomus in mathematical sense [8], [1]. Such systems do not have attractors in the general case. Consequently, attractor neural network models cannot be considered as general frameworks of cortical models.

5.3 Rhythmicity and Synchronization

The existence of single cell level oscillation does not imply the generation of global oscillatory behaviour. To avoid "averaging out" due to irregular phase shifts, some synchronization mechanism should appear. Synchronization phenomena in different neural centers have been the subject of many recent investigations. The appearance of single cell oscillation is, however, not a necessary condition for global oscillations, since the latter may arise as an emergent network property. This was suggested based on combined physiological - pharmacological experiments in the thalamus (3), and also on theoretical studies discovering the connection between network structure and qualitative dynamic behaviour.

Specifically, following the work of Wang and Buzsáki [18] we studied the emergence of different types of synchronized oscillations in the model of networks of hippocampal inhibitory interneurons [11].

6 Population Models

As we have seen, structure-based bottom-up modeling has two extreme alternatives, namely multi-compartmental simulations, and simulation of networks composed of simple elements. There is an obvious trade-off between these two modeling strategies. The first method is appropriate to describe the activity patterns of single cells, small and moderately large networks based on data on detailed morphology and kinetics of voltage- and calcium-dependent ion channels. The second offers a computationally efficient method for simulating large network of neurons where the details of single cell properties are neglected.

To make large-scale and long-term realistic neural simulations there is a need to find a compromise between the biophysically detailed multicompartmental modeling technique, and the sometimes oversimplified network models. Statistical population theories offer a good compromise. Ventriglia [16], [17] introduced a kinetic theory for describing the interaction between a population of spatially fixed neurons and a population of spikes traveling between the neurons. A scale-invariant theory (and software tool) was developed, which gives the possibility to simulate the statistical behavior of large neural populations, and synchronously to monitor the behavior of an "average" single cell [9], [10], [5]. There is a hope that activity propagation among neural centers may be realistically simulated, and schemas can be built from structure-based modeling.

6.1   Incorporated Single Cell Models

Population models follow the conception of bottom up modeling incorporating basic realistic neuroanatomical and physiological evidences. However, whole brain simulations are computationally rather expensive thus simplifications are needed on different hierarchical levels.

The first idea is to simplify the single cell model constituting cortical areas. Large amount of experimental and simulational data are available about morphology, physiology, channel kinetics and other aspects of numerous types of neurons which are possible to take into consideration in models that do not explicitly compute difficult single cell equations. Such single cell models are complex enough to account for biologically realistic behavior but simple enough to enable fast and accurate numerical computation.

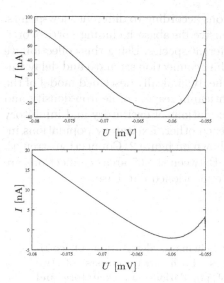

Fig. 1. Membrane potential–membrane current graphs. Figure on the left shows $I(V)$ curve of excitatory cells, while figure on the right the $I(V)$ characteristics of inhibitory ones. Data are acquired using current-clamp simulations of Hodgkin-Huxley neurons. Hard firing thresholds on basis of these graphs for excitatory and inhibitory cells are about -57 mV and -56 mV, respectively.

Considering the notion that neural information is stored in the timing of action potentials and not in the exact shape of spikes the important part of the dynamics is the time between two firings, that is the sub-threshold regime. This way the point of interest in single cell modeling is the description of inter-spike dynamics. This is achieved by the following method. First, detailed simulation of single cell models, acquiring the membrane potential–membrane current $(I(V(t)))$ characteristics (see figure 1.). Second, generating the return map of the membrane potential for a given integration time using the

$$C\frac{dV(t)}{dt} = I(V(t)) \tag{1}$$

Kirchoff-law. Here further modifications are required to corrigate for example the loss of characteristic electrotonic length. Finally, single cells can be integrated into the model framework by adding these rather complex return maps to the population model. Spike emission occurs with a certain membrane potential dependent probability. Action potentials which play critical role in defining activity are represented simply by the refractory period.

In simulations of activity propagation in the cat cortex two types of neurons are used. The prototype of the excitatory cell model is the Traub '91 pyramidal cell model of the hippocampal CA3 region. This is a multicompartmental model in which the membrane potential is described by the Hodgkin-Huxley formalism. Inhibitory cells were derived from Hodgkin-Huxley interneurons of the CA3 region using the above described method.

6.2   Cortical Areas

The cortex can be divided into several regions according to different view-points. Scannell *et al.* [14] have compiled an extensive database including corticocortical and corticothalamic pathways from different species. Using this collection we picked 65 cortical areas and the corresponding connection strength and delay values to construct the connection matrix of the dynamically described model of the cat cortex. Areas are considered to be identical in respect of neuron density and internal connection setup. In each area populations of excitatory and inhibitory cells are present occasionally overlapping each other. Excitatory populations innervate inhibitory ones and vice versa as shown on figure 2. Connection strength here decrease exponentially with distance. Between the 65 areas connections are described by data obtained form the above mentioned database.

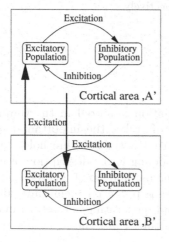

Fig. 2. Schematic description of cortical areas. Each area consists of two kinds of populations: an excitatory and an inhibitory one. Intracortical connections are assumed in both directions, from excitatory to inhibitory and from inhibitory to excitatory populations. Connection strength is an exponentially decreasing function of distance between populations. Intercortical connections are assumed to be excitatory. Excitatory and inhibitory populations mutually innervate each other.

6.3   Mathematical Description of the Model

Neural activity in the population model is described using statistical approach. A probability density function (PDF) is introduced as the number of neurons in a given membrane potential state. The membrane potential of each point in the neural field – the cortex – is given by a PDF. This idea is adopted from earlier works by [16,17]. The time-evolution equation of the PDF is given by the

$$
\begin{aligned}
&\frac{\partial g_s(\mathbf{r}, u, t)}{\partial t} + \frac{\partial}{\partial u}\left(\varepsilon_s(\mathbf{r}, u, t) \cdot g_s(\mathbf{r}, u, t)\right) + \\
&- \frac{D_u}{2} \cdot \frac{\partial^2 g_s(\mathbf{r}, u, \chi, t)}{\partial u^2} = \\
&= b_s(\mathbf{r}, u, t) - n_s(\mathbf{r}, u, t)
\end{aligned}
\tag{2}
$$

equation. The second term of the left hand side – which side gives the interspike dynamics – is a drift term with ϵ velocity, while the third term describes a diffusion process of the membrane potential due to random events influencing the it. The drift velocity is

$$
\varepsilon_s(\mathbf{r}, u, t) = -\frac{1}{C_s}\sum_i I_s^{(i)}(u) - \frac{1}{C_s}\sum_{s'} I_{s's}^{syn}(\mathbf{r}, u, t).
\tag{3}
$$

Here C_s is the membrane capacitance of neurons in the postsynaptic population, $I_{s's}^{syn}$ is the synaptic current between the s and s' populations, $I_s^{(i)}$ stands for all other currents simulations take into consideration. The synaptic current satisfies the

$$
I_{s's}^{syn}(\mathbf{r}, u, t) = -\gamma_{s's}(\mathbf{r}, t) \cdot (u - E_{s's})
\tag{4}
$$

equation. $E_{s's}$ is the reversal potential of the synapse between populations s and s'. $\gamma_{s's}(\mathbf{r}, t)$, the postsynaptic conductance function is of the form

$$
\gamma_{s's}(\mathbf{r}, t) = \int_{R_{s'}} \Phi_{s'}(\mathbf{r}') \int_0^\infty a_{s'}(\mathbf{r}', t - t' - d_{s's}(\mathbf{r}', \mathbf{r})) A_{s's}(k_{s's}(\mathbf{r}', \mathbf{r}), t') dt' d\mathbf{r}'.
\tag{5}
$$

$\Phi_s(\mathbf{r})$ is the density in the given $\mathbf{r}$ point, R_s is the area and $a_s(\mathbf{r}, t)$ the activity of neural population s. $d_{s's}(\mathbf{r}', \mathbf{r})$ gives the axonal delay between the two inter-connected population and $A_{s's}(x, t)$ is the change of postsynaptic conductance of a synapse with strength x t seconds after the beginning of the synaptic action. $A_{s's}(x, t)$ is usually given by the dual-exponential function, $d_{s's}(\mathbf{r}', \mathbf{r})$ is taken from database.

Activity in the model is defined as the number of firing cells:

$$
a_s(\mathbf{r}, t) = \int_0^{t_{ref}} dt' \int_{-\infty}^\infty du' \left(n_s(\mathbf{r}, u', t - t') - b_s(\mathbf{r}, u', t - t')\right),
\tag{6}
$$

where $n_s(\mathbf{r}, u, t)$ is a sink term giving the amount of neurons emitting a spike, $b_s(\mathbf{r}, u, t)$, the source term, the number of neurons returning into the interspike dynamics.

7   (Over)Synchronization in the Cerebral Cortex

Large-scale simulation is a proper tool for studying the relationship between structural and functional connectivity. anatomical data for the connectivity in the cat cerebral cortex was presented [14] and used by a simplified threshold model [12]. Based on these anatomical data and estimated time delays between regions the activity propagation as a spatiotemporal response for (even a local) stimulus can be simulated. Specifically, the generation and propagation of epiletic signals have been studied.

Epilepsy is a typical example of *dynamical diseases*. A dynamical disease is defined that occurs in an intact physiological system yet leads to abnormal dynamics. Epilepsy itself is characterized by the occurrence of seizures (i.e. ictal activities). During epileptic seizures oscillatory activities emerge, which usually propagate through several distinct brain regions. The epileptic neural activities generally displayed in the local field potentials measured by local electroencephalogram (EEG). Epileptic activity occurs in a population of neurons when the membrane potentials of the neurons are "abnormally" synchronized.

Simulation of the generation and propagation of disinhibition induced epiletic seizure is shown in Fig. 3

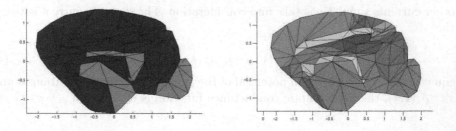

Fig. 3. Epileptiform activity pattern evolving on the cat's cortex. Initial state is depicted on the left image: activity in most regions are low (blue color). Image on the right shows typical epileptiform activity. Results represented are acquired by disinhibition of the prefrontal cortex.

Acknowledgements. This work was supported by the National Scientific Research Foundation (OTKA) No T 025500.

References

1. Aradi I, Barna G, Érdi P, Gröbler T, Chaos and learning in the olfactory bulb. Int J Intel Syst. 10:89-117, 1995

2. Arbib MA, The Metaphorical Brain 2: Neural Networks and Beyond, New York: Wiley-Interscience, 1989
3. Arbib M, Érdi P, Szentágothai J, Neural Organization: Structure, Function and Dynamics. MIT Press, 1997
4. Arbib M and Érdi P, Structure, Function and Dynamics: An Integrated Approach to Neural Organization. Behavioral and Brain Sciences (accepted)
5. Barna G, Gröbler T, Érdi P, Statistical model of the hippocampal CA3 region II. The population framework: Model of rhythmic activity in the CA3 slice. Biol. Cybernetics 79 (309-321) 1998
6. Érdi P, Neurodynamic system theory: scope and limits. Theoret. Medicine 14(137-152)1993
7. Érdi P, Structure-based modeling of schemas. Artificial Intelligence, 101 (341-343) 1998
8. Érdi P, Gröbler T, Tóth J, On the classification of some classification problems. Int. Symp. on Information Physics, Kyushu Inst. Technol., Iizuka, (110-117) 1992
9. Érdi P, Aradi I, Gröbler T, Rhythmogenesis in single cells and populations: olfactory bulb and hippocampus. BioSystems 1997; 40: 45-53
10. Gröbler T, Barna G, Érdi P, Statistical model of the hippocampal CA3 region I. The single cell module: bursting model of the pyramidal cell. Biol. Cybernetics 79 (301- 308)1998
11. Kiss T, Orbán G, Lengyel M, Érdi P, Hippocampal rhythm generation: gamma related theta frequency resonance. Cybernetics and System Research: 2000 (ed. Trappl R), Austrian Society for Cybernetic Studies, Vienna, (330-335)2000
12. Kötter R, and Sommer FT, Global relationship between anatomical connectivity and activity propagation in the cerebral cortex. 2000, Phil. Trans. R. Soc. Lond. pages 127–134
13. Mountcastle VB, The columnar organization of the neocortex, 1997, Brain 120:701-22
14. Scannell JW, Blakemore C, and Young MP, Analysis of connectivity in the cat cerebral cortex. 1995, Journal of Neuroscience 15:1463–1483
15. Szentágothai J, The neuron network of the cerebral cortex: a functional interpretation, Proc.R Soc.Lond.B. 201:219-248(1978)
16. Ventriglia F, Kinetic approach to neural systems I. Bull. Math. Biol. 36:534-544(1974)
17. Ventriglia F, Towards a kinetic theory of cortical-like neural fields. In: Ventriglia F, Ed., Neural Modeling and Neural Networks. pp. 217-249. Pergamon Press, 1994
18. Wang XJ and Buzsáki G, Gamma oscillation by synaptic inhibition in a hippocampal interneuronal network model. J. Neurosci. 16:6402-6413(1996)

Synchronisation, Binding, and the Role of Correlated Firing in Fast Information Transmission

Simon R. Schultz[1], Huw D.R. Golledge[2], and Stefano Panzeri[2]

[1] Howard Hughes Medical Institute & Center for Neural Science, New York
University, New York, NY 10003, USA
[2] Department of Psychology, Ridley Building,
University of Newcastle upon Tyne, Newcastle NE1 7RU, UK
stefano.panzeri@ncl.ac.uk
http://www.staff.ncl.ac.uk/stefano.panzeri/

Abstract. Does synchronization between action potentials from different neurons in the visual system play a substantial role in solving the binding problem? The binding problem can be studied quantitatively in the broader framework of the information contained in neural spike trains about some external correlate, which in this case is object configurations in the visual field. We approach this problem by using a mathematical formalism that quantifies the impact of correlated firing in short time scales. Using a power series expansion, the mutual information an ensemble of neurons conveys about external stimuli is broken down into firing rate and correlation components. This leads to a new quantification procedure directly applicable to simultaneous multiple neuron recordings. It theoretically constrains the neural code, showing that correlations contribute less significantly than firing rates to rapid information processing. By using this approach to study the limits upon the amount of information that an ideal observer is able to extract from a synchrony code, it may be possible to determine whether the available amount of information is sufficient to support computational processes such as feature binding.

1 Introduction

Does synchronization (or more generally temporal correlations) between action potentials from different cells in the central visual system play a substantial role in solving crucial computational problems, such as binding of visual features or figure/ground segmentation? One theory suggests that synchrony between members of neuronal assemblies is the mechanism used by the cerebral cortex for associating the features of a coherent single object [1].

Although several groups have reported compelling experimental evidence from the visual system in support of this theory (for a review see [2]), the role played by synchronous firing in visual feature binding is still highly controversial [3,4,5,6], and far from being understood. In our view, it is possible that one or

S. Wermter et al. (Eds.): Emergent Neural Computational Architectures, LNAI 2036, pp. 212–226, 2001.
© Springer-Verlag Berlin Heidelberg 2001

more methodological factors contribute to the continuing uncertainty about this issue. In fact, almost all the reported neurophysiological evidence in favor or against the temporal binding hypothesis relies upon the assessment of the significance of peaks in cross-correlograms (CCG, [7]) and of their modulation with respect to stimulus configuration. While investigating stimulus modulation of peaks (or of other features) of CCGs can clearly bear some evidence on the role of synchrony in binding, it does not address the crucial issue of *how much* synchrony tells the brain about the configuration of objects in the visual field. This question is particularly important as it is well known that firing rates of individual cells are commonly related to features of the sensory world [8], and even to perceptual judgements (see e.g. [9,10]). Firing rate modulations can potentially contribute to association of features through the use of population codes, or also in other ways[3]. Therefore the specific contribution of synchrony (or in general of correlations between firing of cells) as a coding mechanism for binding should be assessed against the contribution of independent firing rate modulation to the encoding of object configurations in the visual field.

To address these issues, a pure analysis of CCG characteristics is insufficient. In addition to CCG quantification, information theory can be used to address the specific contribution of synchronized or correlated firing to visual feature binding, and to compare the contribution of synchrony against that of firing rates. In fact, Information theory [11] allows one to take the point of view of an ideal observer trying to reconstruct the stimulus configuration just based on the observation of the activity of neuronal population, and to determine how much the presence of correlated firing helps in identifying the stimulus.

In this paper we present and develop a rigorous information theoretic framework to investigate the role of temporal correlations between spikes. We first discuss how information theory could overcome the limitations of the pure CCG analysis. We then present a mathematical formalism that allows us to divide the information into components which represent the information encoding mechanisms used by neuronal populations – i.e. it determines how many bits of information were present in the firing rates, how many in coincident firing by pairs of neurons, etc., with all of these adding up to the overall available information. The mathematical approach developed here is valid for timescales which are shorter than or of the order of a typical interval between spike emissions, and it makes use of a Taylor series approximation to the information, keeping terms up to the second order in the time window length. This approximation is not merely mathematically convenient; short timescales are likely to be of direct relevance to information processing by the brain, as there is substantial evidence that much sensory information is transmitted by neuronal activity in very short periods of time [12,13,14]. Therefore the mathematical analysis is relevant to the study of the computations underlying perceptual processes. In particular, it enables the quantitative study of the rapidly appearing correlational assemblies that have been suggested to underlie feature binding and figure/ground segmentation.

2   Problems with Conventional Cross-Correlogram Analysis

The CCG represents a histogram of the probability of a spike from one neuron at a given time relative to a reference spike of a second neuron [7]. Whilst cross-correlation is capable of identifying synchrony between neurons, several aspects of the analysis of CCGs present problems or are incomplete. First, CCG analysis itself does not provide a criterion to choose which periods of a response epoch should be analysed. Since, in many cases, moving stimuli are employed, the response varies with time and it may be that correlations are present or are stimulus modulated for only a short part of the response [15]. This short period is not necessarily related simply to the response peak, although some studies have analysed only the period in which the peak of the response is made to a moving stimulus [16]. Second, the width of the time window over which correlations should be assessed is arbitrary. CCG analysis does not entirely address over which time scales correlations contribute most information about object configuration. Using long windows (e.g. much larger than the width of CCG peaks) may "wash out" transient correlations. Narrow windows centered upon the PSTH peak may ignore the part of the responses that contains most of the information about the stimuli (e.g. in firing rate modulations). Third, if the window length used to assess correlations is varied between stimulus conditions (e.g [16]) then an undesirable extra source of variation is introduced when the stimulus conditions are compared. Information theory can mitigate some of these problems by providing a criterion for the selection of time windows, by identifying the windows in which most information is actually transmitted.

Many previous studies also differ in the methods used to quantify the temporal structure in CCGs. Some studies rely on the fitting of a damped sine wave to the CCG (e.g. [5,17]). Other methods quantify solely the likelihood that the peak in the CCG did not arise by chance [18]. Analysis of the significance of a peak, or of structure in the CCG must be made in relation to the flanks of the CCG. What length of flank is chosen will affect the significance of peaks. However, downstream neurons are unlikely to be able to compare the likelihoods of spikes occurring at lags of tens of milliseconds against the likelihood of simultaneous spikes.

The parameters of a CCG do not themselves quantify the informational contribution of synchronous firing. Conventional CCG analysis techniques attempt to assess correlation in a manner independent of the firing rate in order to disambiguate synchronous modulations from firing rate variations. It is unlikely, though, that any downstream detector of the synchronous discharge of neurons would be capable of assessing the significance of correlation independent of the firing rate. It is more likely that it would make use of the actual number of coincident spikes available in its integration time window. Therefore cross-correlation peaks and firing rate modulation are probably intrinsically linked in transmitting information to downstream neurons, and an analysis of the functional role of synchrony should be able to take this into account. Most studies that appear to show stimulus-dependent synchrony have employed relatively strongly stimulated cells. An important prediction of the temporal correlation hypothesis is that synchrony should encompass the responses of sub-optimally stimulated neurons [19]. A thorough test of this hypothesis requires the study of cells that fire very few spikes. The number of spikes included in the calculation of a CCG of course affects the precision with which correlations can be detected [20]. Variations in the number of evoked spikes, rather than a true change in correlation between the neurons, could thus affect comparisons between optimal and sub-optimal stimuli. While analysing cells firing at low rates may be a challenge for CCG analysis, it is tractable for analyses developed from information theory, as we shall see in Section 4.

3   Information Theory and Neuronal Responses

We believe that the methodological ambiguities that attend studies purely based on quantification of spike train correlograms can be greatly reduced by employing in addition methods based upon information theory [11], as we describe in this Section.

Information theory [11] measures the statistical significance of how neuronal responses co-vary with the different stimuli presented at the sensory periphery. Therefore it determines how much information neuronal responses carry about the particular set of stimuli presented during an experiment. Unlike other simpler measures, like those of signal detection theory, which take into account only the mean response and its standard deviation, information theory allows one to consider the role of the entire probability distributions. A measure of information thus requires sampling experimentally the probabilities $P(r|s)$ of a neuronal population response r to all stimuli s in the set, as well as designing the experimental frequencies of presentation $P(s)$ of each stimulus. The information measure is performed by computing the distance between the joint stimulus-response probabilities $P(r, s) = P(r|s)P(s)$ and the product of the two probabilities $P(r)P(s)$, ($P(r)$ being the unconditional response probability) as follows:

$$I(S; R) = \sum_s \sum_r P(s, r) \log_2 \frac{P(s, r)}{P(s)P(r)} \tag{1}$$

If there is a statistical relationship between stimuli and responses (i.e. if $P(r, s)$ is dissimilar from $P(r)P(s)$) , our knowledge about what stimulus was presented increases after the observation of one neuronal spike train. Eq. (1) quantifies this fact. The stronger the statistical relationship between stimuli and responses, the higher is the information value. Eq. (1) thus quantifies how well an ideal observer could discriminate between different stimulus conditions, based on a single response trial. There are several advantages in using information theory to quantify how reliably the activity of a set of neurons encodes the events in the sensory periphery [21,22]. First, information theory puts the performance of neuronal responses on a scale defined at the ratio level of measurement. For example, an increase of 30% of on the peak height of a cross-correlogram does not tell us how this relates to synchrony-based stimulus discrimination, but values of information carried by synchronous firing have a precise meaning. Information theory measures the reduction of uncertainty about the stimulus following the observation of a neuronal response on a logarithmic scale. One bit of information corresponds to a reduction by a factor of two in the stimulus uncertainty. A second advantage of information theory in this context is that it does not require any specific assumption about the distance between responses or the stationarity of the processes, and it can therefore lead to objective assessments of some of the hypotheses.

In the above discussion we have mentioned that we are calculating information 'about a stimulus'. In fact, more generally it can be information 'about' any quantifiable external correlate, but we shall continue to use the word stimulus in an extended sense. If we were studying information about the orientation of a grating, we would define our stimulus to be which of a number of different orientations the grating appeared at on any given experimental trial. If we wish to study problems such as binding or figure-ground segregation within this framework, we have to specify our stimulus description accordingly. An illustration of this is shown in the scene of Fig. 1, which contains two objects in front of a background. Also shown are a number of receptive fields, which are taken to be those of cells from which we are simultaneously recording the activity ('spike trains'). We can define our stimulus, or external correlate, as a multidimensional variable representing the object to which each receptive field is associated. The dimensionality of our stimulus is in this case the number of cells from which we are recording at once. By quantifying the information contained in the spike trains about this variable, and breaking it down into individual components reflecting firing rates and correlations (or synchronisation), we can determine the aspects of the spike train which best encode the figure-ground (or object-object-ground) segregation. Furthermore, by examining how this relationship scales with the stimulus dimensionality (number of receptive fields recorded from), it may be possible to determine whether enough information is present in correlations to support binding in perceptually realistic environments.

It is worth noticing that information values are always relative to the stimulus set used, and that testing a neuron with different stimuli may lead to rather different information values. This has some interesting implications. On the one hand, it allows us to characterise neuronal selectivity by searching for a stimulus set that maximises the neuronal information transfer, a more rational characterisation strategy than searching for stimuli eliciting sustained responses. On the other hand, the intrinsic dependency of mutual information values on the nature of stimulus set allows us to test whether different encoding strategies are used by visual cortical neurons when dealing with external correlates of a different nature. The last property is of interest because one of the predictions of the binding-by-synchrony hypothesis is that synchrony is particularly important when stimulus configurations requiring some kind of associations are included, and less important in other situations. Information theory thus provides a natural framework to test this theory.

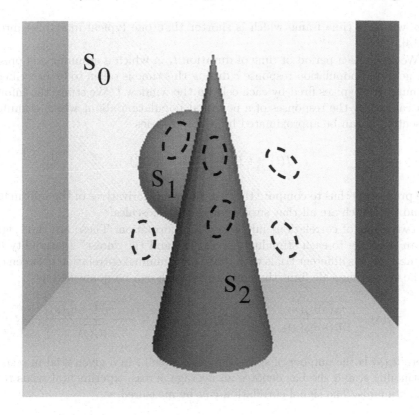

Fig. 1. An illustrative segregation problem in which there are two objects in front of a background. The background is labeled by s_0, and the objects by s_1 and s_2 respectively. The dashed ellipses represent the receptive fields of visual cortical cells which we are recording the responses of simultaneously. This situation can be examined in the framework of information theory by considering the 'stimulus' to be a multidimensional variable indicating which object (s_0, s_1 or s_2) is associated with each receptive field. The problem is thus to determine which response characteristics are most informative about the visual configuration.

4   Series Expansion of the Mutual Information

Although information theory is, as explained, a natural framework to address the role of correlated firing on e.g. binding of visual features, some work is needed to separate out of the total information contained in a population response r into components, each reflecting the specific contribution of an encoding mechanism. We perform this separation in the limit in which the relevant window for information encoding is short enough that the population typically emits few action potentials in response to a stimulus. As discussed in the introduction and in [23], there is evidence that this is a relevant limit for studying the computations underlying perception, as cortical areas in several cases perform their computa-

tions within a time frame which is shorter than one typical interspike interval
[13,14].

We examine a period of time of duration t, in which a stimulus s is present.
The neuronal population response r during this time is taken to be described by
the number of spikes fired by each cell[1] in the window t. We study the information carried by the responses of a neuronal population about which stimulus is
presented. It can be approximated by a power series

$$I(t) = t\, I_t + \frac{1}{2}t^2\, I_{tt} + O(t^3). \tag{2}$$

The problem is thus to compute the first two time derivatives of the information,
I_t and I_{tt}, which are all that survive at short timescales.

Two kinds of correlations influence the information. These are the "signal"
(mean response to each stimulus) correlation and the "noise" (variability from
the mean across different trials with the same stimulus) correlation between cells.
In the short timescale limit the noise correlation can be quantified as

$$\gamma_{ij}(s) = \frac{\overline{n_i(s)n_j(s)}}{(\overline{n}_i(s)\overline{n}_j(s))} - 1, (i \neq j) \qquad \gamma_{ii}(s) = \frac{(\overline{n_i(s)^2} - \overline{n_i(s)})}{\overline{n_i(s)}^2} - 1, \tag{3}$$

where $n_i(s)$ is the number of spikes emitted by cell i in a given trial in response
to stimulus s, and the bar denotes an average across experimental trials to the
same stimulus. The signal correlation can be measured as

$$\nu_{ij} = \frac{< \overline{r}_i(s)\overline{r}_j(s) >_s}{< \overline{r}_i(s) >_s < \overline{r}_j(s) >_s} - 1, \tag{4}$$

where $\overline{r}_i(s)$ is the mean firing rate of cell i to stimulus s, and $< \cdots >_s$ denotes
an average across stimuli. These are scaled correlation densities ranging from -1
to infinity, which remain finite as $t \to 0$. Positive values of the correlation coefficients indicate positive correlation, and negative values indicate anti-correlation.

Under the assumption that the probabilities of neuronal firing conditional
upon the firing of other neurons are non-divergent, the t expansion of response
probabilities becomes an expansion in the total number of spikes emitted by the
population in response to a stimulus. The probabilities of up to two spikes being
emitted are calculated and inserted into the expression for information. This
yields for the information derivatives

$$I_t = \sum_{i=1}^{C} \left\langle \overline{r}_i(s) \log_2 \frac{\overline{r}_i(s)}{\langle \overline{r}_i(s') \rangle_{s'}} \right\rangle_s \tag{5}$$

$$I_{tt} = \frac{1}{\ln 2} \sum_{i=1}^{C} \sum_{j=1}^{C} \langle \overline{r}_i(s) \rangle_s \, \langle \overline{r}_j(s) \rangle_s \left[\nu_{ij} + (1 + \nu_{ij}) \ln(\frac{1}{1 + \nu_{ij}}) \right]$$

---

[1] The additional temporal information contained in the spike times is studied in [24]

$$+ \sum_{i=1}^{C} \sum_{j=1}^{C} \left[\langle \overline{r}_i(s) \overline{r}_j(s) \gamma_{ij}(s) \rangle_s \right] \log_2 \left(\frac{1}{1+\nu_{ij}} \right)$$

$$+ \sum_{i=1}^{C} \sum_{j=1}^{C} \left\langle \overline{r}_i(s) \overline{r}_j(s)(1+\gamma_{ij}(s)) \log_2 \left[\frac{(1+\gamma_{ij}(s)) \langle \overline{r}_i(s') \overline{r}_j(s') \rangle_{s'}}{\langle \overline{r}_i(s') \overline{r}_j(s')(1+\gamma_{ij}(s')) \rangle_{s'}} \right] \right\rangle_s . \quad (6)$$

The first of these terms is all that survives if there is no noise correlation at all. Thus the *rate component* of the information is given by the sum of I_t (which is always greater than or equal to zero) and of the first term of I_{tt} (which is instead always less than or equal to zero). The second term is non-zero if there is some correlation in the variance to a given stimulus, even if it is independent of which stimulus is present; this term thus represents the contribution of *stimulus-independent noise correlation* to the information. The third component of I_{tt} is non-negative, and it represents the contribution of *stimulus-modulated noise correlation*, as it becomes non-zero only for stimulus-dependent correlations. We refer to these last two terms of I_{tt} together as the correlational components of the information.

In any practical measurement of these formulae, estimates of finite sampling bias must be subtracted from the individual components. Analytical expressions for the bias of each component are derived in the online appendix of [23].

5   Correlations and Fast Information Processing

The results reported above for the information derivatives show that the instantaneous rate of information transmission (related to the leading order contribution to the information) depends only upon the firing rates. Correlations contribute to information transmission, but they play only a second order role. This has interesting implications for the neural code, that we develop further here.

It was argued [25] that, since the response of cortical neurons is so variable, rapid information transmission must imply redundancy (i.e. transmitting several copies of the same message). In other words, it should necessary to average away the large observed variability of individual interspike intervals s by replicating the signal through many similar neurons in order to ensure reliability in short times. Our result, eq. (6), shows that to have a high information rate, it is enough that each cell conveys some information about the stimuli (because the rate of the information transmitted by the population is the sum of all single cell contributions); therefore we conclude that it is not necessary to transmit many copies of the same signal to ensure rapidity.

Also, since firing rates convey the main component of information, correlations are likely to play a minor role in timescales of the order of 20-50 ms, in which much information is transmitted in the cortex. As an example, Fig. 2 shows the information conveyed by a population of simulated Integrate and Fire neurons, which share a large proportion (30%) of their afferents (see [23] for details of the neuronal model used). It can be seen that, despite the strong

correlations between the cells in the ensemble correlations play only a minor role
with respect to firing rates.

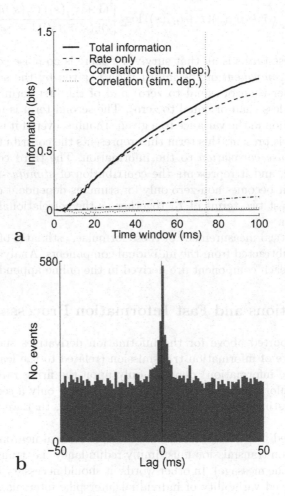

Fig. 2. (a) The short-timescale information components for a set of 5 simulated neurons sharing 30% of their inputs. (b) A typical example of the cross-correlogram between two of the simulated neurons. Despite of the strong correlation between neurons,
the impact of the cross correlations on information transmission is minimal.

To model a situation where stimulus dependent correlations conveyed information, we generated simulated data using the Integrate-and-Fire model for
another quintuplet of cells which had a stimulus dependent fraction of common
input. This might correspond to a situation where transiently participate in different neuronal assemblies, depending on stimulus condition. This is therefore a
case that might be found if the binding-by-synchrony theory is correct. There

were ten stimuli in the sample. The spike emission rate was constant (20 Hz) across stimuli. One of the stimuli resulted in independent input to each of the model cell, whereas each of the other nine stimuli resulted in an increase (to 90%) in the amount of shared input between one pair of cells. The pair was chosen at random from the ensemble such that each stimulus resulted in a different pair being correlated The change in responses of one such pair of cells to changes in the amount of common input is shown in Fig. 3a. The upper panel of Fig. 3a shows the fraction of shared connections as a function of time; the central and lower panel of Fig. 3a show the resulting membrane potentials and spike trains from the pair of neurons. This cross-correlation is also evident in the cross-correlograms shown in Fig. 3b. The results for the information are given in Fig. 3c: all terms but the third of I_{tt} are essentially zero, and information transmission is in this case almost entirely due to stimulus-dependent correlations. This shows that the short time window series expansion pursued here is able to pick up the right encoding mechanisms used by the set of cells. Therefore it is a reliable method for quantifying the information carried by the correlation of firing of small populations of cells recorded from the central nervous system *in vivo*. Another point that is worth observing is that Fig. 3c also shows that the total amount of information that could be conveyed, even with this much shared input, was modest in comparison to that conveyed by rates dependent on the stimuli, at the same mean firing rate. This again illustrates that correlations typically convey less information than what can be normally achieved by rate modulations only. Therefore they are likely to be a secondary coding mechanism when information is processed in time scales of the order of one interspike interval.

6   Optimality of Correlational Encoding

In the preceding sections we have shown that the correlational component is only second order in the short time limit, essentially because the probability of emission of pairs of spikes, and the reliability of this process, are much smaller than the corresponding quantities for single spikes. For this reason, one can expect a correlational code to carry appreciable information only when it is *efficient*, i.e. when each correlated spike pair carries as much information as possible.

In this section we investigate the statistical conditions that have to be satisfied by a correlational code in order to be efficient in the short time limit. If the population code is purely correlational (i.e. the firing rates are not modulated at all by the stimuli), then it is possible to show that the mean information per coincidence Ψ carried by a pair of cells (obtained dividing the total information by the mean number of observed coincidences) is bounded only by the sparseness of the distribution of coincident firing across stimuli α:

$$0 \leq \Psi \leq \log_2(1/\alpha) \tag{7}$$

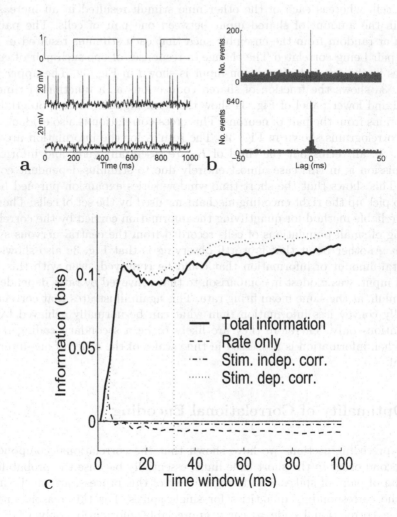

Fig. 3. A situation in which the stimulus dependent correlational component dominates: with a fixed mean firing rate, two of the five simulated cells (chosen randomly for that stimulus) increase their correlation by increasing the number of shared connections while the other two remained randomly correlated. The effect of this on cell spiking activity is shown in (a): upper panel shows the fraction of shared connections, while central and lower panels of (a) show the membrane potential and spike emission of the simulated cells. (b) shows the cross-correlograms corresponding to the low and high correlation states. The result of this is seen in (c): information due to correlations, although modest in magnitude, in this demonstration dominates the total information.

The maximal (most efficient) value of information per coincidence $\Psi_{\max} = \log_2(1/\alpha)$ is reached by a binary distribution of correlations across stimuli, with a fraction of stimuli α eliciting positively correlated firing, and the other $1 - \alpha$ stimuli eliciting fully anti-correlated firing (i.e. coincident firing is never observed when presenting one of the latter stimuli). Nearly uniform, or strongly unimodal distributions of correlations across stimuli would give poor information, $\Psi \sim 0$.

By analyzing Eq. (7), it is easy to realize that there are two statistical requirements that are necessary to achieve high values of information per coincidence. The first one is that the correlational code should be em sparse (i.e. the fraction of stimuli leading to a "high correlation state" should be low). The sparser the code, the more information per coincidence can be transmitted. The second important factor for fast and efficient transmission of correlational information, is that the low correlational state must be strongly *anti-correlated* in order to achieve an information per coincidence close to its maximum $\log_2(1/\alpha)$. In fact correlations in short times have fluctuations that may be big compared with their mean value, and therefore for any observer it is difficult to understand in less than one ISI if an observed coincidence is due to chance or neuronal interaction. This is why low correlational states with no coincidences are so helpful in transmitting information. We note here that states of nearly complete anticorrelation have never been observed in the brain. Therefore the "low state correlational state" of a realistic correlational assembly should be the "random correlation state" (i.e. the state in which the number of coincident spikes is on average that expected by chance).

We have quantified the reduction in the information per coincidence, compared to its maximum $\Psi_{\max}$, that arises as a consequence of the presence of the random correlation state. Fig. 4 plots the ratio between the information per coincidence carried when the "low correlation state" is random and the optimal amount of information per coincidence $\log_2(1/\alpha)$ obtained when the low correlation state is totally anticorrelated. The plot is shown as a function of the fraction α of stimuli eliciting a "highly correlated state". Fig. 4 shows clearly that, if the "low correlation state" of the assembly elicits uncorrelated firing, then the information per coincidence is far from its maximum, unless the correlation in the "high state" is extremely high and the correlational code is not sparse at all. However, in which case the information per coincidence is very low anyway (see eq. (7)).

Therefore correlational assemblies in the brain, if any, are likely to be inefficient in the short time limit. This consideration further limits the possible role played by correlations in fast information encoding.

7   Discussion

If cells participate in context-dependent correlational assemblies [2], then a significant amount of information should be found in the third component of I_{tt} when analysing data obtained from the appropriate experiments. The series expansion approach thus enables the testing of hypotheses about the role of correlations in

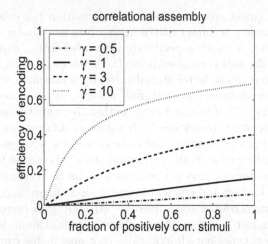

Fig. 4. The ratio between the information per coincidence carried by a binary correlational encoder with a fraction of stimuli eliciting positive correlation γ and the other stimuli eliciting no correlation, and the optimal information per coincidence carried in the same situation, but with full anticorrelation in the "low correlation state". This ratio is plotted, for different values of the strength γ of the "high correlation state", as a function of the fraction of stimuli eliciting positive correlation.

solving the binding problem, as opposed to other solutions, and about information coding in general. Data analyses based on the time-expansion approach have the potential to elucidate the role of correlations in the encoding of information by cortical neurons.

It is worth noticing that the formalism presented here evaluates the the information contained in the neuronal responses themselves, it does not make about the system that is going to read the code. For this reason, the information computed "directly" from neuronal responses is an upper bound to what any type of decoder can extract from the responses themselves. Therefore it is termed the information that an "ideal observer" can extract [22]. Of course, the relative contribution of rate and synchrony modulations to information transmission will depend on the specific read-out mechanism used by a downstream neural system that listens to the neuronal code. However, if the information that an ideal observer is able to extract from the synchrony code is small, as the mathematical analysis indicates for the fast information processing limit, one can be sure that any decoding device cannot extract more information in the synchrony than that small amount evaluated from the responses.

Whether this small amount is sufficient to support computational processes such as figure-ground segregation remains to be seen, and depends upon how it scales empirically with the number of receptive fields examined. Ultimately, as suggested by [3], the only way we will achieve any degree of confidence in a proposed solution to the binding problem will be to study recordings made from a monkey trained to make behavioural responses according to whether individual features are bound to particular objects. The information theoretic approach described here would be a natural way to analyse such data.

In conclusion, the methodology presented here can provide interesting and reliable bounds on the role of synchrony on cortical information encoding, and we believe that its application to neurophysiological experiments will advance our understanding of the functional interpretation of synchronous activity in the cerebral cortex.

Acknowledgements. We would particularly like to thank Alessandro Treves and Malcolm Young for useful discussions relating to this work.

References

1. C. von der Malsburg. Binding in models of perception and brain function. *Current Opinion in Neurobiology*, 5:520–526, 1995.
2. W. Singer, A. K. Engel, A. K. Kreiter, M. H. J. Munk, S. Neuenschwander, and P. Roelfsema. Neuronal assemblies: necessity, signature and detectability. *Trends in Cognitive Sciences*, 1:252–261, 1997.
3. M. N. Shadlen and A. J. Movshon. Synchrony unbound: a critical evaluation of the temporal binding hypothesis. *Neuron*, 24:67–77, 1999.
4. G. M. Ghose and J. Maunsell. Specialized representations in visual cortex: a role for binding? *Neuron*, 24:79–85, 1999.
5. M. P. Young, K. Tanaka, and S. Yamane. On oscillating neuronal responses in the visual cortex of the monkey. *J. Neurophysiol.*, 67:1464–1474, 1992.
6. H. D. R. Golledge, C. C. Hilgetag, and M. J. Tovée. Information processing: a solution to the binding problem? *Current Biology*, 6(9):1092–1095, 1996.
7. A. M. H. J. Aertsen, G. L. Gerstein, M. K. Habib, and G. Palm. Dynamics of neuronal firing correlation: modulation of "effective connectivity". *J. Neurophysiol.*, 61:900–917, 1989.
8. E. D. Adrian. The impulses produced by sensory nerve endings: Part I. *J. Physiol. (Lond.)*, 61:49–72, 1926.
9. K.H. Britten, M. N. Shadlen, W. T. Newsome, and J. A. Movshon. The analysis of visual-motion - a comparison of neuronal and psychophysical performance. *J. Neurosci.*, 12:4745–4765, 1992.
10. M. N. Shadlen and W. T. Newsome. Motion perception: seeing and deciding. *Proc. Natl. Acad. Sci. USA*, 93:628–633, 1996.
11. C. E. Shannon. A mathematical theory of communication. *AT&T Bell Labs. Tech. J.*, 27:379–423, 1948.
12. M. J. Tovée, E. T. Rolls, A. Treves, and R. P. Bellis. Information encoding and the response of single neurons in the primate temporal visual cortex. *J. Neurophysiol.*, 70:640–654, 1993.
13. S. Thorpe, D. Fize, and C. Marlot. Speed of processing in the human visual system. *Nature*, 381:520–522, 1996.
14. E. T. Rolls, M.J. Tovee, and S. Panzeri. The neurophysiology of backward visual masking: Information analysis. *J. Cognitive Neurosci.*, 11:335–346, 1999.
15. C. M. Gray, A. K. Engel, P. König, and W. Singer. Synchronization of oscillatory neuronal responses in cat striate cortex: Temporal properties. *Visual Neuroscience*, 8:337–347, 1992.

16. A. K. Kreiter and W. Singer. Stimulus-dependent synchronization of neuronal responses in the visual cortex of the awake macaque monkey. *J. Neurosci.*, 16:2381–2396, 1996.
17. P. König, A. K. Engel, and W. Singer. Relation between oscillatory activity and long-range synchronization in cat visual cortex. *Proc. Natl. Acad. Sci. USA*, 92:290–294, 1995.
18. S. C. deOliveira, A. Thiele, and A. Hoffmann. Synchronization of neuronal activity during stimulus expectation in a direction discrimination task. *J. Neurosci.*, 17:9248–9260, 1997.
19. P. König, A. K. Engel, P. R. Roelfsema, and W. Singer. How precise is neuronal synchronization? *Neural Computation*, 7:469–485, 1995.
20. Y. Hata, T. Tsumoto, H. Sato, and H. Tamura. Horizontal interactions between visual cortical neurones studied by cross-correlation analysis in the cat. *Journal of Physiology*, 441:593–614, 1991.
21. F. Rieke, D. Warland, R. R. de Ruyter van Steveninck, and W. Bialek. *Spikes: exploring the neural code*. MIT Press, Cambridge, MA, 1996.
22. A. Borst and F. E. Theunissen. Information theory and neural coding. *Nature Neuroscience*, 2:947–957, 1999.
23. S. Panzeri, S. R. Schultz, A. Treves, and E. T. Rolls. Correlations and the encoding of information in the nervous system. *Proc. R. Soc. Lond. B*, 266:1001–1012, 1999.
24. S. Panzeri and S. Schultz. A unified approach to the study of temporal, correlational and rate coding. *Neural Comp.*, page submitted, 1999.
25. M. N. Shadlen and W. T. Newsome. The variable discharge of cortical neurons: implications for connectivity, computation and coding. *J. Neurosci.*, 18(10):3870–3896, 1998.

Segmenting State into Entities and Its Implication for Learning

James Henderson

University of Exeter, Exeter EX4 4PT, United Kingdom,
J.B.Henderson@ex.ac.uk,
http://www.dcs.ex.ac.uk/~jamie

Abstract. Temporal synchrony of activation spikes has been proposed as the representational code by which the brain segments perceptual patterns into multiple visual objects or multiple auditory sources. In this chapter we look at the implications of this neuroscientific proposal for learning and computation in artificial neural networks. Previous work has defined an artificial neural network model which uses temporal synchrony to represent and learn about multiple entities (Simple Synchrony Networks). These networks can store arbitrary amounts of information in their internal state by segmenting their representation of state into arbitrarily many entities. They can also generalize what they learn to larger internal states by learning generalizations about individual entities. These claims are empirical demonstrated through results on training a Simple Synchrony Network to do syntactic parsing of real natural language sentences.

1 Sequences, Structures, and Segmenting with Synchrony

Neural Network models have been successful at pattern analysis using just a holistic representation of the input pattern. There is not usually any need to partition the problem or impose strong prior assumptions. This approach has been extended to processing sequences and to processing structures, but with much less success. The storage capacity of a network which is appropriate for short sequences or shallow structures is soon overwhelmed when faced with long sequences or deep structures, leading to much relevant information being lost from the network's representation of state. The solution advocated here is to segment the representation of state into multiple components. The number of components can grow with the length of the sequence or the size of the structure, thereby allowing the capacity required for each component to remain constant. Given this approach, the questions are how to represent these components, on what basis to segment the state information, and what are the implications for learning and computation.

In cognitive neuroscience one proposal for pattern segmentation is that the brain does this segmentation and represents the result using the temporal synchrony of neuron activation spikes [18]. Any two features of the input which are associated with the same entity are represented using synchronous (or more

S. Wermter et al. (Eds.): Emergent Neural Computational Architectures, LNAI 2036, pp. 227–236, 2001.
© Springer-Verlag Berlin Heidelberg 2001

generally, correlated) activation spikes, and any two features which are associated with different entities are represented with non-synchronous (uncorrelated) activation spikes [17,1,4,15,16]. This approach has been applied to visual object recognition [20] and auditory source recognition [19]. In both domains there are multiple entities in the world which are together producing the sensory input, and the brain needs to segment the sensory input according to which entity is responsible for which aspects of it.

This neuroscientific proposal suggests answers to two of our questions. First, that the synchrony of spiking neurons can be used to represent the components of the segmented state. Second, that this segmentation should be done on the basis of the entities in the domain. The work discussed in the remainder of this abstract looks at our third question, the implications of these two proposals for learning and computation. We argue that these two proposals are intimately related because of their implications for learning.

2   Temporal Synchrony for Artificial Neural Networks

To investigate the implications of using temporal synchrony to segment state into entities we have developed an artificial neural network model of computation and learning with temporal synchrony, called Simple Synchrony Networks [12,9]. This architecture extends Simple Recurrent Networks [2] with an abstract model of temporal synchrony.

2.1   An Abstract Model of Temporal Synchrony

To apply the idea of using the synchrony of neuron activation spikes to a computational architecture, we first need to define an abstraction of the neurobiology which captures only those properties which are relevant to our computational investigation. For this work we use the following abstract model of temporal synchrony:

1. Time is divided into discrete periods, and each period is divided into discrete phases.
2. The ordering between phases within a period cannot be used to represent information.
3. Units are divided into two types, those which process each phase independently (called pulsing units) and those which output the same activation to all phases (called nonpulsing units).
4. There is no bound on the number of periods and no bound on the number of phases in a period.

A computational abstraction of temporal synchrony was first defined in the SHRUTI model of reflexive reasoning [15,14]. The SHRUTI model was hand built to do resource-constrained logical inference, although it has been extended to include some trainable weights [13]. The above model of temporal synchrony is based on the model used in SHRUTI. In SHRUTI the periods correspond to the steps of the logical inference and the phases correspond to instantiations

Segmenting State into Entities   229

of the variables in the logical rules. As in SHRUTI, we abstract away from the problem of creating and maintaining synchrony between neurons by assuming discrete phases. Two unit activation pulses are either synchronous and firing in the same phase, or not synchronous and firing in different phases. There is no problem with noise when propagating this synchrony from one period to the next. However by organizing activation pulses into discrete phases we have introduced the potential for representing information in the ordering between the phases within a period. This organization is not present in the biological system, so in the second feature above we explicitly exclude exploiting this ordering within the computational model. In the third feature above, the two types of units correspond to those which utilize temporal synchrony to represent information (pulsing units) and those which do not (nonpulsing units). Thus the pulsing units represent information about individual entities, and nonpulsing units represent information about the situation as a whole.

The last feature listed in the above model of temporal synchrony is a departure from that used in SHRUTI. Based on biological constraints, SHRUTI places a bound on the number of phases per period. We remove this bound in order to more easily demonstrate the computational advantages of segmenting state into multiple entities.[1] We also do not address the problem of multiple instantiations of predicates, which requires single pulsing units to be active in multiple phases during the same period.[2] As is commonly done, a unit is intended to correspond to a collection of neurons. We assume that these neurons are able to allocate themselves to as many phases as are necessary.

2.2   Simple Synchrony Networks

The artificial neural network architecture which we use to investigate temporal synchrony differs significantly from the SHRUTI model in that it is primarily designed to support learning. For this reason Simple Synchrony Networks (SSNs) [12,9] are an extension of Simple Recurrent Networks (SRNs) [2]. SRNs have been demonstrated to be effective at learning about sequences. A sequence of inputs are fed into the network one at a time, and a sequence of outputs are produced by the network after each input. The outputs are computed on the basis of both the current input and the SRN's internal state, which records information about the previous sequence of inputs. At each position in the sequence this state information is represented as a fixed-length vector of unit activations, known as a holistic representation because of its unstructured nature. SSNs use this same form of representation in their nonpulsing units. The positions in an SRN's input/output sequence correspond to the periods in an SSN's temporal pattern of activation, and during each period the state of the nonpulsing units can be represented as a fixed-length vector.

---

[1] But see [11] for a discussion of reintroducing this bound.

[2] Although we don't explicitly introduce a constraint, the nature of temporal synchrony imposes a form of multiple instantiation constraint on predicates with two or more arguments. See [7] for a discussion of this constraint and its linguistic implications.

The difference between SRNs and SSNs is in SSN's use of pulsing units. During each period, pulsing units have a different activation value for each phase within the period. Thus the state of the pulsing units must be represented with a set of fixed-length vectors, where there is no fixed bound on the number of vectors in the set. This allows an SSN to segment the representation of its internal state into an unbounded number of components. Each component is one of the vectors of activation values. Because the segmentation is done on the basis of entities, each of these vectors is a representation of the state of an entity, which records information about the previous sequence of inputs which is relevant to that entity.

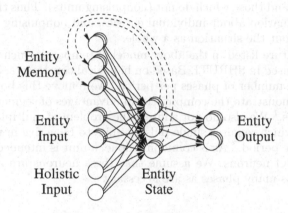

Fig. 1. An example of an SSN network. Nonpulsing units are shown as circles, pulsing units as stacks of circles, and links as solid arrows. The dashed arrows represent the copying of activation from the previous period.

By combining the SRN's ability to learn about sequences with temporal synchrony's ability to represent entities, SSNs can learn about sequences of entities. As illustrated in Figure 1, the pattern of links in an SSN is the same as that for an SRN. The pattern of activation in the state layer is computed from the pattern of activation from the previous state plus the input pattern. The output activation is computed from the state. The difference is in the inclusion of pulsing units in the layers.[3] The pattern of activation for each entity's state is computed from the entity's previous state, the current input about that entity, and the input about the situation as a whole (the holistic input). The output activation for an entity is computed from the state of that entity. The links which perform these computations can be trained using a simple extension of Backpropagation Through Time [12].

---

[3] Some versions of the SSN architecture have nonpulsing units in their state and memory layers as well as the pulsing units shown in Figure 1. Without loss of generality, we will restrict attention to this simpler architecture, called type A in [12].

3   Generalization with Temporal Synchrony

The motivation for segmenting state information into components was that this would allow us to vary the number of components, and thus the capacity of the state, without having to vary the capacity requirements of each individual component. This requirement is met by using temporal synchrony for segmenting state information because the number of phases can vary without requiring additional training. These additional phases are simply processed with the same link weights as all the other phases, as is done in SSNs. This processing strategy has significant implications for learning. Anything learned about a computation performed in one phase will have to be applied to computations in all other phases. This generalization of learned information across phases is an inherent property of using temporal synchrony, because link weights do not change at the time scale of phases.

Now we can see the relationship between using temporal synchrony to segment state and segmenting state on the basis of entities. If we use temporal synchrony to segment state then learned info will be generalized across the segmented components. But if the segmented components correspond to entities in the domain, then that is exactly the kind of generalization which we want. For example, if we learn to recognize that one snake is poisonous, then we want to generalize that knowledge to other examples of snakes. And if we encounter a situation with several snakes, then we want to generalize that knowledge to each snake individually (rather than generalizing it to the situation as a whole). In language processing, if we learn that a noun phrase can begin with "the", then we want to generalize that to all noun phrases, even in sentences with several noun phrases.

Generalizing learned information over entities is crucial for many domains. Hadley [5] argues, in the context on natural language constituents, that this kind of generalization is required for a neural network to demonstrate "systematicity", a property closely related to compositionality of representations [3]. Given this definition of systematicity, the above arguments can be used to show that neural networks which use temporal synchrony to represent entities have systematicity as an inherent property [6]. Even beyond natural language and compositional structures, it is hard to imagine any domain where we would want to say that particular entities exist in the domain but would not want to say that there are regularities across those entities. If such regularities are a part of what we mean by "entity", then temporal synchrony is applicable to representing any set of entities. Perhaps even the presence of entities in our higher level cognitive representations is a result of the evolution of temporal synchrony as a neurological mechanism for segmenting the representation of state. But that question is beyond the scope of this chapter!

4   An Application to Language Processing

Parsing natural language is an application which imposes widely varying capacity requirements on a network's representation of state. There is wide variation in the length of sentences and in the size of their syntactic structures. As the length of a sentence grows so does the size of its syntactic structure, and so does the amount of information about that structure which might be relevant to each parsing decision. In order for a parser to store all this information in its state, its representation of state must have sufficient storage capacity. In theory the capacity required can be arbitrarily high, but in practice it is sufficient to argue that there is wide variation between the capacity required for small structures and that required for large structures.

Many attempts have been made to apply holistic neural network representations to parsing, but in one way or another they all run into the problem of a network's limited storage capacity [10]. As advocated above, this limitation has been overcome using Simple Synchrony Networks [11]. The SSN's phases are used to store the information about syntactic constituents,[4] thereby segmenting the parser's state into an arbitrarily large set of constituents and allowing each constituent's representation to be of constant size. It also means the network has the linguistically desirable property of generalizing over constituents, as well as simplifying the parser's output representation [9].

Recent experiments demonstrate that an SSN parser can handle varying sentence lengths and generalize appropriately, even with the complexities of real natural language. Henderson [8] compares the performance of an SSN parser to a simple version of the standard statistical parsing technique of Probabilistic Context Free Grammars (PCFGs). Both models were trained on the same set of sentences from a corpus of naturally occurring text, and then tested on a previously unseen set. The results of this testing are shown in Table 1. Constituent recall (percentage of correct constituents which are output) and precision (percentage of output constituents which are correct) are the standard measures for parser performance, and F_1 is a compromise between these two.

Table 1. Testing results.

	Sentences		Constituents		
	Parsed	Correct	Recall	Precision	F_1
SSN	100%	14.4%	64%	65%	65%
PCFG	51%	3.3%	29%	54%	38%

The biggest difference between the SSN parser and the PCFG is in its robustness. For about half the sentences the PCFG did not find any non-zero probability parses. The SSN is always able to find a parse, and that parse is exactly correct more than four times as often as the PCFG. Because the PCFG's lack of robustness makes it difficult to make a meaningful comparison with the SSN parser on the other performance measures, we also calculated their perfor-

---

[4] By "syntactic constituent" we mean the substrings of the sentence which are the leaves of some subtree in the syntactic structure of the sentence. These substrings are labeled with the label of the root of their subtree. For example, the sentence "The dog ran" with the structure [S [NP the dog] ran] would have two constituents, a noun phrase [NP the dog] and a sentence [S the dog ran].

mance on the subset of sentences which the PCFG was able to parse, shown in Table 2. Even on these sentences the SSN parser outperforms the PCFG on the standard parsing measures by about 10%.

Table 2. Testing results on the sentences parsed by the PCFG.

	Sentences Correct	Constituents		
		Recall	Precision	F_1
SSN	16.3%	65%	67%	66%
PCFG	6.5%	57%	54%	56%

These results show that an SSN is able to learn effectively despite the fact that parsing real text places widely varying capacity requirements on the SSN's representation of state. The test set contained sentences ranging from 2 words to 59 words in length, and sentence structures ranging from 2 levels to 16 levels in depth. The SSN parser processes a sentence incrementally, one word at a time, so by the time the SSN parser reaches the end of the sentence, all the words and all the constituents in the sentence could possibly be relevant to the final parsing decision.

To show that the SSN parser indeed achieves this performance by generalizing what it learns to more complex representations, we analyzed the performance of the SSN parser on constituents of varying length and depth. Figure 2 plots the performance of the SSN and the PCFG as a function of the length of the constituent, and Figure 3 plots their performance as a function of the depth of the constituent. The points on each plot are the F_1 performance on the subset of constituents which include that many words, or that many levels, respectively. These results are all for the sentences which the PCFG can parse, and only results for subsets that include at least 10 constituents are shown. By comparing the performance of the SSN to that of the PCFG we can see whether the SSN has the same ability to generalize to more complex representations as statistical methods based on compositional symbolic representations.

Figure 2 shows that the SSN is generalizing successfully across constituent length. The difference in performance between the SSN and the PCFG stays consistently high up to length 17, at which point the results become more noisy due to the small number of constituents involved. Figure 3 shows that the SSN also generalizes successfully across constituent depth.[5] The difference in performance between the SSN and the PCFG stays consistently high up to depth 7, at which point the results become more variable.

---

[5] To be correct in this measure a constituent must have the correct words and label, but the structure of the subtrees associated with the output and target constituents can be different. For these results recall was calculated on the basis of the depth of the target constituent structure, precision was calculated on the basis of the depth of the output constituent structure, and F_1 was calculated from these values.

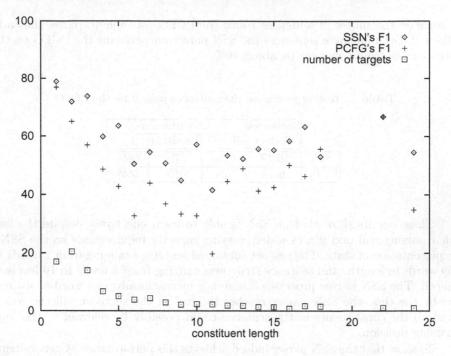

Fig. 2. F_1 results of the network and the PCFG on the PCFG's parsed sentences, plotted by constituent length. The lower curve shows the percentage of target constituents at each length.

5 Conclusions

These empirical results on parsing natural language show that segmenting state information on the basis of entities is an effective way to allow neural networks to handle wide variations in state capacity requirements. They also show that using temporal synchrony to represent this segmentation results in learning which generalizes appropriately to these larger states. This generalization is achieved because temporal synchrony results in learning generalizations about individual entities, which is appropriate for complex domains such as language processing. Thus the neuroscientific proposal of using the synchrony of activation spikes to segment patterns according to entities has been shown to have significant computational implications for artificial neural networks. Simple Synchrony Networks are an artificial neural network architecture which demonstrates the promise of this proposal.

References

[1] Eckhorn, R., Bauer, R., Jordan, W., Brosch, M., Kruse, W., Munk, M., Reitboeck, H.J.: Coherent oscillations: a mechanism of feature linking in the visual cortex? *Biological Cybernetics* **60** (1988) 121–130

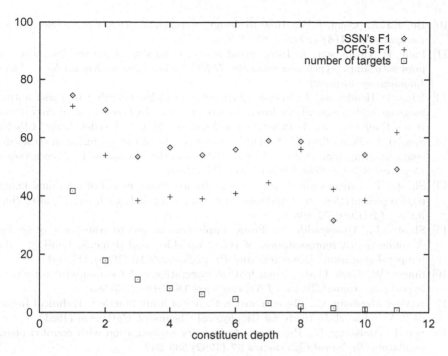

Fig. 3. F_1 results of the network and the PCFG on the PCFG's parsed sentences, plotted by constituent depth. The lower curve shows the percentage of target constituents at each depth.

[2] Elman, J.L.: Distributed representations, simple recurrent networks, and grammatical structure. *Machine Learning* **7** (1991) 195–225

[3] Fodor, J.A., Pylyshyn, Z.W.: Connectionism and cognitive architecture: A critical analysis. *Cognition* **28** (1988) 3–71

[4] Gray, C.M., Koenig, P., Engel, A.K., Singer, W.: Oscillatory responses in cat visual cortex exhibit intercolumnar synchronization which reflects global stimulus properties. *Nature* **338** (1989) 334–337

[5] Hadley, R.F.: Systematicity in connectionist language learning. *Mind and Language* **9(3)** (1994) 247–272

[6] Henderson, J.: A connectionist architecture with inherent systematicity. In *Proceedings of the Eighteenth Conference of the Cognitive Science Society*, La Jolla, CA (1996) 574–579

[7] Henderson, J.: Constituency, context, and connectionism in syntactic parsing. In Matthew Crocker, Martin Pickering, and Charles Clifton, editors, *Architectures and Mechanisms for Language Processing*, Cambridge University Press, Cambridge UK (2000) 189–209

[8] Henderson, J.: A neural network parser that handles sparse data. In *Proceedings of the 6th International Workshop on Parsing Technologies*, Trento, Italy (2000) 123–134

[9] Henderson, J., Lane, P.: A connectionist architecture for learning to parse. In *Proceedings of COLING-ACL*, Montreal, Quebec, Canada (1998) 531–537

[10] Ho, E.K.S., Chan, L.W.: How to design a connectionist holistic parser. *Neural Computation* **11(8)** (1999) 1995–2016

[11] Lane, P., Henderson, J.: Incremental syntactic parsing of natural language corpora with simple synchrony networks. *IEEE Transactions on Knowledge and Data Engineering* (in press)

[12] Lane, P, Henderson, J.: Simple synchrony networks: Learning to parse natural language with temporal synchrony variable binding. In *Proceedings of the International Conference on Artificial Neural Networks*, Skovde, Sweden (1998) 615–620

[13] Shastri, L., Wendelken, C.: Seeking coherent explanations – a fusion of structured connectionism, temporal synchrony, and evidential reasoning. In *Proceedings of Cognitive Science 2000*, Philadelphia, PA (2000)

[14] Shastri, L.: Advances in shruti – a neurally motivated model of relational knowledge representation and rapid inference using temporal synchrony. *Applied Intelligence* **11** (1999) 79–108

[15] Shastri, L., Ajjanagadde, V.: From simple associations to systematic reasoning: A connectionist representation of rules, variables, and dynamic bindings using temporal synchrony. *Behavioral and Brain Sciences* **16** (1993) 417–451

[16] Singer, W., Gray, C.M.: Visual feature integration and the temporal correlation hypothesis. *Annual Review of Neuroscience* **18** (1995) 555–586

[17] von der Malsburg, C.: The correlation theory of brain function. Technical Report 81-2, Max-Planck-Institute for Biophysical Chemistry, Gottingen (1981)

[18] von der Malsburg, C., Buhmann, J.: Sensory segmentation with coupled neural oscillators. *Biological Cybernetics* **67** (1992) 233–242

[19] Wang, D.L.: Primitive auditory segregation based on oscillatory correlation. *Cognitive Science* **20** (1996) 409–456

[20] Wang, D.L., Terman, D.: Image segmentation based on oscillatory correlation. *Neural Computation* **9** (1997) 805–836 (For errata see Neural Computation **9** (1997) 1623–1626)

Temporal Structure of Neural Activity and Modelling of Information Processing in the Brain

Roman Borisyuk[1,2], Galina Borisyuk[2], and Yakov Kazanovich[2]

[1]University of Plymouth, Centre for Neural and Adaptive Systems, School of Computing
Plymouth, PL4 8AA, UK
Borisyuk@soc.plym.ac.uk
[2]Institute of Mathematical Problems in Biology, Russian Academy of Sciences, Pushchino,
Moscow Region, 142290, Russia
{Borisyuk, Kazanovich}@impb.psn.ru

Abstract. The paper considers computational models of spatio-temporal patterns of neural activity to check hypotheses about the role of synchronisation, temporal and phase relations in information processing. Three sections of the paper are devoted, respectively, to the neuronal coding; to the study of phase relations of oscillatory activity in neural assemblies; and to synchronisation based models of attention.

1 Introduction

The theoretical study of the temporal structure of neural activity is very important. Despite the great complexity and variability of electrical activity in the brain, constantly increasing experimental data reveal consistent temporal relations in the activities of single neurons, neural assemblies and brain structures. Without a proper theoretical background, it is very difficult to guess how these relations can appear and what their role could be in information processing. This is especially important in the situation when detailed knowledge of the mechanisms of neural activity and neural interactions have not led to a significant progress in understanding how the information in the brain is coded, processed, and stored. What are the general principles of information processing in the brain? How can they be discovered through the analyses of electrical activity of neural structures? Which part of available experimental data reflects these general principles and which is related to the peculiarities of biological implementation of these principles in different brain structures? Are the observed temporal relations in neural activity related to information coding or they are the artefacts generated by special experimental conditions? These questions still wait their answer in future theoretical studies.

Computational neuroscience is one of the promising directions in developing the brain theory. The mathematical and computer models provide the possibility: to form general concepts and to apply them to the analysis of experimental data; to extract essential variables and parameters of neural systems which determine their information processing capabilities; to analyse the role of different mechanisms (biophysical, biochemical, etc.) in neural system functioning; to propose new hypotheses and to check their validity by comparing the modelling results with

S. Wermter et al. (Eds.): Emergent Neural Computational Architectures, LNAI 2036, pp. 237-254, 2001.

experimental data, to make suggestions about the further progress of neuroscience, to formulate the main ideas of new experiments and possible drawbacks.

In this paper, we consider several hypotheses that have been put forward to explain the role of temporal structure of neural activity in information processing. We describe neural networks that have been developed in support of these hypotheses and whose analysis reveals what kind of model neurons or neural assemblies are suitable and how their interaction should be organised to implement different types of information coding and processing.

2 Neuronal Coding

Traditionally, a neuron is considered as a device that transforms a changing sequence of input spikes into discrete action potentials that are transmitted through the axon to the synapses of other neurons. What are the properties of the neural spike trains that provide the possibility to carry the information or take part in information processing? Until recently, the most popular hypothesis has stated that this property is the rate of spikes in the train. Rate coding explains how the presentation and intensity of the stimulus can influence neural activity but this coding neglects the temporal organisation of spike trains.

Experiments show that temporal patterns of neural activity can be very complex, and it is natural to admit that there should be some information encoded by the moments of spike generation. For example, different stimuli or tasks can elicit different patterns of activity that have the same firing rate. Experimental data obtained in the last years show that in the slowly changing surrounding the rate code might be useful, but its efficiency drops abruptly if stimulation conditions change quickly. In the latter1. case, the fine temporal structure of spike trains should play a much more important role [1].

If we agree that the temporal pattern of activity carries the information about the stimulus, which features of this pattern are important? The popular hypothesis is that the stimulus-driven oscillatory activity of a neuron is a code of a significant stimulus [2] (more recent discussion of this problem can be found in [3]). The essence of oscillatory coding can be reduced to two basic ideas:

- place coding (the stimulus is coded by the location of a neuron that shows the oscillatory activity) and
- binding (the integration of local stimulus representations that can be realised through impulse synchrony).

The other approach takes into account the fine temporal structure of spike trains. The approach is based on the evidence that under some conditions of multiple stimulus presentation a neuron can reply, reproducing the moments of spikes with precision of 1 msec [1]. Note that the general pattern of activity is far from being regular in these experiments. Hence, the temporal structure of neural activity might be irregular and very complex. Therefore, the understanding of neural coding requires a very detailed study of temporal relations in multi-dimensional sets of spike trains. A powerful approach in describing the complex neural activity is based on the synchronisation principle [4]. The experimental evidence shows synchronous neural activity in the form of very precise and coherent spiking of many neurons relating to some specific stimulation [5].

Neural information coding can be implemented not only at the level of individual neurons but also at the level of the activity of neural assemblies. The ideas about information coding by neural populations: are similar to those of individual neurons: 1) rate coding, where the rate/activity at time t is determined as the average number of firing neurons in the given population per a time unit (or as the average membrane potential of neurons at the moment t); 2) spatio-temporal coding, where some characteristics of spatio-temporal patterns of neural activity are determined.

Thus, a general theoretical approach to neural coding systems can be formulated in the following way. A coding system is a network of dynamically interacting elements supplied by a set of inputs (which deliver the information about a stimulus to the network) and a set of outputs (which transmit the results of coding to other information processing systems). Depending on the type of coding, the elements of the network simulate in different detail the functioning of individual neurons, neural populations (excitatory and inhibitory), or neural structures. As the input signals, both constant and changing continuous signals can be used, as well as stochastic or deterministic sequences of spikes. The signals at different inputs can be identical to each other or not. The output signals are spike trains or the average activity of the coding elements (in these two cases, the adequate mathematical descriptions .of coding system dynamics are multidimensional stochastic processes and ordinary differential equations, respectively). Typically, the coding system has the capability to learn by experience, that is the parameters of the system elements and the strength of connections are modified, which may leads to the change of dynamics of the system while different stimuli are supplied to it.

The above definition emphasizes the importance of studying the dynamics of activity of encoding networks that arises as a result of the presentation of different stimuli. Not only the final state of the system after stimulation but its whole dynamics properly reflects what kind of coding is done [6,7]. Many investigations confirm that oscillatory dynamics of neural activity and its synchronisation should play a key role in the models of information processing in the brain (see, e.g., [5]). The synchronisation mechanism can be considered as a basis for modelling associative memory [8], feature binding [5], and attention [5, 9]. An oscillatory neural network formed by interacting neural oscillators is a powerful paradigm in the description of the dynamics and synchronisation of neural activity [10-12]. Different synchronous patterns that can appear in oscillatory neural network models include regular oscillations, chaotic oscillations, and quasiperiodic (envelope) oscillations with multiple frequencies. In the paper [13] we described how multifrequency oscillations could be used for information coding. In the case of two-frequency oscillations, we consider the synchronisation at a high frequency and lack of synchronisation at a low frequency as a specific code that is different from the one when synchronisation is spread to both high and low frequencies. Note that multi-frequency oscillations drastically improve the encoding capabilities of the system because different frequencies play the role of independent coordinates. Another fruitful approach to modelling the dynamics of neural activity in different brain structures is represented by stimulus-dependent persistent (metastable) states [6].

An interesting example of complex dynamics in an oscillatory neural network is represented by the chain of locally coupled oscillators with different connection types.

Such chain-like architecture is important for different applications including artificial neural networks and computational neuroscience (e.g., the chain-like architecture is natural for the sensory-motor cortex [14], also the 3D structure of the hippocampus may be considered as a chain of oscillators allocated along the septo-temporal axis [15-16]).

Figure 1a shows the wave propagation in a chain of 101 locally coupled Wilson-Cowan oscillators (see Appendix 1 for model description). To start the wave propagation, we choose a low value of external input to all oscillators, which corresponds to the low level background activity of a single oscillator [17]. Only the external input of the 51st oscillator is capable to induce oscillations. Thus, the 51st oscillator becomes the source of the propagating wave. Figure 1a shows the pattern of neural activity with propagating waves for a moderate value of the connection strength. The wave propagation is very limited for weak coupling because after propagating for a short distance the wave decays and disappears. A greater value of the connection strength allows the wave propagation along the chain. As a result, some spatio-temporal pattern of neural activity arises.

Computational experiments with wave propagation are related to the study of stimulus dependent neural activity. The source of a propagating wave appears as a result of constant simulation of an oscillator. Another possibility would be to present a stimulus during some limited time and to study the resulting stimulus-dependent pattern of neural activity. Figure 1b shows the result of stimulations with the formation of the stimulus-dependent activity pattern. In this computational experiment we chose the parameter value of the external input equal to $P = 0.8$ (this value corresponds to the low level background activity of a single oscillator) for all oscillators of the chain except one oscillator in the middle of the chain whose external input is $P_{51} = 1.5$ (the oscillatory mode of an isolated oscillator). We present the stimulus for five time units (keeping the values of external input parameters fixed), after that we stop the stimulation making $P_{51} = 0.8$ (all oscillators receive the same external input). Despite the termination of stimulation, a complex pattern of neural activity arises and exists for a long time. Figure 1b shows a spatio-temporal pattern related to the parameter values $\beta = 1.6$, $\alpha = 0.6$ that determine the coupling strength of an oscillator with its neighbours of the first and second order, respectively. These parameter values are minimal in the sense that coupling is strong enough to support the propagation of the wave but the wave rapidly decays for any smaller coupling strength.

3 Phase Relations of Neural Activity

In this section we consider the oscillatory dynamics of neural activity. This case is important for modelling since many EEG recordings show various rhythms (alpha, beta, gamma, etc.) during background activity (without presentation of an external stimulus) and in some functional states. While spectral analysis has been a dominating

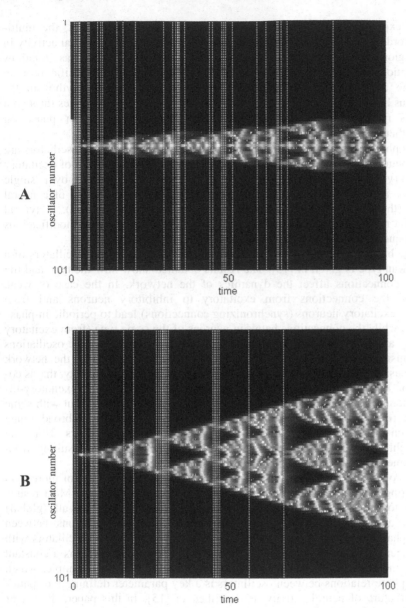

Fig. 1. Wave propagation in a chain-like neural network for the case of local excitatory-to-inhibitory connections. (A) Spatio-temporal pattern of neural activity for the permanent source of wave propagation. (B) Spatio-temporal pattern of neural activity for the case of a short stimulus application. The activity of 101 oscillators is shown in grades of grey, changing from black (low) to white (high).

method of experimental data processing under one-channel recordings, the multi-channel recordings can give the information about phase relations of neural activity in different regions of the brain. As an example of the important role of phase relations for information processing we refer to the paper [18], where a specific relation between the activity of place cells and the phase of the theta rhythm in the hippocampus has been shown. The key observation is that as the rat moves through a place cell's receptive field, the cell fires with progressively earlier phase on successive theta cycles (the so-called "phase-advance").

To study phase relations in oscillatory dynamics, networks of neural oscillators are helpful. A neural oscillator can be formed by two interacting populations of excitatory and inhibitory neurons. For simplicity, a population is approximated by a single excitatory or inhibitory element, which represents the average activity of a neural population (the term "neuron" is kept to denote this element as well). A typical example of a neural oscillator is the Wilson-Cowan oscillator [19], its modifications are most frequently used in oscillatory neural networks.

Complete bifurcation analysis of the system of two coupled neural oscillators of a Wilson-Cowan type is given in [13]. The results obtained show how the type and the strength of connections affect the dynamics of the network. In the case of weak connections, the connections from excitatory to inhibitory neurons and from inhibitory to excitatory neurons (synchronizing connections) lead to periodic in-phase oscillations, while the connections between neurons of the same type (from excitatory to excitatory and from inhibitory to inhibitory) lead to periodic anti-phase oscillations (desynchronising connections). For intermediate connection strengths, the network can enter quasiperiodic or chaotic regimes; it can also exhibit multistability, that is co-existence of several types of oscillatory dynamics. In the case of excitatory-to-excitatory connections, the oscillators can run with the same frequency but with some phase shift (out-of-phase oscillations). The phase shift varies in a broad range depending on the coupling strength and other parameters of oscillators. Thus, the analysis highlights the great diversity of neural network dynamics resulting from network connection architecture and strengths.

There is no general theory of oscillatory neural networks in the case of more than two interacting oscillators and an arbitrary architecture of connections. Most results are related to the following important types of architectures: all-to-all (global) connections and local connections. In both cases non-trivial relations between oscillation phases can appear. For example, in a network of identical oscillators with global connections, a so-called splay-phase state can exist, when there is a constant time lag in the phase dynamics of oscillators (see, [20]). Another example, which shows that phase relations between oscillators is a key parameter defining the spatio-temporal structure of neural activity, is described in [15]. In this paper, the set of models of hippocampal oscillatory activity is presented that takes into account some important features of the three-dimensional structure of the hippocampus. Two external periodic inputs are considered, one from the entorhinal cortex and another from the septum. Both inputs are engaged in slow theta-rhythm oscillations with some phase shift between them. It is shown that the phase shift is a crucial parameter that determines the spatio-temporal pattern of hippocampal neural activity.

Let us illustrate this phenomenon using an oscillatory neural network composed of integrate-and-fire units. The mathematical description of activity of integrate-and-fire neuron is given in Appendix 2. Each oscillator is formed by two units: an excitatory pyramidal neuron and an inhibitory inter-neuron. Oscillators are located at the nodes of the three-dimensional grid of size NxMxM. Here the first coordinate represents a position along the septotemporal axis of the hippocampus and N squares of size MxM are located on the planes orthogonal to the septotemporal axis (Fig 2A). We consider each plane as a hippocampal segment ("lamella"). Thus, the model consists of N segments allocated along the septo-temporal axis. The total activity of pyramidal neurons of a segment is an important characteristic corresponding to the hippocampal theta rhythm. Another important feature of the hippocampal activity is the level of synchrony (coherency) between different segments. The integrate-and-fire model allows one to study both spike trains of neurons and the average activity of each segment.

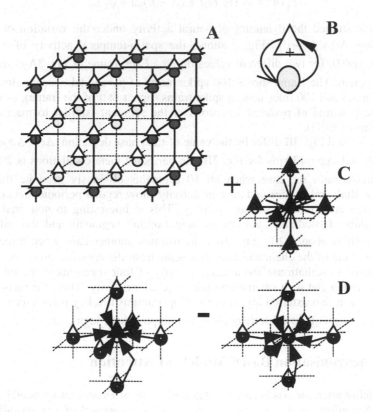

Fig. 2. Architecture of interconnections in the hippocampal model. (A) 3D structure of the model. (B) Each node contains an excitatory pyramidal neuron and an inhibitory inter-neuron. (C) Excitatory connections incoming to the pyramidal neuron. (D) Inhibitory connections incoming to the pyramidal neuron and inter-neuron.

The connections between neurons are shown in Fig. 2B-D. Each pyramidal neuron receives 6 excitatory connections from nearest pyramidal neurons and 7 inhibitory connections from nearest inter-neurons (6 from the nearest nodes and one from the inter-neuron at the same node). Each inter-neuron receives only one excitatory connection from the pyramidal neuron at the same node and 6 inhibitory connections from nearest inter-neurons. The input I_C from the entorhinal cortex is distributed along the septotemporal axis so that all neurons in the same segment (a transverse plane) have got the same input value

$$I_C(n) = a_C \sin(\omega_0 t + (n-1)\Delta t + \phi_C). \tag{1}$$

Here n is the segment number $(n = 1,...,N)$.

The input I_S from the septum is also distributed along the septotemporal axis but it propagates from the opposite side. All neurons in the segment n have got the same input value

$$I_S(n) = a_S \sin(\omega_0 t + (N-n)\Delta t + \phi_S). \tag{2}$$

We have studied the dynamics of neural activity under the variation of the phase deviation $\Delta\phi = \phi_C - \phi_S$. Fig. 3 shows the spatio-temporal activity of N segments (N=10, M=10) for two different values of $\Delta\phi$. Each frame of Fig. 3A corresponds to one segment. The frame shows 100 spike trains of pyramidal neurons (lower part of each frame) and 100 inter-neuron spike trains (upper part of the frame), as well as the average potential of pyramidal neurons of the segment (similar to macro electrode recording of EEG).

Fig. 3A and Fig. 3B differ by the value of the phase deviation $\Delta\phi$ ($\Delta\phi = 30ms$ for Fig. 3A and $\Delta\phi = 130$ ms for Fig. 3B, the period of theta oscillations is 200ms). Fig. 3A demonstrates the case when all 10 segments obviously generate theta-rhythm activity. Both spike trains and average activity show regular periodic theta oscillations with large amplitude of average activity. This is interesting to note that there is a phase shift of oscillatory activity of neighbouring segments and the value of this phase shift is about $120°$. Fig. 3B demonstrates another case when three segments from one side of the chain and three segments from the opposite side generate regular theta-rhythm oscillations. The average activity of four segments in the middle of the chain is rather flat with small amplitude of regular activity. Thus, the phase deviation $\Delta\phi$ between two external inputs to the hippocampus is a key parameter that controls the distribution of partial activity along the hippocampal segments.

4   Synchronisation Based Models of Attention

A model of attention has been formulated in terms of an oscillatory neural network in [21]. According to this model, the network is composed of the so-called central oscillator (CO), which is coupled with a number of other oscillators, the so-called peripheral oscillators (PO), by feedforward and feedback connections. Such network architecture facilitates the analysis of network dynamics and interpretation of network elements in terms of brain structures. For example, it is presumed that the septo-hippocampal system plays the role of the CO, while POs are represented by cortical columns which are sensitive to particular features. This concept is in line with Damasio's hypotheses that the hippocampus is the vertex of a convergent zones pyramid [22] and the ideas of Miller [23], who formulated the theory of representation of information in the brain based on cortico-hippocampal interplay.

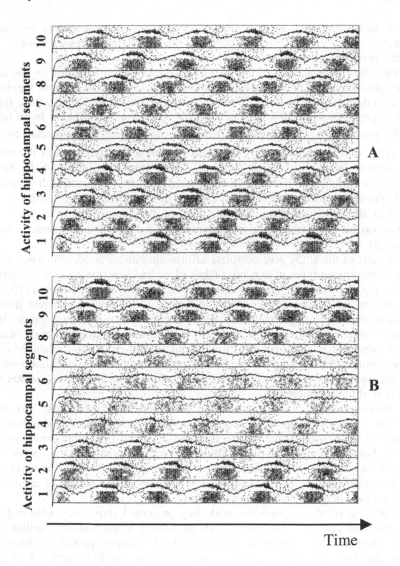

Fig. 3. Spatio-temporal pattern of neural activity for different values of the phase deviation. Each frame shows the activity of hippocampal segment: 100 spike trains of pyramidal neurons (bottom) and 100 spike trains of interneurons as well as the total potential of pyramidal neurons belonging to the segment versus time. (A) Phase deviation equals 30 ms. (B) Phase deviation equals 130 ms.

Attention is realised in the network in the form of synchronisation of the CO with some POs. Those POs that work synchronously with the CO are supposed to be included in the attention focus (here the synchronisation implies nearly equal current frequencies of corresponding oscillators). The parameters of the network that control attention focus formation are coupling strengths between oscillators and natural frequencies of oscillators (the frequency of an oscillator becomes natural if all connections of this oscillator are cut off).

Let the set of POs be divided into two groups, each being activated by one of two stimuli simultaneously presented to the attention system. Different oscillators of a group are interpreted as coding different features of a stimulus. We suppose that the natural frequencies of oscillators of group k $(k = 1, 2)$ are distributed in an interval Δ_k. The intervals Δ_k are supposed to be non-overlapping and separated by some empty space on the frequency axis. The following types of dynamics of the network are of interest for attention modelling: (a) **global synchronisation** of the network (this mode is attributed to the case when the attention focus includes both stimuli); (b) **partial synchronisation** of the CO and a group of POs (this mode is attributed to the case when the attention focus includes one of two competing stimuli); (c) **no-synchronisation** mode (this mode is attributed to the case when the attention focus is not formed).

The results of the study give complete information about conditions when each of the above-mentioned types of dynamics takes place and describe possible scenarios of transition from one mode to another under the variation of some parameters [9,24]. In particular, the model shows that when the focus of attention is switched from one stimulus to the other an intermediate state of attention focusing must appear. In this state the focus of attention is either absent or both stimuli are included in the attention focus. Another result of modelling is the formulation of conditions, when decreasing the interaction of the CO with the oscillators representing one of two stimuli included in the attention focus may lead not to focusing attention on the other stimulus but to destruction of the attention focus.

An important type of attention focus formation in experimental and natural conditions is a spontaneous attention shift between several stimuli simultaneously presented at the sensory input. There is some evidence that spontaneous attention switching between several complex stimuli (a complex stimulus is the one described by more than one feature) may be a general way of item-by-item examination during pop-out experiment [25]. Let us describe how spontaneous switching of attention can be implemented in an oscillatory network with a central element.

Note that the results on attention modelling presented above were obtained under fixed values of network parameters, such as natural frequencies of oscillators and coupling strengths. The current frequency of the CO during partial synchronisation depends on these parameters and determines the set of POs included in partial synchronisation. It may happen that this set is empty, or includes only a small part of oscillators of a given group or even combine some number of POs from different groups (in the latter case the focus of attention will contain a "monster" formed from incomplete sets of features of different stimuli). Such effects may appear in the model due to improper location of the attention focus in the frequency range. To overcome this difficulty, we consider a simple algorithm of adaptation of the natural frequency of the CO. By computer simulation we show that this adaptation results in automatic movement of the attention focus to an optimal position, where a maximal number of oscillators of one group corresponding to one stimulus are synchronised.

The fact that oscillators in the brain may tune their frequency to the frequency of the input stimulus is well known from the works of Ukhtomsky and his colleagues [26] and Thatcher & John [27]. Ukhtomsky suggested that frequency adaptation is one of general mechanisms of information processing in the brain and that it is used for forming stable excitation of neural assemblies (the so-called dominanta). A neural network implementation of frequency adaptation mechanism has been worked out in [28]. The idea that the central executive of the attention system is an oscillator and that it may change its natural frequency during attention focus formation belongs to Kryukov [21]. This assumption is in compliance with significant variation of theta-frequency activity observed in the hippocampus in the experiments on orienting response.

A mathematical formulation of the model of attention is presented in Appendix 3. Equation (A3.1) describes the dynamics of the CO and equation (A3.2) describes the dynamics of POs. The elements of the network are the so-called phase oscillators and the synchronisation is described in terms of a phase locking procedure. (Phase oscillator networks are under intensive investigation in the last years as a powerful instrument in neurophysiological modelling, see, e.g., [10,29,30,31]). The main new feature of the model is that the natural frequency of the CO changes according to equation (A3.3), which implies that the natural frequency of the CO tends to its current frequency. To avoid abrupt jumps of CO's natural frequency during transitional stages of synchronisation, the dynamics of this frequency is made slow relative to the rate of phase locking.

We suppose that r stimuli are presented simultaneously to the attention system. The stimulus k is coded by the activity of a group of POs with q oscillators in each group whose natural frequencies are randomly distributed in an interval Δ_k on the frequency axis. The intervals are supposed to be non-overlapping. For simplicity, all intervals have the same size 1 and are separated by empty spaces of length 1. Thus, the range $(\omega_{min}, \omega_{max})$ of distribution of the frequencies of all POs has the size $2r - 1$. In computation experiments we put

$$\Delta_k = (2k - 1, 2k), \ k = 1, 2, ..., r. \tag{3}$$

(Note that a shift of all values of natural frequencies in the equations for phase dynamics is equivalent to a change of phase variables, therefore, without loss of generality, one can choose any value for ω_{min}).

Let the period $(0, T)$ of stimulation be divided into equal intervals $t_1, t_2, ..., t_s$. The procedure of spontaneous attention focusing and switching works in the following way. At the initial moment, all oscillators start with phases randomly distributed in the range $(0, 2\pi)$. The initial value of the natural frequency ω_0 of the CO is 0. At the initial moment of each time interval t_j $(j = 1, ..., s)$ the natural frequency ω_0 of the CO is set to be equal to a randomly chosen value in the range $(\omega_{min} - 1, \omega_{max} + 1)$. During all other moments the dynamics of network variables is determined by equations (A3.1-A3.3).

We fix the value $w = 0.5$ of coupling strength between oscillators. This value is large enough to synchronise oscillators of one group under a proper value of ω_0 (near the middle of an interval Δ_k) but it is too small to synchronise all oscillators of more than one group, because the range of phase locking of a PO is not more than $d\omega_0 / dt \pm w$. During different time intervals t_j, different groups of POs will be synchronised by the CO. Which group will be synchronised depends on the random value of ω_0 at the initial moment of t_k. Starting from this value, the natural frequency of the CO will move in the direction of natural frequencies of the nearest group of POs (the word "near" is used in the sense of the distance on the frequency axis). The final location of ω_0 will be around the average value of natural frequencies of POs in the synchronised group.

5 Conclusions

The following parameters of the network have been used in computation experiments: the number of groups $r = 5$, the number of oscillators in each group $q = 20$, duration of all t_j is 500 units of time. The period of stimulation is $T = 10000$.

A typical result of computation experiments is presented in Fig. 4. The graphics show time evolution of the current frequencies of POs and of the natural frequency of the CO. During the whole period of stimulation, the groups of oscillators were synchronised in the following order: group1, group 2, group 5, group 1, group 1, etc. As can be seen from the figure, during each time segment t_j similar behaviour of oscillators is observed. The natural frequency of the CO tends to some value, which is more or less optimal for the synchronisation of oscillators of one group. The current frequencies of oscillators of this group tend to the natural frequency of the CO. As a result, after some transitional period nearly all oscillators of one group are synchronised by the CO. Exceptions can be observed for those POs whose natural frequencies are at the edges of their frequency range Δ_k. The synchronisation of such oscillators is most difficult and sometimes they may be out of synchronisation with the CO. Such situation can be observed, for example, during time segments $t_4=(1500, 2000)$, $t_5=(2000, 2500)$, $t_7=(3000, 3500)$, when one or two oscillators of a group do not work synchronously with the CO.

The results described above can be generalised to the case when the attention focus includes several stimuli. In this case one should assume that w is not constant but for each t_k the value of w is randomly chosen from some range (w_1, w_2), where w_1 is large enough to synchronise oscillators of one group and w_2 is large enough to synchronise oscillators of, say, two groups. (Such changes of coupling strength may be attributed to random changes of synchronising activity of the CO in different time periods). The result of synchronisation during time interval t_j will depend on the position of ω_0 at the initial moment of t_j and also on the value of w during t_j. Depending on these values, ω_0 may float to the average of natural frequencies of POs of one group (in this case one group of POs will be synchronised by the CO) or to the average of natural frequencies of POs in two neighbouring groups (in this case two groups of POs will be synchronised by the CO and the focus of attention will consist of two stimuli).

The model of attention considered above is based on synchronisation of regular oscillations. The following example shows that regularity is not an obligatory condition to obtain synchronous oscillations of neural activity, also synchronised chaotic oscillations can be generated by the networks of integrate-and-fire neurons. This gives the possibility to use chaotic oscillations in binding and attention models.

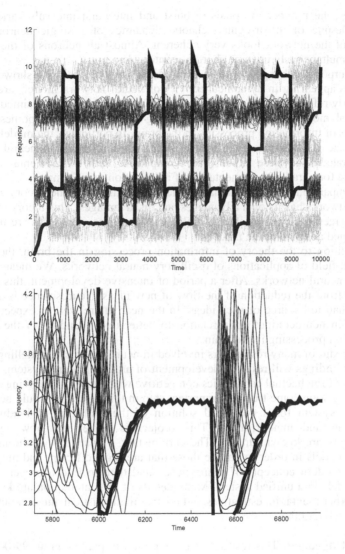

Fig. 4. Above: time evolution of current frequencies of POs and of the natural frequency of the CO. Period of stimulation is T = 10000. Below: zoomed fragment of synchronisation of oscillators in group 2 (the range of natural frequencies in the group is between 3 and 4). The thick trajectory shows the natural frequency of the CO, thin trajectories show the current frequencies of POs.

An example of a neural network that combines chaotic dynamics with synchronisation is presented in [4]. The authors have developed a neural network of excitatory and inhibitory integrate-and-fire neurons with global connections that can show spatially coherent chaotic oscillations. A specific property of this regime is that the periods of neuron bursting activity alternate with irregular periods of silence.

Moreover, the number of spikes in burst and interburst intervals varies in a broad range. Despite of the irregular, chaotic dynamics of a single neuron, the global activity of the network looks very coherent. Almost all neurons of the network fire nearly simultaneously in some short time intervals.

The purpose of this paper was twofold. First, we were going to show that temporal structures appearing in dynamical activity of neural network models are rich enough to properly reflect the basic neurophysiological data. Second, we aimed to show that dynamical models are helpful for checking the validity of hypotheses about the principles of information processing in the brain. In particular, the models can be used to elucidate the possible significance of temporal relations in neural activity. We demonstrated that these models are compatible with experimental data and are promising for parallel implementation of information processing.

In comparison with traditional connectionist theories, the theory of oscillatory neural networks has its own advantages and disadvantages. This theory tries to reach a better agreement with neurophysiological evidence, but this results in more complicated analysis of the models.

In addition to the theory of information processing in the brain, there is another important field of applications of oscillatory neural networks. We mean the theory of artificial neural networks. After a period of intensive development, this theory seems to suffer from the reduction of the flow of new ideas. Neuroscience is a reliable and never exhausted source of such ideas. In the near future we can expect a significant progress in neurocomputing in relation to better understanding of the principles of information processing in the brain.

The dream of many researchers involved in neurobiological modelling is that some day their findings will result in development of artificial neural systems with a broad spectrum of intellectual capabilities competitive with those of the living systems. This dream may soon come true. An important step in this direction would be to develop a computer system for a combined solution of the problems of binding, attention, recognition and memorisation. This project is quite real now. All necessary components are known already. The current task is just to scan through the set of existing models in order to choose those that are most efficient and most compatible with the modern concepts of neuroscience and then to find a proper interaction of these models in a unified system. Many details of this interaction are known already, further experimental investigations and relative mathematical and computer modelling should discover others.

Acknowledgement. This research was supported in part by grant 99-04-49112 from the Russian Foundation of Basic Research and UK EPSRC grant GR/N63888/01.

References

1. Mainen, Z.F., Sejnowski, T.J.: Reliability of spike timing in neocortical neurons. Science 268 (1995) 1503-1506
2. Borisyuk, G.N., Borisyuk, R.M., Kirillov, A.B., Kryukov, V.I., Singer, W.: (1990) Modelling of oscillatory activity of neuron assemblies of the visual cortex. In: Proc. of Intern. Joint. Conf. on Neural Networks - 90, Vol. 2. San-Diego, (1990) 431-434
3. Singer, W.: Putative functions of temporal correlations in neocortical processing. In: Koch, C., Davis, J.L. (eds.): Large scale neuronal theories of the brain. MIT Press, Cambridge MA (1994)

4. Borisyuk, R., Borisyuk, G.: Information coding on the basis of synchronization of neuronal activity. BioSystems 40 (1997) 3-10
5. Borisyuk, G.N., Borisyuk, R.M., Kazanovich, Y.B., Strong, G.: Oscillatory neural networks: Modeling binding and attention by synchronization of neural activity. In: Levine, D.S., Brown, V.R., Shiry, V.T. (eds.): Oscillations in Neural Systems. Lawrence Erlbaum Associates, Inc. (2000) 261-284
6. Kryukov, V.I., Borisyuk, G.N., Borisyuk, R.M., Kirillov, A.B., Kovalenko, Ye.I.: Metastable and unstable states in the brain. In: Dobrushin, R.L., Kryukov, V.I., Toom, A.L. (eds.): Stochastic Cellular Systems: Ergodicity, Memory, Morphogenesis, Part III. Manchester University Press, Manchester (1990) 225-358
7. Arbib, M., Erdi, P., Szentagothai, J.: Neural organization. Structure, function and dynamics. MIT Press (1998)
8. Borisyuk, R., Hoppensteadt, F.: Memorizing and recalling spatial-temporal patterns in an oscillator model of the hippocampus. BioSystems 48 (1998) 3-10
9. Kazanovich, Y.B., Borisyuk, R.M.: Dynamics of neural networks with a central element. Neural Networks 12 (1999) 441-453
10. Hoppensteadt, F., Izhikevich, E.M.: Weakly Connected Neural Networks. Springer-Verlag, New York (1997)
11. Ermentrout, B.: The analysis of synaptically generated travelling waves. J. Comput. Neurosc. 5 (1998) 191-208
12. Abarbanel, H., Huerta, R., Rabinovich, A., Rulkov, N., Rowat, P., Selverston, A.: Synchronized action of synaptically coupled chaotic model neurons. Neural Computatation 8 (1996) 1567-1602
13. Borisyuk, G., Borisyuk, R., Khibnik, A., Roose, D.: Dynamics and bifurcations of two coupled neural oscillators with different connection types. Bulletin of Mathematical Biology 57 (1995) 809-840
14. Abeles, M.: Corticonics: Neural circuits of the cerebral cortex. Cambridge University Press, NY (1991)
15. Borisyuk, R., Hoppensteadt, F.: Oscillatory model of the hippocampus: A study of spatio-temporal patterns of neural activity. Biol. Cybern. 81 (1999) 359-371
16. Amaral, D., Witter, M.: The three dimensional organization of the hippocampal formation: a review of anatomical data. Neuroscience 31 (1989) 571-591
17. Borisyuk, R.M., Kirillov, A.B.: Bifurcation analysis of a neural network model. Biol. Cybern. 66 (1992) 319-325
18. O'Keefe, J., Recce, M.: Phase relationship between hippocampal place units and EEG theta rhythm. Hippocampus 3 (1993) 317-330
19. Willson, H.R., Cowan, J.D.: Excitatory and inhibitory interactions in localized populations of model neurons. Biophys. Journal 12 (1972) 1-24
20. Swift, J.W., Strogatz, S.H., Wiesenfeld, K.: Averaging in globally coupled oscillators. Physica D 15 (1992) 239-250
21. Kryukov, V.I.: An attention model based on the principle of dominanta. In: A.V. Holden., V.I. Kryukov (eds.): Neurocomputers and Attention I. Neurobiology, Synchronization and Chaos. Manchester University Press, Manchester (1991) 319-352
22. Damasio, A.: The brain binds entities and events by multiregional activation from convergent zones. Neural Computations 1 (1989) 123-132
23. Miller, R.: Cortico-hippocampal interplay and the representation of contexts in the brain. Springer-Verlag, Berlin (1991)
24. Kazanovich, Y.B., Borisyuk, R.M.: Synchronization in a neural network of phase oscillators with the central element. Biol. Cybern. 71 (1994) 177-185
25. Horowitz, T.S., Wolfe, J.M.: Visual search has no memory. Nature 394 (1998) 575-577
26. Ukhtomsky, A.A.: Collected Works. Nauka, Leningrad (1978) 107-237 (in Russian)
27. Thatcher, R.W., John, E.R.: Functional Neuroscience (Foundations of Cognitive Processes, v.1), Lawrence Erlbaum, New York (1977)

28. Torras, C.: Neural network model with rhythm assimilation capacity. IEEE Trans. System Man, and Cybernetics, SMC-16 (1986) 680-693
29. Schuster, H.G., Wagner, P.: A model for neuronal oscillations in the visual cortex. 2. Phase description of the feature dependent synchronization. Biol. Cybern. 64 (1990) 83-85
30. Ermentrout, B., Kopell, N.: Learning of phase lags in coupled neural oscillators. Neural Computation 6 (1994) 225-241
31. Hoppensteadt, F.: An introduction to the mathematics of neurons (2nd edition, 1997), Cambridge Univ Press, Cambridge (1986, 1997)

Appendix 1:   A Chain-Like Network of Neural Oscillators with Local Connections

We consider a Wilson-Cowan neural oscillator consisting of an excitatory neural population $E_n(t)$ and an inhibitory neural population $I_n(t)$ [8]. Let us suggest that identical oscillators are arranged in a chain with local connections between nearest oscillators. The dynamics of the network is described by the equations:

$$\frac{dE_n}{dt} = -E_n + (k_e - E_n) * S_e(c_1 E_n - c_2 I_n + P_n + V_n),$$

$$\frac{dI_n}{dt} = -I_n + (k_i - I_n) * S_i(c_3 E_n - c_4 I_n + W_n),$$ (A1.1)

$$n = 1,2,...,N.$$

Here c_1, c_2, c_3, c_4 are the coupling strengths between the populations constituting an oscillator; P_n is the value of the external input to the excitatory population;

$$k_p = 1/S_p(+\infty);$$ (A1.2)

$$S_p(x) = S_p(x; b_p, \theta_p), \quad p \in \{e, i\}$$

where $S_p(x)$ is a monotonically increasing sigmoid-type function given by the formula

$$S_p(x) = 1/(1 + \exp(-b_p(x - \theta_p))) - 1/(1 + \exp(b_p \theta_p)).$$ (A1.3)

For indexes n outside the range $[1,N]$ we use the reflection symmetry relative to the ends of the chain.

The parameter values are:

$$c_1 = 16, \quad c_2 = 12, \quad c_3 = 15, \quad c_4 = 3, \quad \theta_e = 4,$$ (A1.4)

$$b_e = 1.3, \quad \theta_i = 2, \quad b_i = 2, \quad P_n = 1.5, \quad N = 101.$$

Coupling between oscillators is described by the terms V_n and W_n. We consider four different connection types and study the dynamical behaviour of the chain of oscillators in four different cases corresponding to different connection types. Each oscillator (with the index n) receives four connections: two connections with the connection strength β from the nearest oscillators $(n-1)$ and $(n+1)$ and the other two connections with the connection strength α from the oscillators $(n-2)$ and $(n+2)$.

Appendix 2: Dynamics of Integrate-and-Fire Units

The dynamics of an integrate-and-fire neuron is governed by the following equations:
1. Threshold:

$$r(t+1) = (r_{max} - r_\infty)\exp(-\alpha_{th}(t-t_{sp})) + r_\infty ,$$
(A2.1)

where r_{max} is the maximum value of the threshold; r_∞ is the asymptotic threshold value when $t \to \infty$; α_{th} is the threshold decay rate; t_{sp} is the last spike moment before t.

2. Post-synaptic potential for the input of the neuron:

$$PSP^j(t+1) = PSP^j(t)\exp(-\alpha_{PSP}^j) + a ,$$
(A2.2)

$$a = \begin{cases} w^j, \text{if } t_{sp}^j + \tau^j = t+1 \\ 0, \quad \text{otherwise} \end{cases}$$

where w^j is the connection strength, positive for the excitatory connection and negative for the inhibitory one ; τ^j is the time delay; α_{PSP}^j is the jth neuron PSP decay rate; t_{sp} is the last spike moment of the jth neuron before t.

3. Noise:

$$N(t+1) = N(t)\exp(-\alpha_N) + \xi ,$$
(A2.3)
$$\xi \in N(0,\sigma) ,$$

where α_N is the noise decay rate; ξ is a random variable with the normal distribution.

4. Somatic membrane potential:

$$V(t+1) = V_{AHP}\exp(-\alpha_V(t-t_{sp}))$$
(A2.4)

where V_{AHP} is the value of after spike hyperpolarisation; α_N is the somatic membrane potential decay rate; t_{sp} is the last spike moment before t.

5. Total potential:

$$P(t+1) = \sum_j PSP^j(t+1) + N(t+1) + V(t+1) + I_{ext}(t+1) ,$$
(A2.5)

where I_{ext} is the value of the external input.

6. Spike generation:

$$\text{if } P(t+1) > r(t+1), \text{ then } t_{sp} = t+1 .$$
(A2.6)

Appendix 3: Phase Oscillator Model of Attention

The model of attention is formulated in terms of a network of phase oscillators. It is described by the following equations:

$$\frac{d\theta_0}{dt} = \omega_0 + \frac{w}{n}\sum_{i=1}^{n}\sin(\theta_i - \theta_0) \qquad (A3.1)$$

$$\frac{d\theta_i}{dt} = \omega_i + w\sin(\theta_0 - \theta_i),\ i = 1,2,...,n\ , \qquad (A3.2)$$

where θ_0 is the phase of the CO, θ_i is the phase of a PO, ω_0 is the natural frequency of the CO, ω_i is the natural frequency of a PO, n is the number of POs, w is the coupling strength between oscillators, $\dfrac{d\theta_0}{dt}$ is the current frequency of the CO, $\dfrac{d\theta_i}{dt}$ is the current frequency of a PO.

The adaptation of the natural frequency of the CO is described by the equation

$$\frac{d\omega_0}{dt} = -\gamma\left(\omega_0 - \frac{d\theta_0}{dt}\right), \qquad (A3.3)$$

where the parameter γ controls the rate of natural frequency adaptation to the value of its current frequency.

Role of the Cerebellum in Time-Critical Goal-Oriented Behaviour: Anatomical Basis and Control Principle

Guido Bugmann

Centre for Neural and Adaptive Systems
School of Computing, University of Plymouth, Plymouth PL4 8AA, UK
http://www.tech.plymouth.ac.uk/soc/Staff/GuidBugm/Bugmann.htm

Abstract. The Brain is a slow computer yet humans can skillfully play games such as tennis where very fast reactions are required. Of particular interest is the evidence for strategic thinking despite planning time being denied by the speed of the game. A review of data on motor and cognitive effects of cerebellar lesions leads to propose that the brain minimizes reaction time during skilled behaviour by eliminating on-line planning. Planning of the next action is concurrent with the execution of the current action. The cerebellum plays a role in preparing fast visuo-motor pathways to produce goal-specific responses to sensory stimuli.. Anatomically, the cerebellum projects to all extra-striate components of the fast sensory-motor route: Posterior parietal cortex (PPC), Premotor cortex (PM) and Motor Cortex (M1). Indirect evidences suggest that the cerebellum sets up stimulus-reaction (S-R) sets at the level of the PPC. Among unresolved issues is the question of how S-R mappings are selected in PPC, how planning is performed and how the cerebellum is informed of plans. Computationally, the proposed principle of off-line planning of S-R associations poses interesting problems: i) planning must now define both the stimulus and the action that it will trigger. ii) There is uncertainty on the stimulus that will appear at the time of execution. Hence the planning process needs to produce not a single optimal solution but a field of solutions. It is proposed here that problem i) can be solved if only learned S-R associations are involved. Problem ii) can be solved if the neural network for S-R mapping has appropriate generalization properties. This is demonstrated with an artificial neural network example using normalized radial basis functions (NRBF). Planning a single optimal trajectory enables to generate appropriate motor command event for initial states outside of the optimal trajectory. Current implementations include a simulated robot arm and the control of a real autonomous wheelchair. In terms of control theory, the principles proposed in this paper unify purely behaviour-based approaches and approaches based on planning using internal representations. On one hand sensory-motor associations enable fast reactions and on the other hand, being products of planning, these associations enable flexible goal adaptation.

S. Wermter et al. (Eds.): Emergent Neural Computational Architectures, LNAI 2036, pp. 255-269, 2001.
© Springer-Verlag Berlin Heidelberg 2001

1 Introduction

In the game of tennis, the time for the ball to cross the court is almost exactly equal to the reaction time of the player added to its displacement time[1]. There does not seem to be time for planning and it is unclear how players manage to produce strategic shots under these conditions. The cerebellum is known to be involved in the co-ordination of action and its lesion results in slower reactions [1,2]. Can we learn something from studying the cerebellum that may shed some light on the computational strategies used by tennis players? Can we learn something that may help design faster robots despite limitations in computational resources? These are the questions underlying this paper.

How can the cerebellum accelerate reactions? The answer is not straightforward, because the cerebellum does not initiate actions, its inputs arriving usually to the motor cortex around the time when the action starts [3]. Several experiments will be described in section 2 that suggest that the cerebellum helps selecting actions to be executed in response to sensory stimuli. In section 3, a possible biological substrate is suggested. In section 4, a conceptual model based on these observations is proposed that combines planning and reactive behaviour. In section 5, a neural network implementation is described to illustrate the principle. The conclusion follows in section 6.

2 Role of the Cerebellum in Planning

Several experiments show that the cerebellum is involved in planning. A more precise view on its role in that respect emerges from an analysis of the following experiments.

- In one experiment, human subjects are asked to walk a path of irregularly placed stepping stones as fast as possible [4]. Subjects usually fixate the next stone just before starting to lift the foot that will be placed on the stone. To assess the role of visual sampling, tests were also performed in the dark, with stone positions indicated by LED's turned off at different times. Visual deprivation had no effect when the goal was not illuminated during fixation time and swing phase of the movement. However, deprivation towards the end of the swing affected sometimes the accuracy of the next step. This indicated that subjects were planning the move to the next stepping stone, while fixating the current stone an executing the current stepping action. Interestingly, shift of fixation from the current to the next stone took also place in the dark, indicating that the fixation was more a reflection of a pre-existing plan than a source of information needed for its execution. The intermittent effect of visual deprivation

---

[1] The length of the court is approximately 24 m and its width 8m. A ball flying at 100 km/h takes 860ms to travel 24m. Choice reaction time is approximately 270ms and the displacement time, to reach a corner of the court starting from the middle (at half the speed of a 100m world record, i.e. 5m/sec, and moving only 3m due to arm-length of 1m), is estimated at 600ms. Planning normally takes several seconds [5]. The problem is that the approach run is dependent on the intended type of hit. Therefore, the decision must be made almost a the start of the run.

could be due to compensation by spatial memory of the layout, which, in another experiment, has be shown to persist for several seconds in the dark [6]. The stepping-stone experiment done with cerebellar patients showed a reduced accuracy and intermittent delays (prolonged stance) correlated with inaccurate fixating saccades to the next stone [7]. Other experiments described below may allow narrowing down the numerous possible reasons for these effects.

- In another experiment, subjects were told to touch, upon hearing a GO sound, one of the elements of a display that is indicated by a light lit for a few seconds before the GO signal [8]. The reaching movements of healthy subjects had the same characteristics in the dark or in the light, indicating the execution of a pre-planned movement, or as noted above, the existence of an internal spatial memory (world model) that is replacing actual visual input. In contrast, cerebellar patients showed increased variability in initial movement direction and, for the final portion, performed worse in the dark, indicating their reliance on visual feedback. However, the overall correct initial direction suggests that the problem was not one of spatial memory (not remembering the position of the target), but one of precise planning or plan execution.

- In another experiment, a monkey was trained to touch in the correct order the two elements lit simultaneously in a succession of displays [9]. In such experiments, a trained animal shows anticipatory saccades to the next element, indicating knowledge of the sequence, and as suggested above, the existence of a plan as for which element to touch next. Lesions in the dorsolateral dentate nucleus, that relays cerebellar outputs to the cortex, reduced the number of anticipatory saccades and caused more errors. This could indicate a loss of sequence memory, or errors in translating sequence knowledge into a correct plan or action. The next experiment shows that the latter is more likely.

- A patient with cerebellar lesions showed spatial dysgraphia characterised by a large number of stroke omissions or stroke repetitions when writing in upper case, even when copying a word [10]. Errors occurred despite a memory of the sequence of strokes not being required. The authors suggested that this was a case of neglected visual feedback (e.g. informing that all strokes of a letter have been produced), but as the performance worsened in darkness, visual feedback was being used. Somehow, cerebellar lesions prevented available knowledge (visible example or memory of sequence) to be correctly converted into a motor plan. If the patient was unfamiliar with writing in upper cases, his task becomes similar to the stepping stone experiment where, while writing one letter, he has to decide which one to write next. Here the problem is clearly one of programming this next action.

- Patients with cerebellar lesions tended to do more illegal moves in a Tower of Hanoi test which is a test of planning capability [11]. This effect was neither due to not understanding the instructions nor to some error in motor control. It was more like an inadequate plan was being selected and executed from start to end.

- In a target detection task, patients with cerebellar lesions have difficulties in setting up rapidly a response program, as shown by longer response times when the target appears shortly (50ms) after the start of the experiment. It takes approximately 800ms before evidence for readiness to respond is seen [2].

In summary, healthy subjects or animals animals plan ahead while acting, which is revealed by anticipatory saccades and short reaction times. The symptoms of cerebellar patients can be described as resulting from an inappropriate and late setup of a response rule or action program. The role of the cerebellum in fast responses is seen here as one of preparing the sensory-motor system for producing a task-specific response to a subsequent sensory stimulus.

A similar although less specific hypothesis was formulated in [12] to explain the participation of the cerebellum in attention orienting. However, the view put forward here differs from earlier suggestions that the cerebellum may be directly involved in action generation via learned stimulus-response (S-R) mappings, or so-called "context-response linkage" [13,14,15]. In a later extension to these ideas, it was suggested [16] that the cerebellum may also learn context-plan linkages, i.e. inform areas of the cortex involved in planning of a plan - that may or may not be executed - evoked by a context. Specification of an action plan, as proposed in [16], or enabling given S-R sets as proposed here are probably similar concepts (see section 4).

Applying these ideas to the problem of fast yet strategic responses by a tennis player, it can be assumed that the tennis player plans his or her next shot during the flight of the ball towards the other side of the court and during his own shot. The notion of concurrent planning is supported by the stepping stone experiment described above showing that planning and action can be performed in parallel. The cerebellum would then setup a S-R set that triggers a response as soon as the ball is returned. The stimulus S could be defined as a view of the court when the ball is fixated. One problem with that approach is that the response of the opponent is unpredictable. Therefore a series of responses needs actually preparing. We refer to that as a "response field". In the next section, we will examine briefly the network connecting the cerebellum to other brain areas that may support the preparatory role of the cerebellum. The question of response field will be addressed in section 5.

3   Anatomical Basis for the Role of the Cerebellum in Response Pre-selection

In this section the neural network supporting the preparatory role of the cerebellum are analysed. First, the fastest paths from primary visual areas to the motor output are identified. Then the most likely target areas for cerebellar control signals are discussed.

Several paths with very short latencies link visual input to motor output (figure 1A, B). A series of paths devoted to the control of eye movements comprise areas V1, PO, LIP, FEF and SC [17]. Another group of paths is devoted to reaching movements and comprises V1, PO, MIP, PM and M1 [18,19,20, 21].

Average visual response latencies in FEF are only 75ms [19] and are close to 100ms in PM [22]. Total eye saccade reaction time (from stimulus onset to movement onset) is approximately 120 ms [23]. Simple arm reaching reaction time is 250ms for a single fixed visual target, and 270 ms for a target that can appear in one of two positions [24]. This includes a 50-100 ms delay between the issue of the motor command in M1 and the initiation of movement [25,20]. When action selection depends on stimulus properties evaluated in the ventral visual path, reaction times are clos to 360 ms [26].

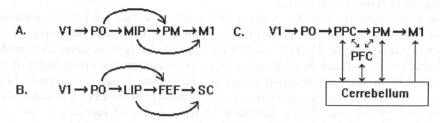

A. V1 → PO → MIP → PM → M1   **C.** V1 → PO → PPC → PM → M1

B. V1 → PO → LIP → FEF → SC

Fig. 1. A. Fastest path from visual input to motor output for arm movements. B. Fatst paths for eye movements. C. Cerebello-cortical connections. V1: Primary visual area, PO : Parieto Occipital Cortex, MIP and LIP are the medial and lateral intraparietal areas, parts of Posterior Parietal Cortex (PPC). The frontal eye field (FEF) is a pre-motor area for eye movements, M1: Primary motor cortex. SC: Superior Colliculus, PFC: Prefrontal cortex.

The main outputs[2] of the cerebellum target areas M1, PM and PPC in the fast paths [27,28,29,30]. There is also a projection to the prefrontal cortex (figure 1C). These connections are reciprocal. Other inputs to the cerebellum include higher visual areas devoted to the peripheral visual field, SC, and the cingulate gyrus [31,32].

The cerebellum is thus in position to influence information processing along fast sensori-motor routes. The more likely targets for an action on stimulus-response mapping are areas PPC and PM. Area M1 is activated by the cerebellum after motor commands have been issued [3]. Therefore the projections to M1 are more likely to be involved in controlling ongoing movements [33] than initiating or preparing them. Areas PO and V1 have no documented cerebellar inputs.

The posterior parietal cortex (PPC) contains maps of intended movements relative to objects in the field of view. When given hand, arm or eye movements are intended, specific subgroups of neurones become active [34]. It has been proposed that area PPC rather represents "potential motor responses" that are communicated to the PM where the "intended response" would be selected in a task dependent way [35]. Finally the PM would activate neurons in M1 which initiate the "actual response" [36].

Electrophysiological data give a more complex image of these three processing stages [37,38]. Neurons coding for "potential responses" have been observed in PM [39] and even in M1 [26]. It has been suggested that visuo-motor transformations are a gradual rather than a clear-cut multistage process [21].

In an experiment by [26], a monkey was presented with two colored LED placed on either side of a fixation point. Its task was to point either in the direction of the lit led, or in the opposite direction, depending on the color of the LED. It was found that the activity in a population of cells in M1 reflects initially the cue location (latency after stimulus onset 80-200 ms), then the S-R rule (LED color) (latency 180-240ms) then the response direction (latency 236-286 ms). On the one hand, such observations are consistent with a multistage sequential process whereby fast position information from

---

[2] Different sub-areas in the cerebellum are linked to different cortical areas (see. e.g. [12, 27].

the PPC is combined with slower color information from the ventral visual stream to finally determine the response. On the other hand, the fact that area M1 has access to information from intermediate processing stages indicates a distributed rather than hierarchical network organization.

In this experiment [26], the position of the stimulus does not provide complete information on the action to perform and combination of ventral and dorsal visual information is required. The PFC and PM are the more likely areas where multimodal information is combined. Brain imaging experiments confirm the involvement of the dorsal PM in response selection based on non-spatial visual information [40]. However, these experiments also suggest an involvement of the PPC. Clinical studies show that PPC lesions disable voluntary or memory-guided saccades to the side contralateral to the lesion [23]. There is no clear reason for the involvement of the PPC in these tasks, as there are direct connections between prefrontal memory areas and pre-motor areas. Similarly it is unclear why the PFC and PPC are coactive when a response needs to be memorized in a delayed response task [41]. These data suggest that the PPC rather than the PM could be last stage before the motor cortex. However, transmagnetic stimulation of the PPC impairs the accuracy of memory guided saccades only if applied in the first 50ms after visual cue offset [42]. This suggests that PPC only provides movement specifications that are then memorised and executed in further stages.

These contradictions indicate that further research is needed. However whatever the interpretation, the PPC stands out as a crucial element of the fast sensori-motor path. It is therefore an ideal target for modulation by the cerebellum. [43] suggest that the cerebellum initiates actions commands in M1 via the PPC relay. Imaging experiments show that the PPC and Cerebellum are more active when the S-R mapping rule is frequently changed (e.g. detect a color or a shape in the stimulus)[44]. The same is observed during motor learning of finger flexion sequences by trial and error [45]. In this experiment, a new finger has to be selected for execution at each pacing tone. Thus trial-and-error learning may involve frequent setting of new S-R rules.

Other issues also need to be resolved before a complete picture of the preparatory role of the cerebellum can be drawn. For instance, it is not known how planning is performed in the brain. It is known that planning requires PFC, anterior cingulate and caudate areas but also the PM, an areas in common with fast sensory-motor paths [46]. This complicates further the question of how planning can occur concurrently with action execution. It is also unclear by which route the cerebellum is informed of which plan to implement in the form of S-R sets. Finally, is unclear how a S-R rule can be projected to the PPC. Projections from the cerebellum to PPC terminate in superficial layers [47] and could be better placed for a modulatory role than for a driving role. It is therefore plausible that the cerebellum operates by pre-selecting previously learnt S-R rules rather than by projecting new rules. This would restrict fast responses to learnt movements, highlighting the importance of training.

In summary, there is still some way to go before we will be able to produce a definitive biological model of the role of the cerebellum in fast responses. However in designing artificial systems, anatomical details do not need copying. Only the computational principle needs to be reproduced.

4 Conceptual Model of the Role of the Cerebellum

The review of behavioural and neurophysiological data in sections 2 and 3 lead to propose following model of response generation (figure 2).

1. The context C elicits a plan P, via a mapping C-P that could be located in the cerebellum [16] but possibly also elsewhere, e.g. in PFC.
2. The plan P is communicated to the cerebellum that implements a mapping between plan P and stimulus response set S-R.
3. The S-R set corresponding to the plan is pre-selected in the PPC via cerebellar projections.
4. The stimulus S activates the response R.

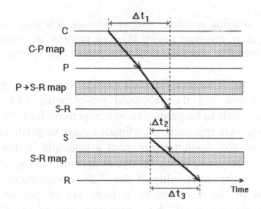

Fig. 2. Timing of the processes leading to the pre-selection of a S-R set in PPC and response selection. The delay $\Delta t1$ is the time for information from cortex to reach the cerebellum and from there the PPC (estimated 100-200ms). The delay $\Delta t2$ is the time for visual information to reach the PPC (approx. 70-100ms). The delay $\Delta t3$ is the reaction time (approx. 270ms). C: Contextual information, P: Plan, S: Stimulus, R: Response. The C-P map links context to plan. Possibly located in Cerebellum. The P to S-R map converts a plan into a stimulus-response (S-R) selection. This map is possibly located in the cerebellum. The S-R selection is conveyed to the S-R map that will compute the response upon arrival of the stimulus. This map is probably located in the PPC.

Observing the diagram in figure 2, one may wonder if the mapping from context C to S-R set needs to be a two-stage process. It may be possible that the cerebellum realizes the mapping directly from context C to S-R set. In that case, the cerebellum would need accessing contextual information which comprise sensory and motivational information. It remains to be verified if the necessary projections exist.

In the proposed system, the S-R set can be adapted dynamically to changes of the context. Hence the response R is appropriate for a context not more than $\Delta t = \Delta t1 - \Delta t2 + \Delta t3$ old. This may be not more than 200-300 ms, depending on the circuits involved and the nature of the relevant contextual information. In the tennis example described above, it would be the game situation 30-130ms before the opponent hits the return that determines the selection of a shot. This time window contains the last moments of the opponent's swing.

From a control perspective, the key notion in the hypothesis is that actions are *selected*. This suggest that time can be saved if, instead of planning on the basis of current sensory information, pre-calculated sets of sensori-motor responses (a "plan") were available and appropriate responses were selected on the basis of current sensory information.

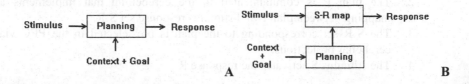

Fig. 3. A. Serial model of action selection: Sensory information is used for planning and the resulting plan is executed. B. Proposed model: Planning precedes the stimulus and sets up a stimulus-response (S-R) map. The stimulus selects the response in the S-R map.

Figure 3 illustrates the difference between the classical serial approach to goal-directed action generation and the proposed pre-planning model. In the serial approach, the response needs to be predictive to compensate for computational delays in planning. In the proposed approach, the stimulus must be predicted. An important characteristic of a controller implementing such a principle is that a plan must be reflected in a variety of response being set up, to account for the variability of ones action execution and the unpredictability of the of the environment. Choosing a plan therefore corresponds to the selection of a large set of possible sensory-motor associations. Another solution would be a representation of the S-R mapping that generalizes over variants of S and produces variants of R that compensate for deviations from expectations. In the next section an artificial neural network with this property is outlined.

5   Neural Network Implementation

The neural network architecture described bellow was initially proposed in [48]. The principle of the implementation is: i) Use the output of a planner to train a sensory-motor (SM) association network; ii) Let the SM network control the sequence of actions.

As a planner, a neuro-resistive grid was used [49] that approximates Laplacian planning. In principle, any other techniques could be used, directed graphs, etc. For encoding sensory-motor associations, a network of Normalized Radial Basis Functions (NRBF) was used.

This section focuses on the details the operation of the NRBF network and its generalization properties. This will be illustrated with S-R maps coding for the trajectory of a 2-jointed robot arm.

5.1 The Robot-Arm Planning Problem

The task that the planner had to solve was to find a sequence of arm configurations allowing the gripper to reach the object marked by the black circle in figure 4A. The result obtained using a resistive grid planner is a trajectory in configuration space shown in figure 4B [48].

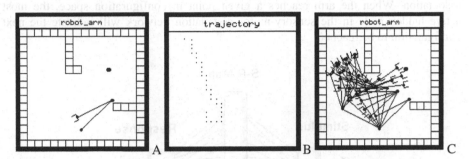

Fig. 4. A. Robot arm in its initial configuration in the workspace. B. Trajectory in configuration space. The x-axis is the shoulder joint angle. The y-axis is the elbow joint angle. The initial configuration is at the top left. C. Sequence of arm positions corresponding to the trajectory in B.

The main advantage of the resistive-grid method is that it guarantees finding a solution to the planning problem, if there is one. However, it has also the limitations of grid-based methods: a limited resolution. Underlying figure 4B is a 25 x 25 grid that represent all possible combinations of the two joint angles. As the joints can rotate by a full 360 degree, there is a 14.4 degree angular step change from one node to the next. This has two consequences, i) the planned movement is very saccadic (Figure 4C), ii) the goal is not reached (Figure 4C) because its exact location does not necessarily correspond to the centre of a node in the grid. It is possible to solve that problem [48] by using a variant of the neural network described in the next section. However, this will not be detailed here.

5.2 A Normalized RBF Network Representing Sensory-Motor Associations

A normalized RBF net has the same architecture as a standard RBF net (fig 5). The only difference is the normalized sum done in the output nodes. The input nodes code for the stimulus and the output nodes code for the desired response. Any number of outputs can be used. Each node in the hidden layer (S-R Map) encodes one association between a stimulus and a response.

In this simplified example, sensory inputs are measurements of the joint angles α_1, α_2 and motor outputs are desired joint angles α'_1, α'_2. This input corresponds to proprioceptive information rather than visual input, but the network could operate in the same way with appropriately encoded visual information[3]. A complete trajectory of the robot arm is encoded here rather than a single response. Inputs and outputs are defined by the trajectory (fig 4B) produced by the planner. Each point in the trajectory is successively used as input (corresponding to the current values α_1, α_2 of joint angles) and the next point is used as desired output (corresponding to the next set of values α'_1, α'_2). One new node in the hidden layer is recruited for each S-R association. When the arm reaches a given point in configuration space, the most active hidden node in the sensory-motor association network will indicate the next configuration to be reached.

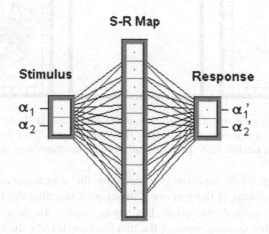

Fig. 5. Neural network for Sensory Motor Association.

The sensory motor association network has two active layers, as depicted in Figure 5. Nodes in the hidden layer have a Gaussian radial basis transfer function. Their output ϕ_j is given by

$$\phi_j = \exp\left(-\frac{1}{2\sigma_j^2} \sum_{i=1}^{2} (\alpha_i - \alpha_{ij})^2\right) \qquad (1)$$

Hidden layer nodes respond maximally when the input angles α_i are equal to the prefered input values α_{1j} and α_{2j} set during learning. These define the centre of the node's "receptive field" in the configuration space. The value of σ_j sets the width of the tuning curve (or size of the receptive field). Hidden layer nodes project to the output "Response" layer with weights W_{j1} and W_{j1}. The output layer comprises here two nodes with outputs α'_1 and α'_2. These outputs are calculated as follows:

---

[3] The representation of visual information needs to generalises in such a way that the activity at each input of the net diminuishes gradually as the image is transformed. This is satisfied by brain cells, but is difficult to achieve with artificial vision systems.

$$\alpha'_i = \frac{\sum_j W_{ij}\phi_j}{\sum_j \phi_j} \tag{2}$$

This is a weighted average over the input weights, where the input activities ϕ_j play the role of weights. The weights W_{ji} are actually the desired output angles. This can be checked by assuming that only one hidden node is active (e.g. because the current input falls into its receptive field). In that case the output is equal to W_j. Interestingly, the actual value ϕ_j does not influence the output value. This is the basis of the generalisation property described in the next section.

It may be possible to approximate the normalization function with biological neurones, e.g. assuming feedforward inhibition.

Training the network consists of setting the weights from hidden to output layer to the desired values of the next angles. This is a fast one-shot learning procedure.

During replay of the learned sequence, the output of is used to control the arm movement, i.e. modify it towards the desired configuration. The actual values of joint angles are used as input. As soon as one configuration has been reached, the network provides the next values of joint angles. As NRBFs have overlapping receptive fields, a small number of nodes in the hidden layer are simultaneously active which may each point to different future configurations. However, the function (2) of the output nodes allows the most *active* of the inputs to have the largest *weight* in the decision.

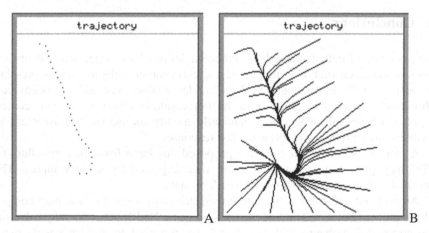

Fig. 6. A. Trajectory in configuration space produced by the Sensory Motor Association network trained on data from figure 4B. B. Example of generalization: Trajectories in configuration space generated by the sensory-motor association network for various initial arm configurations. It should be noted that the network was trained solely with the trajectory data shown in figure 4B.

Figure 6A shows the trajectory in configuration space produced by a movement controlled by the sensori-motor association network. The sequence is not replayed with the same number of steps or speed as the sequence used for the training (Figure

4B). The steps in the replay depend on the size of the receptive fields. If these are large, many hidden nodes respond for any given position, and the next value of the angles is an average over the values set for each of the active hidden nodes. This average can be close to the current angle and only a small displacement occurs. If the receptive fields are too small, the smoothness of the trajectory is lost, but the original sequence of steps is reproduced. For the results shown here $\sigma = 2\pi/25$ was used.

5.3 Generalisation

NRBFs nets behave as nearest neighbor classifier [50]. This has the consequence that inputs far from any input value used during training produce the response ("next position") corresponding to the nearest input in the training set. Figure 6B illustrates the trajectories in configuration space generated by the sensory-motor association network for various initial positions.

The particular type of sensory-motor association network used here needs to be trained only on *one trajectory* produced by the planner, and can lead an arm to the goal from *any* initial configuration. Hence, the need for planning a field of sensory-motor associations that is mentioned in section 4 might not be a practical problem. This property was used for a robust control of the trajectories of an autonomous wheelchair [51]. In that application, additional context input was used to select among several subsequences.

6 Conclusions

Observations of deficits caused by cerebellar lesions have suggested that one of the roles of the cerebellum is to prepare fast sensory-motor paths to express pre-planned or learned S-R associations. Such a model enables fast skilled behaviour by eliminating the planning time from the perception-to-action cycle. Yet, contextual input or off-line planning can dynamically modify the active S-R association and enable strategic thinking to influence fast responses.

As for the underlying circuit, it is proposed that input from the cerebellum to the PPC may pre-select responses that are then triggered by sensory inputs. This is speculative but not inconsistent with available data.

An artificial network has been described that implements the S-R map component of the proposed model. The S-R map design using NRBF networks may solve one of the conceptual problems with the model, i.e. the need to predict a wide range of possible sensory stimuli S and calculate the corresponding responses R. Thus, implementation of the complete model with artificial neural networks may not be of overwhelming complexity.

From a control theory perspective, the proposed model unifies i) behaviour-based control [52] that is fast but with limited flexibility in terms of goals   and ii) classical plan-based control which is flexible but slow due to the computational overhead.

Further work is needed to evaluate the relative contributions of Basal Ganglia and Cerebellum in setting up S-R sets and the potential of the model for encoding complex schemas and their management [53].

References

1. Stein J.F. & Glickstein M.G. (1992) "Role of the cerebellum in visual guidance of movement", Physiological Reviews, 72:4, pp. 967-1017.
2. Townsend J., Courchesne E., Covington J., Westerfield M., Singer Harris N., Lynden P., Lowry T.P. and Press G.A. (1999) "Spatial attention deficits in patients with acquired or developmental cerebellar abnormality", J. Neuroscience, 19:13, pp. 5632-6543.
3. Butler E.G., Horne M.K. & Hawkins N.J. (1992) "The activity of monkey thalamic and motor cortical-neurons in a skilled ballistic movement", J. of Physiology-London, 445, pp. 25-48.
4. Hollands M.A. & Marple-Horvat D.E. (1996) "Visually guided stepping under conditions of step cycle-related denial of visual information", Exp. Brain Res., 109, pp. 343-356.
5. Owen A.M., Downes J.J., Sahakian B.J., Plkey C.E. and Robbins T.W. (1990) Planning and Frontal Working Memory Following Frontal Lobe Lesions in Man. Neuropsychologia. 28:10, pp. 1021-1034.
6. Thomson J.A. (1980) "How do we use visual information to control locomotion", Trends in Neuroscience, October issue, pp. 247-250.
7. Marple-Horvat D.E., Hollands M.A., Crowdy K.A. & Criado J.M (1998) "Role of cerebellum in visually guided stepping", J. of Physiology, 509P, pp. 28S.
8. Day B.L., Thompson P.D., Harding A.E. & Marsden C.D. (1998) "Influence of vision on upper limb reaching movement in patients with cerebellar ataxia", Brain, 121, pp. 357-372.
9. Lu X., Hikosaka O. & Miyach S. (1998) "Role of monkey cerebellar nuclei in skill for sequential movement", J. of Neurophysiology, 79:5, pp 2245-2254.
10. Silveri M.C., Misciagna S., Leggio M.G. & Molinari M. (1997) "Spatial dysgraphia and cerebellar lesion: A case report", Neurology, 48:6, pp. 1529-1532.
11. Grafman J., Litvan I., Massaquoi S., Stewart M., Sirigu A. & Hallet M. (1992) "Cognitive planning deficits in patients with cerebellar atrophy", Neurology, 42:8, pp. 1493-1496.
12. Allen G., Buxton R.B., Wong E.C. and Courchesne E. (1997) Attentional Activation of the Cerebellum Independent of Motor Involvement., Science, 275, pp. 1940-1943.
13. Brindley, G.S. (1969) The use made by the cerebellum of the information that it receives from the sense organs. Int. Brain. Res. Org. Bull. 3:80.
14. Marr, D. (1969) A theory of cerebellar cortex. J. Physiol. 202:437- 470.
15. Albus, J.S. (1971) A theory of cerebellar function, Math. Biosci. 10:25-61.
16. Thach, W. T. (1996). On the specific role of the cerebellum in motor learning and cognition: Clues from PET activation and lesion studies in man. Behavioral and Brain Sciences 19(3): 411-431.
17. Bullier J., Schall J.D. and Morel A. (1996) "Functional stream in occipito-frontal connections in the monkey", Behavioural Brain Research, 76, pp. 89-97.
18. Rushworth M.F.S., Nixon P.D. and Passingham R.E. (1997) Parietal Cortex and Movement. 1. Movement Selection and Reaching. Experimental Brain Research, 117:2, pp. 292-310.
19. Schmolesky M.T., Wang Y., Hanes D.P., Thompson K.G., Leutgeb S., Schall J.D. and Leventhal A.G. (1998) "Signal timing across the visual system", J. Neurophysio., 79:6, pp. 3272-3278.

20. Johnson P.B., Ferraina S., Bianchi L. & Caminiti R. (1996) "Cortical networks for visual reaching: Physiological and anatomical organization of frontal and parietal lobe arm regions", Cerebral Cortex, 6, pp. 102-119.

21. Caminiti R. (1996) From vision to movement: Combinatorial Computation in the Dorsal Stream. In: Caminiti R., Hoffmann K.P., Lacquaniti F. and Altmann J. "Vision and Movement Mechanisms in the Cerebral Cortex", HFSP, Strassbourg pp. 42-49.

22. Shimodozono M., Mikami A. and Kubota K. (1997) Visual receptive fields and movement fields of visuomovement neurons in the monkey premotor cortex obtained during a visually guided reaching task. Neuroscience Research 29:1, pp. 55-71.

23. Braun D., Weber H., Mergner TH. and Schulte-Mönting J. (1992) "Saccadic reaction times in patients with frontal and parietal lesions", Brain, 115, pp. 1359-1386.

24. Bonnefoi-Kyriacou B., Trouche E., Legallet E. and Viallet F. (1995) "Planning and execution of pointing movements in cerebellar patients". Movement Disorders, 10:2, pp. 171-178.

25. Schwartz A.B., Kettner R. and Georgopoulos A.P. (1988) Primate motor cortex and free arm movements to visual targets in three-dimensional space. I. Relation between single cell discharge and direction of movement. J. of Neuroscience, 8, pp. 113-117.

26. Zhang Z., Riehle A., Requin J. and Kornblum S. (1997) "Dynamics of Single Neuron Activity in Monkey Primary Motor Cortex Related to Sensorimotor Transformation", J. Neuroscience, 17:6, pp. 2227-2246.

27. Middleton F.A. & Strick P.L. (1998) "Cerebellar output: motor and cognitive channels", Trends in Cognitive Science, 2:9, pp. 348-354.

28. Kakei S., Yagi J., Wannier T., Na J and Shinoda Y. (1995) Cerebellar and cerebral input to corticocortical and corticofugal neurons in areas 5 and 7 in the cat. J. of Neurophysiology, 74, pp. 400-412.

29. Matelli M., and Luppino G. (1996) "Thalamic Input to Mesial and Superior Area 6 in the Macaque Monkey", J. of Comparative Neurology, 372, pp. 59-87.

30. Fanardzhyan V.V. and Papoyan Y.V. (1995) synaptic Effects Induced in efferent Neurons of the Parietal Associative Cortex of the Cat by Stimulation of the Cerebellar Nuclei. Neurophysiology, 27:3, pp. 148-155.

31. Schmahmann J.D. and Pandya D.N. (1991) "Projections to the basis pontis from the superior temporal sulcus and superior temporal region in the Rhesus monkey". J. of Comparative Neurology, 308, pp. 224-248.

32. Schmahmann J.D. and Sherman J.C. (1998) "The cerebellar cognitive affective syndrome", Brain, 121, pp. 561-579.

33. Aizawa H. and Tanji J. (1994) "Corticocortical and thalamocortical responses of neurons in the monkey primary motor cortex and their relation to a trained motor task", J. of Neurophysiology, 71:2, pp. 550-560.

34. Snyder L.H., Batista A.P. & Andersen R.A. (1997) "Coding of intention in the posterior parietal cortex", Nature, Vol. 386, no 6621, pp. 167-170.

35. Kalaska J.F. & Crammond D.J. (1995) "Deciding not to GO: Neuronal Correlates of Response Selection in a GO/NOGO Task in Primate Premotor and Parietal Cortex", Cerebral Cortex, 5:5, pp.410-428.

36. Riehle A. (1991) "Visually induced signal-locked neuronal activity changes in precentral motor areas of the monkey: hierarchical progression of signal processing". Brain Research, 540, pp. 131-137.

37. Shen L.M. & Alexander G.E. (1997a) "Neural correlates of a spatial sensory-to-motor transformation in primary motor cortex", J. of Neurophysiology, 77:3, pp.1171-1194.

38. Shen L.M. & Alexander G.E. (1997b) "Preferential representation of instructed target location versus limb trajectory in dorsal premotor area", J. of Neurophysiology, 77:3, pp.1195-1212.

39. Murata A., Fadiga L., Fogassi L., Gallese V., Raos V. and and Rizzolatti G. (1997) "Object representation in the ventral premotor cortex (Area F5) on the monkey", J. of Neurophysiology, 78, pp 2226-2230.

40. Grafton S.T., Fagg A.H. and Arbib M.A. (1998) Dorsal premotor cortex and conditional movement selection: A PET functional mapping study. J. Neurophysiology. 79, pp. 1092-1097.

41. Toni I., Schluter N.D., Joseph O., Friston K. and Passingham R.E. (1999) "Signal-, set- and movement-related activity in the human brain: An event-related FMRI study", Cerebral Cortex, 9:1, pp. 35-49.

42. Brandt S.A., Ploner C.J., Meyer B.-U., Leistner S. and Villringer A. (1998) "Effects of repetitive transcranial magnetic stimulation over dorsolateral prefrontal and posterior parietal cortex on memory guided saccades", Experimental Brain Research, 118:2, pp. 197-204.

43. Lalonde R. and Botez-Marquard T. (1997) The neurobiological basis of movement initiation. Reviews in the Neurosciences, 8, pp. 35-54.

44. Le T.H., Pardo J.V., and Hu X.(1998) "4 T-fMRI Study of Nonspatial Shifting of Selective Attention: Cerebellar and Parietal Contributions ", J Neurophysiol, 79, pp. 1535-1548.

45. Jueptner M., Frith C.D., Brooks D.J., Frackowiak R.S.J. and Passingham R.E. (1997) Anatomy of motor learning. II. Subcortical structures and learning by trial and error. J. Neurophysiology, 77, pp. 1325-1337.

46. Dagher A., Owen A.M., Boecker H. and Brooks D.J. 1999. Mapping the network for planning: a correlational PET activation study with the Tower of London task. Brain, 122:Pt.10, pp. 973-1987.

47. Okuda B. (1997) "Early morphological changes in the thalamocortical projection onto the parietal cortex following ablation of the motor cortex in the cat", Brain Research Bulletin, 44:3, pp. 281-287.

48. Althoefer K. & Bugmann G. (1995) "Planning and Learning Goal-Directed Sequences of Robot-Arm movements", Proc. of ICANN'95, Paris, Vol. 1, 449-454.

49. Bugmann, G., Taylor, J.G. and Denham M. (1995) "Route finding by neural nets" in Taylor J.G (ed) "Neural Networks", Alfred Waller Ltd, Henley-on-Thames, p. 217-230

50. Bugmann, G. (1998) "Normalized Radial Basis Function Networks", Neurocomputing (Special Issue on Radial Basis Function Networks), 20, pp. 97-110.

51. Bugmann G., Koay K.L., Barlow N., Phillips M. and Rodney D. (1998) "Stable Encoding of Robot Trajectories using Normalised Radial Basis Functions: Application to an Autonomous Wheelchair", Proc. 29th Intl. Symp. Robotics (ISR'98), 27-30 April, Birmingham, UK, pp. 232-235.

52. Brooks R.A. (1986) A robust Layered Control System for a Mobile Robot. IEEE Journal of Robotics and Automation. RA-2:1, pp. 14-23.

53. Shallice, T. (1988) "From neuropsychology to mental structure", New York: Cambridge University Press

Locust Olfaction
Synchronous Oscillations in Excitatory and Inhibitory Groups of Spiking Neurons

David C. Sterratt*

Institute for Adaptive and Neural Computation, Division of Informatics,
University of Edinburgh,
dcs@cogsci.ed.ac.uk,
http://www.cogsci.ed.ac.uk/~dcs

Abstract. During odour recognition, excitatory and inhibitory groups of neurons in the second stage of the locust olfactory system, the antennal lobe (AL), fire alternately. There is little spread in the firing times within each group. Locust anatomy and physiology help to pin down all parameters apart from the weights in a coarse spiking neuron model of the AL. The time period and phase of the group oscillations do however constrain the weights; this paper investigates how.
I generalise the spiking neuron locking theorem [3] to derive conditions that allow stable synchronous firing within multiple groups. I then apply the general result to the AL model. The most important result is that for a general form of postsynaptic potential (PSP) function the excitatory and inhibitory neuronal populations cannot fire alternately at certain time periods and phases, regardless of the size of the weights between and within groups.

The dynamics of groups of spiking neurons connected by purely excitatory or purely inhibitory synapses and pulsatile connections have been well studied (for example [3,7,8,9]). The Wilson-Cowan oscillator [11] has inspired a number of models of the dynamics of the mass activity of reciprocally connected groups of inhibitory and excitatory neurons. However, the dynamics of reciprocally connected groups of excitatory and inhibitory spiking neurons do not appear to have been studied, perhaps due to lack of biological motivation.

A system that provides some motivation is the peripheral olfactory system of the locust *Schistocerca americana*, which has been studied by Laurent and co-workers [4]. Its anatomy and mass activity is summarised in Fig. 1. Each antenna contains about 50 000 excitatory primary olfactory sensory neurons, called receptor neurons (RNs), which project into the ipsilateral AL. The AL contains ~830 excitatory projection neurons (PNs) and ~300 inhibitory local neurons (LNs). There appear to be connections from RNs to PNs and LNs and reciprocally within and between the PN and LN groups. The connections are

---

* This work was carried out under the supervision of David Willshaw and Bruce Graham, with the financial support of the Caledonian Research Foundation. I would like to thank Leslie Smith for comments on an earlier draft.

S. Wermter et al. (Eds.): Emergent Neural Computational Architectures, LNAI 2036, pp. 270–284, 2001.
© Springer-Verlag Berlin Heidelberg 2001

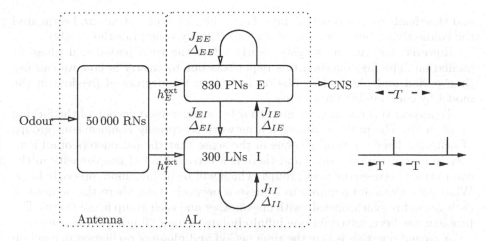

Fig. 1. The basic anatomy and activity of the locust AL. The left hand part of the figure shows the anatomy as described in the text and the symbols used later on in the paper. Open arrowheads denote excitatory connections and filled ones indicate inhibitory ones. The right hand part shows the group activity of the PNs (top) and LNs (bottom).

organised in 1000 groups of synapses called glomeruli, and the mean number of each type of neuron projecting into a glomerulus is known.

When no odour reaches the antenna, the RNs fire spontaneously at a low rate. When odour is present, their firing rate either increases, decreases or remains the same, depending on the identity of the RN and the odour. The PNs are conventionally spiking neurons, whereas the LNs are axon-less neurons that produce graded action potentials. With odour present at the antenna, subsets of the PNs and LNs fire; otherwise they are mainly quiescent. The overall activity of the both sets of neurons oscillates at about 20 Hz and the excitatory PNs lead the inhibitory LNs by quarter of a cycle [5]. This frequency and phase seem to be fairly independent of the concentration of the odour. Furthermore, it appears that there is usually only one spike per PN per cycle [6].

Experimental evidence suggests that the spatiotemporal pattern of activation of the PNs over a period of about 1 s encodes the identity of odours presented to the system [4,5]. The oscillations seem to be important for refining the representation of the odour, as when they are pharmaceutically abolished honeybees find it harder to discriminate between similar odours.

In order to understand the mechanism of the oscillations and spatiotemporal patterns I have been modelling the RNs and AL. Although the LNs are not true spiking neurons, one way of modelling them is as spiking neurons since it appears that their effect is similar to inhibitory postsynaptic potentials (IPSPs) from synchronously-firing inhibitory neurons [5]. As there is not enough data to produce a biophysical model of each type of neuron, I used the spike response model (SRM) [2] and fitted the PSPs, refractory curves, axonal delays and thresholds to experimental data. From the glomerular structure I estimated the connectivity, but there is no data that directly constrains the weights.

However, the (mean) weights clearly affect the time period and phase of oscillation. The first question this paper sets out to answer is how we can use time period and phase of oscillation to reduce the degrees of freedom in the model by constraining the weights.

To answer this question we need to solve a generalised version of the locking problem [3]. The problem concerns a network comprising homogeneous groups of neurons. They are homogeneous in the sense that the parameters of all neurons in a group are the same and there is a certain kind of homogeneity in the connections between different groups, which will be defined more precisely later. What conditions are required to achieve a network state where the neurons of each group fire synchronously with one another and each group fires in turn? The problem has been solved for one infinitely-large group [3] using SRM neurons.

A second question is how the time period and phase of oscillation depend on the input strength to the system.

In Sect. 1, I briefly introduce the SRM. In Sect. 2, I give the mathematical form of the constraints for multiple groups and confirm that the locking theorem can be extended. In Sect. 3, I present the implications of the constraints for the excitatory-inhibitory oscillator. In Sect. 4, a network simulation with one set of parameters is consistent with most of the analytical and numerical predictions but shows some differences, possibly due to the simulation method. Finally, in Sect. 5, I discuss the conclusions and suggest some further work.

1 The Spike Response Model

The SRM [2] provides a framework for describing and analysing networks of spiking neurons. It comes in two forms, both of which can be derived by integrating the integrate-and-fire (IF) neuron equations. I use the simpler SRM model as it can describe adaptation.

In a network of N neurons, the membrane potential $h_i(t)$ at time t of each neuron i is given by

$$h_i(t) = h_i^{\text{syn}}(t) + h_i^{\text{ref}}(t) + h_i^{\text{ext}}(t) \tag{1}$$

where $h_i^{\text{syn}}(t)$ is the *postsynaptic* potential of neuron i, $h_i^{\text{ref}}(t)$ is the *refractory* potential and $h_i^{\text{ext}}(t)$ is the *external* (input) potential. When the membrane potential reaches a threshold θ_i the neuron fires.

The postsynaptic potential comprises PSPs caused by spikes from the neurons in the network:

$$\sum_{i=1}^{N} \sum_{f=1}^{F} h_i^{\text{syn}}(t) = J_{ij} \epsilon_{ij}\left(t - t_j^{(f)}\right) , \tag{2}$$

where J_{ij} is the weight from neuron j to neuron i, $\epsilon_{ij}(s)$ is the PSP function from neuron j to neuron i, $t_i^{(f)}$ represents the fth last firing time of neuron i and

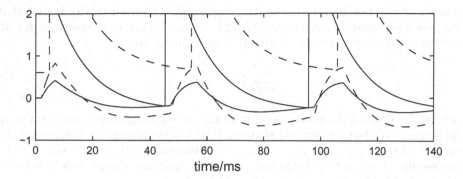

Fig. 2. An example of the synaptic potentials (lower traces) and effective threshold $\theta - h^{\text{ext}}(t) - h^{\text{ref}}(t)$ (upper traces) of neurons in the excitatory (solid line) and inhibitory (dashed line) groups. Each neuron fires when its membrane potential crosses the effective threshold. The parameters are: τ_E =10 ms, τ_I =15 ms, Δ = 2 ms, J_{EE} = 0.5, J_{EI} = 0.5, J_{IE} = 1, J_{II} = 1, F = 3, θ = 0, h_E^{ext} = 0.3 and h_I^{ext} = −0.6. The resulting time period and phase are T = 51.3 ms and ϕ = 0.200. For the meaning of the parameters see Sect. 3.

F is the number of spikes remembered. A common form for the PSP function is the alpha function $\epsilon_{ij}(s) = (s - \Delta_{ij})/\tau e^{-(t-\Delta_{ij})/\tau}$ where Δ_{ij} is the delay from neuron j to neuron i.

The refractory potential depends on the firing time of the neuron:

$$h_i^{\text{ref}}(t) = \sum_{f=1}^{F} \eta_i\left(t - t_i^{(f)}\right) \ . \tag{3}$$

Here the refractory function $\eta(s)$ derives from the reset of an IF neuron without an absolute refractory period, but can be used to simulate absolute refractory periods and after-potentials. When derived from the IF model its derivative is always positive $(d\eta_i/ds > 0)$. This property is important for the locking properties later on, and is called the *standard dynamics* [3]. In this paper I will use an exponential function derived from recordings from locust AL PNs [10]:

$$\eta(s) = \begin{cases} 0 & s \leq 0 \\ \exp(1.5 - s/12) & s > 0 \end{cases} \ . \tag{4}$$

Figure 2 shows an example of the synaptic potential and effective threshold of an excitatory and inhibitory neuron in a model of the AL.

2   Constraints on Stable Synchronous Oscillations

Suppose we have a network comprising N neurons organised into M groups defined by the connectivity, delays, thresholds, and PSP and refractory functions.

I denote the set of neurons in the lth group by $\mathcal{G}_l$. The connectivity is such that the total presynaptic connection strength from each group is the same for all neurons in a group. We can express this mathematically by

$$\sum_{j \in \mathcal{G}_m} J_{ij} = \tilde{J}_{lm} \ , \tag{5}$$

where $\tilde{J}_{lm}$ denotes the total weight from the neurons of group m onto a neuron in group l. We further demand that the PSP functions and the weights are the same ($\epsilon_{ij}(s) = \tilde{\epsilon}_{lm}(s)$ and $\Delta_{ij} = \tilde{\Delta}_{lm}$) for all neurons $i \in \mathcal{G}_l$ and $j \in \mathcal{G}_m$. Finally, we require that the thresholds and refractory functions for neurons in a group be equal: $\theta_i = \tilde{\theta}_l$ and $\eta_i(s) = \tilde{\eta}_l(s)$ for $i \in \mathcal{G}_l$.

In a multi-group locked state the neurons of each group fire synchronously and periodically with a time period T. We assume that group l fires at $(k + \phi_l)T$ for integer k where ϕ_l is a phase variable in the range $[0,1)$. It is convenient to arrange the phases in ascending order and set $\phi_1 = 0$ so that $0 = \phi_1 < \phi_2 < \ldots < \phi_M$.

Now assume that the $l - 1$th group has fired and that we are waiting for the lth group to fire (or that the Mth group has fired and we are waiting for the first group to fire). Although time is arbitrary, we will assume that we are in the first cycle after $t = 0$, that is that $\phi_{l-1} < t < \phi_l$ (or $\phi_l < t < T$ if we are waiting for the first group). For any neuron $i \in \mathcal{G}_l$ the expected synaptic and refractory potentials are

$$h_i^{\mathrm{syn}}(t) = \sum_{l=1}^{m-1} \tilde{J}_{lm} \sum_{k=0}^{1-F} \tilde{\epsilon}_{lm}(t - (k + \phi_m)T) + \sum_{l=m}^{M} \tilde{J}_{lm} \sum_{k=-1}^{-F} \tilde{\epsilon}_{lm}(t - (k + \phi_m)T) \tag{6}$$

and

$$h_i^{\mathrm{ref}}(t) = \sum_{k=-1}^{-F} \tilde{\eta}_m(t - (k + \phi_l)T) \ . \tag{7}$$

Group l should next fire when the membrane potential reaches the threshold θ_l. If the threshold has been set correctly this will be at $t = \phi_l T$, leading to the self-consistency or *threshold* condition:

$$\tilde{\theta}_l = h_i(\phi_l T) \ . \tag{8}$$

A more precise requirement for the existence of coherent solutions is that for all $i \in \mathcal{G}_l$,

$$\phi_l T = \inf(t > (\phi_{l-1})T | h_i(t) = \tilde{\theta}_l) \ . \tag{9}$$

As $h_i(t)$ reaches $\tilde{\theta}_l$ from below, this implies that a necessary condition for stability is

$$\dot{h}_i(\phi_l T) > 0 \ . \tag{10}$$

I refer to this as the *rising potential condition*.

Although these conditions show that synchronised solutions exist, they do not show that they are stable. In order to show this, we have to show that a small perturbation to the firing time of a neuron from the group firing in a cycle time will decrease in the next cycle. This linear perturbation analysis applied to a homogeneous network of neurons with standard dynamics yields the *locking theorem* [3]:

Theorem 1. *In a spatially homogeneous network of spiking neurons with standard dynamics, a necessary and, in the limit of a large number n of presynaptic neurons (n → ∞), also sufficient condition for a coherent oscillation to be asymptotically stable is that firing occurs when the postsynaptic potential arising from all previous spikes is increasing in time.*

In a system with standard dynamics this is a stricter condition than the rising potential condition since the positive refractory derivative contributes to making the derivative of the membrane potential positive.

This result does not apply to more than one group of simultaneously firing neurons. However, it is straightforward to extend the proof to prove part of the the *infinite-limit extended locking theorem*:

Theorem 2. *In a network of spiking neurons with standard dynamics with M homogeneous groups of neurons such that all neurons in group l receive the same total postsynaptic weight from a large number of presynaptic neurons N_m in each group m a necessary and sufficient condition for constant phase, periodic, coherent oscillation to be asymptotically stable is that firing occurs when the postsynaptic potential arising from all previous spikes is increasing in time.*

Appendix A contains the proof. The mathematical expression of the extended locking theorem is that

$$h_i^{\text{syn}}(\phi_l T) > 0 \quad \text{for } i \in \mathcal{G}_l \ . \tag{11}$$

3 Applying the Constraints to Excitatory and Inhibitory Groups

It is now a straightforward task to apply the extended locking theorem to one excitatory and one inhibitory group of neurons, a system that approximates the AL.

To make the notation clearer, I will refer to the excitatory group as E and the inhibitory one as I. Later it will be convenient to work with positive weights, so I refer to the group connection weights which should be called $\tilde{J}_{\text{EE}}$, $\tilde{J}_{\text{EI}}$ and so on as J_{EE}, $-J_{\text{EI}}$, J_{IE} and $-J_{\text{II}}$. (I drop the tildes from here on as there is no further need to refer to the weights between individual neurons.) I take the delays to be uniformly Δ. I assume that PSP functions do not depend on the postsynaptic neurons and denote the excitatory postsynaptic potential (EPSP) function by $\epsilon_{\text{E}}(t)$ and the IPSP function by $\epsilon_{\text{I}}(t)$. I will assume that they are zero until $t = \Delta$, rise monotonically to a maximum at time $t = \tau_{\text{E}}$ or $t = \tau_{\text{I}}$ and then fall monotonically back to zero. The refractory functions, denoted $\eta_{\text{E}}(t)$ and $\eta_{\text{I}}(t)$, both have positive derivatives so the network has standard dynamics.

To simplify the analysis I will assume that neurons only remember their last firing times, that is $F = 1$. This is reasonable if the membrane potential due to a second spike is much smaller than the membrane potential due to the first, which is true when the time period is long compared to the PSP time constant. In any case, it is straightforward to generalise to longer spike memories. I will also assume that the refractory function complies with the formal version of the threshold condition (9). This means the absolute refractory period is less than the desired period T and that it prevents multiple firing during a cycle.

Since with the standard dynamics the stability condition is stronger than the rising potential condition, we can ignore the rising potential condition. Of the two remaining conditions, the stability condition is the more fundamental, since it constrains only the weights, whereas the threshold condition also constrains the input levels. I therefore explore the consequences of the stability condition first, in Sect. 3.1, before moving onto the consequences of the threshold condition in Sect. 3.2.

3.1 Consequences of the Stability Constraint

In the alternating state, network activity oscillates with a time period of T and the excitatory group leads the inhibitory group by ϕT, where ϕ is a phase variable with the range $[0,1)$. Applying the stability constraint (11) yields

$$J_{EE}\dot{\epsilon}_E(T) - J_{EI}\dot{\epsilon}_I\left((1-\phi)T\right) > 0 \tag{12}$$

for the excitatory group and

$$J_{IE}\dot{\epsilon}_E(\phi T) - J_{II}\dot{\epsilon}_I(T) > 0 . \tag{13}$$

for the inhibitory group.

The consequence of these inequalities is that there are combinations of time periods and phase for which there can be no stable firing, regardless of the size of the weights. This is because either inequality is not satisfied when the excitatory derivative is less than or equal to zero and the inhibitory derivative is greater than or equal to zero. For the condition on the excitatory group (12) this happens when $T < \Delta$ or $T \geq \tau_E + \Delta$ and $0 \leq (1-\phi)T \leq \tau_I + \Delta$. The condition on the inhibitory group (13) is violated when $T \leq \tau_I + \Delta$ and $\phi T \geq \tau_E + \Delta$ or $\phi T < \Delta$. As it is enough for one group to become unstable to plunge the whole system into instability, the forbidden combinations of T and ϕ are given by the union of the regions defined by the above inequalities. They can be displayed in a plot of ϕ versus T (referred to as T-ϕ space from now on) as in Fig. 3.

The second consequence of the inequalities (12) and (13) is that for those combinations of T and ϕ for which activity is not inherently unstable, stability depends on appropriate weights. Thus for any possible pair of T and ϕ there will be a region of the four-dimensional weight space that allows stable firing. We can make a major simplification by noting that the inequalities concern the *ratios* $\hat{J}_{EE} := J_{EE}/J_{EI}$ and $\hat{J}_{IE} := J_{IE}/J_{II}$ and finding the regions of a two-dimensional weight-ratio space that allow stable firing. Since the weights are by definition positive, the weight ratios are always greater than or equal to zero.

The stability condition (12) imposes different constraints on T, ϕ and the weight ratio $\hat{J}_{EE}$ depending on the sign of $\dot{\epsilon}_E(T)$, and hence on the size of T compared to Δ and $\tau_E + \Delta$.

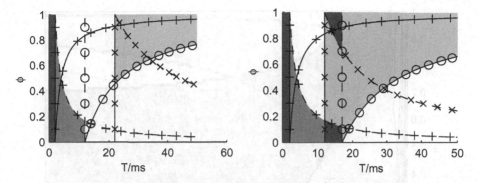

Fig. 3. Forbidden regions of the T-ϕ plane for two sets of parameters. The light regions are unstable due to the excitatory group becoming unstable, medium-grey areas unstable due to the inhibitory group and the darkest areas unstable due to both groups. Dashed lines pertain to the excitatory group, dotted lines to the inhibitory group and dash-dotted lines to both. Plus signs indicate where condition (14) and its inhibitory counterpart become true. Crosses indicate zeros in the denominators and circles zeros in the numerators of equations (15) and (16) and their inhibitory counterparts. Left: $\Delta = 2\,\text{ms}$, $\tau_\text{E} = 20\,\text{ms}$ and $\tau_\text{I} = 10\,\text{ms}$. Right: $\Delta = 2\,\text{ms}$, $\tau_\text{E} = 10\,\text{ms}$ and $\tau_\text{I} = 15\,\text{ms}$ (estimated locust parameters).

1. For $T < \Delta$ or $T = \tau_E + \Delta$, $\dot\epsilon_\text{E}(T) = 0$ and from (12) we can deduce that

$$\dot\epsilon_\text{I}((1 - \phi)T) = 0 \ , \tag{14}$$

   so that stability condition is not satisfied. This is the region $T < \Delta$ where both conditions are violated (see Fig. 3).
2. If $\Delta < T < \tau_\text{E} + \Delta$, $\dot\epsilon_\text{E}(T) > 0$ and

$$\hat{J}_\text{EE} > \frac{\dot\epsilon_\text{I}((1 - \phi)T)}{\dot\epsilon_\text{E}(T)} \ . \tag{15}$$

   When $\dot\epsilon_\text{I}((1 - \phi)T) \leq 0$ (for $(1 - \phi)T \leq 0$ or $T \geq \tau_\text{I} + \Delta$) this equation is satisfied for any weight ratio $\hat{J}_\text{EE}$ (and thus any weights J_EE and J_EI) since weight ratios are always positive.
3. When $T > \tau_\text{E} + \Delta$, $\dot\epsilon_\text{E}(T) < 0$ and equation (15) does not hold, since we have to reverse the sign of the inequality when rearranging (12). Instead,

$$\hat{J}_\text{EE} < \frac{\dot\epsilon_\text{I}((1 - \phi)T)}{\dot\epsilon_\text{E}(T)} \ . \tag{16}$$

   When $\epsilon_\text{I} \geq 0$ (for $0 < (1 - \phi)T \leq \tau_\text{I} + \Delta$) the inequality cannot be satisfied for positive weight ratios. This corresponds to the forbidden region where $T > \tau_\text{E} + \Delta$.

We can derive similar inequalities for the inhibitory group from (13).

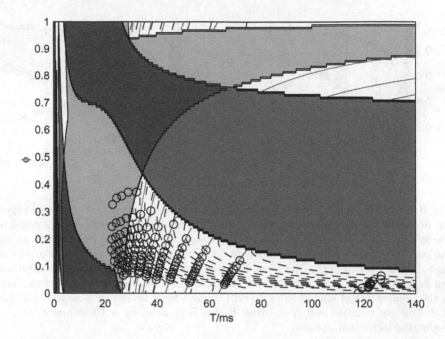

Fig. 4. Unstable regions of T-ϕ space and mapping from h_E^{ext}-h_I^{ext} space onto T-ϕ space for a pair of weight ratios. The shaded regions are unstable, with the shade representing the groups whose stability conditions are violated: the light region corresponds to the excitatory group, the intermediate region to the inhibitory group, and the darkest region to both groups. The unshaded regions with lines plotted in them are stable given the correct input. The lines represent the levels of external excitation required to produce the values of T and ϕ. Lighter shaded lines correspond to higher values. The solid lines represent constant h_E^{ext} and the dashed lines represent constant h_I^{ext}. The circles represent simulation values obtained by setting the input levels to those of the lines. They agree well with the theoretical results, although they appear in the theoretically unstable area too. Parameters: $\tau_E = 10$ ms, $\tau_I = 15$ ms, $\Delta = 2$ ms, $J_{\text{EE}} = 0.5$, $J_{\text{EI}} = 0.5$, $J_{\text{IE}} = 1$, $J_{\text{II}} = 1$ and $F = 3$.

The third consequence of the inequalities (12) and (13) complements the second. Any pair of weight ratios limits the stable combinations of T and ϕ even more than the inherent stability. This follows from setting $\hat{J}_{\text{EE}}$ in the inequalities (14), (15) and (16) and solving for T and ϕ. By contrast to the inherently unstable areas, the solution does depend on the exact form of the PSP functions; we cannot simply consider the signs of the PSP gradients.

With alpha function PSPs (defined in Sect. 1) there appears to be no analytical solution to the inequalities (14), (15) and (16) and their inhibitory counterparts. We must solve both sets of inequalities numerically to find the regions each allow. The intersection of these regions is the allowable region. Figure 4 shows the regions of stability and instability for a particular pair of weight ratios.

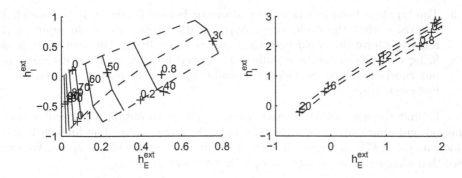

Fig. 5. The mapping from T-ϕ space into h_E^{ext}-h_I^{ext} space for two sets of parameters. Solid lines correspond to constant T and dashed lines to constant ϕ; numbers between 0 and 1 label some of the constant-T lines with the phase and multiples of 10 label some of the constant=ϕ lines. Left: $\tau_E = 10$ ms, $\tau_I = 15$ ms, $\Delta = 2$ ms, $J_{EE} = 0.5$, $J_{EI} = 1$, $J_{IE} = 0.5$, $J_{II} = 1$, $F = 3$. Right: $\tau_E = 20$ ms, $\tau_I = 10$ ms, $\Delta = 3$ ms, $J_{EE} = 1$, $J_{EI} = 0.5$, $J_{IE} = 1$, $J_{II} = 0.5$, $F = 3$.

3.2   Consequences of the Threshold Constraint

The threshold condition (8) means that

$$\theta_E = J_{EE}\epsilon_E(T) - J_{EI}\epsilon_I((1-\phi)T) + \eta_E(T) + h_E^{\text{ext}} \tag{17}$$

holds for the excitatory group and that

$$\theta_I = J_{IE}\epsilon_E(\phi T) - J_{II}\epsilon_I(T) + \eta_I(T) + h_I^{\text{ext}} \tag{18}$$

holds for the inhibitory group.

These equations tell us directly how much activation is required to sustain activity in a stable region of T-ϕ space. Figure 5 shows the mapping of a grid in T-ϕ space into h_E^{ext}-h_I^{ext} space for a particular set of weights and thresholds. There are a few points to notice here.

1. As the excitatory or inhibitory external potential is increased, the time period decreases. It seems intuitive that more excitatory input should speed up the oscillations, but not that more inhibitory input should. The reason is this happens is that more excitatory input makes the inhibitory neurons fire earlier, which in turn release the excitatory neurons from their inhibition earlier.

2. More input to the excitatory neurons means they increase their lead over the inhibitory neurons. We would expect more excitatory input to make the excitatory neurons fire earlier, but without doing calculations we would not know whether it brings forward the inhibitory neurons' firing by more or less. By contrast, more input to the inhibitory neurons reduces the excitatory neurons' lead.

3. The input to both groups can be above or below threshold. It is somewhat surprising that the excitatory group can have below-threshold input as it requires above-threshold external input to start firing. However, once it is firing the self-excitation is sufficient to keep it firing and the external input can be reduced. So below-threshold input to the excitatory group is a hysteresis state.

To find the amount of external input required to obtain a particular time period and phase we must solve (17) and (18) numerically. The lines in Fig. 4 show a grid in h_E^{ext}-h_I^{ext} space mapped into allowed regions of T-ϕ space. We can see that changes to either h_E^{ext} or h_I^{ext} change both T and ϕ.

4   Simulations

To test whether the analysis is correct, I simulated a network with 100 excitatory and 100 inhibitory neurons and the parameters in the caption of Fig. 4 for varying values of external input. I analysed the results to find whether there was stable firing, and if so, the time period and phase of oscillations. In order to make sure that the network did not stay in a stable state by chance, I added some noise to the firing times by making the firing probabilistic [2]. The simulation results are shown in Fig. 4, along with the theoretical results.

In the allowed regions they fit well with the theoretical predictions. However, the network appears to fire stably even when it should not according to the theory. It is not clear why this happens, though it is possible that the noise may stabilise the network in these areas.

5   Discussion

This paper started with a model of the first stage of olfactory processing in the locust, the AL. This model raised the problem of choosing weights in a network of excitatory and inhibitory spiking neurons that produce the right time period and phase of oscillation. In order to do this I had to extend the locking theorem [3] slightly. By applying the extended locking theorem to a network comprising excitatory and inhibitory homogeneous groups I showed:

- that the excitatory and inhibitory neurons can only fire synchronously and alternately for certain time periods and phases of oscillation;
- how to find weights that produce a certain, allowed, time period and phase of oscillation;
- that oscillations are possible for below-threshold input to inhibitory neurons and, when the network is in a hysteresis state, for below-threshold input to excitatory neurons; and
- how the time period and phase depend on the input and vice versa.

Simulations confirm most of these results, but, puzzlingly, seem to be stable when the theory predicts they should not. This may be due to the symmetry-breaking noise included in the simulations.

I am not aware of any model of the locust AL at a similar level to mine, nor am I aware of the problem of the time period and phase of networks of spiking excitatory and inhibitory spiking neurons having been posed or analysed before. Although there is a sizable literature on the dynamics of networks of spiking neurons, none of them appear to deal with networks comprising excitatory and inhibitory neurons [7,9]. While there has been much work on excitatory and inhibitory oscillators based on the population model of [11], this does not relate to spiking neurons, which allow different dynamics to continuous-valued population models [1]. By contrast, my extension to the locking theorem is trivial — indeed it has been hinted at [3]. Nevertheless, I believe that it has been a worthwhile accomplishment to have put it on a firmer footing.

Perhaps the most surprising result is the forbidden values of time period and phase, independent of the values of the weights. These regions possibly act to limit the time periods at which neural systems can work, and so are perhaps responsible for the time periods observed in the locust. The results also indicate that it should be rather difficult for the locust to maintain the experimentally-observed constant oscillation time period, even if the excitatory and inhibitory inputs are proportional to each other. This implies that either the input is relatively constant or that the oscillation time period does actually change in the locust, perhaps as a function of odour concentration. It may be possible that there are parameter regimes that do give the desired characteristics, however. It may also be possible that a more complex neuron model with active channels is required to achieve a fairly constant oscillation time period.

In deriving the results I have assumed that the refractory potential is strong enough to prevent PNs firing before they are inhibited. If this is not true, firing may not be "correct" as neurons may fire more than once on each cycle. Another, more limiting assumption is that of the homogeneity of the groups, a condition unlikely to be found in the nervous system. Less homogeneous groups will introduce differences in the firing times of units of a group, thus making the locking less precise. However, an extension of the one-group locking theorem to noisy neurons suggests that moderate amounts of noise still allow reasonably precise, stable locking and there are arguments as to why noisy neurons might be equivalent to inhomogeneous connections [1].

Several questions remain. In particular the effect of noise and inhomogeneity on the locking is not well understood. There may be ways of setting parameters to make large stable regions or to minimise the dependence of the time period on the inputs. The behaviour as the external inputs are increased in proportion to each other is relevant to the locust and should be studied.

In conclusion, this simple model of excitatory and inhibitory spiking neurons is a useful tool in predicting and understanding the activity of real neural systems without going to complex simulations. It may have application beyond the locust in some of the many excitatory-inhibitory subnetworks of nervous systems that show oscillatory activity.

A   Proof of the Extended Locking Theorem

To show whether the group firing times are stable to small perturbations, we can perform a linear stability analysis. Suppose that the firing times before $\phi_l T$ are perturbed by small times $\delta_i^{(k)} \ll T \min_l \phi_l$. The firing times are now $(k + \phi_l)T + \delta_i^{(k)}$ for $k = 0, -1, -2, \ldots$ for a neuron in group $m < l$ and $(k + \phi_l)T + \delta_i^{(k)}$ for $k = -1, -2, \ldots$ for a neuron in group $m \geq l$.

The slight changes in firing times will change the membrane potential slightly. An increase in membrane potential when it is near the threshold makes a neuron fire earlier. The change in firing time is greater when the rate of change of the membrane potential is small. Mathematically,

$$\delta_i^{(0)} = -\frac{\delta h_i(\phi_l T)}{\dot{h}(\phi_l T)} \tag{19}$$

where $\delta h_i(\phi_l T)$ is the change in membrane potential due to the perturbations in the previous firing times.

We can linearise so that:

$$\eta_l(\phi_l T - (k + \phi_l)T - \delta_i^{(k)}) \approx \eta_l(-kT) - \dot{\eta}_{l,k}\delta_i^{(k)} \tag{20}$$

and

$$\epsilon_{lm}(\phi_l T - (k + \phi_m)T - \delta_i^{(k)}) \approx \epsilon_{lm}((\phi_l - \phi_m - k)T) - \dot{\epsilon}_{lm,k}\delta_i^{(k)} \tag{21}$$

where

$$\dot{\eta}_{l,k} = \dot{\eta}_l(-kT) \text{ and } \dot{\epsilon}_{lm,k} = \dot{\epsilon}_{lm}((\phi_l - \phi_m - k)T) .$$

To find the small change in membrane potential $\delta h_i(\phi_l T)$, we substitute the perturbed times for the unperturbed times in (6) and (7) and substitute the linearised expressions for the refractory, synaptic and membrane potential terms. Substituting the result in a rearranged version of (19) yields

$$\dot{h}_i(\phi_l T)\delta_i^{(0)} = \tag{22}$$

$$\sum_{k=-1}^{-F} \dot{\eta}_{l,k}\delta_i^{(k)} + \sum_{m=1}^{l-1} \sum_{j \in \mathcal{G}_m} J_{ij} \sum_{k=0}^{1-F} \dot{\epsilon}_{lm,k}\delta_j^{(k)} + \sum_{m=l}^{M} \sum_{j \in \mathcal{G}_m} J_{ij} \sum_{k=-1}^{-F} \dot{\epsilon}_{lm,k}\delta_j^{(k)} .$$

By the law of large numbers it is reasonable to assume that mean time shifts vanish, that is $(1/N) \sum_i \delta_i^{(k)} \approx 0$. In order to simplify the problem to make it tractable we must further assume that

$$\sum_{j \in \mathcal{G}_m} J_{ij}\delta_j^{(k)} \approx 0 . \tag{23}$$

Note that this is a stricter condition than (5) as it implies that the weights and perturbations are uncorrelated and that the weights sample the perturbations fairly well. For example, $J_{ij} = J_0 \delta_{ij}$ (where δ is here the Dirac delta) satisfies (5) but does not necessarily lead to (23).

The assumption (23) allows us to neglect the last two summations in (22) so that the only delays influencing the 0th delay of neuron $i \in \mathcal{G}_l$ are the previous delays of neuron i itself:

$$\delta_i^{(0)} = \sum_{k=-1}^{-F} A_l^{(k)} \delta_i^{(k)} \quad \text{where} \quad A_l^{(k)} = \frac{\dot{\eta}_{l,k}}{\sum_{k=-1}^{-F} \dot{\eta}_{l,k} + \dot{h}_i^{\text{syn}}(\phi_l T)} . \tag{24}$$

We can recast (24) in matrix form:

$$\begin{pmatrix} \delta_i^{(0)} \\ \vdots \\ \delta_i^{(1-F)} \end{pmatrix} = \mathbf{F}_l \begin{pmatrix} \delta_i^{(-1)} \\ \vdots \\ \delta_i^{(-F)} \end{pmatrix} \quad \text{where} \quad \mathbf{F}_l = \begin{pmatrix} A_l^{(-1)} & A_l^{(-2)} & A_l^{(-3)} & \cdots & A_l^{(-F)} \\ 1 & 0 & 0 & \cdots & 0 \\ 0 & 1 & 0 & \cdots & 0 \\ \vdots & \vdots & \vdots & \ddots & \vdots \\ 0 & 0 & 0 & \cdots & 0 \end{pmatrix} .$$

The condition for stability is that as $t \to \infty$, $\delta_i^{(k)} \to 0$. In the matrix notation, this is equivalent to $\lim_{n\to\infty} \mathbf{F}_l^n(\boldsymbol{\delta}) = 0$ for an arbitrary delay vector $\boldsymbol{\delta}$. A sufficient condition for this to happen is that the eigenvalues of $\mathbf{F}_l$ all lie within the unit circle. [3] use various matrix theorems to prove that $\sum_k A_l^{(k)} < 1$ is a necessary and sufficient condition for the modulus of the maximum eigenvalue to be less than one. Substituting (24) into this condition yields

$$\frac{\sum_{k=-1}^{-F} \dot{\eta}_{l,k}}{\sum_{k=-1}^{-F} \dot{\eta}_{l,k} + \dot{h}_i^{\text{syn}}(\phi_l T)} < 1 . \tag{25}$$

As, in the standard dynamics, $\dot{\eta}_{l,k} > 0$, the second part of the denominator must be positive, yielding the stability condition (11).

References

[1] Gerstner, W.: Population dynamics of spiking neurons: Fast transients, asynchronous states, and locking. Neural Computation **12** (2000) 43–89
[2] Gerstner, W., van Hemmen, J. L.: Associative memory in a network of 'spiking' neurons. Network: Computation in Neural Systems **3** (1992) 139–164
[3] Gerstner, W., van Hemmen, J. L., Cowan, J. D.: What matters in neuronal locking. Neural Computation **8** (1996) 1689–1712
[4] Laurent, G.: Dynamical representation of odors by oscillating and evolving neural assemblies. Trends in Neurosciences **19** (1996) 489–496
[5] Laurent, G., Davidowitz, H.: Encoding of olfactory information with oscillating neural assemblies. Science **265** (1994) 1872–1875
[6] Laurent, G., Naraghi, M.: Odorant-induced oscillations in the mushroom bodies of the locust. Journal of Neuroscience **14** (1994) 2993–3004
[7] Mirollo, R. E., Strogatz, S. H.: Synchronisation of pulse-coupled biological oscillators. SIAM Journal of Applied Mathematics **50** (1990) 1645–1662
[8] Smith, L. S., Cairns, D. E., Nischwitz, A.: Synchronization of integrate-and-fire neurons with delayed inhibitory connections. In *Proceedings of ICANN94*. Springer-Verlag (1994) pp. 142–145.
URL ftp://ftp.cs.stir.ac.uk/pub/staff/lss/icann94.ps.Z

[9] van Vreeswijk, C., Abbott, L. F., Ermentrout, G. B.: When inhibition not excitation synchronises neural firing. Journal of Computational Neuroscience **1** (1994) 313–321

[10] Wehr, M., Laurent, G.: Relationship between afferent and central temporal patterns in the locust olfactory system. Journal of Neuroscience **19** (1999) 381–390

[11] Wilson, H. R., Cowan, J. D.: Excitatory and inhibitory interactions in localized populations of model neurons. Biophysical Journal **12** (1972) 1–24

Temporal Coding in Neuronal Populations in the Presence of Axonal and Dendritic Conduction Time Delays

David M. Halliday

Department of Electronics, University of York, Heslington, YORK, YO10 5DD, UK.
dh20@ohm.york.ac.uk

Abstract. Time delays are a ubiquitous feature of neuronal systems. Synaptic integration between spiking neurones is subject to time delays at the axonal and dendritic level. Recent evidence suggests that temporal coding on a millisecond time scale may be an important functional mechanism for synaptic integration. This study uses biophysical neurone models to examine the influence of dendritic and axonal conduction time delays on the sensitivity of a neurone to temporal coding in populations of synaptic inputs. The results suggest that these delays do not affect the sensitivity of a neurone to the presence of temporal correlation amongst input spike trains, and point to a mechanism other than electrotonic conduction of EPSPs to describe neural integration under conditions of large scale synaptic input. The results also suggest that it is the common modulation rather than the synchronous aspect of temporal coding in the input spike trains which neurones are sensitive to.

1. Introduction

Recent experimental evidence has emerged that temporal coding on a millisecond time scale may underlie the representation and transmission of information between neurones within the central nervous system [1,2,3]. Modelling studies have highlighted both rate coding and temporal coding as candidate mechanisms which may subserve neural integration between populations of spiking neurones [4-10].

An important aspect of neuronal systems is the presence of both axonal and dendritic conduction time delays. This study investigates whether these delays impact on temporal coding in neurones, by determining if they alter the ability of a neurone to respond to temporal coding in pre-synaptic inputs. In this study, temporal coding is carried as temporal correlation amongst a population of input spike trains. We consider a realistic correlation structure amongst the spike trains in the population. This correlation is both weak and stochastic, and is based on common rhythmic modulation of spike trains, which induces a tendency for synchronised firing. The resultant temporal correlation is manifest as a rhythmic modulation of inter spike intervals in individual spike trains and a tendency for synchronised firing between sample pairs of spike trains [11].

This paper addresses the question of propagation of temporal coding within neurones under conditions of large scale synaptic input, and in particular how temporal coding on a millisecond time scale can be achieved in the presence of

S. Wermter et al. (Eds.): Emergent Neural Computational Architectures, LNAI 2036, pp. 285-295, 2001.
© Springer-Verlag Berlin Heidelberg 2001

dendritic and axonal conduction delays of a similar order of magnitude. We use simulations of biophysical neurone models, based on a compartmental model to study how the spatial and temporal aspects of large scale synaptic input interact and shape the output discharge. The simulations are based on a class of neurones which have been extensively studied and possess dendritic trees which have electrotonic conduction time delays of the order of several milliseconds, spinal motoneurones. The simulations capture the large degree of convergence present in neural systems, where each neurone is acted on by a large number of inputs which are distributed over the surface area of the dendritic tree and cell body.

The results of the simulations show that, under conditions of large scale synaptic input, temporal coding can be preserved in the output discharge when the temporal code is carried by a subset of the total synaptic input in the form of temporally correlated spike trains. Dendritic conduction time delays do not affect this process, similar results are obtained with temporally correlated inputs which have a high degree of spatial correlation (clustered) or a low degree of spatial correlation (distributed). The inclusion of random axonal conduction delays similarly does not alter the results. This suggests that a mechanism other than electrotonic conduction of EPSPs is responsible for synaptic integration and the propagation of temporal coding under conditions of large scale synaptic input.

2. Model Neurone

The model motoneurone is based on morphological studies of cat spinal α motoneurones [12] and consists of a spherical soma from which radiate 12 tapered dendrites of different lengths. These 12 dendrites consist of three types, which we call short, medium and long, each cell contains four of each type. The membrane parameters are based on complementary electophysiological studies of the same motoneurones [13,14], these are a membrane resistivity, R_m, of 11,000 $\Omega \cdot$cm^2, a ctyoplasmic resistivity, R_i, of 70 $\Omega \cdot$cm, and a membrane capacitance, C_m, of 1.0 μF/cm^2. The initial diameters at the soma of the three dendrite types are 5.0 μm (short), 7.5 μm (medium) and 10.0 μm (long). The taper used to estimate dendritic diameters is 0.5 μm reduction per 100 μm length. The physical lengths of the three dendrite types are 766 μm (short), 1258 μm (medium) and 1904 μm (long), the equivalent passive electrotonic lengths are L=0.7, 1.0 and 1.5, respectively. The dendritic tree and soma are modelled using a compartmental model [15], with the dendrites represented as a sequence of connected cylinders, each of electrotonic length L=0.1, with an additional compartment for the soma. Thus a total of 129 compartments are used to model each motoneurone. The total membrane areas represented in the compartmental model for each dendrite type are 8100 μm^2 (short), 18500 μm^2 (medium) and 33500 μm^2 (long). The compartmental model for the complete cell represents a total membrane area of 248300 μm^2, 97% of this area is taken up by the 12 dendrites. The model has a steady state input resistance of 4.92 MΩ, and a time constant of 9.7 ms, estimated from the response to hyperpolarizing current injected at the soma.

Individual excitatory synaptic inputs use a time dependent conductance change, specified by an alpha function: $g_{syn}(t)=G_s\, t/\tau \exp(-t/\tau)$ [16,17], where τ is the time constant, and G_s a scaling factor which defines the peak conductance change. In the present study we use values of $\tau=0.2$ ms, and $G_s=1.19\times10^{-8}$ S, this results in a peak conductance change of 4.38 nS at t=0.2 ms [17]. A value of –70 mV is used for the membrane resting potential, and the reversal potential for excitatory post synaptic potentials (EPSPs) is –10 mV. This conductance change, when activated from the resting potential at the soma, results in an EPSP with a peak magnitude at the soma of 100 μV at t=0.525 ms, and a rise time (10% to 90%) and half-width of 0.275 ms, 2.475 ms [18]. The time course of this EPSP is illustrated in Fig. 1 (top trace), as is the time course of the EPSP which is generated at the soma when the same conductance change is activated in the most distal compartments of each of the three dendrite types. The magnitudes of these EPSPs are 52.8 μV (short dendrite), 40.4 μV (medium) and 27.1 μV (long). The delays to the peak of voltage response are 2.0 ms, 3.08 ms and 4.75 ms, the rise times are 0.90 ms, 1.35 ms and 2.08 ms, and the half widths are 7.05 ms, 9.35 ms and 11.38 ms, respectively. Comparison of the four traces in Fig. 1 illustrates the reduction in magnitude and increased conduction times associated with electrotonic conduction of localised voltage changes in passive dendrites. The resultant EPSPs generated at the soma are reduced in magnitude and spread out in time, reflecting the low pass filtering characteristics of dendrites [19].

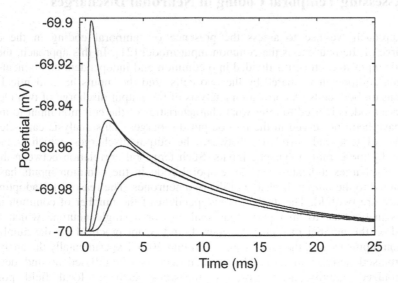

Fig. 1. Voltage response at soma to individual synaptic events activated at the soma (top trace) and in the most distal compartments of the short, medium and long (bottom trace) dendrite. These responses show a decreasing magnitude and increased time to peak voltage, effects due to electrotonic conduction in passive dendrites.

The threshold parameters and afterhyperpolarization (AHP) conductances associated with repetitive firing are based on the work of [20]. The firing threshold is –55 mV, 15 mV above resting potential, [20] proposes a three term model for a time dependent potassium conductance, $g_K(t)$. In the present study, we incorporate only the dominant F2 term (see [20]) in the AHP conductance model, $g_K(t)= G_K \exp(-t/\tau_K)$, with G_K =6.0×10^{-7} S, and τ_K=14 ms. The reversal potential for this potassium conductance is –75 mV. The other fast acting terms in the AHP conductance model [20] have no influence at the low firing rates in the present study, and are omitted. The AHP trajectory during repetitive firing at 12 spikes/sec, induced by constant current input to the soma, has a peak depression of 12 mV. An action potential is taken to occur when the somatic membrane potential exceeds the threshold, and $g_K(t)$ is activated in the somatic compartment to simulate the AHP.

The results in figure 1 illustrate the dendritic conductions delays present in the model, these can be up to several milliseconds for more distal inputs to the dendritic tree. Thus, the spatial dimension converts to a temporal delay before the effects of synaptic input arrive at the soma by electrotonic conduction. The traces in Fig. 1 suggest that electrotonic conduction of EPSPs is the underlying mechanism by which sparse synaptic input to a dendritic tree acts to modulate the output discharge. The objective of this study is to assess if these time delays impact on the sensitivity of the neurone to temporal patterns in the timing of synaptic inputs under conditions of large scale synaptic input.

3. Assessing Temporal Coding in Neuronal Discharges

The approach we use to assess the presence of temporal coding in the output discharge of the neurone is the common input model [21]. In this approach, the total synaptic input to each cell is divided into common and independent components. The common component is shared by the two cells, and the inputs have identical input locations on both cells. A correlation analysis of the output discharge of two identical neurones models is used to infer what characteristics of the common input to the two cells have been preserved in the two output discharges. This analysis can detect the presence of temporal correlation between the output discharges, which is induced entirely by the common synaptic inputs. Such temporal correlation between the two output discharges indicates that the temporal code in the common inputs has been transmitted to the output discharge of the two neurones. The reasons for adopting this approach are twofold. The method is independent of the number of common inputs. This is in contrast to an input/output analysis for a single neurone which would depend on the number of inputs considered, and is not practical for the numbers of common inputs used in the present study (around 200). Experimentally the analysis of synchronised activity in ensembles of neurones is often based around detecting synchronized activity in macroscopic measures such as local field potential recordings. The common input simulation approach is closer to the experimental situation, where access to input signals are rarely possible in vivo.

We use the frequency domain analogue of time domain correlation, coherence functions [22, 23], to assess the temporal correlation between the output discharges of the paired neurone model. This has some advantages over a time domain approach, in particular the estimated coherence between the two output discharges can be used to directly infer the frequency content of the common inputs to the two cells [24], and thus indicates what frequency components in the temporal code are transmitted to the output discharge of the neurones. Details of the derivation, estimation and interpretation of coherence functions can be found in the above references. Briefly, coherence functions provide a normative measure of linear association between two stationary stochastic signals (spike trains or time series data) on a scale from 0 to 1. Zero (more correctly values below a confidence limit) indicates no correlation between the signals, one indicates a perfect linear association. A coherence value is estimated for each frequency of interest over a range of frequencies, usually the Fourier frequencies resulting from the use of an FFT algorithm to calculate the discrete Fourier transform of the data. A confidence limit can be set based on the null hypothesis of independent signals, values of coherence below this can be taken as evidence of uncorrelated signals at a particular frequency. The model for interpreting coherence estimates between the output discharge of two neurones acted on by common synaptic input is described in [24]. The estimated coherence can be shown to be proportional to the spectrum of the common input signal for a single common input. We use this interpretation as the basis to interpret the result from the present simulations. Coherence is used to infer which frequency components in the common synaptic inputs are effective in modulating the output discharge of the two cells, providing an indication of preserved temporal coding in the two output discharges.

4. Results

The aim of this study is to investigate spatial temporal interactions in a dendritic structure, in other words how spatial and temporal correlations amongst a subset to the total synaptic input interact to shape the output discharge of the neurone. Spatial correlation refers to clustering of inputs about a particular location on the dendritic tree, measured as the passive electrotonic length from the soma. Temporal correlation refers to the correlation structure present in the population of input spike trains, measured as correlation between two sample input spike trains. The pre-synaptic inputs are distributed over the soma and all dendrites uniformly by surface area. The number of pre-synaptic inputs and the mean firing rate of each input spike train are adjusted such that each compartment receives the same average number of EPSP's/sec in all simulations. In the basic configuration each cell has a total of 996 inputs, 32 at the soma and 33, 74 and 134 inputs to each short, medium and long dendrite, with inputs distributed over all dendritic cylinders. The desired output rate for repetitive firing was chosen as 12 spikes/sec, to achieve this requires a total synaptic input of 31,872 EPSPs/sec, this can be achieved with each of the 996 inputs activated by a random (Poisson) spike train firing at 32 spikes/sec. During repetitive firing the membrane potential is dominated by the large AHP, approximately 12 mV in amplitude, following each output spike. Smaller fluctuations, approximately 1 mV peak-to-peak amplitude, are superimposed on top of this trajectory. These fluctuations are a result of spatial temporal interactions amongst the synaptic inputs, and are larger in magnitude than individual EPSPs [25]. It is these fluctuations and the process by which they modulate the output discharge which this study is concerned with.

Only the common subset of the total synaptic carries the temporal coding which is considered to provide the input "signal" to the two cells. In this study, the temporal coding is carried by two sets of synaptic inputs, each of which consist of 5% of the total synaptic input, measured as EPSPs/sec input to each neurone. These two populations consist of 160 spike trains firing at 10 spikes/s and 64 spike trains firing at 25 spikes/s. These inputs are periodic and have a coefficient of variation (c.o.v.) of 0.1. Each population of spike trains has a weak stochastic correlation structure. The spike timings are generated by a population of integrate-and-fire encoders, with mean rates of 10 spikes/s or 25 spikes/s, respectively, within each population. Correlated spike times are generated by applying common pulse inputs to each population of encoders, these pulses are derived from a single spike train of rate 10 or 25 spikes/sec and a c.o.v. of 0.1. This approach generates spike trains which have a weak stochastic correlation structure between pairs of discharges, the aim is to model accurately the pattern of correlation found in mammalian central nervous systems. The strength of correlation within the two populations can be quantified using coherence analysis, for the present study the coherence between sample pairs of spike trains from each population is 0.015 at 10 Hz or 25 Hz, respectively. For details see [11]. This weak stochastic correlation links the timing of individual spikes across the two population of discharges to a single "parent" discharge of 10 spikes/s or 25 spike/s which is the source of the temporal coding in each population. The time course of the temporal correlation between inputs is a few ms (not shown). Thus, temporal coding is carried by a tendency for synchronized spikes occurring within a few ms.

The coherence results in this study are based on analysis of simulated output spike trains of duration 100 seconds with spike timings defined to an accuracy of 1 ms. The Fourier frequencies have a spacing of 0.977 Hz, this results from using a discrete Fourier transform length of 1024 points to construct spectra and coherence estimates.

Fig. 2 shows coherence estimates between the output discharges of the two model neurones with the two populations of common inputs applied in different input locations. The numbers of independent inputs and their activation rate are adjusted in each simulation so that each compartment receives the same average number of EPSPs/sec as the control configuration above. A Poisson discharge is used to activate all independent inputs to the two cells. Fig. 2a is a control case, when two populations of uncorrelated spike trains are applied as common input. These have the same firing rate and c.o.v. as the correlated spike trains, except there is no temporal correlation between the spike times within each population, and thus no temporal code. Figures 2b, 2c and 2d show the effects of applying the correlated spike train inputs with a high degree of spatial correlation (clustering) at different electrotonic lengths from the soma, concentrated around L=0.0 to 0.1 (Fig. 2b), L=0.5 to 0.6 (Fig. 2c) and L=0.8 to 1.1 (Fig. 2d). This last case has no common inputs to the 4 short dendrites, which have a passive electrotonic length of 0.7. Fig. 3a shows the estimated coherence between the two output discharges with the common correlated inputs uniformly distributed over the entire dendritic tree, this represents the smallest spatial correlation amongst input locations.

The coherence estimates in Fig. 2b, 2c, 2d and 3a are similar in form. Both have distinct peaks at 10 Hz and 25 Hz, indicating that the output discharges of the two model neurones are correlated only at the frequencies at which the two populations of common inputs are correlated. This correlation does not reflect the firing rates of the two populations of input spike times, since the coherence estimate for the simulation with uncorrelated periodic inputs at 10 and 25 spikes/sec is not significant (Fig. 2a). The strength of correlation between the two output discharges is an order of magnitude larger than the strength of correlation amongst the two populations of correlated inputs (see above). We interpret the coherence between the two output discharges to indicate that the temporal coding in each population of correlated input spike trains is preserved in the output discharges of the two neurones. These results suggest that the location of the synaptic inputs on the dendritic tree does not influence the sensitivity of the neurone to the presence of temporal coding in the common synaptic inputs.

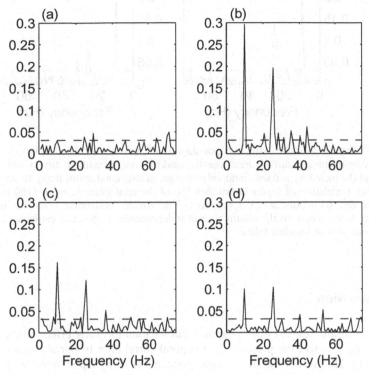

Fig. 2. Coherence estimate between the output discharges of the two cell model with (a) two populations of uncorrelated spike train input applied to L=0.0, 0.1, and two populations of weakly correlated spike trains applied to (b) L=0.0, 0.1, (c) L=0.5, 06 and (d) L=0.8 to 1.1. For both the correlated and uncorrelated inputs, each population constitutes 5% of the total synaptic input (160 inputs at 10 spikes/sec and 64 inputs at 25 spikes/sec). The dashed horizontal lines are upper 95% confidence limits based on the assumption of independence. Coherence estimates constructed from 100 seconds of simulated data.

We next consider the effect of adding an axonal conduction delay to the configuration with uniformly distributed common inputs (Fig. 3a). In Fig. 3b is a coherence estimate between the two output discharges with the addition of a fixed, randomly chosen, delay to both branches of each common synaptic input. The delay was normally distributed with a value of 2.5 ± 2.3 ms (mean ± 2 SD). This coherence estimate still has distinct peaks at 10 Hz and 25 Hz, indicating that the random axonal delay also has a minimal effect on the sensitivity of the neurones to the temporal correlation amongst the populations of common input spike trains.

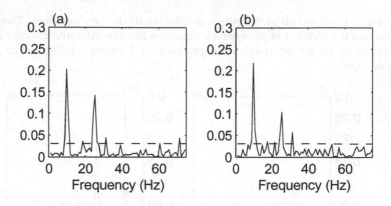

Fig. 3. Coherence estimate with two populations of weakly correlated spike trains applied uniformly by membrane surface area over the dendritic tree, (a) without any axonal conduction delay, and (b) including a fixed, randomly chosen, axonal conduction delay to each synaptic input. Each population of inputs constitutes 5% of the total synaptic input (160 inputs at 10 spikes/sec and 64 inputs at 25 spikes/sec). The dashed horizontal lines are upper 95% confidence limits based on the assumption of independence. Coherence estimates constructed from 100 seconds of simulated data.

5. Discussion

The above results suggest that synaptic input location has little effect on the ability of the neurone to detect the presence of temporal correlation in populations of synaptic inputs (Fig. 2b, 2c, 2d, 3). In contrast, populations of synaptic input with the same rhythmic components, but which are uncorrelated, have no significant effect on the output discharge (Fig. 2a). The results are obtained despite the considerable electrotonic conduction time delay present for the effects of individual synaptic inputs to propagate to the somatic spike generation site, which are the order of several ms (Fig. 1). This time delay is similar to the time course of the temporal correlation amongst inputs. In light of these results it seems reasonable to suggest that a mechanism other than electrotonic conduction of EPSPs is the dominant mechanism whereby temporal coding amongst input spike trains modulates the output discharge of each neurone.

The temporal code in this study is carried by rhythmic components which result in weak stochastic temporal correlation between pairs of spike trains within the population of inputs. In the temporal domain this is manifest as a tendency for synchronous discharge. An interesting question is whether it is the tendency for synchronous discharge within a time scale of a few ms, or the presence of common rhythmic modulation which the neurone is sensitive to. The inclusion of a random axonal conduction delay prior to the activation of synaptic events has little effect on the results (Fig. 3b). These delays results in post synaptic events which are separated in time by a fixed (random) amount between 0 and 5 ms. This suggests that it is the common rhythmic modulation which the neurone is sensitive to rather than the tendency for synchronous inputs. The effect on the output discharge of the paired neuron model is, however, to restore the tendency for synchronous spikes which have the same common rhythmic modulation as the population of inputs. Thus, although the tendency for synchronous events does not appear to be necessary for the neurone to respond to temporal correlation, the effect within a population of post synaptic neurones is to preserve both aspects of the temporal code. In addition, the strength of correlation within the post synaptic discharges appears to be enhanced, although this may be a feature of the specific parameters in the present simulations. A recent study using a random walk model of cortical neurones found that weakly synchronized inputs did not cause any increase in synchronization in output neurones [7]. Experimental evidence has been presented that input synchrony is an important variable in determining the characteristics of the output discharge of cortical neurones [26].

Similar simulation studies have demonstrated that the presence of weak stochastic correlation can control the bandwidth of the neurone, defined as the frequency components transmitted from pre-to post synaptic discharges, and led to the concept of dynamic modulation of neuronal bandwidth [27]. The present results suggest that this process can also preserve the temporal coding in the correlated spike trains, and further, that the mechanism does not depend on dendritic input location.

The present simulations are designed to model accurately the large degree of convergence in central and peripheral mammalian neural systems [28, 29]. Under conditions of large scale synaptic integration many hundreds of synaptic events contribute to the excitation which generates each output spike [30,31]. In the above simulations there are, on average, 2666 input spikes for each output spike, where each input spike contributes a small increment of excitation to the somatic membrane potential. Studies on the mechanisms underlying neural integration in neurones have proposed that temporal coding in input spike trains will not be preserved under conditions of temporal integration (defined as a large ratio of input to output spikes) [5, 7, 10]. However, as the above simulations demonstrate, the temporal coding carried by each population of 5% of the total synaptic input is transferred to the output discharge of the neurone.

The study of the mechanisms by which spiking neurones encode information in their output discharge, relates to the fundamental questions of how information is represented and processed in the central nervous system [32]. There is currently considerable interest and debate about the underlying mechanisms. Rate coding, coincidence detection and temporal integration have been considered as fundamental mechanisms underlying neural coding [4-10]. Despite the considerable research effort in this area, as yet the basic mechanisms which underlie neural coding are poorly understood [33]. Simulation studies, such as those reported above, are useful tools with which to investigate neural integration. The present simulations have demonstrated that temporal coding, in the form of weak stochastic temporal correlation, can be preserved by spatial temporal integration in passive dendrites. This may be an important mechanism in large scale synaptic integration across different classes of neurone.

References

1. Singer, W., Gray, C.: Visual feature integration and the temporal correlation hypothesis. Annual Review of Neuroscience 18 (1995) 555-586
2. MacKay, W.A.: Synchronized neuronal oscillations and their role in motor processes. Trends in Cognitive Sciences 1 (1997) 176-183
3. Riehle, A., Grün, S., Diesmann, M., Aertsen, A.: Spike synchronization and rate modulation differentially involved in motor cortical function. Science 278 (1997) 1950-1953
4. Abeles, M.: Role of cortical neuron: Integrator or coincidence detector? Israel Journal of Medical sciences 18 (1982) 83-92
5. Shadlen, M.N., Newsome, W.T.: Noise, neural codes and cortical organization. Current Opinion in Neurobiology 4 (1994) 569-579
6. Shadlen, M.N., Newsome, W.T.: Is there a signal in the noise? Current Opinion in Neurobiology 5 (1995) 248-250
7. Shadlen, M.N., Newsome, W.T.: The variable discharge of cortical neurons: Implications of connectivity, computation and information coding. Journal of Neuroscience 18 (1998) 3870-3896
8. Softky, W.R.: Simple versus efficient codes. Current Opinion in Neurobiology 5 (1995) 239-247
9. Softky, W.R., Koch, C.: The highly irregular firing of cortical cells is inconsistent with temporal integration of random EPSPs. The Journal of Neuroscience 13 (1993) 334-350
10. König, P., Engel, A.K., Singer, W.: Integrator or coincidence detector? The role of the cortical neuron revisited. Trends in Neurosciences 19 (1996) 130-137
11. Halliday, D.M.: Generation and characterization of correlated spike trains. Computers in biology and medicine 28 (1998) 143-152
12. Cullheim, S, Fleshman, J.W., Glenn, L.L., Burke, R.E.: Membrane area and dendritic structure in type identified triceps surae alpha motoneurons. Journal of comparative neurology 255 (1987) 82-96
13. Fleshman, J.R., Segev, I., Burke, R.E.: Electrotonic architecture of type-identified α-motoneurones in the cat spinal cord. Journal of Neurophysiology 60 (1988) 60-85
14. Rall, W., Burke, R.E., Holmes, W.R., Jack, J.J.B., Redman, S.J., Segev, I.: Matching dendritic neuron models to experimental data. Physiological Reviews 72 (1992) (Suppl.) S159-S186
15. Segev, I., Fleshman, J.R., Burke, R.E.: Compartmental models of complex neurons. In Methods in neuronal modeling: From synapses to networks, eds Koch C, Segev I. (1989) 63-96. MIT Press
16. Jack, J.J.B., Noble, D., Tsien, R. W.: Electrical Current Flow in Excitable Cells. 2nd Edition Clarendon Press, Oxford (1975)
17. Segev, I., Fleshman, J.R., Burke, R.E.: Computer simulations of group Ia EPSPs using morphologically realistic models of cat a-motoneurones. Journal of Neurophysiology 64 (1990) 648-660
18. Cope, T.C., Fetz, E.E., Matsumura, M.: Cross-correlation assessment of synaptic strength of single Ia fibre connections with triceps surae motoneurones in cats. Journal of Physiology 390 (1987) 161-188
19. Rinzel, J., Rall, W.: Transient response to a dendritic neuron model for current injected at one branch. Biophysics 14 (1974) 759-789
20. Baldissera F., Gustafsson B.: Afterhyperpolarization conductance time course in lumbar motoneurones of the cat. Acta physiol. scand 91 (1974) 512-527
21. Moore, G.P., Segundo, J.P., Perkel, D.H., Levitan, H.: Statistical signs of synaptic interaction in neurones. Biophysical Journal 10 (1970) 876-900

22. Rosenberg, J.R., Amjad, A.M., Breeze, P., Brillinger, D.R., Halliday, D.M.: The Fourier approach to the identification of functional coupling between neuronal spike trains. Progress in Biophysics and molecular Biology 53 (1989) 1-31

23. Halliday, D.M., Rosenberg, J.R., Amjad, A.M., Breeze, P., Conway, B.A., Farmer, S.F.: A framework for the analysis of mixed time series/point process data - Theory and application to the study of physiological tremor, single motor unit discharges and electromyograms. Progress in Biophysics and molecular Biology 64 (1995) 237-278

24. Rosenberg, J.R., Halliday, D.M., Breeze, P., Conway, B.A.: Identification of patterns of neuronal activity - partial spectra, partial coherence, and neuronal interactions. Journal of Neuroscience Methods 83 (1998) 57-72

25. Calvin, W.H., Stevens, C.F.: Synaptic noise and other sources of randomness in motoneuron interspike intervals. Journal of Neurophysiology 31 (1968) 574-587

26. Stevens, C.F., Zador A.M.: Input synchrony and the irregular firing of cortical neurons. Nature Neuroscience 1 (1998) 210-217

27. Halliday, D.M.: Weak, stochastic temporal correlation of large scale synaptic input is a major determinant of neuronal bandwidth, Neural Computation 12 (2000) 737-747

28. Braitenberg, V., Schüz, A.: Cortex: Statistics and geometry of neuronal connectivity. Springer-Verlag, Berlin (1997)

29. Henneman, E., Mendell, L.M.: Functional organization of motoneuron pool and its inputs. In Handbook of Physiology. Section 1 Vol 2 Part 1 The nervous system: Motor control (eds Brokkhart, J.M. & Mountcastle, V.B.) American Physiological Society, Bethesda, MD, (1981) 423-507

30. Bernander, Ö., Douglas, R.J., Martin, K.A.C., Koch, C.: Synaptic background activity influences spatiotemporal integration in single pyramidal cells. Proceedings of the National Academy of Sciences 88 (1991) 11569-11573

31. Rapp, M., Yarom, Y., Segev, I.: The impact of parallel fiber background activity on the cable properties of cerebellar purkinje cells. Neural Computation 4 (1992) 518-533

32. Marr, D.: Vision. WH Freeman & Company, San Francisco (1982)

33. Abbott, L. Sejnowski, T.J.(eds): Neural codes and distributed representations. MIT Press (1999)

The Role of Brain Chaos

Péter András

Department of Psychology, Ridley Building, University of Newcastle upon Tyne,
Newcastle upon Tyne, NE1 7RU, United Kingdom

Abstract. A new interpretation of the brain chaos is proposed in this
paper. The fundamental ideas are grounded in approximation theory. We
show how the chaotic brain activity can lead to the emergence of highly
precise behavior. To provide a simple example we use the Sierpinski
triangles and we introduce the Sierpinski brain. We analyze the learning
processes of brains working with chaotic neural objects. We discuss the
general implications of the presented work, with special emphasis on
messages for AI research.

1 Introduction

There is growing evidence of chaotic dynamics within biological neural activity. It
is known that individual neurons produce chaotic firing in certain conditions [17].
Other experimental results show that chaotic dynamics is characteristic of some
simple biological neural networks, like those, which form the stomato-gastric nucleus of crustaceans [12][20] or the central pattern generators in the breath controlling nuclei of mammals [14]. Freeman and his colleagues have shown chaotic
activity in large scale biological neural networks in the olfactory bulb [7] and
primary visual areas of small mammals [7]. Other researchers found chaotic dynamics in short-term behavioral transitions (e.g., change of gait) [13] and in the
eye movement patterns [2] of mammals.

Research results suggest that the observed chaotic dynamics has functional
roles in the biological neural systems. Schiff and his co-workers found that the
change in the dynamic nature of the EEG can be used to predict the occurrence of epileptic seizures [22]. The results of Freeman suggest that the chaotic
dynamics within the olfactory bulb is crucial for the emergence of olfactive sensation [7]. Recent experimental findings indicate that neurons in various visual
areas are specifically active during involuntary eye movements [16], which are
characterized by chaotic dynamics [2].

There were several attempts to give an interpretation and explanation for
the role of chaotic dynamics in the context of biological brains. Two recent
reviews are provided by Tsuda [25] and Érdi [6]. Freeman proposed that the high-
dimensional chaotic attractors characterizing the activity of the olfactory bulb
in the absence of olfactive stimuli, contain low-dimensional embedded attractors,
which may code the distinguishable olfactive sensations [7]. Rabinovich, Tsuda
and others proposed that the role of chaotic activity is mainly to provide a kind
of efficient random search over the space of meaningful encodings [18][20]. This
pathway of search is called chaotic itinerancy [18][25]. Tsuda [25] merged the

S. Wermter et al. (Eds.): Emergent Neural Computational Architectures, LNAI 2036, pp. 296–310, 2001.

these proposals by suggesting that the high-dimensional attractor is formed by a chaotic itinerancy pathway over so-called 'attractor ruins' [25], which correspond to the concept of embedded attractors of Freeman. Schiff formulated a similar proposal, in which the role of 'attractor ruins' is played by the quasi-periodic orbits [22].

Two main critical points can be raised against the mentioned explanations and interpretations. The first is that although the proposed chaotic systems have the ability to encode many meaningful sensorial or behavioral objects and to search efficiently over their space, this can be done by other, less complicated systems too, like fixed-point attractor neural networks [11]. The second point is that these proposals cannot account for the link between the observed brain chaos and the high precision of behavioral actions (e.g., gaze direction, reaching to an object). In our view the above mentioned proposals explain only partially the role of the brain chaos.

A new interpretation and explanation of the brain chaos is proposed in this paper. This proposal is grounded in nonlinear approximation theory. The general role of biological brains is to classify the incoming stimuli and / or to make predictions based on these. Both tasks can be understood as approximation of discrete or continuous valued functions [10]. Multiple-goal, high precision approximation can be performed efficiently by the right choice of basis functions from an appropriate pool of such functions [10]. Traditionally it is supposed that individual neurons in the brain represent these basis functions (e.g., Gabor functions in the primary visual cortex) [24]. The key element of the present proposal about the role of brain chaos is that such basis functions can be formed by combinations of chaotic attractors, which emerge in the behavioral space (i.e., the space of spatio-temporal firing patterns) of neural assemblies [8]. We call chaotic neural objects the chaotic attractors emerging in neural assemblies. We suggest that the inputs to the brain are encoded as chaotic neural objects and such objects are combined at the pre-actuator level (e.g., motor neurons controlling directly the muscles) to provide the right selection of basis functions necessary for high precision behavior. The argument in favor of coding with chaotic neural objects is that they share the stability properties of other dynamically stable configurations (e.g., fix-point attractors) and at the same time they have superior information encoding capacity [15]. The paper describes specifically, the use of chaotic attractors to generate classes of basis functions, the interpretation of these chaotic attractors in neural context and the learning in such neural systems. The presented proposal for the role of chaotic neural activity explains the advantages of using chaos in neural computations and the link between chaotic neural activity and high precision behavior. At the same time it suggests new directions for neural coding related artificial intelligence (AI) research.

The structure of the paper is as follows. Section 2 describes the way of using chaotic attractors for the generation of classes of basis functions. In Section 3 a neural interpretation is given for the previously presented method of basis function generation. Section 4 discusses the issues of learning in neural systems working with chaotic neural objects. In Section 5 we discuss some implications of our proposal. Section 6 contains the conclusions of the paper.

2   Computation with Chaos

In this section we discuss first the main properties of chaotic objects (e.g., chaotic attractors), and next we use the Sierpinski triangles to show how to generate pools of basis functions by interaction of chaotic objects.

2.1   Chaotic Objects

We call chaotic objects all those static or dynamic systems, patterns, configurations, which can be described in terms of mathematical chaos. Examples of such terms of mathematical chaos are the various non-integer chaotic dimensions (e.g., Hausdorff dimension [3]) or the positive Lyapunov exponents [3]. A critical evaluation of such mathematical chaos measures in neural context can be found in [9].

The chaotic objects usually can be described as invariant sets of some non-linear transformations of the mathematical space where these objects exist. The invariant sets are stable under these transformations in the sense that a member point of the invariant set is transformed into a member point of the same invariant set. In general the invariant sets may have simple (e.g., point) or complex (e.g., chaotic attractor, for example see Figure 1) shapes.

The stability of chaotic attractors is the same as that of the other simpler attractors, like the fix-point attractors. This means that a system that works with fix-point attractors (e.g., the Hopfield neural networks [11]) can work with chaotic attractors, too, by implementing the necessary modifications in the dynamics of the system [18]. Although this is not an argument in favor of chaotic attractors, it neutralizes the critique, which suggests that the apparently disorganised nature of these attractors may imply reduced stability in comparison with simpler attractors (e.g., fix-point attractors).

A dramatic feature of chaotic objects is their enormous variability in their details. To give an impression we present two images (see Figure 1) of the border of a famous chaotic object, the Mandelbrot set [3]. This high variability can be used for information encoding. It is known that a very simple chaotic dynamical system, the one driven by the logistic map:

$$x_{n+1} = \lambda x_n (1 - x_n) \qquad (1)$$

for appropriate λ (e.g., $\lambda = 4$), can be used to generate all non-periodic binary sequences (i.e., the binary sequences, which correspond to irrational numbers) [3]. By means of symbolic dynamics [15] we can systematically evaluate the information encoding capacity of various dynamical systems. In general it is true that chaotic dynamical systems have a very large information encoding capacity, due to their variability [15]. With respect to the dynamical systems characterized by simpler non-chaotic attractors, we note that their information encoding capacity is reduced, due to the limited variability of their attractors.

Summarizing, the chaotic objects have similar stability properties as their non-chaotic counterparts, while the information encoding capacity of chaotic objects is much higher than the same ability of the non-chaotic objects. A simple way to use these properties of chaotic objects is to encode members of a basis

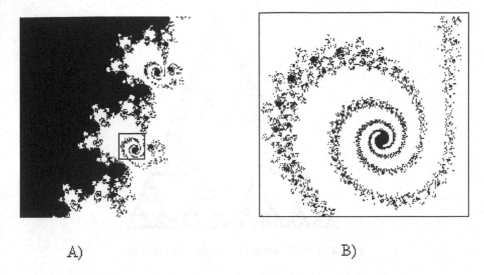

A)   B)

Fig. 1. The border of the Mandelbrot set, B) is the refined view of the inset in A).

function family, which is suitable for approximation of large classes of functions. An example is given to this task in the next sub-section.

2.2 Sierpinski Basis Functions

We use the Sierpinski triangle, a simple chaotic object, for demonstration purposes. The Sierpinski triangle is defined in the following way. Let S^0 be a triangle. Construct the geometric object S^1 by adding the three midlines of S^0 to the S^0 triangle. Construct S^2 by adding the midlines of the three corner triangles of S^1 to S^1. Continue this procedure, by adding to S^n the midlines of the corner triangles of S^n. Figure 2 shows S^6 for an example triangle. The Sierpinski triangle is the limit of this procedure, i.e., it is $S = \lim_{n \to \infty} S^n$. A different way of obtaining the Sierpinski triangle is the stochastic generation method. Start with the three corners of the triangle S^0, denoted by a, b, and c, and a random point p_0 in the plane of the triangle. Consider three functions $f_a(p) = \frac{1}{2}(p+a)$, $f_b(p) = \frac{1}{2}(p+b)$, $f_c(p) = \frac{1}{2}(p+c)$. Randomly apply the three functions and generate the series of points p_t, $p_t = f_x(p_{t-1})$, where x randomly takes one of the values from $\{a, b, c\}$. The geometric object made by the union of the p_t points, when $t \to \infty$, converges to the Sierpinski triangle $S = \lim_{n \to \infty} S^n$.

In order to generate the Sierpinski basis functions we impose some restrictions over our Sierpinski triangles. We denote by S_w the Sierpinski triangle generated from the triangle, which has as vertices the points $(0,0)$, $(0,1)$ and $(w,1)$. We denote by T_x the Sierpinski triangle generated from the triangle, which has as vertices the points $(1,0)$, $(1,1)$ and $(x,0)$. Let $c(w,x,n)$ be the number of intersections between the n-th levels of the two triangle, i.e. between S_w^n and T_x^n. We associate a Sierpinski basis function to the triangle S_w, denoted by s_w.

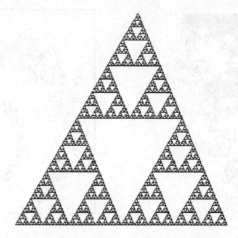

Fig. 2. The S^6 stage of a Sierpinski triangle.

The values of the basis function s_w are calculated as the weighted sum of the $c(w, x, n)$ values, i.e.,

$$s_w(x) = \sum_{n=0}^{\infty} \gamma_n \cdot c(w, x, n) \qquad (2)$$

Practically we use the weight set $\gamma_0 = \ldots = \gamma_6 = 1$, $\gamma_n = 0$ for $n > 6$. Two typical examples of the Sierpinski basis functions are shown in Figure 3

As we can note in Figure 3, the first type (case A in the figure) resembles the combination of a sigmoidal function with two radially symmetric functions, while the second type of the basis functions (case B in the figure) resembles a sigmoidal function. By more accurate analysis the Sierpinski basis functions can be decomposed into a combination of nine functions which have similar structural characteristics as the Gaussian functions, e.g., $e^{-\frac{x^2}{2r^2}}$, their complementary functions, e.g., $\frac{r}{x}(1 - e^{-\frac{x^2}{2r^2}})$, and the sigmoidal functions, e.g., $\frac{1}{1+e^{-ax}}$. It is known from neural network approximation theory [1][5][10] that these families of functions have the universal approximation property with respect to continuous and smooth functions (i.e., linear combinations of them can approximate with arbitrary precision the latter functions). Based on these results and on the fact that the Sierpinski basis functions can be considered as linear combination of such functions we conjecture that the family of Sierpinski basis functions has the universal approximation property. To prove this formally we would need to have the analytical form of the Sierpinski basis functions, which is unavailable, due to their complex nature.

Based on our conjecture we conclude that it is possible to use simple chaotic objects to generate basis functions, which can approximate a large range of functions and can be used efficiently to perform complex predictions and classifications.

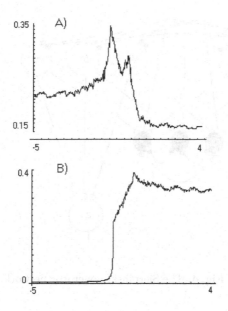

Fig. 3. Two typical Sierpinski basis functions. The values are normalized to be in the interval $[0, 1]$.

3   The Sierpinski Brain

In this section we give a neural interpretation for the Sierpinski triangles and we design the Sierpinski brain, which is able to control simple actions. We emphasize that we do not propose that Sierpinski triangles exist in the brains of animals, but we intend to give a suggestive example of how the brain might work by using chaotic neural objects.

We use the stochastic generation method of the Sierpinski triangles in order to build simple neural networks, which exhibit a behavior that can be described in terms of such chaotic objects. The Sierpinski brain has two key neural structures. The first is responsible for the generation of Sierpinski triangle behavior and the second is responsible for the calculation of the interaction of such behaviors.

The Sierpinski generator network consists of eleven neurons. These neurons are organized as shown in Figure 4. The neurons in the upper row have a stable periodic firing frequency, which corresponds to the coordinates of the three vertices of a triangle. The role of the inhibitory neurons, shown as black filled circles, is to select one of the three pairs of the periodically firing neurons. Due to the competition mediated by the inhibitory neurons, only one of the ax, bx, cx, respectively ay, by, cy neurons fire. They are grouped in such a way that the member neurons of the groups (ax, ay), (bx, by), (cx, cy) fire in the same time periods. The zx and zy neurons integrate the incoming firing frequencies for a longer time period. Due to the use of the synapses this integration process is a weighted integration, i.e., the later the incoming signal, the smaller its contri-

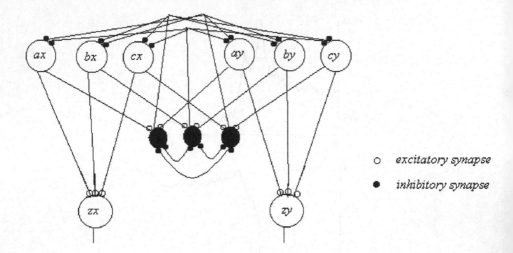

Fig. 4. The Sierpinski generator network.

bution to the final sum. The integration process is formulated mathematically
as

$$f_{zx} = \sum_{n=1}^{N} \lambda_n \cdot f_n \qquad (3)$$

where $\lambda_n > \lambda_{n+1}$, a good choice being $\lambda_n = \frac{1}{2^n}$. The f_n-s randomly take the
values f_{ax}, f_{bx}, f_{cx}, due to the competition mediated by the inhibitory neurons.
The formula is similar for the zy neuron, with the difference that f_n randomly
takes the values f_{ay}, f_{by}, f_{cy}. The Sierpinski generator network produces the
Sierpinski triangle through the firing rates of the zx and zy neurons. Figure 5
shows the distribution in the plane of the points with coordinates (zx, zy) for
5000 integration periods at the level of zx and zy neurons. Each integration
period consisted of the integration of ten consecutive incoming firing rates (i.e.,
f_n-s, for $n = 1, \ldots, 10$).

The Sierpinski brain uses the Sierpinski basis functions for the computation of
its actions. Each basis function is represented by a Sierpinski generator network.

The input to the Sierpinski brain comes from sensory cells and is represented
in the form of partially variable Sierpinski network. In this case one of the pairs
of neurons with fixed firing rate changes its firing rate depending on the input.
The input is quantified as a real number u and this is represented by the generation of the Sierpinski triangle T_u through a Sierpinski network. The single
input dependent component of the input representing network is the cx neuron,
which adjusts its firing rate to represent the value u. The other components have
genetically fixed values, namely $ax = 0$, $bx = 1$, $ay = 1$, $by = 1$, $cy = 0$.

The second main structural component of the Sierpinski brain is the interaction calculation network. This network is shown in Figure 6. The C neuron
receives excitatory input from two Sierpinski generator networks, i.e., the zx,

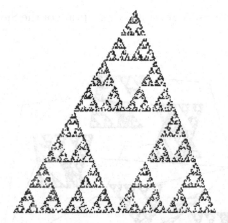

Fig. 5. The output of a Sierpinski generator network: the points with the coordinates (zx, zy).

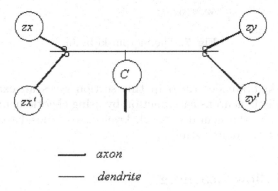

——— *axon*

——— *dendrite*

Fig. 6. The interaction calculation network.

zy, respectively zx', zy' neurons are the output neurons of these networks. The C neuron detects the coincident firing of the zx and zx', respectively zy and zy' neurons. If both pairs fire coincidentally they signal an intersection point of the two Sierpinski triangles. The C neuron integrates the incoming signals, by counting the simultaneous coincident firings of the (zx, zx') and (zy, zy') neuron pairs for a time period. The output of the C neuron represents the normalized result of this counting (i.e., the number of counts is divided by an upper limit number). In this way the output of the C neuron gives the value of a Sierpinski basis function, which has as its parameter w the cx value of the (zx, zy) network and as its argument the u value of the (zx', zy') network.

The integrated output of several interaction calculation networks gives the right answer to the input. The Sierpinski brain consisting of many Sierpinski generator and interaction calculation networks can perform classification and prediction tasks, due to the above mentioned approximation properties of the

Sierpinski basis functions. A graphical description of the Sierpinski brain is given in Figure 7.

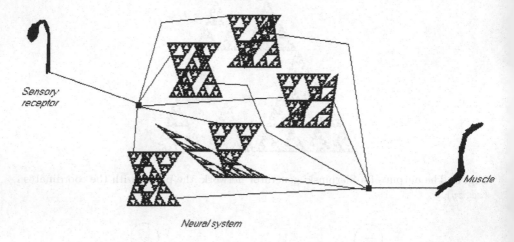

Fig. 7. The Sierpinski brain.

The Sierpinski brain described in this section gives an example of how a brain having realistic neurons can function by using chaotic neural objects. This example shows how the neural network level chaos is used to produce a very precise response to incoming stimuli.

4   Fast and Slow Learning

In this section we discuss the learning processes of Sierpinski brains. The case of the Sierpinski brain provides an example of how a real brain that uses chaotic neural objects might work during learning.

First we consider five populations of neurons at the level of Sierpinski generator networks. These populations are the following: P_0 consisting of neurons with fix genetic specification that represent the value 0 and have regular periodic firing, P_1 a similar population of neurons representing the value 1, P_x that consists of neurons which have modifiable regular periodic firing (i.e., these represent the non-0, 1 coordinate values for the various Sierpinski triangles), P_{inh} consisting of inhibitory neurons participating in the generator networks, and P_z the population of neurons that give the output value of generator networks. In this context, the input to the Sierpinski brain is represented by neurons from the P_x population and the learning process affects the distribution of the represented non-0, 1 values in the P_x population.

The individual learning is performed by the Sierpinski brain through a trial-and-error process. The objective of this learning is to optimize the weights of the basis functions such that the output of the Sierpinski brain matches closely

the ideal input-output relationship (i.e., this can be a classification of the input
or a continuous functional relationship between the input and output). The
weight of a basis function is given by the number of generator networks, which
encodes the corresponding Sierpinski triangle. Thus, by learning the number of
Sierpinski networks encoding a certain basis function is changed. We term the
individual learning *fast learning*, because it happens fast, during the lifetime of
an individual.

The learning in the Sierpinski brain is governed by two principles, these are
the energy minimization principle and the principle of the replication of the
fittest.

The energy minimization principle means that the activity of the Sierpinski brain during functioning is minimized. This is equivalent to minimizing the
number of activity at the level of interaction calculation networks, because the
activity in the Sierpinski generator networks is fixed. Thus, we minimize the
number of intersections between all possible combination of Sierpinski triangles
represented by using neurons of population P_x.

Let us formally note the number of interactions as

$$K(x,y) = s_x(y) = s_y(x) \tag{4}$$

where s_x and s_y are the Sierpinski basis functions with parameter x, respectively
y. The minimization of interactions is formulated as

$$\min_{\mathbf{a}} \frac{1}{2}\sum_{i=1}^{N}\sum_{j=1}^{N} a_i a_j K(x_i, x_j) \tag{5}$$

where N is the number of represented different Sierpinski triangles, x_t, $t = 1,\ldots,N$, are the parameters of these Sierpinski triangles (i.e., they are the value
of the x-coordinate of the third vertex), a_t is the number of representations of
the Sierpinski triangles with parameter x_t, and $\mathbf{a} = (a_1,\ldots,a_N)$.

The replication of the fittest is an evolutionary optimization principle and
its interpretation in neural systems context is rooted in the work of Edelman [4].
The fundamental idea behind this principle is that the neurons and assemblies
of neurons participate in a competition for resources within the living brain.
The competition selects those assembly structures which are the most active
and which use most of the resources. The individual neurons tend to participate
in the winning assembly structures. This happens by changing synaptic weights
and redirecting axonal and dendritic connections (i.e., making new connections
and retracting from existing connections) [4][24]. In our case this optimization
principle means that the existing neural networks (i.e., the Sierpinski generating
networks) support the recruiting of neurons in similar neural networks. In other
words, if the Sierpinski brain has a network that represents the Sierpinski basis
function s_x than the existence and functioning of this network supports the
formation of other similar networks that represent the same basis function s_x.
Formally we write the effect of this replication pressure as

$$\max_{\mathbf{a}} \sum_{i=1}^{N} a_i \tag{6}$$

The two counteracting optimization principles are joined as follows

$$\min_{\mathbf{a}} \frac{1}{2} \sum_{i=1}^{N} \sum_{j=1}^{N} a_i a_j K(x_i, x_j) - \sum_{i=1}^{N} a_i \qquad (7)$$

where we turned the maximization into minimization by applying the minus sign in the front of the sum.

The optimization process is activated only when the output of the Sierpinski brain is not adequate. The failure in providing the right response triggers the learning process that is formulated by the equation (7).

The optimization in equation (7) corresponds to finding a support vector machine [23], which solves a classification problem. By taking into account more information about the actual and desired output, the equation (7) can be transformed into optimization corresponding to support vector machines that solve more sophisticated problems (e.g., prediction by function approximation).

As we know from the theory of support vector machines, these represent the minimal complexity faithful model of the given data. Consequently, our Sieprinski brain is able to obtain a minimal complexity good model of the observed reality. This model is robust (i.e., has little sensitivity to noise) and provides good decisions on the basis of input information. This indicates that a neural system working with chaotic neural objects and subject to the above mentioned natural optimality principles is able to build high precision input-output models of its environment similar to real animals.

The individual learning happens truly fast if the initial pool of basis function given by the P_x population is good enough to allow fast optimization of the representational strength of basis functions. Otherwise, if the initial pool of basis functions is far from the optimal the individual learning is slowed down. Good initial pools of basis functions can be achieved by evolution within a population of Sierpinski brains. We name this process the *slow learning*.

During slow learning the individual Sierpinski brains compete in terms of their classification and/or prediction tasks. New populations of Sierpinski brains are generated by evolutionary selection. The effect of this evolutionary selection is that the initial basis function pools of the individual Sierpinski brains will be sufficiently close to the optimal. As the task set of Sierpinski brains may contain several different tasks, the unilateral focusing of the initial basis function pool on performing only one or some of these tasks may have negative effects on the overall performance of such Sierpinski brains. Consequently, by slow learning we obtain Sierpinski brains that are specialized enough, but not over specialized. The presence in small number of generalist Sierpinski brains (i.e., with little specialization) in the population allows the turn of adaptational directions to new targets as new tasks appear. On the other hand highly specialized Sierpinski brains may perform very well in their task specific niches.

The combined process of slow and fast learning shows how neural systems working with chaotic neural objects may evolve and learn individually in order to adapt to their environmental challenges.

5   Discussion

In this section we discuss two issues related to the presented work. First we discuss the how brain may work in our view. Second we discuss how the presented work can give useful hints for artificial intelligence implementations.

5.1   How the Brain May Work ?

The presented work leads to interesting ideas about how the brain may work. It suggests that the inputs to the brain are encoded as ensembles of dynamic neural activities that form chaotic neural objects. These neural objects interact with already existing neural objects. The interactions lead to the creation and transformation of the processed objects. After several such transformations the active neural objects encode the right basis functions. These are used to generate precise motor actions, which are at the basis of behaviors. The high information encoding capacity of chaotic neural objects allows for efficient multi-purpose coding of the same information in a compact way. Every line of processing uses that part of the chaotic object code, which matches its particular informational requirements. The result is the complex pattern of behaviors executed in parallel by the animal, in response to multi-channel inputs coming from the environment.

As we indicated, the chaotic neural objects have the same stability properties as the non-chaotic stable dynamic configurations and they have superior information encoding capabilities than the latter. Still, the question remains, how such connective structures, which give rise to chaotic neural objects could form. A hint in this respect is provided by studies on the formation of patterns during morphogenesis [4]. These studies show that the so called 'Turing patterns' [4] are very frequent during morphogenesis, a typical example being the spot and stripe pattern of various animals [4]. Such patterns are the result of chaotic reaction - diffusion processes, where the individual characteristics of the cells are determined by various biochemical gradients. These observations suggest that similar mechanisms may guide the formation of the connective structure between the neurons. Recent results about axonal growth [19] and neuron migration [21] support these ideas about the role of chemical gradients in the neural system development. Thus the development of the connective structure that supports the emergence of chaotic neural objects may be very natural and expectable.

5.2   Messages for AI Research

The AI research besides intending to copy how the natural brain works, is directed towards successful practical applications of 'intelligent' technologies. From the point of view of the latter, this work offers guidelines for new research directions.

The natural chaotic neural objects are coded by the genes through the system of protein interactions and morphological development of the brain. Similarly to this, we may code the artificial chaotic neural objects by short sequences of artificial genes that determine mathematically the encoded chaotic objects (e.g., simple fractals encoded in the form of L-systems [3]). Having an efficient coding mechanism for our artificial brain, we can evolve it using the ideas of slow and fast learning described above. Thus we can evolve simulated artificial brains that can solve many problems and can live successfully in an environment made by various tasks.

The presented work indicates a new way of finding the right choice of basis function family for the solution of prediction and classification problems. As we pointed out above, the family of optimal basis functions can be obtained through the evolution of artificial brains working with chaotic neural objects. In this way we may even obtain natural combinations of independent families of basis functions that arise by applying the constraints of the task environment to the population of simple artificial brains.

The thinking about neural coding in terms of chaotic objects reveals a possible new direction in the realization of hardware systems with high computational power. Chaotic objects, like those described in this paper, can be obtained as natural forms of organization of various substances (e.g., chaotic diffusion patterns [4]). Realizing such objects at the nanometer scale can lead to the building of high efficiency, reconfigurable, molecular level computation machines that can be used in intelligent appliances (e.g., set-top box, wearable computer).

Besides the noted practical application lines, an interesting theoretical issue is to investigate the approximation properties of the various basis function families that are generated by classes of artificial chaotic neural objects. Having a priori knowledge about the specific advantages and disadvantages of classes of chaotic objects in certain task environments, the evolutionary process can be speeded up by directed evolution (i.e., by imposing additional restrictions on the evolutionary process to avoid the undesired gene sequences).

6   Conclusions

A new interpretation of brain chaos is introduced in this paper. This is based on the idea that chaotic objects can be used for generation of families of basis functions with general approximation capabilities. The use of chaotic neural objects in information processing within the brain is favored by the stability and information encoding properties of such objects. We also indicated the possible developmental background for the formation of chaotic neural objects.

We have shown through a simple example, how to use chaotic objects to generate a family of basis functions. We introduced the Sierpinski basis functions that are generated by the Sierpinski triangles. We presented the Sierpinski brain that uses these basis functions to generate responses to inputs. The Sierpinski brain shows how simple chaotic objects can emerge in realistic neural networks and how they are used to perform input-output mapping.

We analyzed the learning processes in Sierpinski brains. We have shown that the fast, individual learning leads to a robust optimal model of the input-output map. We discussed the role of the slow, population level learning, which provides the right pool of basis functions for the fast learning.

Our interpretation of brain chaos leads to a new view of how the brain may work. According to this the inputs to the brain are represented as chaotic neural objects, which interact with other similar objects, and at the end these interactions lead to the formation of combinations of basis functions that are used for the control of movements and other actions of the organism.

With respect to AI implications we noted four future research directions. These are: the evolution of multi-purpose intelligent agents (i.e., simulated brains) using chaotic objects, fast and slow learning; the generation of new classes of basis functions using the presented methods; the hardware implementation of the described artificial brains at the nano scale; and the theoretical analyses of the approximation properties of the basis functions that are generated by chaotic objects.

References

1. Andras, P.: RBF neural networks with orthogonal basis functions. To appear in: Howlett, R.J., Jain, L.C. (eds.): Recent Developments in Theory and Algorithms for Radial Basis Function Networks. Physica-Verlag, Heidelberg (2000)
2. Brockmann, D., Geisel, T.: The ecology of gaze shifts. Neurocomputing **32-33** (2000) 643-650
3. Devaney, R.L.: A First Course in Chaotic Dynamical Systems. Theory and Experiment. Addison - Wesley, Reading, MA (1992).
4. Edelman, G.M.: Neural Darwinism. Basic Books, New York, NY (1987)
5. Ellacott, S.W.: Aspects of numerical analysis of neural networks. Acta Numerica **3** (1994) 145-202
6. Erdi, P.: On the 'Dynamic Brain' metaphor. Brain and Mind **1** (2000) 119-145
7. Freeman, W.J.: Role of chaotic dynamics in neural plasticity. In: Pelt, J. van, Corner, M.A., Uylings, H.B.M., Lopes da Silva, F.H. (eds.): Progress in Brain Research **102** (1994) 319-333
8. Fujii, H., Ito, H., Aihara, K., Ichinose, N., Tsukada, M.: Dynamical Cell Assembly Hypothesis Theoretical Possibility of Spatio-temporal Coding in the Cortex. Neural Networks **9** (1996) 1303-1350
9. Glass, L.: Chaos in neural systems. In: Arbib, M.A. (ed.): The Handbook of Brain Theory and Neural Networks. MIT Press, Cambridge, MA, (1995) 186-189
10. Haykin, S.: Neural Networks: A Comprehensive Foundation. Macmillan Publishers, Englewood Cliffs, NJ, (1994)
11. Hopfield, J.J.: Neural networks and physical systems with emergent collective computational abilities. Proceedings of the National Academy of Sciences **79** (1982) 2554-2558
12. Hooper, S.L.: Crustacean stomatogastric system. In: Arbib, M.A. (ed.): The Handbook of Brain Theory and Neural Networks. MIT Press, Cambridge, MA, (1995) 275-278
13. Kelso, J.A.S.: Dynamic Patterns : The Self-Organization of Brain and Behavior. Bradford Books, Cambridge, MA (1997)
14. Lieske, S.P., Thoby-Brisson, M., Telgkamp, P., Ramirez, J.M.: Reconfiguration of the neural network controlling multiple breathing patterns: eupnea, sighs and gasps. Nature Neuroscience **3** (2000) 600-607
15. Lind, D.A., Marcus, B.: An Introduction to Symbolic Dynamics and Coding. Cambridge University Press, Cambridge (1995)
16. Masson, G.S., Castet, E.: Motion perception during saccadic eye movements. Nature Neuroscience **3** (2000) 177-183

17. Mpitsos, G.J., Burton, R.M., Creech, H.C., Soinilla, S.O.: Evidence for chaos in spike trains of neurons that generate rhythmic motor patterns, Brain Research Bulletin **21** (1988) 529-538
18. Nara, S., Davis, P.: Learning feature constraints in a chaotic neural memory. Physical Review E **55** (1997) 826-830
19. Van Ooyen, A., Willshaw, D. J.: Development of nerve connections under the control of neurotrophic factors: parallels with consumer-resource systems in population biology. Journal of Theoretical Biology **206** (2000) 195-210
20. Rabinovich, M.I., Abarbanel, H.D.I., Huerta, R., Elson, R., Selverston, Al.I.: Self-regularization of chaos in neural systems: Experimental and theoretical results. IEEE Transactions on Circuits and Systems - I.: Fundamental Theory and Applications **44** (1997) 997-1005
21. Rakic, P., Komuro, H.: The role of receptor-channel activity during neuronal cell migration. Journal of Neurobiology **26** (1995) 299-315
22. Schiff, S.J.: Forecasting brain storms. Nature Medicine **4** (1998) 1117-1118
23. Schölkpof, B., Burges, C.J.C., Smola, A.J. (eds.): Advances in Kernel Methods. Support Vector Learning. MIT Press, Cambridge, MA, (1999)
24. Shepherd, G.M.: The Synaptic Organization of the Brain. Oxford University Press, Oxford (1997)
25. Tsuda, I.: Towards an interpretation of dynamic neural activity in terms of chaotic dynamical systems. To appear in: Behavioral and Brain Sciences (2000)

Neural Network Classification of Word Evoked Neuromagnetic Brain Activity

Ramin Assadollahi[1] and Friedemann Pulvermüller[2]

[1]Department of Psychology. University of Konstanz. Germany.
[2] Medical Research Council. Cognition and Brain Sciences Unit. Cambridge. England.

Abstract. The brain-physiological signatures of words are modulated by their psycholinguistic and physical properties. The fine-grained differences in complex spatio-temporal patterns of a single word induced brain response may, nevertheless, be detected using unsupervised neuronal networks. Objective of this study was to motivate and explore an architecture of a Kohonen net and its performance, even when physical stimulus properties are kept constant over the classes.

We investigated 16 words from four lexico-semantic classes. The items from the four classes were matched for word length and frequency.

A Kohonen net was trained on the data recorded from a single subject. After learning, the network performed above chance on new testing data: In the recognition of the neuromagnetic signal from individual words its recognition rate was 28% above chance (Chi-square = 16.3, p<0.0001) and its accuracy was 44% above chance (Chi-square = 40.8, p<0.0001). The classification of brain responses into lexico-semantic classes was also unexpectedly high (recognition rate 16% above chance, Chi-square = 27.2, p<0.0001, accuracy 20% above chance, Chi-square = 42.0, p<0.0001).

Our results suggest that research on single trial recognition of brain responses is feasible and a rich field to explore.

1 Introduction

Physiological data from psycholinguistic research suggests that words are distinguishable by some of their properties such as word length, frequency and meaning. Assumed that word evoked fields are modulated by these properties and the corresponding MEG signals reflect these distinctions, it might be possible to classify neuromagnetic brain responses to single words automatically.

Recognition of single carefully chosen words seems feasible when some well studied properties of words are taken into account:

- The representation of a word is neurobiologically realised in mostly cortical networks having a topology that depends on the modalities correlating with external aspects of the meaning (Pulvermüller 1999, 1995): the representation of action verbs for example would include neurons from the primary motor cortex in addition to the neurons involved in the phonetic representation located in the persylvian cortex
- Word length and frequency modulate the brain response to words (Osterhout et al. 1997, Rugg 1990) in respect of latency and/or amplitude of certain ERP components: the presentation of short high frequency words leads to smaller

S. Wermter et al. (Eds.): Emergent Neural Computational Architectures, LNAI 2036, pp. 311–319, 2001.

amplitudes or short latencies whereas long low frequency words are often reported to induce larger amplitudes or longer latencies.

In Suppes et al. (1997, 1998) test samples and single trials of auditory evoked potentials and fields were analysed. Under investigation were whole sentences and as well single words. It was not reported whether physical properties of stimuli were matched, e.g. for word length or root mean square of the acoustic signal. The studies were performed with several subjects that were analysed as single subjects. Whole samples consisting of averages of ten trials were analysed. Optimising the parameters of a bandpass that was applied to the data before classification improved the recognition performance.

There are two problems in interpreting the results of these studies: First, psycholinguistic research has demonstrated the influence of word length on brain responses (e.g. Osterhout et al 1998). Word length is a physical property of words and modulates the brain response as such. Second, since the response of the auditory cortex to certain phonetic properties of the auditory signal is very prominent (e.g. plosives) the mere absence in one stimulus or the other might lead to strong brain responses that are easily detectable. In sum, mere physical properties of the stimuli may have been the critical factor for word recognition performance. These influences must be ruled out by carefully matching stimulus words for physical parameters and, where possible, for other psychologically relevant parameters as well (e.g. word frequency).

The objective of our study was to investigate physical properties of words but mainly their psycholinguistic features. We therefore varied word length (physical factor) over the two other factors word frequency and word class (cognitive factors). To surpass problems arising from different RMSs and other acoustic word properties we preferred to present the stimuli visually.

In order to explore the cognitive aspects of word processing semantically distinct words from three different lexical categories were chosen, nouns, verbs and grammatical function words. A questionnaire study was performed on 12 subjects to determine association strength for the words.

No semantic associations for words were reported for function words (e.g., "this"). For other words, there was evidence of strong associations in only one modality, for example only visual or motor associations. This was found for action verbs and for visually related nouns ("get", "sheep"). Finally, there were words with strong multimodal associations, typically nouns referring to entities that can be visually perceived and can also be manipulated with the hands (e.g., "boat" or "newspaper"). Thus we had four stimulus groups in our experiment, (1) function words with no associations, (2) action verbs with motor associations (3) nouns with strong visual associations, and (4) nouns with strong multimodal associations.

When processing the representations of words it is assumed that their phonetic representation in the perisylvian fissure will be activated. The activation of function words might be confined to this cortex area. The representation of the other words should include additionally the visual cortex and/or motor cortex.

Since cortical representations of the word classes in our stimulus set have brain areas in common (e.g. the visual cortex for visual nouns and multimodal nouns), their neuromagnetic brain responses should have partially similar topologies. Other aspects of similarity in the brain response include the influence of word length and frequency that are not specific to the meaning. For construction of an automated recognition, it is therefore well motivated to implement a type of neuronal networks that can model similarities and employ them for recognising new samples:

- In Kohonen's model the development of cell responses is organised in an unsupervised and self-organising fashion. Neurons automatically develop stimulus specific properties and the network self-organises such that neighbouring cells are tuned to similar stimuli. (Kohonen 1982, 1999)
- Kohonen networks can reduce dimensionality of the input data while preserving the topography of the data in the high dimensional space.

In his classic work Kohonen (1984) introduced a two-dimensional map for Finish phonemes. Digitised acoustic speech signals were submitted to a Fourier-transformation subdivided into 15 bands. The energy of the frequency bands served as 15 dimensional feature vector. Feature vectors were generated for windows of 25.6ms length and were shifted by 10ms over the signal. The overlap in the signal guaranteed the similarity of subsequent features. The feature vectors were fed to a 2D Kohonen map resulting in spatial clusters of the 21 Finnish phonemes. The phonemes in the map were automatically ordered in a way that acoustically similar phonemes were situated in neighbourhood. The network thus collapsed the 15 dimensional feature space as well as temporal information of the phonemes' signals into one representational space while preserving similarity. The network was unable to distinguish stop-phones. Stops are distinguished by the length of silence between two features. Since the net only gets time slices, the number of time frames containing silence is not represented. This leads to the architecture inherent inability to distinguish plosives.

In the present study of neuromagnetic responses to words we chose to use the energy (RMS) of up to eight sites over the cortex in order to provide spatial information to the net. In correspondence to Kohonen (1984) we chose to model the signal representation through time windows that are moved over the signal instead of whole samples. Thus parts of signals that are common across word classes are stored together in some parts of the map whereas signal parts that are typical for a certain word class should occupy larger coherent areas in the network.

Following Suppes et al. (1997), we chose to generate and analyse data from a single subject for the following reasons: (1) The latency of word evoked brain responses related to cognitive processing varies substantially as a function of stimulus familiarity and, consequently, between experiment participants (Osterhout et al. 1997). (2) Structural and functional brain organisation differ considerably between individuals (Damasio 1995), thus causing devastating between-subject variance in both topography and timing of evoked fields.

2 Materials and Methods

2.1 MEG-Experiment

Subject

The participant was a 21 year old right handed monolingual female student with 15 years of formal education. She had normal eyesight and no history of neurological disease or drug abuse. She was paid for her participation.

Stimuli

Words from four word classes were presented frequently in the experiment (recurrent words): (1) function words, (2) action verbs, (3) nouns with strong visual associations and (4) nouns eliciting both visual and action associations.

Words were 1 and 2 syllables long (4 to 7 letters). Half of the words had high (range: 123-329 average: 210.375) and low (range: 10-15 average: 12.875) frequency. Word length and frequency were exactly matched between lexico-semantic categories. Each word category included two long (6-7 letters, two syllables) and two short (4-5 letters, one syllable) words. Before the experiment, ratings from 12 subjects confirmed the classification of these words into the four lexico-semantic categories (amodal, visual unimodal, action unimodal, multimodal). Association ratings for function words (1) were low. Strong associations of actions were only reported for the stimuli in classes (2) and (4), whereas strong visual associations were reported for classes (3) and (4). Semantic association scores were computed as the sum of action and visual ratings. Additional ratings confirmed that arousal and valence ratings did not significantly differ between categories.

Additional words were only presented once or twice (new words). These were generated by exchanging one letter of one of the 16 recurrent words forming a correct word.

Procedure

In each block, the 16 recurrent words and two of the set of new words were presented in random order. Before the experiment, the subject was familiarised with the recurrent stimuli and she was told to memorise these words. She was instructed to respond to those stimuli that were not in the memorised set (new words) by pressing a button with her left index finger. No response was required after the recurrent words. The task was applied to assure sustained attention to the stimuli, and to force the participant to keep the set of recurrent words in active memory. One run consisted of 20 of such blocks and 3 runs made one session. Thus, each of the repeated words was presented 60 times in each session. The subject was allowed to pause freely between blocks and between runs. 11 sessions took place within 4 weeks, with a minimum pause of 2 days between subsequent sessions.

All stimuli were presented for 100 ms in white upper case letters (maximum word size 7x3 cm) on a black background 1.4m away from the participant's eyes. A fixation cross appeared in the middle of the screen whenever no word was visible. The asynchrony of stimulus onset was varied randomly between 1.4 and 2.0 sec. An LCD-projector outside the MEG chamber was used to project the stimuli onto the screen inside.

MEG Data Recording

Neuromagnetic signals were recorded continuously with a 148 channel whole head magnetometer (4D NeuroImaging WH2500) using a 0.1-100 Hz band-pass filter and sampled at a rate of 508 Hz. Along with the magnetic signals the vertical and the horizontal EOG together with the ECG were recorded.

Data Preprocessing

MEG data were submitted to a global noise filter subtracting the external, non-biological noise obtained by the MEG reference channels. The data was then split into epochs while discarding all epochs where the button had been pushed by the subject. Epochs with an EOG level > 100 μV or MEG level > 5 pT between minimum and maximum on one or more MEG channels were automatically excluded from further analysis. 11 sessions delivered a maximum of 660 MEG-traces for single word. The epochs were baseline corrected (100ms). Data was split by the ratio of 3 to 7 into testing and training data.

2.2 Neural Net Design and Evaluation

In order to reduce data size foci were defined located over anterio-central (Broca and Wernicke), parietal and occipital sites in both hemispheres and two additional foci over centro-occipital and centro-central sites. Each of these 8 foci comprised 5 adjacent MEG-channels. RMS of these channels was computed resulting in a higher signal-to-noise-ratio and further data reduction to the detriment of spatial resolution. Input-data for the neuronal network was generated for windows of 50ms length with a shift of 30ms.

Thus an 8 dimensional feature vector containing the RMS of several sites on the cortex was mapped into a two dimensional Kohonen-map (Kohonen 1982, 1984). The size of the Kohonen networks was one subject of investigation. Large networks usually adapt very well to training data due to greater storage capacity. Small networks tend to reduce the necessary information which at a particular size leads to worse performance. Other subjects of investigation were the foci (different focus combinations were tested) and the length of the presented data (i.e. number of windows per trial).

Training Procedure

Data windows for a particular word were presented to net, the best matching neuron learned most and its neighbours in the typical Kohonen fashion (Gaussian distribution). Learning in this context means the reduction of the Euclidean Distance between the presented vector in the 8D feature space and the stored vector in the neuron in the network.

For later classification every neuron stored a score table for all words and another one for all word classes. When a best match occurred the score for the presented word and its word class was incremented. Thus every neuron had a score table that represented the order of the words/word classes it represented best. After training a map of the best represented word classes could be rendered (see Results).

Recognition Procedure

In the recognition phase, again, a word was presented window by window and the best matching neuron per window was chosen. The scores for the best matching neurons of all windows were then summed up which resulted in a table containing a score for every word and every word class over all presented windows. The word/word class with the highest score was considered as most probable word and was put out as 'recognised'.

Statistical Testing

First, we defined recognition rate as the number of correctly recognised words/word classes divided by the over all number of presentations for this word / word class:

recognition rate = hits / (hits + misses)
(expectation values: 4 word classes 25%, 16 words 6,25%)

The accuracy was defined as number of correctly recognised words / word classes divided by all outputs of this word irrespective of correct or incorrect classification:

accuracy = hits / (hits + false positives)
(expectation values: 4 word classes 25%, 16 words 6,25%)

Chi-Square-Tests (Pearson) were performed to evaluate the statistical significance of the recognition rate and its accuracy.

3   Results

The performance was best for a time range form 490ms-790ms post stimulus onset. The word classes diverged most in this time range (see figure 1).

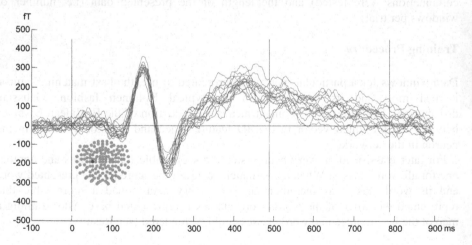

Fig. 1. Plot of all sixteen words over left central channel. The greatest divergence of the words is observable from ~400 to 800 ms.

The best configuration of the network (60x60, timing 490ms-790ms post stimulus onset, all foci except left parietal and centro-occipital) performed as given in tables 1 and 2:

Table 1. The recognition performance on training data (accuracy is given in brackets).

Rates for single words:

	F	A	B	V
1L	78%(52%)	51%(88%)	83%(61%)	68%(78%)
1H	68%(79%)	84%(55%)	77%(67%)	62%(47%)
2L	60%(91%)	54%(85%)	58%(93%)	78%(67%)
2H	63%(81%)	69%(77%)	56%(91%)	74%(58%)

Average over all words: 67%(73%)

Rates for word classes:

F	A	B	V
69%(52%)	58%(67%)	67%(56%)	43%(82%)

Average over all word classes: 59%(61%)

Table 2. The recognition performance on new testing data (accuracy is given in brackets).

Rates for single words:

	F	A	B	V
1L	13%(7%)	2%(9%)	10%(6%)	3%(5%)
1H	4%(5%)	15%(4%)	10%(16%)	27%(16%)
2L	8%(12%)	1%(3%)	6%(16%)	10%(10%)
2H	0%(0%)	8%(10%)	4%(20%)	7%(5%)

Average over all words: 8% (9%)

Rates for word classes:

F	A	B	V
38%(30%)	32%(30%)	38%(29%)	9%(30%)

Average over all word classes: 29% (30%)

Overall word class recognition rate 29% ($\chi2 = 27.2$, p<0.0001) and accuracy 30% ($\chi2 = 42.0$, p<0.0001) as well as recognition rate over all single words 8% ($\chi2 = 16.3$, p<0.0001) and accuracy 9% ($\chi2 = 40.8$, p<0.0001). Therefore results were above chance: individual words the recognition rate was 28% above chance and its accuracy was 44% above chance. The classification into word classes revealed a recognition rate of 16% above chance and an accuracy of 20% above chance.
An example for a net is given in figure 2.

4 Discussion

It seems that a feature based self-organising Kohonen network *can* classify word evoked brain potentials, although similarity (and thus topographic ordering) is only given for signal parts that distinguish word classes.

The performance of the single word recogniser is higher than the performance of the word class recogniser. This may be due to a high interclass variance, which means that members of one class are often recognised as members of another class. One reason for this may be the gross property of word length that has a great impact on signal amplitudes. Thus it appears reasonable that 2 syllable words of different classes are mixed up.

There are still several improvements to make: (1) Improvements on the input feature generation for the net and (2) improvements on the architecture of the net.

Having in mind that a simple technique (RMS of foci) was employed to obtain the features for the network input, it is surprising that a classification occurred at all. To improve recognition rates it is crucial to further investigate methods how to generate better features, e.g. through PCA. A second way to reduce data complexity and to improve the signal-to-noise-ratio would be to find an optimal bandpass filter for the data as was done by Suppes et al. in 1998. Here the EEG gamma band is promising (Lutzenberger et al. 1994).

The current architecture collapses features of different time windows to one representation. If there are different latencies for a component in a given topography of a brain response that depend on a word property and all time windows around this latency shift are presented to the net the latency information would be lost because the same neuron would respond which is maybe one reason for non homogeneous fields in the map (figure 2). The problem is similar to the distinction of plosives discussed in the introduction. Even temporal Kohonen networks using contexts may not be able to address this problem because the context of this component (i.e. the time windows preceding this component) may be the same. Further research on the latency of such components could result in determining the number of context windows necessary to model the signal properly.

```
AA BAA+AFFFF B ABBFF
 AAFAAF      AB+ AA AFF
B  FA +BBF++ A BF BB
FFFABBFF FAAFFABF A+
F BB BA FA BA A FFAF
BFAAF+FBAFBFFFABA FF
  AF A A F  FAAF BBB
BABAAABF BAFB +FBFFA
BABAAAAAB+AFFBFFFFA
+A AAA   V+BF AAF AA A
FFBAA FB+ AABBAAAF+B
+B  BA FFAB+ A +FB
ABAFBFB FB  +AAAAA F
B+ FFB+B BFBFFAAAFBA
 FBF++BB A AFBB AB A
FABBB AA AA+A FFFBB+
FBB+FFAAAFA+F+BFBB
+FBFAAAFB+BABBBBA+A
BBFBABAAB+AF+FBBABFA
BABBA FBA A BB+ +BB+
```

Fig. 2. Example for the Kohonen map representing word classes. ('A' action words, 'B' multimodal words, 'V' visually related words, 'F' function words, '-' no hit by a word class, '+' equally representing more than on word class).

Another solution might be an internal clock that simply counts the number of windows presented so far could be employed as an additional input feature. This feature would have to be weighted according to the amplitude and number of the foci.

Two factors may determine the optimal size of a net: the number of time slices and the number of foci. A larger number of either of the factors might result in the need of a larger network. On the other hand redundancies over time or over foci might reduce the additional complexity of the data and hence the needed storage capacity. Further research on these dependencies is necessary.

Another general architectural issue is the spreading activity in the subject's brain. An incoming visual stimulus causes a wave of activity cortically starting at occipital sites and progressing to more anterior sites like the perisylvian cortex and the central sulcus. Presenting data from all foci for each time window implies that brain activity on some foci is not related to word processing because the spreading activity has not reached the brain area at that time. Data from such foci would not contain any significant information for the net and would deteriorate the overall signal-to-noise-ratio. An architecture that is able model changing foci over time could be a solution to this.

We conclude that automated recognition of single word evoked brain responses is promising and further research on both, input-generation and architectural issues, will increase the performance significantly.

References

Pulvermüller F. *Behav. Brain Sci.*, **22**, 253-336. (1999)
Pulvermüller F, Lutzenberger W, Birbaumer N. *Electroenceph. Clin. Neurophysiol.*, **94**, 357-370 (1995)
Osterhout L, Bersick M, McKinnon R. *Biol. Psychol.*, **46**, 143-168 (1997)
Rugg, M. D. *Mem. Cogn.* **18**, 367-379 (1990)
Suppes P, Bing H, Lu Z-L, *PNAS*, 95, 15861-15866, (1998)
Suppes P, Lu Z-L, Bing H, *PNAS*, 94, 14965-14969, (1997)
Kohonen T , *Biological Cybernetics*, 43, 59-69 (1982)
Kohonen T, Makissra K, Saramaki T, *In: Proc. of the Seventh Int. Conf. on Pattern Recognition, Montreal, Canada*, 182-185 (1984)
Kohonen T, Hari R, *TINS*, 22/3, 135-139 (1999)
Damasio H, *Human brain anatomy in computerized images*, Oxford University Press, New York, Oxford. (1995)
Lutzenberger W, Pulvermüller F, Birbaumer N, *Neuroscience Letters*, 176, 115-118 (1994)

Simulation Studies of the Speed of Recurrent Processing

Stefano Panzeri[1], Edmund T. Rolls[2], Francesco P. Battaglia[3], and Ruth Lavis[1]

[1] Department of Psychology, Ridley Building,
University of Newcastle upon Tyne, Newcastle NE1 7RU, UK
stefano.panzeri@ncl.ac.uk
http://www.staff.ncl.ac.uk/stefano.panzeri/
[2] University of Oxford, Department of Experimental Psychology,
Oxford OX1 3UD, UK
[3] ARL-NSMA - University of Arizona,
Life Sciences North Bldg, Tucson, AZ 85724-5115, USA

Abstract. The speed of processing in the cortex can be fast. For example, the latency of neuronal responses in the visual system increases by only approximately 10-20 ms per area in the ventral pathway sequence V1 to V2 to V4 to Inferior Temporal visual cortex. Since individual neurons can be regarded as relatively slow computing elements, this may imply that such rapid processing can only be based on the feedforward connections across cortical areas. In this paper, we study this problem by using computer simulations of networks of spiking neurons. We evaluate the speed with which different architectures, namely feed-forward and recurrent architectures, retrieve information stored in the synaptic efficacy. Through the implementation of continuous dynamics, we found that recurrent processing can take as little as 10-15 ms per layer. This is much faster than obtained with simpler models of cortical processing that are based on simultaneous updating of the firing rate of the individual units. These findings suggest that cortical information processing can be very fast even when local recurrent circuits are critically involved.

1 Introduction

The speed of processing is a fundamental constraint that must be incorporated into computational models of cortical function. For example, there is evidence that the visual system can operate very fast. An analysis of response latencies indicates that there is sufficient time for only 10-20 ms per processing stage in the visual cortex. In the primate cortical ventral visual system, the difference in latencies between each stage is in the range of 10 to 20 ms [1,2,3]. In the dorsal visual system the difference in latencies between successive stages is even faster when flashed stimuli are used [2]. Information theoretic analysis of responses of single visual cortical cells in primates reveals that much of the information that can be extracted from neuronal spike trains is often found to be present in periods as short as 20-30ms [4,5,6]. Event-related potential studies in humans provide strong evidence that the visual system is able to complete some analyses of complex scenes in less than 150 ms [7]. There is also evidence that the speed of processing may be as fast in other sensory system (e.g. [8]).

The speed of information processing observed in the nervous system is somehow surprising given that individual cortical neurons may be regarded as relatively slow computing elements. The time constant of the neuronal membrane (an indication of the time needed to charge the neuronal membrane to the threshold level for action potential emission) is often estimated to be in the order of 10-30 ms [9]. These observations have led several authors to conclude that such rapid cortical processing may be based, at least in vision, almost entirely on the simplest computational architecture, namely a purely feedforward mechanism [10,11,12,7,13]). In their view, under such a tight time frame there is simply not enough time for more sophisticated computations like recurrent processing.

However, the intuition that recurrent processing is too slow is mainly fostered by results obtained with early models of autoassociative attractor networks. These networks use associative learning in the recurrent collateral connections between the neurons to store information, and to retrieve it through feedback processing [14,15,16]. In these early models the dynamics of retrieval was usually simulated with discrete time steps, with one time-step for each feedback cycle and simultaneous update of the outputs of the neurons, each of which represented some short time average of the neuron's firing rate. With these simulations of associative networks, typically several processing time-steps are needed for the network to perform pattern retrieval through the recurrent processing, and to begin to settle into a dynamical attractor corresponding to a correct memory retrieval state [17]. Discrete time step recurrent processing of this type is therefore far too slow to contribute to rapid visual cognition, given that any meaningful biophysical estimate of the discrete time step of the artificial model should be of the order of the neuronal membrane time constant or the typical inter-spike interval (both of which are in the 10-30 ms range [9]).

More recent studies have shown that such simple models of recurrent processing with discrete dynamics may give rise to artificially slow retrieval dynamics. In fact, more realistic models of recurrent networks of spiking neurons incorporating the dynamics of membrane potential and of synaptic conductance changes in continuous time can process information much faster than the discrete time step models. In particular, networks of leaky Integrate-and-Fire neurons with recurrent connections can respond very quickly to a change in the input current [18,19]. In this paper we want to take this studies a step further, and characterize the dynamics of retrieval of information about patterns that are stored in the associative synapses, in the case when recurrent collaterals are endowed with associative learning. Previous work on this subject has concentrated on purely recurrent architectures representing a single cortical module [20,21]. A full understanding of the contribution of recurrent collaterals to fast cortical processing would also require a study of the dynamics of hierarchically organized networks, in which the signal is transmitted between areas by feedforward connections, and is at the same time processed inside each area by recurrent projections. We therefore present in this paper a simulation study of the dynamics of a multiple-layer network of Integrate-and-Fire neurons, with associative feedforward connections between each layer and the next, and with excitatory recurrent collateral connections in each layer, as well as local inhibition. The main result that we found was that the information latencies in cases where local recurrent processing was important were much smaller than what expected from simultaneous updating of the firing rate of the units. This suggests that the recurrent circuitry may play some role under conditions of fast cortical processing.

2   Methods

2.1   Network Structure and Single Neuron Models

The model consisted of three layers of excitatory and inhibitory populations. Excitatory neurons within each layer were connected by excitatory recurrent collateral synaptic connections, implementing local feedback and denoted as RC in the following. The RC synapses were set up by associative (Hebbian) learning. The excitatory populations within one layer were connected to the excitatory cells in the next layer by Hebbian feedforward (FF in the following) projections. Non-associative inhibition within each layer was also provided. Each layer was composed of $N_E = 800$ excitatory units, and of 400 inhibitory units, divided into $N_S = 200$ "shunting" inhibitory units and $N_H = 200$ "hyperpolarizing" inhibitory units. Each unit was modeled as a leaky Integrate-and-Fire device, and it represented a neuron as a single branch, compartmented, dendrite, and a point-like soma where spikes were generated. The compartmental model was used in order to implement shunting inhibition, which is necessary for stable pattern retrieval in a recurrent system [21]. The current flowing from each compartment to the external medium was expressed as:

$$I(t) = g_{leak}(V(t) - V^0) + \sum_j g_j(t)(V(t) - V_j), \tag{1}$$

where g_{leak} is a constant passive leakage conductance, V^0 the membrane resting potential, $g_j(t)$ the value of the j-th synapse conductance at time t, and V_j the reversal potential of the j-th synapse. $V(t)$ is the potential in the compartment at time t. A list of all the parameter values used is reported in [22]. We just note here that the RC integration time constant of the membrane of excitatory cells was 20 ms long for the simulation presented. Synaptic conductances decayed exponentially in time, obeying the equation

$$\frac{dg_j}{dt} = -\frac{g_j}{\tau_j} + \Delta g_j \sum_k \delta(t - t_k^j), \tag{2}$$

where τ_j is the synaptic decay time constant, and Δg_j is the amount the conductance is increased when the presynaptic unit fires a spike. Δg_j thus represented the (unidirectional) coupling strength between the pre-synaptic and the post-synaptic cell. t_k^j is the time at which the pre-synaptic unit fires its k-th spike.

(The value of the inactivation time constant τ_j was changed across simulations, in order to test its effect on the speed of retrieval). The inhibitory system was set to work faster than the excitatory one, as this was helpful for preventing destabilization of the attractor retrieval dynamics due to neuronal synchronization (see [23] for detailed studies of regimes of inhibitory couplings leading to synchronization).

For each time step of 1 ms, the cable equation for the dendrite was integrated [9] with a finer time resolution of 0.1 ms, and the somatic potential was compared with the spiking threshold V^{thr}. When this was exceeded, post-synaptic conductances were updated and the somatic potential was reset to the after-hyperpolarization value V^{ahp} throughout the neuron.

The intralayer connections from excitatory to inhibitory, from inhibitory to excitatory, and between inhibitory units were taken to be homogeneous, that is, all of the same strength. The intralayer connectivity was 0.25 between excitatory and inhibitory populations, and 0.5 within each inhibitory population. There were no connections between shunting and hyperpolarizing cells. To build the connections, each cell was connected to a fraction of the cells of the receiving population chosen at random according to level of connectivity. In contrast, the excitatory units within a layer were all connected to each other. This high connectivity was necessary to produce sufficient statistical averaging in the synaptic input to each unit because of the small size of the simulated network. The connectivity of the FF excitatory connection between excitatory units of successive layers was set to 0.5. This was to take into account the fact that cortical cells receive more excitatory inputs from neurons in the same region than from axons coming from more distant parts of the cortex.

Both RC and FF excitatory to excitatory connections encoded in their strength p memorized patterns of activity $\eta_i^{L;\mu}$, consisting of binary words (with elements 0 or 1), the fraction of active units in the pattern being $a = 0.1$. The p patterns (indexed by $\mu = 1, \cdots, p$) $\eta_i^{L;\mu}$ to be stored in each layer were chosen at random independently for each layer L. Encoding was implemented through a Hebb rule as follows, with a procedure ensuring that the synaptic weights (conductances g in the model) are always positive. All conductances were initially set to zero and then, for each pattern, the synapse from the i-th to the j-th unit was modified by a covariance term, which is reported below for the RC and FF case (see [22] for the values of the parameters):

$$\Delta g^{RC} = \frac{g_{EE}}{pN_E} \left(\frac{\eta_i^{L;\mu}}{a} - 1 \right) \left(\frac{\eta_j^{L;\mu}}{a} - 1 \right), \tag{3}$$

$$\Delta g^{FF} = \frac{g_{FF}}{pN_E} \left[\left(\frac{\eta_i^{L;\mu}}{a} - 1 \right) \left(\frac{\eta_j^{L+1;\mu}}{a} - 1 \right) \right], \tag{4}$$

If, after summing each pattern, the conductance became negative, it was reset to zero. Memories were therefore stored through a "random walk with one reflecting barrier" procedure. The barrier acted as a "forgetting" mechanism [24, 21], as whenever the conductance value bumped into the barrier, it lost memory about the previously presented patterns. For computational reasons, the network was tested at low memory loading ($p = 10$). The excitatory synapses impinged on the distal end compartment of the post-synaptic dendrite, and they had a positive reversal potential (with respect to the resting membrane potential). Inhibitory synapses of the shunting type were distributed uniformly along the dendritic body, and they had a reversal potential equal to the resting membrane potential. Inhibition of this type leads to a mainly divisive effect on the post-synaptic firing rate [25]. Inhibitory synapses of the hyperpolarizing type were instead located at the distal end of the dendritic tree, colocalized with excitatory inputs, and had a reversal potential lower than the resting membrane potential. This last type of inhibition is predominantly subtractive in nature.

2.2   Simulation Protocol

First, the connection matrix was constructed during the training phase. Then, a test of the retrieval dynamics was performed according to the following protocol:

Layer 1. Layer 1 was a recurrent network responding to a cue current, and was used to set the stage for retrieval in successive layers. L1 provided a pattern-specific cue signal, which initiated retrieval in the successive layers. The structure of L1 was essentially identical to the autoassociative network studied in [21], to which we refer for full details. In brief, in the first 100 ms (for $-100 \leq t < 0$, t being the simulation time), a current was injected into a random fraction $a = 0.1$ of the units of all types. This generated spontaneous, non-specific activity. At simulation time $t = 0$ ms, the retrieval process was started by replacing the random current with a *cue* current, injected in a fraction $a + \varrho(1 - a)$ of the layer 1 units active in the pattern being tested and in a fraction $a(1 - \varrho)$ of the units inactive in the pattern. ϱ was the average correlation between pattern and cue. It was set to 0.3 for all the simulations presented. The cue current lasted for another 300 ms.

Successive layers. Each of the successive layers (layer 2 and layer 3) was activated by the incoming spikes, received by the previous layers through FF excitatory projections. As for L1, spontaneous activity was generated initially (only for $-100 \leq t < 0$) by injecting a current in a random fraction $a = 0.1$ of the units. Then another 300 ms of simulation followed, in which the network processed the pattern specific cue received by the previous layer.

2.3   Quantification of the Performance of Retrieval

The time course of processing and of retrieval of the stored patterns was quantified by using the mutual information about which pattern was presented in L1, carried at different times by the responses of small populations of excitatory neurons in each layer. There are several advantages in using an information theoretic measure. First, mutual information is the only measure of correlation that is consistent with simple and plausible requirements [26]. Second, mutual information naturally quantifies how significant the differences in the responses to different stimuli are with respect to variability or spontaneous activity. Third, unlike simpler measures like pattern overlap, the same information measure can be applied to neurophysiological recordings of small neuronal populations [27, 28], thereby facilitating the comparison between theory and experiment. The mutual information was computed as follows. The retrieval simulation protocol was repeated for 25 "trials" for each stored memory. A population of 30 randomly selected excitatory units was sampled for the number of spikes they fired in a 30 ms long time window. This time window length was chosen because it is of the order of the integration time constant of a downstream neuron reading the population output. The window was moved across the entire simulated time course with steps of 5 ms. The population firing rate vector was constructed at any time step of each trial and was *decoded*. Decoding consisted in [27] predicting the pattern presented in each trial from the average response and its distribution across all the patterns. The predicted memory pattern s^p was the one whose mean response was closest (using an "Euclidean Distance" [28] measure in the response space) to the response on the particular trial. The result of decoding all the trials was a probability table $Q(s^p|s)$ containing the fraction of times in which an actual memory s elicited a response leading to the prediction of s^p. The *mutual information I* between the actual and decoded pattern, given by

$$I(s, s^p) = \sum_s \frac{1}{p} \sum_{s^p} Q(s^p|s) \log_2 \frac{Q(s^p|s)}{Q(s^p)} \tag{5}$$

was calculated and then corrected for limited sampling [29]. To reduce fluctuations, results were averaged, at each time step, over a number of randomly selected populations of 30 excitatory units from the same run. Note that, for the time window examined, the information computed through the decoding procedure is a very close approximation to the total information contained in the population responses about the memory patterns [28]. The additional interest in using such a decoding step to compute the information is that the same decoding procedure can in principle be performed also by a downstream neuron receiving the population responses through appropriate synaptic weights [28]. All the information extracted in this way is thus likely to be readable by downstream cells.

3 Results

The aim of this study is to understand the dynamics of information processing in multiple layer networks with both feedforward and local recurrent connections, and in particular to investigate the speed of the specific RC contribution. For this purpose, it is useful to begin the analysis by looking first at the purely recurrent situation. Hence, we concentrate first on the speed of information retrieval of layer 1 neurons. Layer 1 operated as a recurrent autoassociative network responding to a cue current injected at $t = 0$ ms. The time course of the information transmitted by populations of excitatory neurons in layer 1 is reported in Figure 1. It is evident that the network activity gradually moves from a spontaneous activity state, to a state highly correlated to the activity pattern stored into the synapses that is more similar to the noisy cue injected. This happens also with discrete dynamics based on simultaneous updating [14,15,16,17]. However, Figure 1 shows that the processing is very fast when considering dynamics in continuous time of the neuronal membrane potential. When a synaptic time inactivation constant of 10 ms is used for excitatory synapses, the information about which pattern was presented as the cue becomes very significant within a few tens of ms from cue injection. When the synaptic time constant of 10 ms was used, and an information value of $I = 0.5$ bits was arbitrarily chosen[1], 0.5 bits of useful information about which pattern was presented can be read off from the firing of the neurons within only 25 ms of cue injection.

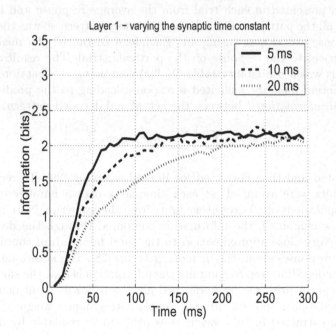

Fig. 1. The speed of pure recurrent processing. The time course of information retrieval in layer 1 is plotted for different values the inactivation time constant of excitatory associative synapses. A value of 20 ms is used for the membrane time constant.

---

[1] The value of 0.5 bits of information about which stimulus was presented is a useful point at which to measure the latency of retrieval because 0.5 bits is a much higher amount than random fluctuations of information values, which corresponded to the amount of information computed in the pre-cue time and were in the range of 0-0.05 bits. Also, 0.5 bits was still far from the $\log 2(10) = 3.3$ bits necessary to discriminate perfectly the 10 patterns, and therefore latency measures were not distorted by "ceiling" effects.

We wanted to investigate which biophysical time constants have the main influence on the recurrent dynamics. We therefore varied the time constant of synaptic inactivation in the range 10 to 40 ms, keeping the time constant of the neuronal membrane fixed to a value of 20 ms. The results are plotted in Figure 1. When the synaptic time constant is made slower, the time course of information retrieval becomes much slower as well. This shows that the inactivation time constant of the excitatory synapses is an important factor in determining the speed of recurrent information retrieval, as predicted by mathematical mean-field analyses [20].

Another biophysical parameter that, according to simple intuition, may be an important factor in determining the speed of recurrent processing is the membrane time constant of the neuron. This is in fact the biophysical parameter that determines the time needed to charge the neuronal membrane to the threshold voltage for firing. We varied the membrane time constant in the range 10 ms to 20 ms, and we measured the information latencies for different values of the membrane time constant. We found that the information latency remained unchanged (25 ms) in the interval considered for the membrane time constant, showing that this parameter is less important for the dynamics of information retrieval. How is this possible? Why is the recurrent dynamics not dominated by the (slow) membrane time constant, but it is shaped by the (faster) synaptic dynamics? In the view of some investigators [20,17,30], this is due to the presence of spontaneous activity and of roughly balanced inhibitory and excitatory inputs. These two properties are not only properties of our simplified network models, but are likely to be present in the cortex during normal operations [31]. The spontaneous activity in the network ensures that the neuronal membrane potentials at the time of cue presentation are not all at the resting level. This makes possible a fast neuronal response for some of the neurons in the network. If the balance of inhibitory and excitatory inputs keeps the average of the distribution of membrane potentials not too far from the firing threshold, then many of the neurons may be able to emit an action potential soon after the cue presentation. Further, since in this case the difference between the typical potential level of the neuron is close to threshold, the dynamics of firing may be driven by fluctuations in the numbers of presynaptic events. Therefore the synaptic time constant is expected to play a major role in this case, as found in our simulations. More details on the dynamics of a single module of recurrent processing can be found in [21].

The analysis of layer 1 (pure autoassociator) has shown that the pattern completion is achieved by the network of spiking neurons with continuous dynamics in a few tens of ms. In particular, 25 ms were necessary to reach an information value of 0.5 bits. However, when considering a sequence of recurrent networks connected by Hebbian excitatory synapses, the dynamics of retrieval in each individual associative module can be even faster. The reason is that one layer, due to the continuous dynamics implemented, can start completing the cue provided by the previous associative layer even before the previous layer has completed retrieval. In this way the signal could be processed and refined by each recurrent layer in the sequence in less than 25 ms per stage.

We have tested this idea in our model consisting of a sequence of three recurrent layers connected by feedforward associative synapses. The resulting information dynamics in each of the three layers is plotted in Figure 2. (Note that a value of 10 ms was used for the inactivation time constant of all the excitatory synapses[2].) The information latencies in layer 1, 2 and 3 respectively are 25, 30 and 45 ms. It is evident that, as predicted above, although the first recurrent layer provides 0.5 bits within 25 ms, the two successive layers need much less time to retrieve the same amount of information. The difference between the information latency of layer 1 and the information latency of layer 3 is 20 ms, with two recurrent layers in-between. Therefore, in conditions of mixed feedforward and recurrent processing, some 10-15 ms per layer may be enough for completing useful computations.

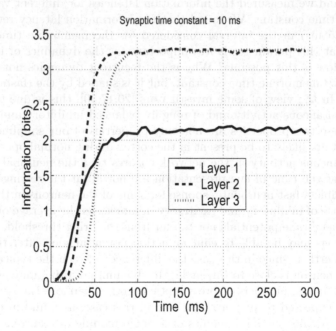

Fig. 2. The speed of mixed feedforward and recurrent processing in a network composed of three layers. The time course of information retrieval in layers one, two and three is plotted. A synaptic time constant of 10 ms and a membrane time constant of 20 ms were used in this simulation.

---

[2] The inactivation time constant of AMPA channels is mainly reported to be in the range 5-8 ms [32]. We used a longer value of 10 ms because we did not include axonal transmission delays. Therefore we chose a time constant value which gave rise to slower retrieval, in order to make sure that the time course reproduced by the model was not artificially fast.

4   Discussion

We simulated the dynamics in continuous time of a multi-layer network of Integrate-and-Fire neurons when it is retrieving the information stored in both the feedforward and within-layer recurrent synaptic pathway.

Given that we used inactivation time constants of 10 ms for the excitatory synapses, and an integration time constant of 20 ms for the membrane of excitatory neurons, the FF processing speed in continuous time is much faster than that predicted by a discrete time-step model. In this latter case at least one time-step would be required for each layer.

We found that when the within-layer excitation involves feedback through Hebbian synapses, it can contribute to substantial information retrieval within some 15 ms per layer. An intuitive explanation of how processing in continuous time can be so fast is that spontaneous activity ensures that some of the neurons are close to threshold when the retrieval cue signal is received. They will, within a fraction of the membrane time constant, start to influence other neurons through their synapses, in which the information is stored. Neuronal ensembles thus achieve fast recurrent processing even with relatively slow computing elements.

Our results indicate that associative recurrent processing may be fast enough to contribute to some particularly fast cortical operations. This possibility has been suggested before [20,33,21]. In this paper we have addressed much more directly the time scale of the RC contribution along a FF multilayer pathway. Note that in our abstract model we did not specify a functional role for the RC, which could be involved in many functions e.g. orientation tuning [34]. Our results indicate that, whatever information is stored in the recurrent synapses, it can influence the propagation of the signal within 15 ms. Also note that we addressed just the speed of local (within area) feedback processing, and not the speed of top-down processing, involving back projections from higher areas. However, by the very same principles of continuous dynamics involved here, one might expect that top-down effects can arise in a much shorter time than expected from discrete time-step dynamics.

We now examine the robustness of the conclusion that the RC can contribute to fast cortical processing in the context of the simplifications of the model. One possible source of overestimation of the speed of recurrent processing might come from the inaccuracy of the neuronal model. Although the Integrate-and-Fire model with a linear dendritic cable used contains the most basic elements of neuronal dynamics, of course several properties of real neurons are missing. One simplification that will have an impact on the dynamics is the lack of axonal conductance delays and synaptic delays in our model. However, they are generally short. Thus the estimates obtained for the information latencies should be realistic enough, especially taking into account that a slightly slow value of 10 ms was used for the inactivation time constant of the excitatory (AMPA) synapses. This is confirmed by the results reported in [22], where we found that a more detailed model taking explicitly into account conduction delays gave very similar results. Other important issues, such as the change of speed in recurrent or feedforward processing in the presence of strong noise, are extensively covered in [22].

If we want to use our model to constrain the functional architecture of the visual system under fast visual processing conditions, we have to keep in mind that, in our abstract and general model, a strictly sequential architecture is used to model successive stages of the system. Naturally, the architecture of the primate visual system presents aspects of further complexity. Although analysis of the anatomical connectivity suggests unequivocally that the primate visual system is organized hierarchically [35,36], there are also direct connections from V1 to several non-primary areas at different level of the visual hierarchy [35]. For this reason, a fraction of the visual processing in each stream might be done in parallel. Indeed, although response latencies in the ventral stream present a precise hierarchical organization, response latencies along the dorsal pathway are largely overlapping and consistent with a more parallel processing [1,2]. We note that, if part of the visual processing is done in parallel, there is more time for each area to process the signal within the 150 ms needed for fast visual cognition [7], and therefore the constraints that the speed of recurrent processing has to satisfy would be much less stringent. Therefore the time scales for recurrent processing met by our sequential model might be taken as a "worst-case" scenario for the case of visual processing.

It is of interest that part of the capability studied in some connectionist networks is obtained by virtue of the discrete time step approach. One example is the connectionist type of recurrent network, which can produce a particular sequence of outputs by virtue of counting through its time-steps (see e.g. [37]). Here we have shown that the continuous time dynamics of networks of spiking neurons seem to be rather different from that of discrete-time models, with no discrete states in 'time-steps' on the way. Hence the issue of how sequences of discrete states are produced becomes more complicated. The way proposed in [37], though interesting, is thus not plausible in biophysical and neurophysiological terms.

Finally, we note that attractor network processes operating by feedback using for example recurrent collateral connections or associatively modifiable forward and back-projections between cortical areas may operate in many brain areas. These areas may include the cerebral cortex, for example to implement short term memories, and the CA3 cells of the hippocampus to store and later retrieve information rapidly [17].

Acknowledgments. We thank A. Treves, M. P. Young and F. Petroni for helpful discussions. This research was supported by the Wellcome Trust and the EC (SP); and by the MRC PG9826105, the MRC IRC in Cognitive Neuroscience, and the Human Frontier Science Program (ETR).

References

1. J. Bullier and L.G. Nowak. Parallel versus serial processing: new vistas on the distributed organization of the visual system. *Current Opinion in Neurobiology*, 5:497–503, 1995.
2. M. T. Schmolesky, Y. Wang, D. P. Hanes, K. G. Thompson, S. Leutgeb, J. D. Schall, and A. G. Leventhal. Signal timing across the macaque visual system. *J. Neurophysiol.*, 79:3272–3277, 1998.
3. Y. Sugase, S. Yamane, S. Ueno, and K. Kawano. Global and fine information coded by single neurons in the temporal visual cortex. *Nature*, 400:869–873, 1999.
4. M. J. Tovée, E. T. Rolls, A. Treves, and R. P. Bellis. Information encoding and the response of single neurons in the primate temporal visual cortex. *J. Neurophysiol.*, 70:640–654, 1993.
5. J. Heller, J. A. Hertz, T. W. Kjaer, and B. J. Richmond. Information flow and temporal coding in primate pattern vision. *J. Comp. Neurosci.*, 2:175–193, 1995.
6. E. T. Rolls, M.J. Tovee, and S. Panzeri. The neurophysiology of backward visual masking: Information analysis. *J. Cognitive Neurosci.*, 11:300–311, 1999.
7. S. J. Thorpe, D. Fize, and C. Marlot. Speed of processing in the human visual system. *Nature*, 381:520–522, 1996.
8. R. S. Petersen and M.E. Diamond. Spatial-temporal distribution of whisker-evoked activity in rat somatosensory cortex and the coding of stimulus location. *Journal of Neuroscience*, 20:L6135–6143, 2000.
9. C. Koch and I. Segev, editors. *Methods in Neuronal Modelling*. MIT Press, Cambridge, MA, 2nd edition, 1998.
10. S. J. Thorpe and M. Imbert. Biological constraints on connectionist models. In R. Pfeifer, Z. Schreter, and F. Fologelman-Soulie, editors, *Connectionism in Perspective*, pages 63–92, Amsterdam, 1989. Elsevier.
11. M. W. Oram and D. I. Perrett. Time course of neuronal responses discriminating different views of face and head. *J. Neurophysiol.*, 68:70–84, 1992.
12. M. W. Oram and D. I. Perrett. Modeling visual recognition from neurobiological constraints. *Neural Networks*, 7:945–972, 1994.
13. M. Riesenhuber and T. Poggio. Are cortical models really bound by the "binding problem"? *Neuron*, 24:87–93, 1999.
14. J. J. Hopfield. Neural networks and physical systems with emergent collective computational abilities. *Proceedings of the National Academy of Sciences of the USA*, 79:2554–2558, 1982.
15. D. E. Rumelhart and J. L. McClelland, editors. *Parallel Distributed Processing*. MIT Press, Cambridge, MA, 1986.
16. D. J. Amit. *Modeling Brain Function*. Cambridge University Press, Cambridge, UK, 1989.
17. E. T. Rolls and A. Treves. *Neural networks and brain function*. Oxford University Press, Oxford, U.K., 1998.
18. M. V. Tsodyks and T. J. Sejnowski. Rapid state switching in balanced cortical models. *Network*, 6:111–124, 1995.
19. W. Gerstner. Population dynamics of spiking neurons: fast transients, asynchronous states and locking. *Neural Comp.*, 12:43–89, 2000.
20. A. Treves. Mean-field analysis of neuronal spike dynamics. *Network*, 4:259–284, 1993.
21. F.P. Battaglia and A. Treves. Rapid stable retrieval in high-capacity realistic associative memories. *Neural Comp.*, 10:431–450, 1998.

22. S. Panzeri, E. T. Rolls, F. P. Battaglia, and R. Lavis. Speed of feed-forward and recurrent processing in multilayer networks of integrate-and-fire neurons. 2000.
23. C. VanVreeswijk, L. F. Abbott, and G. Bard Ermentrout. When inhibition not excitation synchronizes neuronal firing. *J. Comput. Neurosci.*, 1:313–321, 1994.
24. G. Parisi. A memory which forgets. *Journal of Physics*, A 19:L617–619, 1986.
25. L.F. Abbott. Realistic synaptic inputs for model neural networks. *Network*, 2:245–258, 1991.
26. C. E. Shannon and W. Weaver. *The Mathematical Theory of Information*. University of Illinois Press, Urbana, Illinois, USA, 1949.
27. E. T. Rolls, A. Treves, and M. J. Tovée. The representational capacity of the distributed encoding of information provided by populations of neurons in primate temporal visual cortex. *Exp. Brain Res.*, 114:149–162, 1997.
28. S. Panzeri, A. Treves, S. Schultz, and E. T. Rolls. On decoding the responses of a population of neurons from short time windows. *Neural Comp.*, 11:1553–1577, 1999.
29. S. Panzeri and A. Treves. Analytical estimates of limited sampling biases in different information measures. *Network*, 7:87–107, 1996.
30. S. Son, K.D. Miller, and L.F. Abbott. Competitive hebbian learning through spike-timing-dependent synaptic plasticity. *Nature Neuroscience*, 3:919–926, 2000.
31. M. N. Shadlen and W. T. Newsome. The variable discharge of cortical neurons: implications for connectivity, computation and coding. *J. Neurosci.*, 18(10):3870–3896, 1998.
32. D. A. McCormick. Membrane properties and neurotransmitter actions. In G. M. Shepherd, editor, *The Synaptic Organization of the Brain*, chapter 2, pages 37–75. Oxford University Press, Oxford, U.K., 1998.
33. A. Treves. Local neocortical processing: a time for recognition. *Int. J. of Neuronal Systems*, 3:115–119, 1993.
34. R. Ben-Yishai, R. Bar-Or, and H. Sompolinsky. Theory of orientation tuning in visual cortex. *Proc. Natl. Acad. Sci. USA*, 92:3844–3848, 1995.
35. D. J. Felleman and D. C. VanEssen. Distributed hierarchical processing in the primate cerebral cortex. *Cerebral Cortex*, 1:1–47, 1991.
36. C. C. Hilgetag, M. A. O'Neill, and M. P. Young. Indeterminate organization of the visual system. *Science*, 217:776–777, 1996.
37. M. Jordan. An introduction to linear algebra in parallel distributed processing. In D. E. Rumelhart and J. L. McClelland, editors, *Parallel Distributed Processing*, pages 365–422, Cambridge, MA, 1986. MIT Press.

The Dynamics of Learning and Memory: Lessons from Neuroscience

Michael J. Denham

Centre for Neural and Adaptive Systems
University of Plymouth
Plymouth, UK
mike@soc.plym.ac.uk

Abstract. In the biological neural network, synaptic connections and their modification by Hebbian forms of associative learning have been shown in recent years to have quite complex dynamic characteristics. As yet, these dynamic forms of connection and learning have had little impact on the design of computational neural networks. It is clear however that for the processing of various forms of information, in which the temporal nature of the data is important, eg in temporal sequence learning and in contextual learning, such dynamic characteristics may play an important role. In this paper we review the neuroscientific evidence for the dynamic characteristics of learning and memory, and propose a novel computational associative learning rule which takes account of this evidence. We show that the application of this learning rule allows us to mimic in a computationally simple way certain characteristics of the biological learning process. In particular we show that the learning rule displays similar temporal asymmetry effects which result in either long term potentiation or depression in the biological synapse.

1 Introduction

Learning and memory are fundamental concepts of computation with neural networks. The representation of these concepts in neural computing architectures is almost universal. Memories are represented by patterns of activation of the active elements, or nodes, in the network. These patterns are determined by the stored magnitudes of the weighted connections between the nodes. The learning process, the formation of memories, is represented by the manipulation of these weighted connections in response to the presentation to the network of selected relevant training events. The memory retrieval process is represented by the computational process of allowing the nodes in the network to acquire particular levels of activation (the representation of the memory). This is often in response to exogenous stimulus applied to the network during the retrieval process, in which case what is retrieved is an internal representation (a memory) associated with the stimulus.

It is likely that, in the biological brain, memories are also represented by patterns of activation in ensembles, or populations, of neurons. How these memories are formed, retrieved and reach conscious awareness are unsolved problems. However it is also likely that the patterns of activity which represent memories are determined by, amongst other things, the strength of the connections which exist between the neurons

S. Wermter et al. (Eds.): Emergent Neural Computational Architectures, LNAI 2036, pp. 333–347, 2001.

which are involved in the relevant neuronal ensemble. But it is clear that the biological connection between neurons is a much more complex entity than that which is implemented in most neural computing architectures. In particular they display temporal and spatial characteristics which are strongly determined by the intricate and complex biophysical and biochemical mechanisms which underlie the connection. These characteristics are also heavily modulated by a variety of processes which depend on, for example, our emotional state. Thus a neuronal ensemble, viewed over a period of time, is likely to display a complex set of dynamical behaviours, resulting in an equally complex spatio-temporal pattern of activity. It is therefore unlikely that memories in the brain are represented by static levels of activation. Rather, it is likely that a memory is represented by a complex, dynamically changing pattern of activation, in which there is as much information coded in the time course of activations of the neuronal ensemble as in the pattern of activation across the ensemble at a single instant in time.

If we consider only a single neuron in such an ensemble, this may have 10,000 or more synapses on its dendritic tree, which allow it to participate in the activity of the ensemble. Each of these synapses receives inputs from other cells in the ensemble in the form of sequences of presynaptic action potentials (APs). Any one of these 10,000 or more inputs may be active, ie receiving a sequence of APs, within a given time window. The individual effect on the postsynaptic neuron of each input depends on the level and time course of the resultant excitation or inhibition of the post-synaptic membrane at each synapse, generated by the sequence of presynaptic APs. The nature and form of the postsynaptic response is determined by a highly complex set of biochemical and biophysical mechanisms at work in both the pre- and postsynaptic cells at the synaptic site.

In addition, both the instantaneous level and the time course of the postsynaptic excitation is also determined by the presence in the dendritic tree, at the synaptic site, of a back-propagating action potential. This impulse of electrical activity is generated at the axon hillock of the cell, and propagates backwards through the dendritic tree, with changes in amplitude and time course as it progresses through the various branch points of the dendritic tree. Considerable control can be applied to this back-propagating action potential, during its progress through the dendritic tree, by modifying local membrane current flows. This is likely to be a major mechanism for selectively controlling the effects of each of the 10,000 or more inputs to the neuron from other cells to which it is connected. It can thus effectively control both the degree and time course of the participation of the neuron in any given neuronal ensemble. It can also very effectively control the level of synaptic modification which may take place at each synapse, and in different portions/layers of the tree, in particular the dynamic switching of such modifications between long term potentiation (LTP), depotentiation, and long term depression (LTP) of the synaptic strength.

If we are to achieve with neural computing architectures, the levels of performance, robustness, etc which are manifest in biological networks, one possible approach is to incorporate into these architectures at least some of the more complex characteristics of the biological network, as outlined above. This has already been done to a limited extent in the case of network architectures which employ pulse trains as the mode of communication between neurons, so-called "pulse-coupled networks". This approximates the biological situation in which trains of action potentials ("spikes") provide the means by which neurons communicate with each other [1]. An obvious further extension is to increase the complexity of neuronal connections, from a non-spatial array of simple multiplicative weights to a form which includes some of the temporal and spatial characteristics of biological synapses and their distribution on the dendritic tree, as described above. This is particularly appropriate in the case of pulse-coupled networks, or their more abstract formulation as computational models of networks of "spiking", or "integrate-and-fire" neurons, in which the dynamic and spatial behaviour of neuronal connections can be precisely modelled, and the detailed temporal and spatial structure of the trains of action potentials which form the inputs to a given neuron can be precisely specified.

In this paper, I will review the recent knowledge which has been gained in the field of experimental neuroscience concerning the dynamic (both temporal and spatial) characteristics of biological connections between neurons, and their modification under "pairing", the experimental equivalent of Hebbian learning. I will restrict my remarks to the case of chemical synapses, although it is clear that other connections, eg electrical gap junctions, are also extensive and important in information processing in the biological neural network. I will describe a computational model of the dynamic characteristics of the synapse which has been used in a number of theoretical models of neural network behaviour. I will then describe a novel computational associative learning rule which takes account of the evidence for the manner in which local interactions between pre- and postsynaptically generated signals interact at the synapse to cause modifications to synaptic efficacy. We show that the application of this learning rule allows us to mimic in a computationally simple way certain characteristics of this biological learning process. In particular we show that the learning rule displays similar temporal asymmetry effects which result in either long term potentiation or depression in the biological synapse.

A greater understanding of how to represent and exploit computationally the important features of these biological phenomena is likely to be crucial for understanding how neural computing architectures might achieve the high levels of information processing performance apparent in biological neural networks.

2 The Dynamic Behaviour of Synapses

In several experimental studies of the characteristics of synaptic transmission it has been observed that the efficacy of this transmission can undergo transient variations according to the recent time history of the activity of the presynaptic neuron . This can take the form of either a short-term and short timescale increase (facilitation) or decrease (depression) of synaptic efficacy [2], [3], [4], [5]. Synaptic depression has been particularly widely observed in synapses in the developing brain, and may undergo a switch to facilitation in cortical synapses in adult brains [6] .

A model of the dynamics of synaptic depression was described recently by Tsodyks and Markram [7]. In fact, this model of the postulated presynaptic dynamics of neurotransmitter release had been proposed some years before by Grossberg [8], [9] and used subsequently by him and his colleagues in more recent years, for example to explain a number of important perceptual features involving the visual cortex.

An alternative form of the model was described recently by Abbott *et al* [4]. This can be derived from the differential equation form originally proposed by Grossberg [8], [9] and later employed by Tsodyks and Markram [7]. Abbott *et al* [4] show that their equations can be used to model many aspects of their own experimental data relating to synaptic depression, in particular the sensitivity of the synaptic response to abrupt changes in firing rate of the presynaptic cells. A number of similar models of synaptic facilitation and depression, based on the generic synaptic decoding method of Sen *et al* [10] have been discussed recently by Varela *et al* [11].

The dynamic synapse model characterises the synapse by defining a "resource", eg the amount of neurotransmitter in the synapse, a proportion of which can be in one of three states: *available, effective, inactive*. The dynamical behaviour of the proportions of the resource that are in each of these states is determined by a system of three coupled differential equations (1)-(3) below. In these we use notation similar to that in [9, see equations (58)-(63)]:

$$\frac{dx}{dt} = g(t) \cdot y(t) \cdot I(t) - \frac{x(t)}{\tau_x} \tag{1}$$

$$\frac{dy}{dt} = \frac{w(t)}{\tau_w} - g(t) \cdot y(t) \cdot I(t) \tag{2}$$

$$\frac{dw}{dt} = \frac{x(t)}{\tau_x} - \frac{w(t)}{\tau_w} \tag{3}$$

where $x(t)$ is the amount of *effective* resource, eg activated neurotransmitter within the synaptic cleft, as a proportion of the total resource, $y(t)$ is the amount of *available* resource, eg free neurotransmitter in the synapse, and $w(t)$ is the amount of *inactive* resource, eg neurotransmitter being reprocessed.

The input signal $I(t)$ represents the occurrence of a presynaptic AP and is set equal to one at the time of arrival of the AP and for a small period of time δ thereafter, and otherwise is set equal to 0. The instantaneous efficacy of the synapse is determined by the variable $g(t)$, which can be interpreted as the fraction of available resource released as a result of the occurrence of the presynaptic AP. It takes a value in the range zero to one.

The key idea behind the model is that there is a fixed amount $K = x(t)+y(t)+w(t)$ of total resource available at the synapse, a proportion $g(t) \cdot y(t)$ $(0<g(t)<1)$ of which is activated in response to presynaptic activity, rapidly becomes inactive (at a rate α), and is then subsequently made available again through reprocessing (at a rate β) . Thus, if the synapse is very active, ie it is bombarded by a large number of presynaptic APs occurring over a short period of time, the amount of available resource $y(t)$ is rapidly reduced. There must then follow a period during which the synapse must recover in order to respond fully once more. This process appears to replicate the experimentally observed characteristics of synaptic depression, for example as reported in [3] and [7].

Markram and Tsodyks [3] showed that pairing of the action potential generation in both the pre- and postsynaptic neurons, experimentally equivalent to the condition of Hebbian or associative learning, appears to result in a dynamic change in synaptic efficacy, rather than a simple increase in gain. After pairing, the shape of the EPSP resulting from a high-frequency train of presynaptic APs altered substantially, in that the amplitude of the initial transient part of the EPSP was significantly increased, and the rate of depression of the EPSP to subsequent APs was also greater, whilst the steady state value of the EPSP, on average, remained unchanged, for sufficiently high frequency stimuli (>20 Hz), and increased by up to 70% for low frequency stimuli (<20 Hz).

As Markram and Tsodyks [3] point out, in most observations of biological neural networks during information processing, neurons are firing irregularly and at a wide range of frequencies. The kind of changes in the dynamics of synaptic connections resulting from Hebbian-type pairing that they observed will therefore result in significant modification of the temporal structure of EPSPs generated by such irregular presynaptic spike trains, rather than that which a simple synaptic gain change would elicit.

3 The Dynamics of the Learning Process

As we have seen, the changes which occur in synaptic efficacy as the result of Hebbian pairing of pre- and postsynaptic activity can substantially alter the dynamic characteristics of the synaptic connection. In addition it has been observed that the induction of long-term changes in synaptic efficacy, ie either long-term potentiation (LTP) or depression (LTD), by such pairing depends strongly on the relative timing of the onset of the EPSP generated by the pre-synaptic AP, and the post-synaptic AP [12], [13], [14], [15], [16].

The precise cellular mechanisms which are responsible for the induction of LTP and LTD are not known. However the process seems to involve the initiation in the neuron's axon of an AP which actively and rapidly propagates back into the dendritic tree of the cell. The pairing of this *backpropagating* AP with a subthreshold EPSP results in a amplification of the dendritic AP and a localised influx of Ca^{2+} near the site of the synaptic input. This also induces significant LTP of the synapse and corresponding increases in the average amplitude of unitary EPSPs [17].

Back-propagating APs reduce in amplitude with their distance from the cell body, although pairing with synaptic activity increases AP amplitude, and this increase is greater with increasing distance of the synapse from the cell body. APs are also attenuated or their back-propagation blocked by dendritic hyperpolarisation. Thus the occurrence of synaptic input to the dendritic tree can have the effect of closely controlling the back-propagation of the APs to more distal parts of the tree, thus introducing a complex spatio-temporal dynamics into the synaptic modification process. In particular, synapses are only modified in those parts of the tree which back-propagating APs are allowed to access, which depends on the precise timing and location of EPSPs and IPSPs at the inputs to the tree, relative to the APs.

Markram et al [14] are amongst the most recent researchers (others are cited above) to observe that this relative timing of EPSPs and APs can either result in LTP or LTD of the synapse. They observed, in pairing experiments between pyramidal cells in rat neocortex, that: if the onset of the EPSP occurs around 10 ms before the AP then LTP is induced; if the AP precedes the onset of the EPSP by the same amount of time, LTD occurs. Relative timings of 100 ms either way resulted in no change to the synaptic strength. Guo-qiang Bi & Mu-ming Poo [16] showed a similar effect, although in their experiments the critical time window for a change in efficacy extended from -40 ms to +40 ms [16, Figure 7].

Several authors have suggested different forms of simple Hebbian learning rule which attempt to capture computationally this temporal asymmetry [9], [18], [19], [20]. Most recently Migliore et al [21] have simulated this effect, together with the modulation of the amplitude of the back-propagating AP, in a computational model of a hippocampal CA1 pyramidal neuron, modelled using a total of 202 compartments for the cell's axon., soma and dendrites. They show how the modulation and timing effects may depend on the properties of a transient A-type K^+ conductance that is strongly expressed in hippocampal dendrites.

4   Back-Propagating Action Potentials, Dendritic Calcium Transients and the Role of K^+ Channels

As described above, action potentials which are initiated in the initial segment of the axon propagate rapidly into the soma and dendrites, causing large membrane depolarisations and substantial increases in dendritic intracellular Ca^{2+} concentration [22], [23], [24], [25] . Back-propagation is dependent on the presence of Na^+ channels in the dendrites. The APs decline in amplitude with distance from the cell body and fail to propagate beyond certain distal branch points during repetitive firing [22]. Pairing of axonally initiated APs with sub-threshold EPSPs increases dendritic AP amplitude and Ca^{2+} influx [17]. In Magee and Johnston's experiments [17], a subthreshold EPSP train alone produced a small and highly localised increase in Ca^{2+} (~2%). The unpaired AP train produced a more widespread but also small increase in Ca^{2+} (~5%). Pairing of the EPSP and APs resulted in a nonlinear increase (~10%). The amount of pairing-induced increase in Ca^{2+} influx and AP amplitude increased with distance from the cell body. The increase resulting from pairing was particularly great where the AP amplitude had attenuated to a level almost too small to gate dendritic Ca^{2+} channels (2-fold increase in AP amplitude; 3- to 4-fold increase in Ca^{2+} influx) [17].

As the AP back-propagates into the dendritic tree, as far back as 500µm, its amplitude decreases (~10% per 100µm) and its duration increases (~20% per 100µm) [23]. Markram et al [26] performed whole cell voltage recordings from the soma and apical dendrites of neocortical layer V pyramidal neurons in the rat in order to study the determinants and time course of dendritic Ca^{2+} influx, evoked by a single back-propagating AP. They found that a single back-propagating AP evokes a discrete Ca^{2+} transient which rises to a peak within 2-3 milliseconds of the AP and then decays back to a resting level with a time constant of around 80 ms at 35-37°C. Since the voltage-gated calcium channels will have opened and closed within a few milliseconds of the AP [27], the relatively slow decay of the transient may be due to dendritic Ca^{2+} clearance mechanisms. The Ca^{2+}-induced release of Ca^{2+} from intracellular stores appears to be negligible [26]. The AP evoked transients begin to merge at an impulse frequency of 5 APs per second. Above this rate, trains of APs cause a concerted and maintained elevation of dendritic Ca^{2+}. The rise time of Ca^{2+} influx is consistent with most of the influx occurring during the depolarisation phase of the AP [27].

The peak of the Ca^{2+} influx is dependent on the opening of at least four voltage-gated calcium channels in the dendritic membrane. Three of them are high voltage-activated (HVA) Ca^{2+} channels (L-, N-, and P-type). It appears that L-type channels are located predominantly in the soma and proximal dendrites and N-type channels are located mostly in the distal dendrites [28]. HVA Ca^{2+} channels are not likely to be activated significantly at membrane potentials reached during subthreshold summation of EPSPs (below -40mV). One function of the HVA channels may therefore be to convert the signal conveyed by the back-propagating AP into dendritic Ca^{2+} transients. On the other hand, low voltage activated (LVA) Ca^{2+} channels in dendrites may be activated at membrane potentials of around -50 mV, ie during subthreshold EPSPs [29] and following strong hyperpolarisation [30], [31]. Multiple Ca^{2+} channel subtypes may therefore act to convert both EPSPs and APs into dendritic Ca^{2+} transients depending on the level of sub- or supra-threshold activity [32].

Hoffman et al [33] have shown that the dendrites of CA1 pyramidal neurons have a high density of transient A-type K^+ channels and that this density increases with distance from the soma. The presence of these channels causes the amplitude of back-propagating APs to decrease with distance from the soma, inhibits the ability of dendrites to initiate APs and alters the shape of EPSPs. Also the voltage-dependent properties of these channels regulates dendritic excitability by subthreshold synaptic activity. The A-type K^+ channels therefore play the dominant role in determining the electrical properties of CA1 dendrites.

CA1 and neocortical pyramidal neurons contain voltage-gated Na^+ and Ca^{2+} channels at densities which are uniform over the soma and dendrites. The density of dendritic A-type K^+ channels increases with distance from the soma, thus providing a possible reason why the amplitude of backpropagating APs decreases with distance from the soma, why action potentials do not initiate in the dendrites, and why dendritic Na^+ and Ca^{2+} channels only have a small effect on subthreshold EPSPs. Hoffman et al [33] assessed the effect of this increasing density of A-type K^+ channels, and found that this acts to reduce AP amplitude and limit the back propagation of full amplitude APs to proximal dendrites. It also allows single axonally generated APs to induce bursts or repetitive firing in the dendrites. Dendritic A-type K^+ channels also act to counter additional inward current induced by subthreshold Na^+ channel activation, thus preventing any boosting of EPSPs by Na^+ channels. Moderate membrane depolarisation, as provided by subthreshold EPSPs, can rapidly inactivate A-type K^+ channels, thus allowing the pairing of such EPSPs with back-propagating APs to boost the amplitude of the APs locally, together with the associated Ca^{2+} influx via voltage activated calcium channels. This may also provide a mechanism whereby synaptic activation can control the spread of backpropagating APs into specific, synaptically active, parts of the dendritic tree.

5 A Theoretical Basis for the Learning Process

The above evidence suggests a simple theoretical explanation for the experimental pairing experiments reported by Markram et al [14]. As described above, it was observed in these and other experiments that the induction of long-term changes in synaptic efficacy, ie either long-term potentiation (LTP) or depression (LTD), by pairing of a backpropagating AP and a presynaptically generated EPSP depends strongly on the relative timing of the onset of the EPSP and the post-synaptic AP [12], [13], [14], [15], [16]. In particular Markram et al [14] observed, in pairing experiments between pyramidal cells in rat neocortex, that: if the onset of the EPSP occurs around 10 ms *before* the AP then LTP is induced; but, if the onset of the EPSP occurs *after* the AP by the same amount of time, LTD occurs. Relative timings of 100 ms either way resulted in no change to the synaptic strength. They also considered the effect of varying the frequency of the EPSP/ AP pairing, between 2 and 40 Hz. They showed that the amount of LTP was zero at 2 Hz, showed a small increase at 5Hz, followed by a sharp (~30%) increase at 10 Hz. The amount of LTP induced then increased almost linearly, achieving a further 20% increase at 40 Hz [14].

It is now a widely held view, first proposed by Lisman [34], that the bidirectional control of synaptic strength, ie LTP and LTD, is mediated by the influx of Ca^{2+} ions. In particular, it has been proposed that the magnitude and temporal structure of the Ca^{2+} signal might completely determine the sign of the change in efficacy of a synapse in response to stimulation [35], [36], [37].

Now consider first the case that the backpropagating AP occurs alone, ie the EPSP is absent. In this case, the backpropagating AP is likely to have a low amplitude and duration, due to the action of the A-type K^+ channels. Although the resultant membrane depolarisation may be sufficient to release the Mg^- block on the NMDA channel and thus to allow influx of Ca^{2+}, the level of influx is presumably insufficient to reach that necessary for either LTD or LTP induction. Similarly, if the EPSP occurs alone, any influx in Ca^{2+} thereby induced is also likely to be insufficient to induce either LTD or LTP.

Assume now that the AP and the EPSP are both present, and that the onset time of the pre-synaptically generated EPSP precedes the peak of the backpropagating AP by around 5 to 10 ms. The effect of the arrival of the EPSP is to inactivate the A-type K^+ channels. These inactivate with a time constant of ~15 ms for a step to +10 mV. This inactivation of the A-type K^+ channels will allow the backpropagating AP's amplitude and duration to increase and cause a correspondingly large influx of Ca^{2+} via voltage activated calcium channels, closely temporally correlated with the AP. The time to peak Ca^{2+} influx is about 2 ms, with a decay time constant of about 80 ms. We might therefore postulate that the influx of Ca^{2+} under these conditions is sufficiently large for the induction of LTP.

An increase in the frequency of the presynaptic AP train will cause synaptic depression of the amplitude of the EPSPs generated. This will result in less inactivation of the A-type K^+ channels and a corresponding limitation in the amplitude of the backpropagating APs and the associated Ca^{2+} influx. Also an increase in the frequency of the presynaptic AP train causes temporal summation of the Ca^{2+} influx above about 5 Hz [26, Figure 11], showing a sharp increase in dendritic calcium influx at around 10 Hz. As a result of the above, we would expect the induction of

LTP to depend on the AP input frequency in the following way: first, an initial induction of LTP at an AP input frequency of about 5Hz, followed by a sharp increase in the amount of LTP induced at an input frequency around 10 Hz, followed again by a more gradual, linear increase as the frequency increases beyond 10 Hz. This is precisely the relationship between LTP induction and the AP input frequency which was observed in the pairing experiments of Markram et al [14, Figure 2C].

Now consider the case when the backpropagating AP precedes the EPSP, by about 5 to 10 ms. Initially the presence of the AP cause sufficient depolarisation to allow influx of Ca^{2+} , but due to the action of the A-type K^+ channels, the backpropagating AP will have a low amplitude and duration and the level of calcium influx is presumably insufficient to reach that necessary for LTP induction. However the decay time of this Ca^{2+} transient is of the order of 80 ms, so an EPSP arriving within around 30 ms of the AP will cause a further calcium influx which will summate with the AP induced influx. The resultant influx, it is proposed, may be insufficient to reach the level necessary to induce LTP, but sufficient to reach the level necessary to induce LTD.

As mentioned already, it has been proposed that the magnitude and temporal structure of the Ca^{2+} signal might completely determine the sign of the change in efficacy of a synapse in response to stimulation [35], [36], [37]. Physiologically, it appears that an additional influx is the trigger for the postulated post-synaptic mechanism for LTP, based on activation of Ca^{2+}-calmodulin-dependent protein kinase II (CaMKII) and a resultant increase in the responsiveness to glutamate of AMPA and other receptors and their associated channels. Reduced influx is the trigger for LTD, based on the preferential activation, at low CA^{2+} ion levels, of calcineurin (protein phosphatase 2B (PP2B), and the resultant postulated effect on CaMKII or AMPA receptors.

6 A Computational Model of the Learning Process

The above physiological account of the LTP/ LTD process can be captured in a simple computational model, which nevertheless has similar behaviour, in respect of the temporal asymmetry of the learning process, to the experimental observations, in particular those experiments described in [14] and [16]. The model requires the computation of the EPSP at the synapse, $e(t)$, and the backpropagating action potential $a(t)$. We then construct a learning rule which depends on the integration over time of the *product* of these two signals, ie their cross-correlation with zero time shift. This product we equate to the instantaneous calcium influx at any time, and in the learning rule we modify the efficacy of the synapse, $g(t)$, in equations (1) and (2), in a positive or negative direction dependent on the level of this instantaneous influx. The efficacy $g(t)$ is increased differentially if the influx level is *above* that required for LTP, it is decreased differentially if the influx level is *below* that for LTP, but *above* that required for LTD. If the influx level is below that required for LTD, the synaptic efficacy is unchanged. The resultant change in efficacy of the synapse over time is thus the integrated effect of these differential changes at each instant of time.

The EPSP at the synapse, $e(t)$, can be computed from the effective synaptic current $x(t)$ in equation (1) using the following equation for the passive membrane mechanism [7]:

$$\tau_{EPSP} \cdot \frac{de(t)}{dt} = \gamma \cdot x(t) - e(t) \qquad (4)$$

The backpropagating AP, $a(t)$, is computed from a simple, single compartment neuron model, which is a modified version of the model described in [38]:

$$\tau_E \cdot \frac{dE}{dt} = -E(t) + I_S(t) + G_K(t) \cdot (E_K - E(t)) \qquad (5)$$

$$\tau_{GK} \cdot \frac{dG_K}{dt} = -G_K(t) + \eta \cdot s(t) \qquad (6)$$

$$\tau_a \cdot \frac{da}{dt} = -\mu \cdot G_K(t) \cdot a(t) + \nu \cdot s(t) \qquad (7)$$

where $E(t)$ is the variation of the somatic membrane potential of the cell, relative to its resting potential; $I_s(t)$ is the stimulating current injected into the soma of the cell in order to elicit firing, as in the experiments of Markram et al [14]; $G_K(t)$ is the membrane potassium conductance, normalised by the sum of all the voltage-dependent ionic membrane conductances; E_K is the potassium equilibrium potential of the membrane, relative to the membrane resting potential, typically -10 mV; $a(t)$ is the backpropagating action potential; and $s(t)$ is the variable which denotes firing of the cell, defined by:

$$s(t) = 1, \quad \text{if } E(t) \geq \theta \qquad (8)$$
$$= 0, \quad \text{otherwise}$$

where θ is the somatic membrane potential threshold for firing of the cell. Finally, τ_E, τ_a, and τ_{GK} are time constants, and η, μ and ν are constant parameters.

The typical behaviour of the cell model to a train of stimulating current pulses is shown in Figure 1.

The learning rule is modelled as a function of the instantaneous level of calcium influx, $\lambda(t)$, where:

$$\lambda(t) = a(t) \cdot e(t) \qquad (9)$$

The differential change in synaptic efficacy, $g(t)$, is then computed as:

$$\tau_g \cdot \frac{dg}{dt} = \lambda(t) \cdot (1 - g(t)), \text{ if } \lambda(t) > \lambda_{LTP} \tag{10}$$

$$= -\lambda(t) \cdot (1 - g(t)), \text{ if } \lambda(t) \leq \lambda_{LTP} \text{ and } \lambda(t) \geq \lambda_{LTD}$$

$$= 0, \text{ otherwise}$$

where λ_{LTP} and λ_{LTD} are the calcium influx levels necessary for initiating the induction of LTP and LTD respectively.

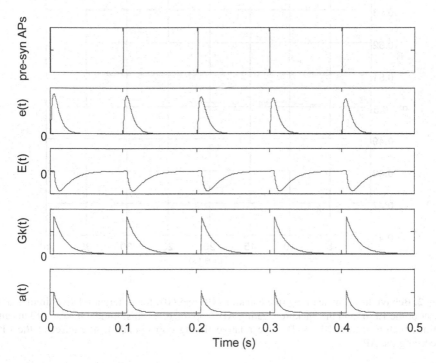

Fig. 1. Behaviour of the dynamic synapse model equations (1)-(4) and the neuron model equations (5)-(7), for a regular presynaptic AP train at 10Hz.

In the simulation experiments shown in Figure 2, the neuron model was stimulated by a random train of current spikes I_s of mean frequency equal to 10 Hz. The resultant backpropagating AP, a(t), generated by equations (5) - (8), was paired in the learning rule (equations (9) and (10)) with the EPSP, e(t). The latter was generated by applying the same train of current spikes I_s to the dynamic model of the synapse (equations (1) - (4)). The effect on the synaptic efficacy, *g(t)*, of introducing a set of time differences of -40 ms, -10 ms, +10 ms, and +40 ms between the onset of the EPSP and the AP is illustrated in Figure 2. Here a positive time difference indicates the onset of the EPSP occurring *before* the AP. As can be seen, the +10 ms difference resulted in a positive increase in synaptic efficacy (LTP), whereas the -10 ms difference resulted in a correspondingly large decrease in efficacy. Both the -40 ms and the +40 ms time differences resulted in almost zero change in efficacy. This result closely mimics the characteristic changes in synaptic efficacy with respect to temporal differences in the onset of the EPSP and the occurrence of the backpropagating AP observed by several experimenters [12], [13], [14], [15], [16]. The values of the constant parameters used in the models described above, for the simulations in Figures 1 and 2, are given in Table 1.

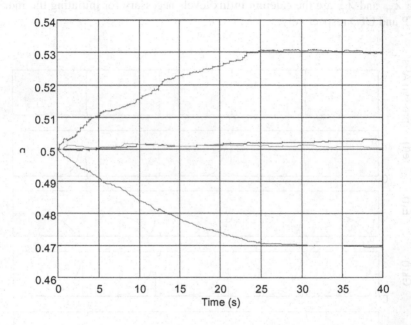

Fig. 2. Behaviour of the learning rule equations (9) and (10) for different relative timings of the onset of the EPSP and the backpropagating AP. Top trace: +10ms; middle traces: +40 ms and -40 ms; bottom trace: -10 ms. Positive relative timing corresponds to the onset of the EPSP preceding the AP.

Note that the learning rule described here directly modifies the synaptic efficacy parameter $g(t)$, which in the dynamic synapse model is interpreted as the proportion of available neurotransmitter released in response to a presynaptic AP. This can be justified physiologically, since postsynaptic Ca^{2+} ion influx, in addition to initiating and regulating postsynaptic LTP/LTD processes, as described above, also induces the synthesis of retrograde messengers, such as arachidonic acid, platelet-activating factor, carbon monoxide (CO) and nitric oxide (NO). These chemicals act by diffusing back to the presynaptic terminal where they are able, in cells which demonstrate LTP or LTD, to modify magnitude of the release of neurotransmitter in response to subsequent stimuli [39], [40]. The synthesis of NO is induced by Ca^{2+}-calmodulin-dependent protein kinase II (CaMKII), which also plays a role in the post-synaptic mechanism for LTP. The learning rule implies however that the differential activation of retrograde messengers can produce both LTP and LTD presynaptically, and that this is directly related to the level of postsynaptic Ca^{2+} influx, in the same manner as the postsynaptic mechanisms for LTP and LTD. In fact, it is likely that several pre- and postsynaptic factors, including different intracellular Ca^{2+} levels and the activities of different second messenger cascades, are involved in integrating the pre- and postsynaptic activities which result in LTP or LTD, as discussed by [36].

Table 1. Values for the constant parameters used in the model simulations

Constant parameter	Value	Constant parameter	Value
τ_x	0.003	τ_{EPSP}	0.007
τ_w	0.45	τ_E	0.005
γ	4.0	τ_{GK}	0.015
η	25	τ_a	0.01
μ	10	E_K	-10
ν	1000	θ	0.4

References

1. Gerstner, W.: Spiking neurons. In: W. Maass and C.M. Bishop (eds): Pulsed Neural Networks. MIT Press, Cambridge (1998) 3-54.
2. Thompson, A.M., Deuchars, J.: Temporal and spatial properties of local circuits in neocortex. Trends in Neuroscience 17 (1994) 119-126.
3. Markram, H., Tsodyks, M.: Redistribution of synaptic efficacy between neocortical pyramidal neurons. Nature 382 (1996) 807-810.
4. Abbott, L.F., Varela, J.A., Sen, K., Nelson, S.B.: Synaptic depression and cortical gain control. Science 275 (1997) 220-224.
5. Ali, A.B., Thompson, A.M.: Facilitating pyramid to horizontal oriens- hippocampus. Journal of Physiology (London) 507 (1998) 185-199.
6. Reyes, A, Sakmann, B.: Developmental switch in the short-term modification of unitary EPSPs evoked in layer 2/3 and layer 5 pyramidal neurons of rat neocortex. Journal of Neuroscience 19 (1999) 3827-3835.
7. Tsodyks, M.V., Markram, H.: The neural code between neocortical pyramidal neurons depends on neurotransmitter release probability. Proceedings of the National Academy of Science USA 94 (1997) 719-723.
8. Grossberg, S.: Some physiological and biochemical consequences of psychological postulates. Proceedings of the National Academy of Science USA 60 (1968) 758-765.
9. Grossberg, S.: On the production and release of chemical transmitters and related topics in cellular control. Journal of Theoretical Biology 22 (1969) 325-364.
10. Sen, K., Jorge-Rivera, J.C., Marder, E., Abbott, L.F.: Decoding synapses. Journal of Neuroscience 16 (1996) 6307-6318.

11. Varela, J.A., Sen, K., Gibson, J., Fost, J., Abbott, L.F., Nelson, S.B.: A quantitative description of short-term plasticity at excitatory synapses in layer 2/3 of rat primary visual cortex. Journal of Neuroscience, 17 (1997) 7926-7940.
12. Levy, W.B., Steward, O.: Temporal contiguity requirements for long-term associative potentiation/depression in the hippocampus. Neuroscience 8 (1983) 791-797.
13. Debanne, D., Shulz, D.E., Fregnac, Y.: Temporal constraints in associative synaptic plasticity in hippocampus and neocortex. Canadian Journal of Physiological Pharmacology 73 (1995) 1295-1311.
14. Markram, H., Lubke, J., Frotscher, M., Sakmann, B.: Regulation of synaptic efficacy by coincidence of postsynaptic APs and EPSPs. Science 275 (1997) 213-215.
15. Debanne, D., Gahwiler, B.H. & Thompson, S.M.: Long-term synaptic plasticity between pairs of individual CA3 pyramidal cells in rat hippocampal slice cultures. Journal of Physiology (London) 507.1 (1998) 237-247.
16. Guo-qiang Bi, Mu-ming Poo.: Synaptic modifications in cultured hippocampal neurons: dependence on spike timing, synaptic strength, and postsynaptic cell type. Journal of Neuroscience 18 (1998) 10464-10472.
17. Magee, J.C. & Johnston, D.: A synaptically controlled associative signal for Hebbian plasticity in hippocampal neurons. Science 275 (1997) 209-212.
18. Senn, W., Tsodyks, M., & Markram, H.: An algorithm for synaptic modification based on exact timing of pre- and post-synaptic action potentials. In: Lecture Notes in Computer Science, Vol. 1327. Springer-Verlag, Berlin Heidelberg New York (1997) 121-126.
19. Denham, M.J., McCabe, S.L.: A dynamic learning rule for synaptic potentiation. Research Report CNAS-98-01, School of Computing, University of Plymouth, (1998).
20. Kempter, R., Gerstner, W., & van Hemmen, J.L.: Spike-Based Compared to Rate-Based Hebbian Learning. In: Advances in Neural Information Processing Systems 11. MIT Press, Cambridge (1999).
21. Migliore, M, Hoffman, D.A., Magee, J.C., Johnston, D.: Role of an A-type K^+ conductance in the back-propagation of action potentials in the dendrites of hippocampal pyramidal neurons. Journal of Computational Neuroscience 7 (1999) 5-15.
22. Spruston, N., Schiller, Y., Stuart, G., Sakmann, B.: Science 268 (1995) 297
23. Stuart, G.J., Sakmann, B.: Active propagation of somatic action potentials into neocortical pyramidal cell dendrites. Nature 367 (1994) 69-72.
24. Jaffe, D.B., Johnston, D., Lasser-Ross, N., Lisman, J.E., Miyakawa, H., Ross, W.N.: The spread of Na+ spikes determines the pattern of dendritic Ca2+ entry into hippocampal neurons. Nature 357 (1992) 244-6.
25. Regehr, W.G., Tank, D.W.: Calcium concentration dynamics produced by synaptic activation of CA1 hippocampal pyramidal cells Journal of Neuroscience 12 (1992) 4202-4223
26. Markram H, Helm PJ, Sakmann B.: Dendritic calcium transients evoked by single back-propagating action potentials in rat neocortical pyramidal neurons. Journal of Physiology 485 (1995) 1-20.
27. Llinas, R., Steinberg, I.Z., Walton, K. : Pre-synaptic calcium currents in squid giant synapse. Biophysical Journal 33 (1981) 289-322.
28. Westenbroek, R.E., Ahlijanian, M.K., Catterall, W.A.: Clustering of L-type Ca2+ channels at the base of major dendrites in hippocampal pyramidal neurons. Nature 347 (1990) 281-4.
29. Markram, H., Sakmann, B.: Calcium transients in apical dendrites evoked by single sub-threshold excitatory post-synaptic potentials via low voltage-activated calcium channels. Proceedings of the National Academy of Sciences USA 91 (1994) 5207-5211.
30. Sayer, R.J., Schwindt, P.C., Crill, W.E.: High- and low-threshold calcium currents in neurons acutely isolated from rat somatosensory cortex. Neuroscience Letters 120 (1990) 175-178.

31. Diesz, R.A., Fortin, G., Zieglgansberger, W.: Voltage dependence of excitatory postsynaptic potentials of rat neocortical neurons. Journal of Neurophysiology 65 (1991) 371-382.
32. McCobb, D.P., Beam, K.G.: Action potential waveform voltage clamp commands reveal striking differences in calcium entry via low and high voltage-activated calcium channels. Neuron 7 (1991)119-127.
33. Hoffman, D.A., Magee, J.C., Colbert, C.M., Johnston, D.: K^+ channel regulation of signal propagation in dendrites of hippocampal pyramidal neurons. Nature 387 (1997) 869-875.
34. Lisman, J.: A mechanism for the Hebb and anti-Hebb processes underlying learning and memory. Proceedings of the National Academy of Science USA 86 (1989) 9574-9578.
35. Artola, A., Singer, W.: Long-term depression of excitatory synaptic transmission and its relation to long-term potentiation. Trends in Neuroscience 16 (1993) 480-487.
36. Lisman, J.: The CaM kinase II hypothesis for the storage of synaptic memory. Trends in Neuroscience, 17 (1994) 406-412.
37. Malenka, R.C.: Synaptic plasticity in the hippocampus: LTP and LTD. Cell 78 (1994) 535-538.
38. MacGregor, R.J.: Neural and Brain Modeling. Academic Press, San Diego (1989).
39. Bolshakov, V.Y., Siegelbaum, S.A.: Hippocampal long-term depression: arachidonic acid as a potential retrograde messenger. Neuropharmacology 34 (1995) 1581-1587.
40. Fitzsimonds, R.M., Mu-ming Poo.: Retrograde signaling in the development and modification of synapses. Physiological Reviews 78 (1998) 143-170.

Biological Grounding of Recruitment Learning and Vicinal Algorithms in Long-Term Potentiation

Lokendra Shastri

International Computer Science Institute, Berkeley CA 94704, USA,
shastri@icsi.berkeley.edu,
http://icsi.berkeley/~shastri

Abstract. Biological networks are capable of gradual learning based on observing a large number of exemplars over time as well as of rapidly memorizing specific events as a result of a single exposure. The focus of research in neural networks has been on gradual learning, and the modeling of one-shot memorization has received relatively little attention. Nevertheless, the development of biologically plausible computational models of rapid memorization is of considerable value, since such models would enhance our understanding of the neural processes underlying episodic memory formation. A few researchers have attempted the computational modeling of rapid (one-shot) learning within a framework described variably as *recruitment learning* and *vicinal algorithms*. Here it is shown that recruitment learning and vicinal algorithms can be grounded in the biological phenomena of long-term potentiation and long-term depression. Toward this end, a computational abstraction of LTP and LTD is presented, and an "algorithm" for the recruitment of *binding-detector* (or *coincidence-detector*) cells is described and evaluated using biologically realistic data.

1 Introduction

Biological neural networks are capable of slow gradual learning as well as rapid one-shot memorization. The former involves an exposure to a large number of exemplars and leads to the acquisition of perceptual-motor skills, category formation, language skills, and certain types of semantic knowledge. In contrast, one-shot memorization can result from a single exposure to an example, and underlies, among other things, the acquisition of "episodic memories" of everyday events, and memories of faces.

The primary focus of research in neural network models has been on slow gradual learning, and the modeling of one-shot memorization has received relatively little attention. Nevertheless, the development of biologically plausible computational models of rapid memorization is of considerable value, since such models would enhance our understanding of the neural processes underlying memory formation and retrieval, and could lead to the design of robust episodic

S. Wermter et al. (Eds.): Emergent Neural Computational Architectures, LNAI 2036, pp. 348–367, 2001.

memory modules for autonomous agents, and perhaps, to the development of memory prosthesis for brain injured humans.

A few researchers have attempted the computational modeling of rapid one-shot learning within a framework described variably as *recruitment learning* [10, 22,8,24,11,20] and *vicinal algorithms* [37]. In simple terms, recruitment learning can be described as follows: Learning occurs within a network of randomly connected nodes. Recruited nodes are those nodes in the network that have acquired a distinct "meaning" (or functionality) by virtue of their *strong* interconnections to other recruited nodes and/or other sensorimotor (i.e., input/output) nodes. Nodes that are not yet recruited can be viewed as "free" nodes. Such nodes are connected via weak links to a large number of free, recruited, and/or sensorimotor nodes. These free nodes form a primordial network from which suitably connected nodes may be recruited for representing new items. For example, a novel concept y which can be expressed as a conjunct of existing concepts x_1 and x_2 can be memorized by (i) identifying free nodes that receive links from nodes representing x_1 as well as nodes representing x_2 and (ii) "recruiting" one or more such free nodes by strengthening the weights of links incident on such nodes from x_1 and x_2 nodes.

Feldman [10] showed that conjunctive concepts can be recruited with a high probability if one makes suitable assumptions about network connectivity. He presented a probabilistic analysis of recruitment learning based on the degree of connectivity and the number of intermediate layers in random interconnection networks. Shastri [22] extended the notion of recruitment learning to relational concepts. He treated a concept as a collection of attribute-value bindings and suggested a two-stage memorization process. In the first stage, *binder* nodes are recruited for each attribute-value binding in a concept. In the second stage, these *binder* nodes are joined together by the recruitment of another conjunctive node. Diederich [8] showed how this form of structured recruitment learning can be used to learn new concepts expressed as modifications of existing concepts. Valiant [37] proposed a formal "neuroidal model" and described several algorithms for the recruitment learning of conjunctive and relational concepts. He also presented a quantitative analysis of these algorithms using plausible assumptions about connectivity in the neocortex. Valiant referred to these algorithms as "vicinal algorithms."

While general arguments in support of the neural plausibility of recruitment learning and vicinal algorithms have been presented in the past (see [10,37]), a specific neural correlate of such learning has not been proposed. In this paper it is shown that recruitment learning can be firmly grounded in the biological phenomena of *long-term potentiation* (LTP) and *long-term depression* (LTD) that involve rapid, long-lasting, and highly specific changes in synaptic strength. Toward this end, a computational abstraction of LTP and LTD is proposed, and an "algorithm" for the recruitment of binding-detector (or *coincidence detector*) cells is described and evaluated using biologically realistic data about region sizes and cell connectivity. In the proposed grounding, the specification of a vicinal algorithm amounts to choosing a suitable network architecture and a set of appropriate parameter values for the induction of LTP and LTD.

The rest of the paper is organized as follows: Section 2 briefly reviews the phenomena of LTP and LTD. Section 3 describes a computational abstraction of cells, synapses, LTP, and LTD. Section 4 describes how a transient pattern of activity can lead to the recruitment of binding-detector cells as a result of LTP (and optionally, LTD) within quasi-random network structures. Finally, section 6 presents some concluding remarks.

2   Long-Term Potentiation and Depression

Long-term potentiation (LTP) refers to a long-term increase in synaptic strength[1] resulting from the pairing of presynaptic activity with postsynaptic depolarization [5,14] LTP was first observed in the rabbit hippocampal formation, and has since been observed in synapses along many excitatory pathways in the mammalian brain. Recent evidence strongly suggests that LTP plays a direct causal role in learning and memory formation (e.g., [33,21].

The most extensively studied form of LTP involves the unusual receptor NMDA[2] (N-methyl-D-aspartate) which is activated by the excitatory neurotransmitter glutamate, but only if the postsynaptic membrane is sufficiently depolarized. In the absence of adequate depolarization, NMDA receptor-gated channels remain blocked by magnesium ions in spite of glutamate being bound to the receptor. Adequate depolarization of the postsynaptic membrane, however, expels the magnesium ions and unblocks the channels. Once the channels are unblocked, calcium ions flood into the dendritic spine of the postsynaptic cell and trigger a complex series of biochemical changes that result in the induction of LTP.

The two conditions required for the activation of NMDA receptor, namely, presynaptic activity and strong postsynaptic depolarization, together entail that the LTP requires the concurrent arrival of activity at several synapses of the postsynaptic cell. This is referred to as the *cooperativity* property of LTP.

---

[1] A synapse is the site of communication between two cells. Typically, a synapse is formed when an axonal (output) fiber emanating from a "presynaptic" cell makes contact with the dendrites (input structure) of a "postsynaptic" cell.

A synapse can be excitatory or inhibitory. The arrival of activity at an excitatory synapse from its presynaptic cell leads to a depolarization of the local membrane potential of its postsynaptic cell and makes the postsynaptic cell more prone to firing. In contrast, the arrival of activity at an inhibitory synapse leads to a hyperpolarization of the local membrane potential of the postsynaptic cell and makes the postsynaptic cell less prone to firing. The strength of an excitatory (or inhibitory) synapse determines the degree of depolarization (or hyperpolarization) that will result from a given presynaptic activity. The greater the synaptic strength, the greater the depolarization (hyperpolarization).

[2] Not all forms of LTP are NMDA receptor-dependent. The LTP of synapses formed by mossy-fibers on CA3 pyramidal cells is a case in point [18].

Several properties of LTP make it suitable for serving as the basis for one-shot recruitment learning. First, it is induced rapidly — within a few seconds, and is fully present within 20-30 seconds. Second it is long lasting. Third, the cooperativity property of LTP makes it an ideal mechanism for transforming a *transient* expression of a relationship between two items or more (encoded as the coherent activity of the ensembles representing these items) into a *persistent* expression of this relationship (encoded via long-term changes in the efficacy of synapses linking the ensembles representing these items). Finally, LTP is synapse specific, and hence, it can express highly specific bindings and correlations.

LTP resulting from the arrival of coincident activity along afferent fibers belonging to a single pathway is referred to as *homosynaptic* LTP. If the arrival of coincident activity along two independent pathways, A and B, leads to the LTP of synapses formed by fibers of A, but the arrival of activity along fibers of A alone does not, then the LTP of synapses formed by fibers of A is referred to as *associative* LTP [12,6]. We will, however, distinguish between associative and homosynaptic LTP based on the *representational* distinctiveness of the afferent sources whose cooperative activity leads to LTP. Thus we will use the qualifier "homosynaptic" to refer to LTP resulting from the arrival of coincident activity along afferents fibers emanating from cells representing the *same* item, and we will use the qualifier "associative" to refer to LTP resulting from the arrival of coincident activity along two sets of afferent fibers, with each set emanating from cells representing a distinct item.

In addition to LTP, synapses along key excitatory pathways in the mammalian hippocampal formation have been shown to undergo long-term depression (LTD) [3,13]. A synapse receiving no presynaptic activity can undergo *heterosynaptic* LTD if other synapses of the same postsynaptic cell receive strong presynaptic activity. In other words, the absence of presynaptic activity in the presence of strong postsynaptic activity can lead to heterosynaptic LTD of a synapse. A synapse may undergo *associative* LTD upon receiving presynaptic activity that is out of phase with strong rhythmic activity converging on other synapses of the postsynaptic cell [34]. Finally, prolonged low frequency stimulation of a synapse can lead to its homosynaptic LTD [9].

3   A Computational Abstraction of LTP and LTD

The computational abstraction of LTP and LTD proposed here is an highly simplified idealization of the complex biophysical processes underlying the induction and expression of LTP and LTD. This abstraction is guided by two considerations. First, the abstraction should be rich enough to capture temporal aspects critical for modeling LTP and LTD. Second, the abstraction should be *discrete* and minimal so as to facilitate quantitative analyses and efficient computer simulations of large-scale neuronal networks.

3.1 Cells

A cell is modeled as an idealized integrate-and-fire neuron (e.g., see [16]), and the spatio-temporal integration of activity arriving at a cell is modeled as follows:

Let $a_i(t)$ be a measure of presynaptic activity occurring at synapse s_i of the cell at time t. In biophysical terms, $a_i(t)$ may correspond to the number of spikes arriving at s_i within a unit time interval anchored at t. Thus the arrival of a high-frequency spike-burst at s_i would correspond to a high value of $a_i(t)$. Let $w_i(t)$ refer to the weight of synapse s_i at time t.

The postsynaptic potential, $psp_i(t\,|a_i(t_0))$, resulting from the presynaptic activity at s_i at time t_0 is modeled as a piecewise linear function consisting of a rising (ramp-up) segment, a flat (plateau) segment, and a falling (decay) segment. That is:

$$psp_i(t\,|a_i(t_0)) = \begin{cases} m_r * (t - t_0) & t_0 \leq t < (t_0 + \Delta T_r) \\[2mm] m_r * \Delta T_r & (t_0 + \Delta T_r) \leq t < (t_0 + \Delta T_{rs}) \\[2mm] m_r * \Delta T_r + (t - \Delta T_{rs}) * m_f & (t_0 + \Delta T_{rs}) \leq t < (t_0 + \omega_{int}) \\[2mm] 0 & \text{otherwise} \end{cases}$$

$$(1)$$

where m_r is the slope of the rising segment and is given by $(a_i(t_0) * w_i(t_0))/\Delta T_r$, ΔT_r is the duration of the rising segment, ΔT_s is the duration of the flat segment, ΔT_{rs} equals $(\Delta T_r + \Delta T_s)$, m_f is the slope of the falling segment, and ω_{int} is the window of temporal integration denoting the maximum amount by which two incident activities may lead/lag and still be summated by the postsynaptic cell. Note that

$$\omega_{int} = \Delta T_{rs} + (m_r * \Delta T_r)/m_f \qquad (2)$$

The postsynaptic potential at time t attributable to s_i, $psp_i(t)$, can be obtained by summing the effect of all the activity arriving at s_i during the past ω_{int} time units. Thus

$$psp_i(t) = \sum_{(0 \leq \tau < \omega_{int})} psp_i(t\,|a_i(t - \tau)) \qquad (3)$$

and $pot(t)$, the cell's potential at time t resulting from the combined effect of presynaptic activity at all its synapses equals:

$$pot(t) = \sum_i psp_i(t) \qquad (4)$$

where i ranges over all synapses of the cell.

A cell has a firing threshold, $thresh_f(t)$, with a resting value of θ_f. A cell fires at time t if $pot(t) \geq thresh_f(t)$, and produces an action potential (spike). This spike arrives at synapses downstream from the cell at time $t + d$, where d is the propagation delay.

After a cell fires, it enters a refractory state for a duration ω_{ref}. During this interval, the cell does not fire irrespective of its inputs. That is,

$$thresh_f(t) = \begin{cases} +\infty \text{ if cell has fired during the interval } [t - \omega_{ref}, t - 1] \\ \theta_f \quad \text{otherwise} \end{cases} \quad (5)$$

Some cell-types can have two firing modes: *supra-active* and *normal*. These modes are associated with firing thresholds θ_{sf} and θ_f, respectively, $(\theta_{sf} > \theta_f)$, and output levels O_2 and O_1, respectively, $(O_2 > O_1)$. Neurally, the *supra-active* mode corresponds to a high-frequency burst response such as the complex spike burst response generated by hippocampal pyramidal cells, and the *normal* mode corresponds to a simple spike response consisting of isolated spikes. The proposed abstraction of the distinction between a complex spike burst response and a simple spike response based on firing thresholds and output levels is a gross simplification. But for suitable choices of parameter values, this simple abstraction offers a computationally inexpensive, yet functionally adequate, means of modeling the two distinct response modes of certain cells.

3.2 Projection

A *projection* refers to the set of links emanating from cells in a source region and impinging on cells in a target region. It is assumed that all the synapses formed by a projection are of the same *type* and have similar attributes.

3.3 Synapses

A synapse can be in any one of following three states: *naive, potentiated,* or *depressed.* The state of a synapse signifies its strength (weight). For a given synaptic type, the weights of all synapses in a given state lie within a restricted band. The weight bands associated with different states are disjoint. The weight bands associated with a synaptic state may differ from one synaptic type to another.

3.4 Computational Modeling of LTP

The induction of LTP is governed by the following parameters: the *potentiation threshold* θ_p, the *weight increment* Δw_{ltp}, the *repetition factor* κ, and the *maximum inter-activity interval* τ_{iai}.

Consider a set of neighboring synapses $s_1, \ldots, s_n$ sharing the same postsynaptic cell. Convergent presynaptic activity at $s_1, \ldots, s_n$ can lead to LTP of naive s_i's and increase their weights by Δw_{ltp} if the following conditions hold:

1. $\sum_{1 \leq i \leq n} psp_i(t) \geq \theta_p$
   Note that in order to summate, the presynaptic activity arriving at $s_1, \ldots, s_n$ must be "synchronous", that is, the maximum lead/lag in incident activity at any pair of synapses should be no more than ω_{int}.
2. Such synchronous presynaptic activity recurs (repeats) at least κ times.
3. The interval between two *successive* arrivals of presynaptic activity at a synapse during the above repetition is at most τ_{iai} time units. In other words, successive volleys of synchronous activity should not be more than τ_{iai} apart.

Note that associative and homosynaptic LTP are modeled in an analogous manner. The difference between homosynaptic and associative LTP is simply this: In the case of homosynaptic LTP, the activity leading to LTP emanates from a cell ensemble representing a single item. In the case of associative LTP such activity emanates from multiple cell ensembles representing more than one item.

3.5   Computational Modeling of LTD

Heterosynaptic LTD is also modeled similarly using five parameters. These are: the potentiation threshold θ_p, the weight decrement Δw_{ltd}, the repetition factor κ, the maximum inter-activity interval τ_{iai}, and the *propensity of LTD* ζ $0 \leq \zeta \leq 0$. When naive or potentiated synapses of a postsynaptic cell receive convergent presynaptic activity, neighboring inactive naive synapses of the postsynaptic cell undergo heterosynaptic LTD and their weights decrease by Δw_{ltd}. As in the case of LTP, θ_p dictates the minimum weighted sum of synchronous activity that neighboring synapses of the postsynaptic cell must receive, and κ specifies the number of times such presynaptic activity must recur in order to induce heterosynaptic LTD of naive inactive synapses. Also as before, τ_{iai} specifies the maximum permissible gap between the successive arrival of presynaptic activity. The parameter ζ specifies the fraction of inactive naive synapse that undergo LTD when the above conditions are met. Thus ζ provides a simple computational mechanism for controlling the prevalence of heterosynaptic LTD. A value of $\zeta = 0$ means that there is no heterosynaptic LTD and a value of $\zeta = 1$ means that a single occurrence of LTP can lead to the heterosynaptic LTD of all inactive naive synapses of the postsynaptic cell.

3.6   Modeling Neuromodulation

The effect of neuromodulators on the response of a cell and on the induction of LTP at a synapse is a complex phenomena. In the present proposal, these effects are modeled by positing an additional input (or bias) that modifies the firing thresholds (θ_f and θ_{sf}) of a cell and the potentiation threshold (θ_p) of a synapse.

3.7 Emergence of Cells and Circuits Responsive to Specific Functionalities

LTP and LTD can transform random networks into structures consisting of cells tuned to specific functionalities. Typically, a cell receives a large number of inputs (afferents), and hence, can potentially participate in a large number of functional circuits. If, however, the weights of selected synapses on the cell increase via LTP (and, optionally, the weights of other synapses decrease via LTD) the cell can become more selective and participate in a limited number of functional circuits. Thus LTP and LTD provide a promising neural mechanism for the recruitment of structures with specific functionalities within quasi-random networks.

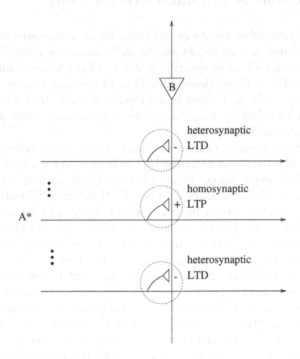

Fig. 1. Cell B becomes linked to ensemble A. The label A* attached to one of the afferents of B indicates that (i) the source of this afferent is a cell in ensemble A and (ii) this source cell is firing. Only a single afferent from ensemble A to B is shown. In general, B may have to receive several afferents from cells in A in order to become linked to A.

Let us consider the simplest of all cases where a cell B (see Fig. 1) becomes *linked* to a cell ensemble A as a result of LTP (cell B is linked to a cell ensemble A if the firing of a significant number of cells in A leads to the firing of B). Under appropriate conditions, the firing of cells in A would lead to the homosynaptic LTP of synapses formed by afferents from A impinging on B and (optionally) to the heterosynaptic LTD of some of the inactive synapses formed by the afferents from other cells impinging on B. The strengthening of synapses from A to B would increase the likelihood that B fires whenever cells in A fire. This would result in cell B becoming linked to cell ensemble A. Additionally, the weakening of B's synapses formed by cells not in A would lower the likelihood that B fires in circumstances when cells in A do not fire.

In the following section we illustrate how transient activity propagating through neural circuits can automatically lead to the recruitment of binding-detector (coincidence detector) cells.

4   Recruitment of Binding-Detector Cells

Our ability to remember events in our daily life demonstrates our capacity to rapidly acquire new memories. Typically, such memories record who did what to whom where and when, or describe states of affairs wherein multiple entities occur in particular configurations. This form of memory is often referred to as episodic memory [36], and there is a broad consensus that the hippocampal formation and neighboring areas in the medial temporal lobes serve a critical role in its formation [19,32,7,35].

The persistent encoding of an event must be capable of encoding role-entity *bindings*. Consider the event *described* by "John gave Mary a book in the library on Tuesday". This event cannot be encoded by simply forming a conjunctive association between "John", "Mary", "a book", "Library", "Tuesday" and "give" since such an encoding would be indistinguishable from that of the event described by "Mary gave John a book in the library on Tuesday". In order to make the necessary distinctions, the encoding of an event should specify the *bindings* between the *entities* participating in the event and the *roles* they play in the event. For example, the encoding of the event in question should specify the following *role*-entity bindings: ($\langle giver$=John$\rangle$, $\langle recipient$=Mary$\rangle$, $\langle give$-$object$=a-Book$\rangle$, $\langle temporal$-$location$=Tuesday$\rangle$, $\langle location$=Library$\rangle$).

As explained in [27], it is possible to evoke a fleshed out representation of an event by "retrieving" the bindings pertaining to the event and activating the web of semantic and procedural knowledge with these bindings. Thus cortical circuits encoding generic "knowledge" about actions such as *give* and entities such as *persons, books, libraries,* and *Tuesday* can recreate the necessary gestalt and details about the event "John gave Mary a book on Tuesday in the library" upon being activated with the above bindings. This view is supported by work on "reflexive reasoning" [28,25,30] and "executing schemas" [4,29].

In view of the above, the recruitment of binding-detectors is expected to be a critical step in the memorization of episodic memory. The following describes how such binding-detector cells can arise spontaneously and rapidly within a biologically motivated network structure as a result of LTP (and optionally, LTD).

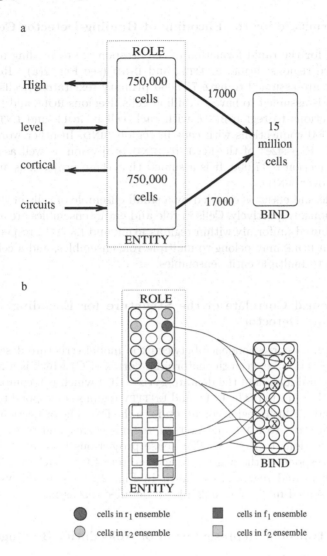

Fig. 2. (a) A structure for the formation of binding-detector cells. Arcs indicate projections and the number on an arc indicates the projective field size. These projections are assumed to be uniformly distributed over the BIND region. Each role and entity is encoded by a small ensemble of cells in the ROLE and ENTITY regions, respectively. Cells in role and entity ensembles are also assumed to be distributed uniformly within regions ROLE and ENTITY, respectively. Binding-detector cells are recruited in region BIND. It is assumed that the ROLE and ENTITY regions lie in the entorhinal cortex and the region BIND corresponds to the dentate gyrus (a part of the hippocampus). The projective field and region sizes are based on [2,40]. (b) A schematic depiction of the ensembles of roles r_1 and r_2 and entities f_1 and f_2. Only links from cells in r_1 and f_1 ensembles to cells in BIND are shown. Cells marked with an "X" are candidates for recruitment as *binder* cells for the binding $\langle r_1 = f_1 \rangle$.

4.1   A Structure for the Encoding of Binding-Detector Cells

A structure for the rapid formation of cells responsive to binding matches consists of three regions: ROLE, ENTITY, and BIND (see Fig. 2(a)). Regions ROLE and ENTITY are assumed to have 750,000 primary (excitatory) cells each, while region BIND is assumed to have 15 million cells. Regions ROLE and ENTITY have dense projections to region BIND, with each cell in ROLE and ENTITY regions making 17,000 connections with cells in region BIND. In other words, the projective field (PF) size[3] of the ROLE to BIND projection as well as the ENTITY to BIND projection is 17,000. It is assumed that these projections are uniformly distributed over BIND.

Each role and entity is encoded by a small ensemble of cells in the ROLE and ENTITY regions, respectively. Cells in role and entity ensembles are also assumed to be distributed uniformly within regions ROLE and ENTITY, respectively. Note that a cell in ROLE may belong to multiple role ensembles, and a cell in ENTITY may belong to multiple entity ensembles.

4.2   A Neural Correlate of the Structure for Encoding Binding-Detectors

There is a direct correspondence between the model structure described above and the interaction between the entorhinal cortex (EC) which is a region in the medial temporal lobe, and the dentate gyrus (DG) which is a component of the hippocampal formation. The ROLE and ENTITY regions correspond to subregions of the EC and the BIND region corresponds to the DG. The projections from high-level cortical areas to ROLE and ENTITY regions correspond to the well known cortical projections to EC [38]. The dense projections from ROLE and ENTITY to BIND correspond to the dense projections from EC to DG [2]. Moreover, the projective field and region sizes shown in Fig. 2(a) are based on anatomical findings presented in [2,40] (see [27] for a detailed discussion).

4.3   The Transient Representation of Role-Entity Bindings

It is assumed that the bindings constituting an event are expressed as a transient pattern of rhythmic activity over distributed high-level cortical circuits (HLCCs) [39,1,28,31]. These HLCCs project to cells in ENTITY and ROLE regions and, in turn, induce transient patterns of rhythmic activity within these regions. Fig. 3 is an idealized depiction of the transient activity induced in ENTITY and ROLE regions by HLCCs to convey the relational instance RI: $(\langle r_1 = f_1 \rangle, \langle r_2 = f_2 \rangle)$. Here r_1 and r_2 are roles, and f_1 and f_2 are entities bound to r_1 and r_2, respectively. Each spike in the illustration signifies the synchronous firing of a cell ensemble. It is shown that cells in the r_1 and f_1 ensembles are firing in

---

[3] The set of cells in the target region that receive links from a cell c in the source region is referred to as the projective field (PF) of c. The PF *size* of c refers to the number of synapses formed by c with cells in the target region.

synchrony, and so are cells in the r_2 and f_2 ensembles. The firing of cells in the r_1 and f_1 ensembles, however, is desynchronized with the firing of cells in the r_2 and f_2 ensembles. This desynchronization is assumed to be $\geq \omega_{int}$ time units. Note that the dynamic encoding of RI can be viewed as a periodic pattern consisting of two *phases*: ρ_1 and ρ_2. Here ρ_1 and ρ_2 are mere labels and the ordering of phases has no significance.

In effect, a role-entity binding is expressed by the synchronous firing of the cell ensembles associated with the bound role and entity [1,28]. In general, the transient encoding of a relational instance with n distinct entities participating as role-fillers involves n interleaved quasi-periodic activities having a period π. It is assumed that $\pi \leq \tau_{iai}$ time units. Such a spatio-temporal encoding enables multiple role-entity bindings to be expressed and propagated concurrently without cross-talk [28].

The following section explains how such a transient encoding of a relational instance may be transformed rapidly into persistent circuits for detecting bindings.

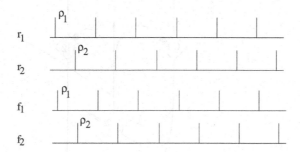

Fig. 3. The transient encoding of a relational instance RI given by the bindings: $(\langle r_1 = f_1 \rangle, \langle r_2 = f_2 \rangle)$. Here r_1 and r_2 are roles, and f_1 and f_2 are entities bound to r_1 and r_2, respectively, in RI. Each spike in the illustration signifies the synchronous firing of a cell ensemble. Cells in the r_1 and f_1 ensembles fire in synchrony and so do cells in the r_2 and f_2 ensembles. The firing of cells in the r_1 and f_1 ensembles, however, is desynchronized with the firing of cells in the r_2 and f_2 ensembles. This desynchronization is assumed to be $\geq \omega_{int}$ time units. Moreover, the period of firing, π, is assumed to be $\leq \tau_{iai}$ time units. The dynamic encoding of RI can be viewed as a periodic pattern consisting of two *phases*: ρ_1 and ρ_2 (the *order* in which these phases appear has no significance).

4.4 Recruitment of Binder Cells for Memorizing Role-Entity Bindings

BIND contains two kinds of cells: principal cells (these correspond to granule cells in the dentate gyrus) and Type-1 inhibitory interneurons.[4] Each principal cell

---

[4] The synapses formed by inhibitory interneurons on other cells have negative weights.

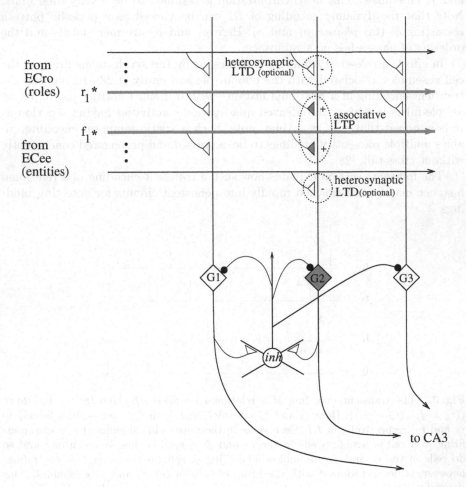

Fig. 4. The BIND region consists of principal cells and inhibitory interneurons (Type-1). In the illustration, G1—G3 are principal cells and *inh* is a Type-1 interneuron. Afferents (incoming links) labeled r_1* and f_1* are from cells in the ensembles for role r_1 and entity f_1, respectively. Since G1 and G2 receive synchronous activity along afferents from r_1 and f_1 cells, they are *candidates* for becoming binding-detector cells for the binding $\langle r_1 = f_1 \rangle$. It is assumed that the inhibition from *inh* prevents the LTP of G1's synapses, and only G2 becomes a binding-detector cell for $\langle r_1 = f_1 \rangle$. Filled blobs denote inhibitory synapses, and the size of a filled blob is meant to convey the strength of the (inhibitory) synapse. See text for additional details.

receives afferents from a number of cells in ROLE and ENTITY regions and makes synaptic contacts on a number of interneurons. The interneurons in turn make contacts on a number of principal cells, thereby forming inhibitory circuits within BIND (see Fig. 4). The significance of inhibitory interneurons will be explained later.

The potentiation threshold, θ_p, of principal cells is sufficiently high, and hence, LTP of a synapse occurs only if multiple synapses of a postsynaptic cell receive coincident presynaptic activity.[5] Moreover, the response threshold, θ_f, of principal cells is such that a cell does not fire unless it receives impulses at multiple potentiated synapses. A set of values for θ_p, θ_f, synaptic weights, and other parameters of LTP and LTD are given below.[6]:

$$\theta_f = 1700; \quad \theta_p = 890;$$

naive weight band = 100-110;

$$\Delta_{ltp} = 100; \quad \Delta_{ltd} = 50;$$

$$\zeta_{ltp} = 1; \quad \zeta_{ltd} = 0;$$

$$\kappa = 4; \quad \omega_{int} = 5; \quad \omega_{ref} = 2.$$

The choice of τ_{iai} is governed by ω_{int} and the number of role-entity bindings in an event. Thus any $\tau_{iai} \geq \omega_{int} * n$, where n is the number of bindings in the event, is appropriate.

The transient encoding of the relational instance RI shown in Fig. 3 leads to the following events in BIND (refer to Fig. 4). The synchronous firing of cells in the r_1 and f_1 ensembles (henceforth, r_1 and f_1 cells) leads to the associative LTP of active synapses of principal cells receiving sufficient afferents from r_1 and f_1 cells. At the same time, depending on the value of ζ_{ltd}, some of the inactive naive synapses of these principal cells may undergo heterogeneous LTD. The LTP of synapses formed by afferents arriving from r_1 and f_1 cells makes these principal cells behave as binding detector cells for the binding $\langle r_1 = f_1 \rangle$ and we will refer to such cells as $binder(\langle r_1 = f_1 \rangle)$ cells.[7]

The claim that $binder(\langle r_1 = f_1 \rangle)$ cells behave as binding-detector cells for $\langle r_1 = f_1 \rangle$ is substantiated quantitatively in Section 5, but let us examine why

---

[5] Here and elsewhere in the paper, "coincidence" is defined with reference to ω_{int}, the window of temporal integration. See Section 3.

[6] LTD does not play a critical role in the recruitment of binding-detector cells described here. It might, however, play an important role in the recruitment of other functional circuits.

[7] To be precise, a cell is deemed to be recruited as a $binder(\langle r_1 = f_1 \rangle)$ cell if during the memorization of $\langle r_1 = f_1 \rangle$, the cell's synapses undergo LTP *and* the cell fires. The firing of the cell at the time of its recruitment is crucial if the cell is to become part of functional circuits lying downstream from the cell.

these cells behave in the desired manner. Note that a $binder(\langle r_1 = f_1 \rangle)$ cell will fire in response to the synchronous firing of r_1 and f_1 cells since the connectivity between r_1 and f_1 cells and a principal cell required for the latter's recruitment as a $binder$ cell during the memorization of $\langle r_1 = f_1 \rangle$ also suffices for the latter's firing during the retrieval of $\langle r_1 = f_1 \rangle$. At the same time, since θ_f is quite high (1700), a $binder(\langle r_1 = f_1 \rangle)$ cell is unlikely to fire as a result of stray impulses arriving at its synapses.

Similar LTP and LTD events occur at the synapses of principal cells that receive coincident activity along afferents from r_2 cells in ROLE and f_2 cells in ENTITY, and lead to their recruitment as $binder(\langle r_2 = f_2 \rangle)$ cells. A $binder(\langle r_2 = f_2 \rangle)$ cell fires whenever r_2 cells in ROLE and f_2 cells in ENTITY fire in synchrony and behaves as a binding detector cell for the role-entity binding $\langle r_2 = f_2 \rangle$.

4.5 Encoding and Response Times

The time required for the recruitment of $binder$ cells is given by $\kappa * \tau_{iai}$. If we assume that the rhythmic activity encoding dynamic bindings corresponds to γ band activity (ca. 40 Hz) we get $\tau_{iai} \simeq 25$ msec. Assuming a plausible value of κ to be 4 suggests that $binder$ cells can be recruited in about 100 msec. The time required for $binder$ cells to respond to a retrieval cue is at most τ_{iai}. Thus both the recruitment and response times of the proposed model are consistent with the requirements of rapid (one-shot) memorization and recognition.

4.6 Potential Problems in the Formation of Binder Cells

The process by which $binder$ cells are formed is susceptible to several problems. First, in order to form $binder(\langle r_i = f_j \rangle)$ cells there should exist cells that receive afferents from both r_i and f_j cells. Given the random connectivity between the regions ROLE and ENTITY and BIND this cannot be *guaranteed*. Second, there may exist cells that receive sufficient activity $(\geq \theta_p)$ along afferents from r_i cells alone. Upon recruitment, such an *ill-formed* cell would produce false-positive responses since it will fire in response to the firing of r_i cells alone, even if there is no coincident activity of f_j cells. Similarly, cells receiving sufficient links from f_j cells alone could also be recruited as ill-formed $binder$ cells. Third, the same cell may get recruited as a $binder$ cell for multiple bindings. Consider a cell that gets recruited as a $binder$ cell for two bindings $\langle r_i = f_k \rangle$ and $\langle r_j = f_l \rangle$. This cell will fire in response to subsequent inputs containing either of these two bindings as well as the bindings: $\langle r_i = f_l \rangle$, and $\langle r_j = f_k \rangle$. Consequently, other cells connected downstream to this cell could receive false-positive binding-match signals in certain circumstances.

Using biologically motivated values of various system parameters it is shown in Section 5 that the probability of not finding cells for recruitment as $binder$ cells is vanishingly small. The problem of too many cells becoming recruited for a binding turns out not to be very serious in the case under consideration, and hence, is not discussed here. In general, however, this problem can be alleviated by inhibitory feedback and feedforward local circuits formed by principal cells and Type-1 inhibitory interneurons. These inhibitory circuits act as soft-WTA and only allow synapses of a limited number of cells to undergo LTP (cf. [15, 17]). Furthermore, it is shown that the ensemble response of binder cells is highly robust, in spite of the possibility that ill-formed and overlapping *binder* cells can be recruited.

5 Quantitative Results

The following quantities have been calculated analytically[8] using the region and projective field sizes described in Section 4.1, the cell, synapse, and LTP parameters described in Section 4.4, and by assuming that each role and entity ensemble contains 600 cells.

1. P_{fail}, the probability that for a given binding *no* cells will be found in BIND (DG) for recruitment as binding detector cells is less than $< 10^{-18}$. The probability that the system would be unable to encode a binding is essentially zero.
2. The expected number of cells in BIND (DG) that will receive appropriate connections and will be candidates for recruitment for a binding is 195. Thus a fairly large number of cells are recruited as *binder* cells for each binding.
3. The expected number of $binder(\langle r1 = f1 \rangle)$ cells that will fire in response to various retrieval *cues* are shown in Fig. 5. Note that $binder(\langle r1 = f1 \rangle)$ cells

   - respond robustly to the matching binding $\langle r1 = f1 \rangle$
   - produce an extremely weak response to erroneous, but partially related bindings of the form $\langle r1 = fx \rangle$ and $\langle rx = f1 \rangle$ (where fx is any entity other than $f1$, and rx is any role other than $r1$)
   - produce essentially no response to unrelated bindings.

   Note that this performance holds irrespective of the number of bindings memorized in BIND.

Since each binding is redundantly encoded by multiple cells, and since these cells are physically dispersed in the BIND region, the probability that a cell loss will destroy many *binder* cells for any given binding remains extremely small. In particular, a loss of $x\%$ of the cells in region BIND will lead to an expected loss of only $x\%$ of the 195 *binder* cells for a given binding. Thus the memorization of binding-detectors is robust with respect to cell loss (e.g., a cell loss of 10% will leave about 175 *binder* cells of any given binding intact).

The results in Fig. 5 are based on $\zeta_{ltd} = 0$ (i.e., no LTD). This condition results in a maximal sharing of *binder* cells among different bindings, and hence, these results provide a measure of the system's performance under conditions of maximal cross-talk. A non-zero value of ζ_{ltd} would reduce cross-talk, but it would also lead to a gradual reduction in the number of cells available for recruitment as more and more bindings are memorized.

---

[8] The bases of these calculation are discussed in [26].

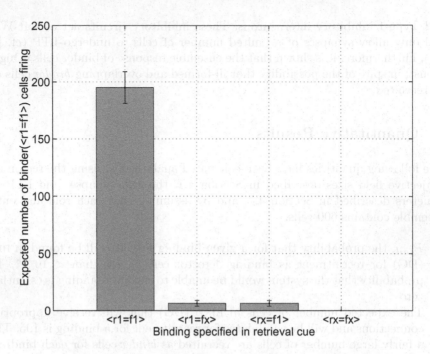

Fig. 5. Response of $binder(\langle r1 = f1 \rangle)$ cells to bindings in a retrieval cue. Here fx refers to an entity other than $f1$ and rx refers to a role other than $r1$.

6 Conclusion

A grounding of recruitment learning and vicinal algorithms in the biological phenomena of LTP and LTD has been described. A realization and specification of a vicinal algorithm using LTP has been illustrated by showing how *binder* cells responsive to specific role-entity bindings can be memorized rapidly in response to a transient pattern of activity encoding the bindings. Using biologically plausible values for the number of cells in the ROLE, ENTITY and BIND regions, and the density of projections from the ROLE and ENTITY regions to BIND, it has been shown that the existence of suitable *binder* cells for encoding arbitrary role-entity bindings is practically certain. It has also been shown that the interference between *binder* cells for different bindings remains extremely low, and that the encoding of *binder* cells is robust with respect to cell loss.

The encoding of binding detectors is just one step in the acquisition of episodic memory. As argued in [23,27] a proper encoding of episodic memory also requires the recruitment of *binding-error detector* circuits, *binding-error integrator* cells, *relational instance match indicator* circuits, and *binding-reinstator* cells. Shastri [24,27] has proposed a model of episodic memory formation that suggests how cells and circuits realizing these functional units can be recruited rapidly via LTP and LTD within quasi-random networks whose architecture and circuitry resembles that of the hippocampal formation. The resulting memory trace can differentiate between highly similar events and respond to partial cues.

Several of the functional units required for encoding an episodic memory trace are more complex than the *binder* cells discussed in this article [24]. The recruitment of these units requires the full-range of features included in the abstraction of LTP and LTD (e.g., spike versus burst firing modes) and further illustrates how interesting vicinal algorithms can arise from a suitable choice of network architecture and parameters for the induction of LTP and LTD.

The investigation of biologically grounded recruitment learning algorithms can increase our understanding of the neural processes underlying memory formation and retrieval. In the long-term, this research may also lead to the design of robust memory modules for autonomous agents, and perhaps, eventually, to the development of memory prosthesis for brain injured humans.

Acknowledgment. This work was partially funded by NSF grants 9720398 and 9970890.

References

1. Ajjanagadde, V., Shastri, L.: Rules and Variables in Neural Nets. *Neural Computation*, 3 (1991) 121–134.
2. Amaral, D.G., Ishizuka, N., Claiborne, B.: Neurons, numbers and the hippocampal network. In *Progress in Brain Research: Understanding the brain through the hippocampus*, J. Storm-Mathisen, J. Zimmer, & O.P. Ottersen (Eds.), 1–11, Elsevier Science, Amsterdam. (1990).
3. Artola, A., Singer, W.: Long-term depression of excitatory synaptic transmission and its relationship to long-term potentiation. *Trends in Neuroscience*, 16 (1993) 480–487.
4. Bailey, D., Chang, N., Feldman, J., Narayanan, S.: Extending Embodied Lexical Development. In the *Proceedings of the 20th Conference of the Cognitive Science Society*, Madison, WI. (1998) 84–89.
5. Bliss, T.V.P., Lomo, T.: Long-lasting potentiation of synaptic transmission in the dentate area of the anaesthetized rabbit following stimulation of the perforant path. *Journal of Physiology*, 232 (1973) 331–356.
6. Brown, T.H, Kairiss, E.W., & Keenan, C.L.: Hebbian Synapses: Biophysical Mechanisms and Algorithms. *Annual Review of Neuroscience*, 13 (1990) 475-511.
7. Cohen, N.J., Eichenbaum, H.: *Memory, Amnesia, and the Hippocampal System.* MIT Press, Cambridge, Massachusetts (1993).
8. Diederich, J.: Instruction and high-level learning in connectionist networks. *Connection Science*, 1 (1989) 161–180.
9. Dudek, S.M., Bear, M.F.: Bidirectional Long-Term Modification of Synaptic Effectiveness in the Adult and Immature Hippocampus. *Journal of Neuroscience*, 13 (1992) 2910-2918.
10. Feldman, J.A.: Dynamic connections in neural networks. *Bio-Cybernetics*, 46 (1982) 27–39.
11. Feldman, J.A., Bailley, D.: Layered hybrid connectionist models for cognitive science. In *Hybrid Neural System*, S. Wermter & R. Sun (Eds.), 14-27, Springer, Heidelberg (2000).

12. Levy, W.B. & Steward, O.: Synapses as associative memory elements in the hippocampal formation. *Brain Research*, 175 (1979) 233-245.
13. Linden, D.J.: Long-term synaptic depression in the mamallian brain. *Neuron*, 12 (1994) 457–472.
14. Malenka, R.C., Nicoll, R.A.: Long-term Potentiation - A Decade of Progress? *Nature*, 285 (1999) 1870-1874.
15. Marr, D.: Simple memory: a theory for archicortex. *Philosophical Transactions of the Royal Society*, B 262 (1971) 23–81.
16. Maass, W., Ruf, B.: On Computation with Pulses. *Information and Computation*, 148 (1999) 202-218.
17. McNaughton, B.L., Morris, R.G.M.: Hippocampal synaptic enhancement and information storage within a distributed memory system. *Trends in Neuroscience*, 10 (1987) 408–415.
18. Nicoll, R.A., Malenka, R.C.: Contrasting properties of two forms of long-term potentiation in the hippocampus. *Nature*, 377 (1995) 115–118.
19. O'Keefe, J., Nadel, L.: *The hippocampus as a cognitive map*, Clarendon Press, Oxford (1978).
20. Page, M.: Connectionist Modeling in Psychology: A Localist Manifesto *Behavioral and Brain Sciences*, 23 (2000), In press.
21. Rioult-Pedotti, M.-S., Friedman, D., Donoghue, J.P.: Learning-induced LTP in Neocortex. *Science*, 290 (2000) 533-536.
22. Shastri, L.: *Semantic Networks: An evidential formalization and its connectionist realization.* (see p. 181–191). Morgan Kaufmann, Los Altos/Pitman Publishing Company, London (1988).
23. Shastri, L.: A Model of Rapid Memory Formation in the Hippocampal System, In the *Proceedings of the 19th Annual Conference of the Cognitive Science Society*, Stanford University, CA, (1997) 680–685.
24. Shastri, L.: Recruitment of binding and binding-error detector circuits via long-term potentiation. *Neurocomputing*, 26-27 (1999) 865–874.
25. Shastri, L.: Advances in SHRUTI — A neurally motivated model of relational knowledge representation and rapid inference using temporal synchrony. *Applied Intelligence*, 11 (1999) 79–108.
26. Shastri, L.: A biological grounding of recruitment learning and vicinal algorithms. Technical Report TR-99-009, International Computer Science Institute, Berkeley, CA. March, 1999.
27. Shastri, L.: From transient patterns to persistent structures: a computational model of rapid memory formation in the hippocampal system. In preparation.
28. Shastri, L., Ajjanagadde V.: From simple associations to systematic reasoning: A connectionist encoding of rules, variables and dynamic bindings using temporal synchrony. *Behavioral and Brain Sciences*, 16 (1993) 417–494.
29. Shastri, L., Grannes, D.J., Narayanan, S., Feldman, J.A.: A Connectionist Encoding of Schemas and Reactive Plans. In *Hybrid Information Processing in Adaptive Autonomous vehicles,* G.K. Kraetzschmar and G. Palm (Eds.), Lecture Notes in Computer Science, Springer-Verlag, Berlin (To appear).
30. Shastri, L., Wendelken, C.: Seeking coherent explanations – a fusion of structured connectionism, temporal synchrony, and evidential reasoning. In the *Proceedings of the Twenty-Second Annual Conference of the Cognitive Science Society,* Philadelphia, PA. (2000) 453-458.
31. Singer, W.: Synchronization of cortical activity and its putative role in information processing and learning. *Annual Review of Physiology* 55 (1993) 349–74.

32. Squire, L.R.: Memory and the hippocampus: A synthesis from findings with rats, monkeys, and humans. *Psychological Review* 99 (1992) 195–231.
33. Tang, Y., Shimizu, E., Dube, G.R., Rampon, C., Kerchner, G.A., Zhuo, M., Liu, G., Tsien, J.Z.: Genetic enhancement of learning and memory in mice *Nature*, 401 (1999) 63 - 69.
34. Stanton, P.K. & Sejnowski, T.J.: Associative long-term depression in the hippocampus induced by hebbian covariance. *Nature*, 339 (1989) 215-218.
35. Treves, A & Rolls, E.T.: Computational analysis of the role of the hippocampus in memory. *Hippocampus*, 4 (1994) 374–391.
36. Tulving, E.: *Elements of Episodic Memory*. Clarendon Press, Oxford (1983).
37. Valiant, L.: *Circuits of the mind*, Oxford University Press, New York (1994).
38. Van Hoesen, G.W.: The primate hippocampus gyrus: New insights regarding its cortical connections. *Trends in Neuroscience*, 5 (1982) 345–350.
39. von der Malsburg, C.: Am I thinking assemblies? In *Brain Theory*, ed. G. Palm & A. Aertsen. Springer-Verlag, Berlin (1986).
40. West, M.J.: Stereological studies of the hippocampus: a comparison of the hippocampal subdivisions of diverse species including hedgehogs, laboratory rodents, wild mice and men. In *Progress in Brain Research: Understanding the brain through the hippocampus* J. Storm-Mathisen, J. Zimmer, & O.P. Ottersen (Eds.) 13–36, Elsevier Science, Amsterdam (1990).

Plasticity and Nativism: Towards a Resolution of an Apparent Paradox

Gary F. Marcus

Department of Psychology, 6 Washington Place, New York University
gary.marcus@nyu.edu

Abstract. Recent research in brain development and cognitive development leads to an apparent paradox. One set of recent experiments suggests that infants are well-endowed with sophisticated mechanisms for analyzing the world; another set of recent experiments suggests that brain development is extremely flexible. In this paper, I review various ways of resolving the implicit tension between the two, and close with a proposal for a novel computational approach to reconciling nativism with developmental flexibility.

1 Introduction: An Apparent Paradox

One strand of contemporary scientific research suggests that human infants are born with sophisticated mechanisms for learning about and analyzing the world. Within the first year of life, human infants can, among other things, anticipate sequences of events [1], keep track of objects that they cannot see [2, 3], discern abstract patterns in artificial languages [4, 5], and discriminate between unfamiliar languages that have different rhythmic properties [6]. In keeping with views advanced by Chomsky [7] and Fodor [8], "nativist" researchers such as Spelke [9] , Pinker [10], Leslie [11] and Crain [12] have taken these studies (and many others like them) to be evidence that the mind is importantly structured in advance of experience.

Another strand of contemporary scientific research suggests that brain development is remarkably flexible (or "plastic") – sizes of some brain areas depends on input [e.g., 13], early in development, some brain cells can be transplanted from one area of the brain to another [14], and certain parts of the brain can even be "rewired" [15, 16].

All this evidence that brain development is flexible has led some to think that nativism is in trouble. How could a newborn be born with language acquisition device if young children with left hemisphere brain injuries can recover language function to a significant extent? If the size of brain regions depends on experience, how could there be a built in module for tracking objects through time? If brain cells are not "born knowing their destinations", how could representations be innate? According to Elman, Bates, Johnson, Karmiloff-Smith, Parisi, & Plunkett [17] "the last two decades of research on vertebrate brain development force us to conclude that innate specification of synaptic connectivity at the cortical level is highly unlikely" (p. 361). Drawing on similar results, Quartz and Sejnowski [18] concluded that experiments in

S. Wermter et al. (Eds.): Emergent Neural Computational Architectures, LNAI 2036, pp. 368-382, 2001.

brain flexibility show that "although the cortex is not a tabula rasa ... it is largely equipotential at early stages" (p. 552) and that "nativist theories [therefore] appear implausible" (p. 555). Elman et al. argue that "Representation-specific predispositions ... may only be specified at the subcortical level *as little more than attention grabbers*" [emphasis added] that ensure the organism will receive "massive experience of certain inputs prior to subsequent learning..." (p. 108).

Do studies of brain development really militate against nativism? Researchers like Elman et al. certainly seem to think so, when they make it clear that the target of their attacks is the nativist positions of researchers like Spelke, Pinker, Leslie, and Crain. In the place of these strong nativist positions, Elman et al. settle for a sort of stripped-down nativism in which "architectural" aspects of brain organization are innate, but "representations " are not.

While I see the appeal in their position, I think it is ultimately untenable. Evidence that brain development is flexible really does challenge some of the simplest ways in which there could be innate mental structure, but, I will argue, it leaves more sophisticated versions of nativism untouched. Moreover, I will argue that the stripped-down nativism of Elman et al. probably relies too much on experience. I will end the paper by sketching a way in which strong nativism might be reconciled with developmental flexibility, proposing a novel computational approach that integrates neural networks simulations with findings in developmental biology.

2   Innateness

Before we can get to arguments about why developmental flexibility might be challenging to innateness, it is worth briefly reviewing some of the reasons for believing that significant aspects of mental structure might be innate. (Given space limitations, I do not aim to be comprehensive here; excellent, recent reviews include Spelke and Newport [19] and Pinker [10].)

2.1   Case Study: Objects

One reason for believing that significant aspects of mental structure might be innate is that recent studies of human infants suggest that they are capable of sophisticated analysis of the world. One case study that has received a great deal of attention is infant's understanding of the idea that objects persist in time. Piaget famously noted that 8-month-olds would cease to show interest in a toy if that toy was covered by a blanket; Piaget argued that the child had to *construct* (de novo) the notion of a persisting object. But dozens of recent studies suggest that infants behave as if they know[1] objects persist in time long before they begin to reach for occluded objects. For example, Spelke and Kestenbaum [2] conducted an experiment in which a four-month-old infant was seated at a stage that initially contained two screens. The infant subject would then see an object, in this case a rod, pass behind a screen, a bit of time

---

[1] I use words like "know" and "understand" loosely here –- I do not mean to say that infants consciously represent knowledge about objects, but rather that their computational systems respond in ways that are consistent with some sort of representation of object permanence.

would pass, and then the infant would then see an identical-appearing rod emerge from behind the other screen. The rod would then go back behind the second screen, some more time would pass, and then the rod would emerge from behind the first screen. This back-and-forth procedure would continue several times until the infant was bored, and then the infant would see the screen lifted, revealing either a single rod or two rods, one behind each screen.   Spelke et al. found that infants look longer when they are shown just one rod. Because infants generally look at longer at novel or unfamiliar outcomes, the results suggest that infants were "expecting" to see two distinct rods. Given that the infant only saw only one rod at any given moment, the result suggests that the infants kept track the rods, even when those rods were occluded. While the exact interpretation of these experiments is still open, it seems likely that at least some of the machinery that infants use in this task is innate.

2.2   Learning

Learning and innateness are often taken to be in opposition, but they need not be: learning mechanisms may themselves be innate. For example, using a variation on the habituation methods of Spelke and others, Saffran, Aslin and Newport [20] recently showed that eight-month-old can detect subtle statistical information from sequences of speech sounds produced in artificial languages. For example, in one experiment Saffran et al. familiarized infants with a two minute long, unbroken string of "familiarization" syllables such as *tibudopabikudaropigolatupabikutibudogolatu-daropidaropitibudopabikugolatu*. In this familiarization, some sounds are always followed by other sounds (e.g., every occurrence of *pa* was followed by *biku*), whereas other sounds are only sometimes followed by a particular sound (e.g., exactly one third the occurrences of *pi* were followed by *gola* ; other occurrences of *pi* were followed by *daro* or *tibu*). Saffran et al. found that infants attended longer during presentations of sequences like *pigola* than during presentations of words like *pabiku*, showing that infants extracted information about how often particular items follow one another. While it is possible that this statistical learning mechanism is learned, I know of no proposal for how it could be learned; instead, my hunch is that the learning mechanism itself is innate, built in prior to experience.

Similarly, my colleagues and I have shown that seven-month-old infants are able to learn "abstract rules" [4]. For instance, we exposed one set of infants to two minutes of "ABA" with sentences like ga ti ga and li na li. After this two-minute familiarization, we exposed infants to test sentences that were made up entirely of novel words that were either consistent with or inconsistent with the familiarization grammar. The prediction was that if infants can distinguish the two grammars and generalize them to new words, they should attend longer during inconsistent items. For example, if infants that were trained on the ABA grammar, we expected them to attend longer during, an ABB test item like *wo fe fe* than during an ABA test item like *wo fe wo*. As predicted, infants looked longer at the inconsistent items, suggesting that infants able to extract the ABA pattern and use it in evaluating new items. (Similar results with twelve-month-old children were reported in [5]). Although I cannot prove that the mechanism for rule-learing is innate, I strongly suspect that it is. A true tabula rasa position would be incoherent -- learning must start somewhere.

2.3   Learning Must Start Somewhere

Generalizing that point – that learning has to start somewhere – a third reason for believing that something is innate is that there may be no other satisfying account for how a given piece of knowledge could arise. So-called "learnability" arguments are perhaps most often made in the context of language acquisition. For example, Gordon [21] asked children to produce compounds such as *mice-eater and rat-eater*. He found that while children often produce compounds that contain irregular plurals (e.g., *mice-eater*) they essentially never produce compounds containing regular plurals (e.g., *rats-eater*). The way that children behave is consistent with a linguistic distinction that holds in English and perhaps cross-linguistically. But plurals inside compounds are so rare that young children are unlikely to have heard any; their inference thus in some sense probably goes beyond the input. From the fact that all children go beyond the data in a consistent way, Gordon argued that there must be some sort of built-in machinery constraining their learning. More general versions of "learnability" arguments have been made in the domain of language acquisition by Wexler and Culicover [22], Pinker [23, 24], and Crain [12], among others.

Similar arguments have been made in other domains; for example, Spelke [9] suggested that the ability to represent objects may be innate:

> If children are endowed with abilities to perceive objects, persons, sets, and places, then they may use their perceptual experience to learn about the properties and behavior of such entities. By observing objects that lose their support and fall, children may learn that unsupported objects fall... it is far from clear how children could learn anything about the entities in a domain, however, if they could not single out those entities in their surroundings.

> ...[in contrast] if children could not represent the object-that-loses-its-support as the *same object* as the object-that-falls (and as a different object from the support itself), they might only learn that events in which something loses support are followed by events in which something falls (the object) and something remains at rest (the support).

3   Developmental Flexibility and DNA as Blueprint

If the mind is indeed importantly structured prior to experience, how did it get that way? I ask this not as a question about evolution, but as a question about developmental biology. To the extent the mind is a product of the brain, how could the brain be organized prior to experience?

It would certainly be convenient for nativists if fertilized eggs contained a blueprint for building the brain. Just as an architectural blueprint might specify exactly where every room and corridor in some new office building might be placed, one might imagine the fertilized egg bearing a neural blueprint that would specify where every neuron and connection in the to-be-born child's brain would be placed. This "DNA-as-blueprint" idea would fit nicely with nativism, but, alas, it clearly cannot be right.

For one thing, there just is not enough information in the human genome to specify exact where each neuron and synapse will go [25]. There are about 10^5 genes which contain about 10^9 nucleotides, as compared with about 10^{10} neurons and about 10^{15} or so synapses.

Moreover, as noted in the introduction, brain development is flexible, and this flexibility seems inconsistent with blueprint idea. For example, if the exact structural organization of some brain region were predestined, its size should not depend on the amount of input received. Yet the size of some brain regions does indeed depend on the amount of input [13].

Similarly, if the DNA provided a blueprint, one would not necessarily expect the brain to be able to adapt itself in response to radical "rewiring", yet experiments by Sur and his colleagues [16] show that when visual thalamic inputs are rewired from their usual destination in visual cortex to a novel destination in auditory cortex, the auditory cortex begins to take on some of the properties of visual cortex.

Plainly the DNA does not specify a point-by-point wiring diagram for the human brain. Other evidence further underscores the view of brain development as flexible. O'Leary and Stanfield [14] showed that when visual cortex neurons are transplanted into somatosensory areas, they develop (at least in some respects) as one would expect for somatosensory neurons rather than for visual neurons, projecting not to the visual cortex, but to the spinal cord. Likewise, somatosensory cells transplanted to visual cortex develop projections that are typical of visual neurons. Furthermore, although recovery from brain injuries that occur in adulthood may be quite minimal (although non-zero), recovery from brain injuries in childhood can be much more substantial, with undamaged areas of the brain taking over some of the functions of damaged areas of the brain [e.g., 26].

Where does this leave us?

4   The Neo-constructivist Synthesis

Scholars such as Quartz and Sejnowski and Elman et al. see the evidence from developmental flexibility as devastating to nativism. Neither set of researchers wishes to dispense with nativism altogether, but both groups direct their criticism towards researchers such as Chomsky, Fodor, Spelke, Pinker, and Crain, and both put the burden of brain organization primarily on learning, stressing "massive experience" over any kind of significant intrinsic organization. For example, since, as they put it, "neurons can't be born knowing their destinations", Elman et al. conclude that strong nativism must be wrong. In its place, they argue that "architectural" aspects of the brain – how many layers there are, how many units are in those layers, and so forth – are organized in advance, but they suggest that the detailed microcircuitry is not; Quartz and Sejnowski make similar points. Collectively, I will call their position the "neo-constructivist synthesis".

In order to make it plausible that weak initial biases could combine with experience in satisfactory ways, both groups of researchers point to a series of connectionist models or neural networks. *Neural networks* are idealized computer simulations that are intended to tell us something important about how the mind/brain works. Typically, they consist of sets of neuron-like *nodes* interacting in parallel.

The models that they describe typically look something like Figure 1. (I assume most readers have at least a passing familiarity with these networks. In brief, a set of *input nodes* represents the input to the network, the set of *output nodes* represent the output from that network. Intervening between the input and nodes is a set of *hidden units* that re-represent the input. The arrows indicate the extent to which different nodes are connected together. Such models are typically trained on the basis of input-output pairs; during this training, connections between nodes are adjusted in way that attempts to minimize error. For a more in-depth introduction, see, for example, [27].)

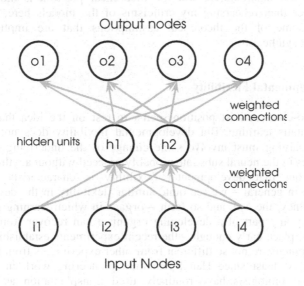

Fig. 1. A simple three-layer neural network

There are, in fact, many ways of arranging nodes and connections, and different arrangements have different implications for basic questions in cognition such as whether the mind is like a symbol-manipulating computer; I have written extensively about such issues [28-30] but will mainly skip them here. Suffice it to say here that the neural networks endorsed by Elman et al. are, by design, among those with the least innate structure – Elman et al. see themselves as providing a computational basis for Piagetian constructivism:

> ... constructivism [considered] development in terms of self-organizing emergent structures arising from the complex interactions between both organism and environment. We believe that the biological-connectionist perspective opens the door to a new framework for thinking about development which embodies some aspects of Piaget's, Werner's, and Vygotsky's constructivist intuitions, but which goes beyond them and provides a formalized framework within which to generate empirically testable questions (p . 114)

5   Discussion of the Neo-constructivist Synthesis

Although some may find the neo-constructivist thesis to be appealing, we are by no means forced to adopt it. One reason is that there are serious limits on the particular *models* that researchers like Elman et al. have advocated [28-32]. Such models are limited in their abilities to generalize, they have difficulties in representing fundamental notions such as the distinction between individuals and kinds, and they are, I think, too unstructured. But the theoretical position is independent of the models; rather than rehearsing my criticisms of the models here, I want to focus instead on some of the theoretical assumptions that are implicit in the neo-constructivist synthesis.

5.1   Developmental Flexibility

First, the neo-constructivist position seems to rest on the idea that developmental flexibility entails learning. But developmental flexibility does not *entail* learning. While any learning must involve some change of the underlying neural substrate, many changes in the neural substrate probably proceed without anything like learning.

It turns out that developmental flexibility is characteristic of mammalian development in general – we see quite similar flexibility in the development of the heart, the kidney, the eye, and so forth – organs in which learning plays little or no role. Virtually any part of a developing organism can recover from damage if that damage takes place early enough – the recent experiments establishing robustness in brain development are not so different from other experiments from the early days of embryology. At least since Han Spemann's pioneering work in the 1920s [33], developmental biologists have routinely used transplantation as a window into embryology; quite often, if those experiments are done early enough, transplanted tissue takes on some or all of the characteristics of its new host region. In the words of noted embryologist Lewis Wolpert [34, p. 42]:

> In general, if cells of vertebrate embryos are moved from one part to another of the early embryo they develop according to their new location and not from where they are taken. Their fate is dependent on their new position in the embryo: they respond to their new address.

For example, if early in development one takes cells from the region of a frog embryo that normally develops into an eye and transplants them into the gut, they develop into gut cells rather than eye cells, much as a transplanted somatosensory cell may take on characteristics of its new home.[2] Such flexibility may even be adaptively advantageous; as Cruz [36] put it

> In a rapidly growing embryo consisting of cells caught in a dynamic flurry of proliferation, migration, and differentiation, it would be desirable for any given cell to retain some measure of developmental

---

[2] Transplants of brain cells, too, seem to be age-dependent, with the chance of a transplanted cell taken on target characteristics greatest earlier in development [35].

flexibility for as long as possible. Such would enable an embryo momentarily disabled by cell cycle delay, for instance, or temporarily compromised by loss of a few cells, to compensate for minor disruptions and resume rather quickly the normal pace of development. It is easy to see how such built-in [flexibility] could contribute to the wide variety of procedural detail manifest in nearly every phase of mammalian embryogenesis (p. 484).

One would not want to say that the eye cell *learns* how to be a stomach cell, and one should not assume that a transplanted somatosensory cell learns how to be a visual cell. None of this rules out learning (and there must be important learning eventually), but it does remind us that developmental flexibility does not on its own entail learning.

5.2 DNA as Blueprint

A second problem with the neo-constructivist synthesis is that it seems to equate nativism with the idea of the DNA as a blueprint. In fact, the DNA rarely if ever serves as literal blueprint in any part of biology, but there is no reason that nativism must depend on such a fantastical view of DNA. One need only look to the heart or the eye to see that nature can build highly intricate structure without depending on learning. The idea of DNA as blueprint is really a strawman that makes little sense in any part of biology. As Richard Dawkins [37] has put it, the DNA is much more like a recipe than a blueprint – the DNA gives a set of instructions for building something, not a diagram of what the finished product will look like. But a recipe is enough – the toolkit of biology is sufficiently powerful that it can build bodies without requiring a whole lot of learning, and I suspect that very same toolkit is powerful to build brains as well. For this reason, I believe that a complete account of brain development must make substantial reference to the toolkit of developmental biology.

5.3 Neural Activity

The third serious problem with the neo-constructivist synthesis is that it rests too heavily on learning and neural activity, attributing virtually all detailed brain organization to neural activity. But a number of recent studies in developmental neuroscience suggest that while neural activity is important to brain development, it may not be essential for early stages of brain development. The idea that an organism's detailed microcircuitry can only be specified on the basis of massive experience simply is not tenable. For example, Crowley & Katz [38] recently demonstrated that the organization of ferret geniculocortical axons into ocular dominance columns could occur even in the complete absence of retinal input. In another set of experiments, Verhage, et al. [39] created "knock-out" mice that lacked the gene *Munc-18*, causing a "complete loss of neurotransmitter secretion from synaptic vesicles throughout development" ; their striking finding was that brain assembly was apparently normal, "including formation of layered structures, fiber pathways, and morphologically defined synapses." Considerations like these led Katz, Weliky, and Crowley [40] to conclude that

"The current emphasis on correlation-based models, which may be
appropriate for later plastic changes, could be obscuring the role of
intrinsic signals that guide the initial establishment of functional
architecture."

6   A New Approach

In a nutshell, what I think is being left out in the neo-constructivist synthesis is the
toolkit of developmental biology. Developing embryos are blessed with an
extraordinary array of techniques for organizing themselves, ways of coordinating the
actions of simple genes into incredibly complex organisms. It is my hunch that a
proper account of brain development should make extensive use of the tools that
biology uses when it builds organisms.

All of which is rather vague. To make it more explicit I would like to borrow an
idea from the neo-constructivists. Like them, I want to use neural networks as a way
of understanding possible mechanisms of development. But unlike them, I want to
build neural networks that *grow*, networks that show a good degree of self-
organization even in the absence of experience.

6.1   Some Principles Drawn from Developmental Biology

In contrast to the neural networks of the neo-constructivists, the neural networks that I
aim to build will integrate ideas about nodes and connections with some of the basic
principles of developmental biology, including the following:

- Basic processes such as cell division, cell migration, and cell death.

- Gene expression. Genes can either be "expressed" or "repressed". What governs
  whether a particular gene is on or off is (among other things) the presence or
  absence of specific *regulatory* proteins that serve as enhancers or repressors for
  that gene [41]. When a gene is on, it sets into motion a transcription process that
  ultimately yields a particular protein. In the simulations, genes are rules with
  preconditions (which correspond to promoter sequences) and actions (which
  correspond, for example, to the construction of various proteins).

- Cell-to-cell communication. Many of the regulatory proteins that serve as triggers
  can pass from one cell to another; in this way, and also by means of electrical
  signaling, cells can communicate with each other. Mechanisms for both chemical
  and electrical signaling are included in the simulations.

- Cascades. Because an expressed gene can yield proteins, and proteins can trigger
  the expression of genes, one gene can trigger the action of another, or even several
  others, each of which in turn might trigger several others, and so forth -- what we
  might call a *cascade* [41-43]. These cascades, sometimes described as *regulatory
  networks* or *gene hierarchies,* are critical, because they provide a way for a

complex coordinated actions to emerge. A particularly vivid example of this comes from the work of Walter Gehring; he and his collaborators have shown that a simple fruit fly gene known as *pax 6* triggers the action of (at least) three other genes, each of which in turns launches the action of still more genes, about 2500 in all [43-45]. What is special about *pax 6*, which Gehring calls a master control gene, is that it sits atop a hierarchy of genes that lead to the construction of an eye. Fruit flies that lack this gene generally do not have eyes; even more striking is the fact that if pax-6 is expressed (turned on) artificially in other parts of the body, eyes may grow in those regions; for example, Gehring and his collaborators were able to induce fruit flies to grow eyes on their antennae. The lesson here is not that there is a gene for building eyes – *pax 6* cannot do this by itself – but rather that the action of a single gene can through the process of cascading snowball into tremendously complex machinery. (Machinery for building cascades emerges in the simulation – and in nature – automatically, in virtue of the mechanisms that control gene expression.)

6.2   The Simulator

Where as most work in developmental biology works bottom-up, by testing what happens if particular genes are "knocked-out" or artificially expressed, I aim to work in a more top-down fashion, asking how brains with particular properties could be assembled by genetic-like processes such as those mentioned above.

It is still very early days for this project, much too early to report any concrete results. Because the project is almost entirely new, I have spent the initial stages developing a prototype simulator that could be used to support this kind of work. A screen shot of the simulator is shown in Figure 2.

The main window shows the current state of a given embryo (what cells there are, what genes are expressed in those cells, etc.); another window shows the "genome" for that organism, a third shows what genes were expressed in a given time step.

In the particular simulation that is illustrated here, the top row depicts one layer of cells (to be thought of as "nodes" or "neurons"), the bottom depicts another layer of

cells; the dark gray cells represent a set of cells that are migrating from top to bottom; the arrows represent "axons" that are growing along gradients attempting to connect the top and bottom layers.

Various buttons allow the user to modify the genome, step forward in time, rebuild the organism, and so forth. Other controls allow the user to display the concentrations of diffusing "morphogens", stain cells according to their patterns of gene expression, and so forth. It is also possible for users to selectively lesion particular cells, transplant cells from one location to another, test the effects of knocking-out particular genes, etc..

Figures 3 and 4 give some more examples.[3] Figure 3 is a sort of time-lapse illustration of the growth of a simple organism, which we might think of as a ladybug. In the ladybug simulation, there is only structure, not function, and nothing neural.

Still, several interesting points are captured. First, the development of the ladybug proceeds in parallel – the *developmental program* that builds the ladybug is like a

---

[3] Further examples may be found on my web site, http://www.psych.nyu.edu/gary.

standard computer program in that it is made up of rules with preconditions and actions (each gene is essentially an IF-THEN rule) but whereas standard computer programs proceed serially, one step at a time, the developmental program (like those in biology) proceeds in parallel – each cell at the same time.

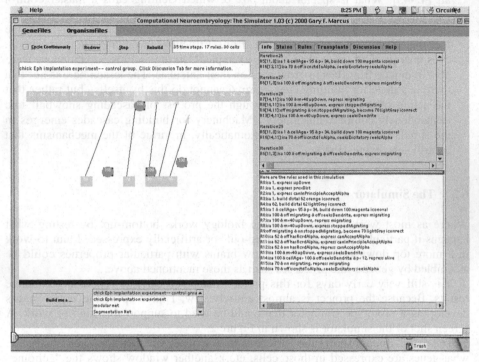

Fig. 2. Simulator

Second, the ladybug program illustrates, in a tiny way, the notion of compression; the complete ladybug has 79 cells, but just 32 rules; the developmental program is thus much more efficient than a literal blueprint would be.

Third, the ladybug program is developmentally robust; portions of the ladybug can be lesioned or amputated, and, salamander-like, they will grow back. The ladybug is thus a primitive illustration of a system that is innately organized (learning plays no role) yet developmentally robust.

Examples in Figure 4 show some simple of the neural structures that can be built in the prototype simulator.

The top left panel illustrates a developmental stage in a two layer network that is topographically organized, such that connections maintain the relative left-right ordering. The top right panel illustrates a more complex network with several layers, and different types of connections. The bottom left panel depicts the core of a network that would solve a very simple visual segmentation task; the bottom right panel depicts a three-dimensional multilayered version of that visual segmentation network. Each of these models is to some degree developmentally robust; none depends on learning for its basic organization.

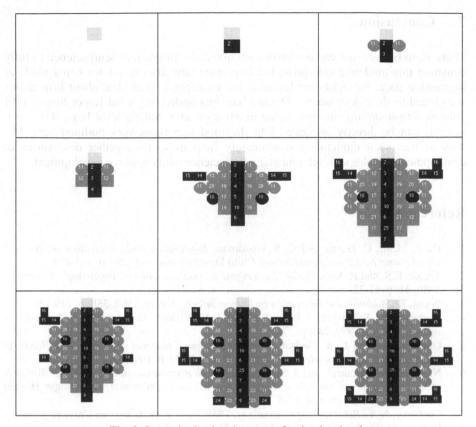

Fig. 3. Stages in the development of a simulated embryo

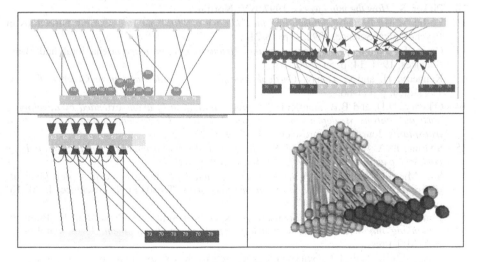

Fig. 4. Four networks "grown" in the simulator. See text for further details

7   Conclusion

There is, to be sure, not enough known yet about developmental neuroscience to fully constrain this modeling enterprise; but important new discoveries are being made at impressive pace; biologists are learning, for example, a great deal about how axons are guided to their destinations. The mechanisms underlying what Roger Sperry [46] dubbed *chemoaffinity* are now being understood at a genetic level [e.g., 47], Such insights can be directly integrated into the modeling framework outlined here. My hope is that such modeling can ultimately help us to tie together discoveries in developmental biology, developmental neuroscience, and cognitive development.

References

1.   Haith, M.M., C. Hazan, and G. S. Goodman, *Expectation and anticipation of dynamic visual events by 3.5-month-old babies.* Child Development, 1988. **59**: p. 467-479.
2.   Spelke, E.S. and R. Kestenbaum, *Les origins du concept d'object.* Psychologie Francaise, 1986. **31**: p. 67-72.
3.   Wynn, K., *Addition and subtraction by human infants.* Nature, 1992. **358**: p. 749-750.
4.   Marcus, G.F., S. Vijayan, S. Bandi Rao, and P.M. Vishton, *Rule learning in 7-month-old infants.* Science, 1999. **283**: p. 77-80.
5.   Gomez, R.L. and L.-A. Gerken, *Artificial grammar learning by 1 year-olds leads to specific and abstract knowledge.* Cognition, 1999. **70**(1): p. 109-135.
6.   Nazzi, T., J. Bertoncini, and J. Mehler, *Language discrimination by newborns: Towards an understanding of the role of rhythm.* Journal of Experimental Psychology: Human Perception and Performance, 1998. **24**: p. 1-11.
7.   Chomsky, N.A., *Rules and representations.* 1980, New York: Columbia University Press.
8.   Fodor, J.A., *The language of thought.* 1975: New York: T. Y. Crowell.
9.   Spelke, E.S., *Initial knowledge: Six suggestions.* Cognition, 1994. **50**: p. 431-445.
10.   Pinker, S., *How the mind works.* 1997, NY: Norton.
11.   Leslie, A.M., *Pretense, autism, and the "Theory of Mind" module.* Current Directions in Psychological Science, 1992. **1**: p. 18-21.
12.   Crain, S., *Language Acquisition in the Absence of Experience.* Behavioral and Brain Sciences, 1991. **14**: p. 597-650.
13.   Kennedy, H. and C. Dehay, *Cortical specification of mice and men.* Cerebral Cortex, 1993. **3**(3): p. 171-86.
14.   O'Leary, D.D. and B.B. Stanfield, *Selective elimination of axons extended by developing cortical neurons is dependent on regional locale: Experiments using fetal cortical transplants.* Journal of Neuroscience, 1989. **9**: p. 2230-2246.
15.   Sharma, J., A. Angelucci, and M. Sur, *Induction of visual orientation modules in auditory cortex [see comments].* Nature, 2000. **404**(6780): p. 841-7.
16.   Sur, M., S.L. Pallas, and A.W. Roe, *Cross-model plasticity in cortical development: differentiation and specification of sensory neocortex.* Trends in Neuroscience, 1990. **13**: p. 227-233.
17.   Elman, J.L., E. Bates, M.H. Johnson, A. Karmiloff-Smith, D. Parisi, and K. Plunkett, *Rethinking innateness: A connectionist perspective on development.* 1996, Cambridge, MA: MIT Press.
18.   Quartz, S.R. and T.J. Sejnowski, *The neural basis of cognitive development: A constructivist manifesto.* Behavioral and Brain Sciences, 1997. **20**: p. 537-56; discussion 556-96.

19. Spelke, E.S. and E.L. Newport, *Nativism, empiricism, and the development of knowledge*, in *Handbook of Child Psychology (5th ed.), Vol. 1: Theories of development*, R.M. Lerner, Editor. 1998, Wiley: NY. p. 275-340.

20. Saffran, J., R. Aslin, and E. Newport, *Statistical learning by 8-month old infants.* Science, 1996. **274**: p. 1926-1928.

21. Gordon, P., *Level-ordering in lexical development.* Cognition, 1985. **21**: p. 73-93.

22. Wexler, K. and P. Culicover, *Formal principles of language acquisition.* 1980, Cambridge, MA: MIT Press.

23. Pinker, S., *Formal models of language learning.* Cognition, 1979. **7(3)**: p. 217-283.

24. Pinker, S., *Language learnability and language development.* 1984, Cambridge, MA: Harvard University Press.

25. Edelman, G.M., *Topobiology : an introduction to molecular embryology.* 1988, New York: Basic Books. xv, 240.

26. Vargha-Khadem, F., D.G. Gadian, K.E. Watkins, A. Connelly, W. Van Paesschen, and M. Mishkin, *Differential effects of early hippocampal pathology on episodic and semantic memory [see comments] [published erratum appears in Science 1997 Aug 22; 277(5329):1117].* Science, 1997. **277**(5324): p. 376-80.

27. Bechtel, W. and A. Abrahamsen, *Connectionism and mind: An introduction to parallel processing in networks.* 1991: Cambridge, MA: Basil Blackwell.

28. Marcus, G.F., *Can connectionism save constructivism?* Cognition, 1998. **66**: p. 153-182.

29. Marcus, G.F., *Rethinking eliminative connectionism.* Cognitive Psychology, 1998. **37**(3): p. 243-282.

30. Marcus, G.F., *The algebraic mind: Integrating connectionism and cognitive science.* 2001, Cambridge, MA: MIT Press.

31. Marcus, G.F., U. Brinkmann, H. Clahsen, R. Wiese, and S. Pinker, *German inflection: The exception that proves the rule.* Cognitive Psychology, 1995. **29**: p. 186-256.

32. Marcus, G.F., S. Pinker, M. Ullman, J.M. Hollander, T.J. Rosen, and F. Xu, *Overregularization in language acquisition.* Monographs of the Society for Research in Child Development., 1992. **57**(4, Serial No. 228).

33. Spemann, H., *Embryonic development and induction.* Yale University. Mrs. Hepsa Ely Silliman memorial lectures [1933]. 1938, New Haven,: Yale University Press. xii, 401.

34. Wolpert, L., *The triumph of the embryo.* Repr. (with corrections) ed. 1992, Oxford England ; New York: Oxford University Press. vii, 211.

35. Levitt, P., *Molecular determinants of regionalization of the forebrain and cerebral cortex*, in *The new cognitive neurosciences*, M.S. Gazzaniga, Editor. 2000, MIT Press: Cambridge, Mass. p. 23-32.

36. Cruz, Y.P., *Mammals*, in *Embryology : constructing the organism*, S.F. Gilbert and A.M. Raunio, Editors. 1997, Sinauer Associates: Sunderland, MA. p. 459-489.

37. Dawkins, R., *The Blind Watchmaker.* 1987, NY: Norton.

38. Crowley, J.C. and L.C. Katz, *Development of ocular dominance columns in the absence of retinal input.* Nature Neuroscience, 1999. **2**(12): p. 1125-1130.

39. Verhage, M., *et al.*, *Synaptic assembly of the brain in the absence of neurotransmitter secretion [In Process Citation].* Science, 2000. **287**(5454): p. 864-9.

40. Katz, L.C., M. Weliky, and J.C. Crowley, *Activity and the development of the visual cortext: New perspectives*, in *The new cognitive neurosciences*, M.S. Gazzaniga, Editor. 2000, MIT Press: Cambridge, Mass. p. 199-212.

41. Jacob, F. and J. Monod, *On the regulation of gene activity.* Cold Spring Harbor Symposium on Quantitative Biology, 1961. **26**: p. 193-211.

42. Gilbert, S.F., *Developmental biology.* 5th ed. 1997, Sunderland, Mass.: Sinauer Associates. 1 v. (various pagings).

43. Gehring, W.J., *Master control genes in development and evolution : the homeobox story.* The Terry lectures. 1998, New Haven: Yale University Press. xv, 236.

44. Halder, G., P. Callaerts, S. Flister, U. Walldorf, U. Kloter, and W.J. Gehring, *Eyeless initiates the expression of both sine oculis and eyes absent during Drosophila compound eye development.* Development, 1998. **125**(12): p. 2181-91.
45. Halder, G., P. Callaerts, and W.J. Gehring, *Induction of ectopic eyes by target expression of the eyeless gene in Drosophila.* Science, 1995. **267**: p. 1788-1792.
46. Sperry, R.W., *Chemoaffinity in the orderly growth of nerve fiber patterns and connections.* Proceedings of the National Academy of Sciences, 1963. **50**: p. 703-710.
47. Brown, A., *et al.*, *Topographic mapping from the retina to the midbrain is controlled by relative but not absolute levels of EphA receptor signaling.* Cell, 2000. **102**(1): p. 77-88.

Cell Assemblies as an Intermediate Level Model of Cognition

Christian R. Huyck

Middlesex University, London, UK,
C.Huyck@mdx.ac.uk,
http://www.cwa.mdx.ac.uk/chris/chrisroot.html

Abstract. This chapter discusses reverberating circuits of neurons or Cell Assemblies (CAs) derived from Hebb's [9] proposal. It shows how CAs can quickly categorise an input and make a quick decision when presented with ambiguous data. A categorisation experiment with a computational model of CAs shows that CAs categorise a broad range of patterns.

This chapter then describes how CAs might be used to implement the primitives of an symbolic cognitive architecture. It also shows how a system based on CAs is theoretically capable of fast learning, variable binding, rule application, integration with emotion and integration with the external environment.

CAs are thus an ideal mechanism for further research into both computational and cognitive neural models. Our medium to long-term plan for exploration of thought via CAs is described.

If humans use CAs as a basis of thought, then studying how biological systems use CAs will provide information for computational models. The reverse is also true; computational modelling can direct our research activity in biological neural systems.

1 Background and Introduction

I want to start this chapter with an apology. This work is about how humans (and other creatures) think. The claims often strike me as arrogant. Discovering how people think has been a major problem of philosophy for thousands of years. For me to say, "we think because of Cell Assemblies (CAs)" is arrogant, and thus we apologise. The statement is also currently backed by little evidence. I once heard Herb Simon say "if you work on an interesting problem, you will get somewhere if you answer it." Understanding thought is important, and CAs are crucial to our understanding of thought. So, I will continue with the claim that we think because of CAs, show how they can provide a sound model of thought, and provide some evidence.

Thought is extremely complex and perhaps the most complex thing that we study. One way to decompose the problem of modelling thought is to model it at different levels of granularity. For instance, Newell divides cognition into levels by time [19]. The biological band is fastest and its primitive operations

S. Wermter et al. (Eds.): Emergent Neural Computational Architectures, LNAI 2036, pp. 383–397, 2001.
© Springer-Verlag Berlin Heidelberg 2001

take place between 10^{-4} and 10^{-2} seconds; the primitives in the cognitive band function from 10^{-1} seconds to 10^1 seconds, the rational band from 10^2 to 10^4 seconds and the social band above that. Models can be built at any level.

Each band is broken into three levels. The neural level, functioning near 10^{-3}, and the neural circuit level, functioning near 10^{-2} seconds, are in the biological band. The deliberate act level near 10^{-1} is in the cognitive band.

While models can be built at different levels, each level depends on those below it. A model at the deliberate act level functions via certain assumed primitive operations. It is assumed that these primitives can be implemented in the lower levels.

Newell and other symbolicists have focused on the cognitive and rational bands [16]. These are the symbol processing levels. Anderson's ACT* [2] system works at this level but also integrates neural models as a means of showing human reaction and performance times.

For instance, Newell bases his architecture on the deliberate act level. The primitives are to assemble operators and operands, apply the operator to the operands, and to store results. Additionally, the system must support structures, input and output. Higher levels are then built upon the deliberate act level.

Why does Newell select these primitives? Can we be sure that these are the right primitives? While Newell has reasons for selecting these primitives, there is no firm basis in the neural circuit level (the level directly below the deliberate act level) that shows these things working. If some basis could be found, then it would strengthen the argument for these primitives. Moreover, it might have further ramifications on higher levels. Currently, Newell's model is a floating island; it is not firmly based on lower level structures.

Aside from the neural level, we have only indirect evidence of the behaviour of all of the other levels. We can look at the firing and activation of neurons, but we cannot see a rule being selected in the operations level. Indirect evidence is useful; reaction times, recognition times and other psychological evidence gives important direction to studies of cognitive models. If a model does not match the psychological evidence it is flawed. Still the size of the space of possible models is huge (and probably infinite).

We are still working on models of neural behaviour but we do know a great deal. Since we have this knowledge, we can use simplified models of neurons to build neural circuits. This can inform higher level models and reduce the size of the search space.

An interesting and somewhat neglected level is the neural circuit level. The neural circuit level can bridge the gap between the neural level and the cognitive band. If neural circuits can implement a model from the cognitive band, then all of the models will be more firmly based.

While we have a good understanding of how neurons work, we have a rather poor understanding of how neural circuits work. Direct analysis of neural circuits is difficult because functioning circuits (in animals) are very complex and our scanning techniques, including fMRI, PET and electrodes are not very good at examining these circuits.

D. O. Hebb proposed a neural circuit model [9], a reverberating circuit of neural cells. Hebb called this reverberating circuit a *Cell Assembly* (CA). Some physiological evidence for CAs exists (e.g. [1,26]).

While there has been a fair amount of work in psychology [21] and neurobiology to show the existence of CAs, there has been little computational modelling of CAs [6,10,11,14,20]. The computational modelling has either been from a more abstract level than neurons [14], has been of a very limited nature focusing on associative memory [6,10,11,20].

The basic CA model is a network of neurons; each neuron connects to other neurons, which in turn may come back (directly or indirectly) to the original neuron. Some neurons (e.g. rods in the eyes) are not parts of circuits; however, physiological studies show that most neurons are in some sort of circuit [24]. That is, the human brain is made mostly of circuits of neurons. Studying these circuits is crucial to our understanding of thoughts.

While there is a considerable theoretical and physical body of evidence for CAs, what exactly do CAs do? CAs have the following (theoretical) properties.

1. CAs recognise concepts.
2. CAs are composed of neurons.
3. Neurons may exist in more than one CA.
4. CAs are in working memory if and only if they are active.
5. A CA is a long-term memory item, and is formed by change in synaptic strength.
6. A CA remains in working memory for a short period (< 5 sec.).
7. CAs interact with other CAs.

This chapter will describe an early model of CAs called CANT. This model is not completed and is not near to being the basis for a model at the deliberate act level; however it is hoped that the model will one day form the basis of such deliberate act level model.

The next section of this chapter explains the CANT model. This is followed by some experiments in categorisation with the model; this should be similar to item 1 in the above list of theoretical properties of CAs. This section is followed by a section showing how CAs can be activated quickly.

At this point this chapter shifts from a functioning program as a model to a theoretical model (we have not written the program yet). The sixth section shows how CAs can quickly resolve ambiguity. The seventh section shows how CAs relate to the other areas of interest for this book. This is followed by a section on CAs as an intermediate level model. The chapter concludes with a section on future work and a section of discussion and conclusions.

2 The CANT Model

A given instance of a CANT model will be a network of neurons. This network may have many CAs. That is, different input patterns will be classified in different ways, and different output patterns may be generated given input patterns.

2.1   Connection Strength and Connectivity

A CANT neuron has connections to other neurons, which are similar to connections inside biological neural systems. Connections are unidirectional (Figure 1). Like most neural net simulations, the connection strength may vary based on a local (Hebbian) learning rule. The connection may have positive or negative strength. Continuous activation is simulated by time steps.

The average biological neuron is activated by about 1000 other neurons, and in turn activates about 1000 other neurons [17]. Current experiments work with small numbers of neurons, so they have a smaller number of connections.

Complex models of neural behaviour (e.g. [4]) consider spread of depolarization, K^+ gating and many many other factors. The CANT model simply has neurons and connections. This simplification makes CANT computationally efficient and hopefully maintains the essential properties of the brain.

Biological neural systems are connected in a distance-biased way [24]. Each neuron is not connected to every other neuron; in the human brain, with 500 billion neurons, this would require each neuron to have 500 billion connections, and as we have seen the average neuron has approximately 1000 connections. If two neurons are closer together they are more likely to be connected. The CANT model adheres to this distance-biased connectivity.

2.2   Activation and Activation Threshold

When the neuron crosses a threshold, it fires and sends activation down each of its axons. The activation of a given neuron i at time t is:

$$h_{i_t} = \frac{h_{i_{t-1}}}{d} + \sum_{j \in V_i} w_{ij} \tag{1}$$

The current activation is the activation from the last time step divided by a decay factor d plus the new activation coming in. This new activation is the weight of the connections of all the active neurons that are connected to i. This weight is the value of the connection from neuron j to neuron i. If a neuron has fired, it loses all activation from the prior time step ($d=\infty$).

There may also be external activation. Theoretically this comes from the environment, but in these experiments it comes from neurons being directly activated. Some experiments allow neurons to spontaneously activate when they have been inactive for a long time.

A neuron fires if and only if it has enough activation to surpass the activation threshold. Each neuron has the same base activation threshold as all other neurons. So if the activation threshold is 5 and a given neuron has an activation of 4, it will not fire and thus will not propagate activation.

2.3   Decay and Fatigue

At each step, the activation of non-firing neurons decays. Of course, new activation may lead to a net gain in activation. Decay is a constant and applies only to non-firing neurons.

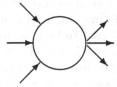

Fig. 1. Neuron with Unidirectional Connections

Active neurons fatigue. When a biological neuron is active for a long time it will fatigue and this will make it less likely to remain active. This is modelled by a fatigue factor, which increases the activation threshold. The threshold is increased by $f_c t_a$. f_c is the fatigue constant and t_a is the time that the neuron has been active. The longer that the neuron is active the larger the threshold becomes, and thus the less likely it is to remain active.

When a neuron becomes inactive, fatigue is reduced. Fatigue is $f_c t_a - t_r R_c$. Fatigue is the time active, minus the time recovering multiplied by the recovery constant. The higher the constant, the faster a neuron recovers from fatigue.

3   The Net and Cell Assemblies

How might a CA look at the cellular level? A CA should consist of a relatively large number of neurons; this could be hundreds or millions[1]. The neurons should have a large connectivity to each other; that is, each neuron should have connections to other neurons in the assembly and the strengths of those connections should be high. Each neuron does not need to be connected to all the other neurons in the CA and may be connected to neurons outside the CA. This large connectivity should lead to mutual activation.

When several neurons in the CA are activated, they should activate other neurons in the CA. When the initial neurons fatigue and cease to be active, the newly activated neurons keep the activation in the CA. The initial neurons, after recovery, may later be reactivated. Thus the CA is a reverberating circuit, and can remain active much longer than a single neuron.

If insufficient evidence is present, a number of cells will still be activated. However, it will not be enough to activate the circuit, and overall activation in the circuit will quickly decline.

Learning in the CANT model is unsupervised. While different learning rules have been tested within the CANT model, all are basically Hebbian. That is, the learning rules are based on the activation of two adjacent neurons.

---

[1] Our current thoughts are that human CAs have on the order of 10^5 neurons.

4   Experimental Results

As Wittgenstein [29] has made clear, there usually are not necessary and sufficient conditions to say an entity is an example of a concept. Generally, dogs have four legs, but three-legged dogs exist. To be a bit more simplistic, a concept is a group of features that tend to travel around together. We have the concept (and thus the word) dog because the features tend to travel around together. We often find wagging tails, fur, wet noses, four legs and other dog features together. Rarely, if ever, do all of the features coexist, but they tend to be together. This type of concept relates to Rosch's [23] prototype theory.

An instance of a concept is present when enough of its features are present. In the CANT model a CA is activated when enough features in a given feature set are present. Neuronal activation is used as a rough equivalent to a feature being present. A concept is made up of a set of features (neurons in CANT terms). So if a sufficient number of those neurons are activated, then the CA should become active. The network will "recognise" that the concept is present in the environment.

The network in this experiment used 400 neurons in a 20x20 matrix with a toroidal topology. The network is presented with input patterns. Each series of runs started from a randomly generated net and used two patterns. The patterns were exhaustive; each neuron was a member of one of the two patterns.

The most basic type of pattern divided the network in half. Pattern A consisted of the neurons 0-199 and pattern B the neurons 200-399. These patterns were entirely local. This is the Local column in Table I.

Fully interleaved patterns had pattern A having all of the even neurons, and pattern B having all of the odd neurons; this is the Interleaved column in Table I. Intermediate patterns combined these; for example pattern A consisted of neurons 0-99 and the even neurons from 100-299, pattern B consisted of neurons 300-399 and the odd neurons from 100-299. This is the Half column in Table II.

The stimulus pattern is presented for 10 cycles. The pattern consists of 20 neurons randomly chosen from the pattern type. Two different A patterns might share no common neurons.

Table 1. Network Correlations Exp. 1

	Local	Half	Interleaved
Number Runs	300	350	1500
A-A Corr.	1	.9286	.9900
B-B Corr.	1	.9492	.9917
A Self Corr.	.9970	.8821	.8621
B Self Corr.	.9734	.9562	.9386
A-B Corr.	-1	-.4973	-.9826

The measurements are Pearson's product correlation coefficient. This shows that the neurons activated when one A pattern is presented is highly correlated with a different A pattern being presented (the A-A row) and when two B patterns are presented (the B-B row). Both are measured 5 cycles after the stimulus stops being presented. Similarly, the A and B patterns are negatively correlated (the A-B row). The pattern of activation is also maintained. The A and B self correlation rows show the correlation between active neurons at cycle 5 and 20 cycles after the end of stimulus presentation.

This experiment shows that this model is capable of learning CAs over the full range of exhaustive patterns. Local patterns, interleaved patterns, and patterns which combined local groups and interleaved groups all formed CAs. These patterns are unique, persist and are reliably activated.

It takes longer to learn interleaved patterns, but they are eventually learned. This is because a distance-biased network helps localised CAs form. Distance-biasing acts as an attractor and when a localised pattern is placed into that attractor, it is easily learned. Patterns that fight the attractor can still be learned but take longer, and are less successful.

Other experiments have been done, but this experiment shows that this model is capable of recognising and categorising different types of patterns. This is the primary function of CAs. From this, more complex behaviour can be generated.

5   Quick Activation of CAs

The theory of CAs states that a CA is quickly activated [14]. This shows that slow neurons can quickly recognise an object.

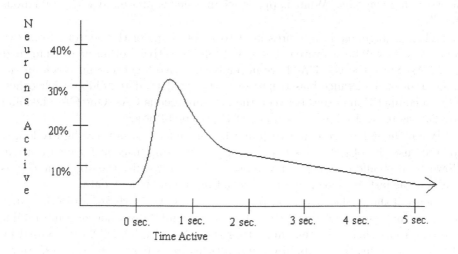

Fig. 2. The Snoopy Curve

Experiments using simple computational models of CAs shows this to be the case [11]. This model shows that activation of 10 neurons can lead to activation of the entire CA of 200 neurons in the next time step. A time step is meant to model 10 ms. of real neural activity. This particular experiment is probably not viable from a neurological standpoint as the size of the CA is very small. However, it does show how quickly a few neurons can lead to activation of many more neurons.

A real CA would probably cross several brain areas. However, it would only take a few time steps (tens to hundreds of milliseconds) to activate the entire assembly. Highly parallel activation compensates for slow neural speed. Practically, it will be easy to get large numbers of neurons activated within the speeds that are needed for normal real-time performance.

Figure 2 shows the activation curve of a CA. At time 0, a stimulus is presented to the network. Rapidly, a large percentage of the neurons in the CA become activated and this is associated with recognition. Stimulus may be removed or remain present. Some neurons in the CA are active for quite some time meaning the CA can spread information, strengthen itself (via the Hebbian learning rule) and remain in short-term memory. Note that no neuron remains active for the whole period; they become active, fatigue, recover and are reactivated by other neurons in the CA.

6   Quick Decision on Ambiguous Data

A human brain must have millions of CAs, but our model networks are much smaller. A given net may have several CAs in it, but only one or a few should be active at a given time. What happens when a net is presented with ambiguous data?

When ambiguous data is presented to a net, some of the neurons from two separate CAs will be activated. These will tend to activate other neurons in their own CAs. Since the two CAs have never been active together and reside in the same area of a distance-biased network, they will tend to inhibit each other. Thus a competition ensues between the two ambiguous CAs. One CA wins, and the data is recognised as an instance of that type of object.

When the net is presented with ambiguous data it does take slightly longer to recognise the object. This is due to the extra time needed for competition. However, it should only take a few more cycles to recognise the object, and thus is almost as fast as recognising an object from unambiguous data.

Figure 3 shows the activation pattern of two CAs. Both get an initial burst of activating neurons. Some of these neurons are inhibitory neurons and inhibit neurons from other CAs, thus inhibiting the other CA. The CA represented by the bold line loses the competition and returns to its base firing rate. Of course this is all theory. We do not know of any model of a CA that handles ambiguous data [2].

---

[2] We are currently working on such an experiment based on the CANT model and the results are promising.

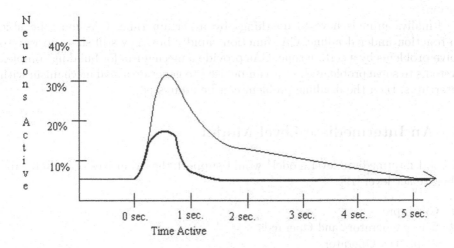

Fig. 3. Competing CA Activation Curves

7 The Other Areas

CAs categorise and they do it quickly. This chapter has described how the CANT model does this. CAs also provide many other properties that are very useful for a computational model of intelligence. They are robust, they learn in context and they enable synchronisation of neural firing. Additionally, they provide an excellent landmark for asking questions about modularity and timing.

Like most ANN architectures, CAs are robust. Loss of neurons might even strengthen a given CA. Since a CA is composed of a suite of neurons, it can handle the loss of neurons; new neurons can be added and incorporated into CAs. The dynamics of CAs recruiting new neurons and fractionating into multiple CAs is not well understood, but the process must be robust.

CAs learn in context. The CA learning mechanism is unsupervised. A given network will spontaneously create CAs based on the input that is seen. Thus all learning is in context. This may lead to unforeseen connections between concepts, but those connections are based on the input that has been presented.

CAs are self synchronising. Neurons in a CA tend to fire in a similar pattern [20]. Thus CAs actually act as a synchronising mechanism. The mechanism also leads to a number of questions about how CAs could work together. There are questions about variable binding [7]; co-operating CAs may be bound together by synchronisation. How is this done? How do CAs combine into larger structures like sequences and cognitive maps [13]?

CAs also provide a landmark for questions of modularity. Braitenberg [5] notes that the brain is largely a uniform mass of neurons. CAs span brain areas [21]. This enables cross modular co-operation, yet intra-modular specialisation.

A working CA model could explore lots of questions about modular communication.

Finally, animals have to do things by a certain time. CAs must be able to function under deadline. CAs function rapidly but they still show no way to solve problems by a certain time. CAs provide a mechanism for building complex systems to solve problems. If CAs can be used to generate a goal mechanism with interrupts, then the deadline problem can be addressed.

8   An Intermediate Level Model

A good intermediate level model[3] would support the primitives of Newell's deliberate act level [19]:

1. Get Input
2. Select Operators and Operands
3. Apply the Operator
4. Store the Results
5. Generate Output
6. Support Structures

The CANT model already has demonstrated support for input, operator and operand selection, and storing results. It holds promise for operator application, output generation and more complex structures.

The CANT model supports direct input. In biological systems, sensory organs activate neurons. In a CANT network, we could connect sensory organs directly to the neurons. The sensory system and underlying neural system are very complex and well studied, and the neural system is largely compatible with the CANT model.

Operators and operands can be selected based on current input and prior input. CA activation can be thought of as operator and operand selection; an operator and operands are selected if and only if their CAs are activated.

Results are stored by either putting them in working memory or in long-term memory. To put them in working memory an existing CA has to be activated. To put them in long-term memory, a new CA must formed. This is done by changing synaptic weights.

Output can be generated by effectors. Like biological input, biological output is done by the body. Muscles move and vocal chords vibrate. The underlying neural control mechanisms, though not entirely understood, are well studied. Again, these are compatible with CA theory and CA based controllers.

Structures are more complex for CAs to generate and have many of the questions in which we are currently interested. How do CAs build, for instance, a verb frame? Weak connections between CAs must be quickly learned via the Hebbian learning mechanism. We do not currently have a functioning computational model of this process, but this is in our medium term plans.

---

[3] a neural circuit model

Operators are more complex in a CA model. Simple operators like *store* have already been implemented. More complex operators like move forward or multiply two numbers are more difficult. These complex functions may be implemented by decomposition into primitives. Primitive operators are done by the output mechanisms and storage mechanisms described above. Still the functions need to compose these primitives via CAs. This composition is done via complex structures like sequences and cognitive maps. As before, these are learned by creating weak links to combine CA primitives.

We like Newell's division into levels and are particularly interested in Soar; thus we have focused our argument in that direction. However, there is nothing sacred about the Soar model. Newell's deliberate act primitives may turn out to be wrong. What is important is that the higher level behaviours that can be built on the deliberate act primitives can be built from CAs. If it turns out that humans build the behaviours differently, CAs should be flexible enough to account for that behaviour. If we can implement Newell's system than CAs will have implemented a strong system.

Additionally, the CANT model may support other aspects of behaviour including emotion, attention, fast learning and forgetting. Emotion can be incorporated using a pleasure/pain mechanism. This interacts particularly well with a Soar-like goal mechanism. Unappealing and dangerous CAs are associated with (connected to) pain, a pleasurable CAs to pleasure. This can direct behaviour.

Attention can be explained with a more global mechanism. CA theory, as presented so far, is entirely localist. There is no global mechanism. However, CAs are connected to other CAs. An attention area or cognitive map could be connected to all other areas. It could focus attention via a global inhibitory mechanism. This would allow only the most active CA to remain active.

Fast learning may be accomplished by reverberation, support from cognitive maps, and attention. These aids can support more neural pair coactivations and thus quick strengthening of connections.

We also forget. CAs can account for this. Forgetting is done by a change in synaptic strengths. Core concepts, CAs, are rarely forgotten (you do not forget what a dog is). This is because the CAs occasionally become active and their synaptic strength is reinforced. Connections between concepts are more likely to be forgotten because they are initially weaker and these connection are also used to connect other CAs.

Finally, the CANT model supports a blackboard like architecture [8,22]. CAs consist of neurons that are closely connected but connections to other CAs exist. This means that information from a wide range of CAs may be brought to bare on a single problem.

A well known example of a blackboard system is the Hearsay speech understanding system [22]. Hearsay consists of several subsystems. Each subsystem can communicate to other subsystems via a blackboard. A special subsystem chooses which subsystem will function next based on which has sufficient information to function. In a CANT blackboard, the architecture would decide which subsystem or subsystems would function based on enough information being available. Similarly information is exchanged via neural activation.

In traditional software systems we break the system into subsystems or into components. This facilitates engineering. In the brain, this really does not happen. Despite physiologists division into brain areas, each area is closely connected to other areas. This whole system is tightly coupled. This tight coupling may not facilitate engineering, but the brain was not engineered.

Existing CA systems are associative memory systems; this includes the experiments described in this chapter. What is needed is a model that can combine CAs in an interesting way [12].

Von der Malsberg [18] proposes dynamical connections as a variant that is much more flexible than standard CA theory. CAs are bound together via synchronous firing. This enables new concepts like *blue square* to be formed quickly and temporarily from primitive concepts like *blue* and *square*.

An unappealing alternative to this is that variable binding is done by the change of synaptic strength. This is more tenable when overlapping set coding [28] is considered. A given neuron participates in more than one CA. So, a given neuron might participate in the *blue* CA and the *square* CA. When the two CAs are coactivated, a new *blue square* CA is formed. This is unappealing because of learning time constraints and forgetting.

However, a middle ground exists. CAs are formed in the traditional way and represent a "strong force" (e.g. *square* and *blue*). Variable binding is done via synchrony and is a short term effect just like CA activation is a short term effect. If the binding is important or repeated, Hebbian learning leads to an association between CAs; this is a "weak force". These weak forces are available for sequences, hierarchies and cognitive maps of CAs. This "weak force" is facilitated by overlapping set coding. This may be similar to the SMRITI system [25].

The important point is that both synchronous firing and overlapping set coding are compatible with CAs in general and the CANT model specifically. Synchronous firing emerges from the activation of neurons [20]. Overlapping set coding also emerges from the input and learning mechanisms.

9   Future Work

The long-term goal of this work is to move from neural models of categorisation to neural models of processing. Currently, we are working on a model of categorisation and associative memory. This work duplicates, and hopefully extends, existing computational CA models.

Next we hope to model interactions of CAs. This includes competition amongst CAs, hierarchies of CAs, sequences of CAs, and maps of CAs. This will be done via overlapping set encoding and synchronous firing patterns.

Variable binding can lead to rule activation. If A and B are active, then complete the pattern and activate C. Rule activation provides the basis for traditional symbolic architectures. One popular symbolic cognitive architecture is Soar. It can be described as an expert system, with goals, operators, operator subgoaling and chunking. If all of these can be implemented in a CA based system, then CAs will have provided an intermediate level for cognitive modelling.

A system like CA-Soar would escape from the symbol grounding criticism of Fodor and Pylyshyn [3]. It would direct psycho-neurological research. It would provide a plausible model for thought and thus for real Artificial Intelligence.

Implementing this set of performances is not straight forward, but one can easily imagine that it is possible. We suspect that this CA-Soar model would be more dependent on categorisation than current Soar programs. This reflects the associative nature of the brain.

10   Discussion and Conclusion

This chapter has shown that CAs can be used to classify a wide range of patterns. It has also argued that CAs can be expanded to account for variable binding, hierarchy and cognitive maps. This functionality will enable CAs to provide the primitives for Newell's deliberate act level. Thus CAs can form a bridge between the relatively solid knowledge that we have of neural behaviour to the highly developed work we have on symbolic cognitive architectures.

One may ask why CAs have not yet been used to build a symbol system. It is likely that the recent interest in connectionist systems [27] has laid new foundations for this work. Additionally, computational speed has only recently enabled large neural models to be run.

CAs can be used to store memories that are based in reality. They can solve the symbol grounding problem. Current computer memory can store the sentence "the cat chased the mouse". When asked "what was being chased" it can respond, but it can not respond to "why was he being chased".

Traditional computer memory is great for number crunching. CA models may be faster than biological CAs but will still be slower than Von Neumann architectures. CAs however function in context and can generate statistically reasonable answers from little data. They will function better than Von Neumann architectures in AI tasks like Natural Language Processing and Decision Support Systems.

This book is derived from the Third International Workshop on Current Computational Architectures Integrating Neural Networks and Neuroscience. The key question for the workshop was *What can we learn from cognitive neuroscience and the brain for building new computational neural architectures?* In relation to this chapter, we would like to break that question into two questions:
1. *What can we learn from cognitive neuroscience and the brain for building CAs?*
2. *What can we learn from CAs for building new computational neural architectures?*

It is simple to say what the brain can tell us about CAs. The way each neuron behaves, and the way they are connected in the brain is the way CAs should behave. Any CA model will be simplified as modelling a single neuron is very complex. Perhaps we can simplify the model by simply modelling changes in neural activation in discrete steps and learning by changes of strength. This is not as robust as modelling neural activation continuously with ionic transfer between neurons, and modelling changes of strength by changes in axonal radius, myelination and geometric properties of the synaptic cleft. This simplification makes the model run faster which should enable us to model the large number of neurons needed for CAs. Still it is a simplification and, as such, may miss important data.

It is more difficult to say what CAs can tell us about building new computational neural architectures. CAs are models for concept storage. They can solve the symbol grounding problem. Perhaps the most interesting question is in what way can CAs work together.

At heart, CAs are a model for data storage. They have several advantages that could be transferred to other models. They remain active for a certain period of time. They are a model for both long and short-term memory. They can be used to choose between equally likely solutions. It is the reverberating activity that makes this model different from most existing ANN models.

Further studies to show how CAs can store more complex structures and process data are important. The class of reverberating circuits is large and we have done little in studying this class. Clearly, the brain is not just a bunch of CAs. For instance, non-local effects [15] probably are important. However, CAs are crucial and we need to understand them and how they work.

Using CAs as a basis for more complex phenomena puts us on reasonably solid ground for future exploration of how the brain works at more complex levels. This in turn leaves us with a better understanding of intelligence and thought.

References

1. Abeles, M., H. Bergman, E. Margalit, and E. Vaadia. (1993) Spatiotemporal Firing Patterns in the Frontal Cortex of Behaving Monkeys. *Journal of Neurophysiology* 70(4):1629-38
2. Anderson, John R. (1983). *The Architecture of Cognition* Cambridge, MA: Harvard University Press.
3. Bechtel, W. and A. Abrhamsen. (1991). *Connectionism and the Mind* Blackwell Publishers.
4. Bower, J. and D. Beeman (1994). *The Book of Genesis* Springer-Verlag.
5. Bratenberg, V. (1989). Some Arguments for a Theory of Cell Assemblies in the Cerebral Cortex In *Neural Connections, Mental Computation* Nadel, Cooper, Culicover and Harnish eds. MIT Press.
6. Fransen, E., A. Lanser, and H. Liljenstrom. (1992) A Model of Cortical Associative Memory Based on Hebbian Cell Assemblies. In *Connectionism in a Broad Perspective.* Niklasson, L. and M. Boden eds. Ellis Horwood Springer-Verlag pp. 165-71
7. Fujii, Hiroshi, I. Hiroyuke, K. Aihara, N. Ichinose and M. Tsukada. 1996. Dynamical Cell Assembly Hypothesis - Theoretical Possibility of Spatio-temporal Coding in the Cortex. In Neural Networks Vol 9:8 pp. 1303-50

8. Hayes-Roth, B. (1985) A Blackboard Architecture for Control. In *Artificial Intelligence* 26 pp. 251-321.
9. Hebb, D.O. (1949) The Organization of Behavior. John Wiley and Sons, New York.
10. Hetherington, P. A., and M. Shapiro. (1993) Simulating Hebb cell assemblies: the necessity for partitioned dendritic trees and a post-net-pre LTD rule. *Network: Computation in Neural Systems* 4:135-153
11. Huyck, Christian R. (2000). Modelling Cell Assemblies. In *Proceedings of the International Conference of Artificial Intelligence* Las Vegas, Nevada
12. Ivancich, J. E., C. Huyck and S. Kaplan. 1999. Cell Assemblies as Building Blocks of Larger Cognitive Structures. In *Behaviour and Brain Science*. Cambridge University Press
13. Kaplan, S., M. Weaver, and R. French (1990) Active Symbols and Internal Models: Towards a Cognitive Connectionism. In *AI and Society* 4:51-71
14. Kaplan, S., M. Sontag, and E. Chown. (1991) Tracing recurrent activity in cognitive elements(TRACE): A model of temporal dynamics in a cell assembly. *Connection Science* 3:179-206
15. Katz, P. S. (1999) What are we talking about? Modes of neuronal communication. In *Beyond Neurotransmission: Neuromodulation and its Importance for Information Processing* Paul Katz ed. Oxford Press. pp. 3-30
16. Laird, John E., A. Newell, and P. Rosenbloom. (1987) Soar: An Architecture for General Cognition. In *Artificial Intelligence* 33,1
17. Lippmann, R. P. (1987) An Introduction to Computing with Neural Nets. *IEEE ASSP Magazine* April 1987.
18. Malsberg, C. von der. (1986) Am I Thinking Assemblies? In *Brain Theory*. Palm, G. and A. Aersten eds. Springer-Verlag pp. 161-76
19. Newell, Allen. 1990. *Unified Theories of Cognition* Cambridge, MA: Harvard University Press.
20. Palm, G. (2000) Robust identification of visual shapes enhanced by synchronisation of cortical activity. In *EmerNet: Third International Workshop on Current Computational Architectures Integrating Neural Networks and Neuroscience*. Wermter, S. ed.
21. Pulvermuller, F. (1999) Words in the brain's language. In *Behavioral and Brain Sciences* 22 pp. 253-336.
22. Reddy, R., L. Erman, R. Fennell, and R. Neely. (1976) The HEARSAY speech understanding system: An example of the recognition process. In *EIII Transactions on Computers* C-25: 427-431
23. Rosch, Eleanor and C. Mervis. 1975. Family Resemblances: Studies in the Internal Structure of Categories. In *Cognitive Psychology* 7 pp. 573-605.
24. Schuz, A. (1995) Neuroanatomy in a Computational Perspective. In *The Handbook of Brain Theory and Neural Networks*. Arbib, M. ed. MIT Press pp. 622-626.
25. Shastri, L. (2000) SMRITI: a computational model of episodic memory formation inspired by the hippocampal system. In *EmerNet: Third International Workshop on Current Computational Architectures Integrating Neural Networks and Neuroscience*. Wermter, S. ed.
26. Spatz, H. (1996) Hebb's concept of synaptic plasticity of neuronal cell assemblies. In *Behavioural Brain Research* 78 pp. 3-7.
27. Sun, R. (1995) Robust reasoning; integrating rule-based and similarity-based reasoning. In *Artificial Intelligence* 75 pp. 241-95.
28. Wickelgren, W. A. (1999) Webs, Cell Assemblies, and Chunking in Neural Nets. *Canadian Journal of Experimental Psychology* 53:1 pp. 118-131
29. Wittgenstein L. (1953) *Philosophical Investigations* Blackwell, Oxford.

Modelling Higher Cognitive Functions with Hebbian Cell Assemblies

Marcin Chady

The University of Birmingham
M.Chady@cs.bham.ac.uk

1 Introduction

Modelling higher cognitive behaviour, namely symbol processing and rule inference, in the connectionist framework is inherently problematic, especially if one strives for biological realism. Problems like compositionality and variable binding, as well as the selection of a suitable representation scheme, are a major obstacle to achieving the kind of intelligent behaviour which would extend beyond simple pattern recognition. Although there are connectionist systems which are capable of advanced inferencing (cf. Shastri and Ajjanagadde [8], or Barnden [3]), they always compromise their flexibility and, most importantly, their capability to learn. All connections in such systems are fixed, having been carefully prearranged by the designer.

Other approaches use backpropagation as the learning mechanism, and rely on arbitrary precision of synaptic efficacies and activity levels (e.g. Chalmers [4], Elman [6] and Chrisman [5]). However, it is now widely accepted that these mechanisms are highly unrealistic in biological systems. Moreover, systems relying on fine-tuning of synapses are sensitive to noise, sacrificing one of the most important properties of connectionist processing, that is graceful degradation.

In this chapter we present ongoing work aimed at developing a self-organising system which is able to learn from a continuous stream of input data and make generalising predictions based on its previous experience. In particular, it addresses the problem of compositionality and context-sensitive learning. We avoid using backpropagation and arbitrary precision of computation. Instead, we use Hebbian learning and a limited number of states to describe each cell and its synapses. Our approach is founded on the basic concept of an auto-associative network, and extends it to consider a large system of such networks interacting with each other. A natural implementation of this model is a network distributed over a two-dimensional plane with local excitation and inhibition. Such system has richer dynamics than a standard globally-connected network. In particular, it is capable of context-sensitive behaviour which, hopefully, can be employed to produce certain forms of structure-sensitive processing in a connectionist system.

2 The Model

The approach presented here is based on the generic concept of associative memory in the form exemplified by the Hopfield network [7]. A large system of such

S. Wermter et al. (Eds.): Emergent Neural Computational Architectures, LNAI 2036, pp. 398–406, 2001.
© Springer-Verlag Berlin Heidelberg 2001

networks interacting with each other is considered. The topology of the system is inspired by the notion of *cortical columns* postulated to exist in the cortex [2], i.e. the networks are arranged in a two-dimensional lattice and the interaction between them is heavily influenced by the nearest neighbourhood (see Figure 1).

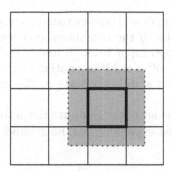

Fig. 1. A lattice of interacting associative networks. The region drawn with the thick line represents the area covered by one of the networks, while the grey region the area it receives its input from.

The interaction with the neighbourhood requires that the standard Hopfield network is extended to include hetero-association of the output with the input received from the neighbourhood:

$$\mathbf{h}' = \tau(\mathbf{W}_H \mathbf{h} + \mathbf{W}_E \mathbf{e})$$

where $\mathbf{h}$ is the internal state of the network (i.e. state of the units inside the thick rectangle in Figure 1), $\mathbf{W}_H$ is the auto-associative matrix of connection weights between the internal units, $\mathbf{e}$ is the state of the neighbouring units with $\mathbf{W}_E$ being the corresponding hetero-associative synaptic matrix; and $\tau : \mathbb{R}^n \to \{0, 1\}^n$ is a thresholding function. The threshold is dynamic and serves the role of local inhibition. It allows to fire only a fixed proportion c of units in a local region. The parameter c is kept low ($c ¡ 0.1$) for two reasons: first, to reflect the typically low activity level in the biological cortex; second, to reduce the interaction between different patterns of activity, and hence eliminate the occurrence of spurious attractors (i.e. false memory traces) as well as increase memory capacity [1].

To allow for the interaction with the environment, a third component is added to the above formula, representing the external input vector $\mathbf{p}$ and the corresponding synaptic matrix $\mathbf{W}_P$:

$$\mathbf{h}' = \tau(\mathbf{W}_H \mathbf{h} + \mathbf{W}_E \mathbf{e} + \mathbf{W}_P \mathbf{p})$$

The input is global, i.e. $\mathbf{p}$ is the same for each network in the lattice, although $\mathbf{W}_P$ is not.

The networks use a simple form of Hebbian learning adapted for sparsely-coded activity patterns, where the efficacy of each synapse in the combined synaptic matrix $\mathbf{W} = \mathbf{W}_H \mid \mathbf{W}_E \mid \mathbf{W}_P$ is given by the formula [1]

$$w_{ij} = \frac{1}{c(1-c)N} \sum_{\mu=1}^{n} (x_j^\mu - c)(h_i^\mu - c)$$

where $\mathbf{x}^\mu$ and $\mathbf{h}^\mu$ are the input and output patterns respectively, N is the number of input units (i.e. the size of the neighbourhood including the internal units, plus the size of the external input vector), and n the number of pattern pairs. Note that $\mathbf{x}^\mu = \mathbf{h}^\mu \mid \mathbf{e}^\mu \mid \mathbf{p}^\mu$ is a concatenation of the internal state $\mathbf{h}^\mu$ (which is also the previous output state), the state of the neighbourhood $\mathbf{e}^\mu$, and the external input vector $\mathbf{p}^\mu$.

For a given input $\mathbf{x}$ the networks compute the activation vector $\mathbf{a} = \mathbf{W}\mathbf{x}$, i.e. $a_i = \sum_{j=1}^{N} w_{ij}x_j$. This last expression can be expanded to

$$a_i = \sum_{j=1}^{N} \left(\frac{1}{c(1-c)N} \sum_{\mu=1}^{n} (x_j^\mu - c)(h_i^\mu - c) \right) x_j$$

$$= \frac{1}{c(1-c)N} \sum_{\mu=1}^{n} (h_i^\mu - c)m(\mathbf{x}^\mu, \mathbf{x})$$

where $m(\mathbf{x}^\mu, \mathbf{x})$ is the so called *overlap* between $\mathbf{x}^\mu$ and $\mathbf{x}$. It is closely related to the inner product of the two vectors

$$m(\mathbf{x}^\mu, \mathbf{x}) = \sum_{j=1}^{N} (x_j^\mu - c)x_j = \mathbf{x}^\mu \mathbf{x} - c \sum_{j=1}^{N} x_j$$

Put this way, the activation vector $\mathbf{a}$ is the interpolation of all output patterns $\mathbf{h}^\mu$ stored in the network, weighed according to the degree of overlap between the input vector $\mathbf{x}$ and the corresponding input patterns $\mathbf{x}^\mu$.

$$\mathbf{a} = \sum_{\mu=1}^{n} (\mathbf{h}^\mu - c)m(\mathbf{x}^\mu, \mathbf{x})$$

In our system it is assumed that the output patterns do not overlap, and that the threshold applied to $\mathbf{a}$ is such that the actual output vector $\mathbf{h}^\eta$ is the pattern for which the corresponding input patterns have the largest overlap with the input:

$$\sum_{\kappa:h^\kappa=h^\eta} m(\mathbf{x}^\kappa, \mathbf{x}) = \max_\mu \sum_{\kappa:h^\kappa=h^\mu} m(\mathbf{x}^\kappa, \mathbf{x})$$

Here, $\kappa : h^\kappa = h^\eta$ is read as "such κ that $h^\kappa = h^\eta$."

Note that computation of the output pattern only requires information about the patterns stored in the network. There is no need to keep the individual synaptic weights and perform expensive summations over large matrices, as long as we keep the history of the input-output patterns. Furthermore, if we take advantage of the dynamics in sparsely-coded networks, which makes the interference between patterns negligible, we can treat them as independent entities represented by symbolic values. Thus, instead of treating $\mathbf{x}$ as a vector of N binary values representing states of the input units, we can decompose it into a tuple of k symbolic values $(x_1, ..., x_k)$, each representing the pattern of activity in one of the input sources. For example, in the context of Figure 1 these values will be $(h, e_{NW}, e_N, e_{NE}, e_W, e_E, e_{SW}, e_S, e_{SE}, p)$, where h represents the internal state of the network, e_{XX} the corresponding neighbouring field, and p the external input vector.

In order for the symbolic representation to fully describe the original state of the network, we need to accompany it with a magnitude vector $\mathbf{g} = (g_1, ..., g_k)$ corresponding to the number of bits occupied by each of the vector's segments. Having this information, the overlap between two vectors in "symbolic" representation will be:

$$m(\mathbf{a}, \mathbf{b}) = \sum_{i=1}^{k} \rho(a_i, b_i) g_i - kc$$

where

$$\rho(a, b) = \begin{cases} 1 \text{ if } a = b \\ 0 \text{ otherwise} \end{cases}$$

In other words, the overlap between two symbolic vectors $\mathbf{a}$ and $\mathbf{b}$ is the sum of magnitudes of the segments in which $\mathbf{a}$ and $\mathbf{b}$ agree, minus kc. The latter is a constant value, which has no effect on the comparison between different overlaps, so it can be safely ignored. Without loss of generality, it is also useful to normalise all overlap measures by expressing g as the proportion of the total vector length, so that $\sum_i^k g_i = 1$. Hence, in the reminder of this paper the overlap will be defined as

$$m(\mathbf{a}, \mathbf{b}) = \sum_{i=1}^{k} \rho(a_i, b_i) g_i$$

and g_i will represent the magnitude of segment i in relation to the whole vector.

Associative systems of the kind discussed here usually exhibit high inertia, i.e. the external input must be very strong to have an effect on the network's state. Strong input, however, tends to erase the network's state completely, so a subtler mechanism is required to ensure sufficient sensitivity to external input and to retain context-sensitivity. In our model this is achieved by introducing fatigue which prevents the network from settling in any global minimum for too long. Fatigue is applied to the value of the overlap between the input pattern $\mathbf{x}$ and patterns $\mathbf{x}^\kappa$ associated with the recently active pattern $\mathbf{h}^\eta$

$$\mathbf{a} = \sum_{\mu=1}^{n} (\mathbf{h}^\mu - c) m(\mathbf{x}^\mu, \mathbf{x}) - f \sum_{\kappa: h^\kappa = h^\mu} (\mathbf{h}^\kappa - c) m(\mathbf{x}^\kappa, \mathbf{x})$$

where $f \in [0, 1]$ is the fatigue factor. In the original connectionist network this mechanism corresponds to the weakening of the response of the output units to pattern $\mathbf{x}^\eta$. In other words, as the active units in the output pattern $\mathbf{h}^\eta$ continue to fire, they become less and less responsive. The value of f will grow asymptotically from 0 to 1, eventually leading to the situation when

$$\sum_{\kappa:h^\kappa=h^\upsilon} m(\mathbf{x}^\kappa, \mathbf{x}) = \max_{\mu \neq \eta} \sum_{\kappa:h^\kappa=h^\mu} m(\mathbf{x}^\kappa, \mathbf{x})$$

where h^υ is the next output vector becoming active in the network. f is reset to 0 when η changes to υ.

In view of the above, learning entails simply adding every encountered pair of vectors $(\mathbf{x}, \mathbf{h})$ to the list of memory traces. However, in a practical application, such approach would give an unfair advantage to patterns in which the system settles for longer periods of time, leading to catastrophic interference. Therefore, a refined learning rule is used, by which the new trace $(\mathbf{x}, \mathbf{h}')$ is remembered only if the postsynaptic cells change state, i.e. when $\mathbf{h}' \neq \mathbf{h}$.

There also remains the question of learning representations for novel input. For this we need to extend the underlying connectionist model by lateral inhibition, effectively making it a competitive, or feature extracting network. At the symbolic level we model this behaviour by a fixed discrimination threshold γ. If the highest activation is below the threshold, i.e. $\max_\mu \sum_{\kappa:h^\kappa=h^\mu} m(\mathbf{x}^\kappa, \mathbf{x}) < \gamma$, the network is assumed to generate a new representation, and $\mathbf{h}'$ is assigned a newly generated random pattern, uniquely identifying the new input.

Another aspect needing attention within our framework is the issue of synchronousness. Namely, once the state of all units in a network is described by a single value, it is no longer possible to perform network update in a fully asynchronous manner, which would be preferable when biological realism is considered. In our approach, we assume that the transient states which a connectionist network goes through on the way from one fixed point to another, are insignificant to the overall result. This assumption is especially safe in sparsely-coded networks, where spurious attractors are eliminated. However, because settling in a fixed point is a local process, its timing is independent from other networks in the lattice. Therefore, a compromise was arrived at, in which individual network states are discrete and their transitions are synchronous, while the state change of the whole lattice is performed in an asynchronous manner, i.e. only one network can change state at any given time.

3   Context-Sensitivity and the Word-Prediction Task

Let us now demonstrate some of the properties of the model described above. In particular, we are going to see that it is capable of context-sensitive processing, which is the first step towards achieving temporal compositionality and systematicity. The demonstration is based on a simple walk-through example.

Consider a minimal system of two networks A and B, as shown in Figure 2. Each network receives input from its peer, its internal state, and the global external input, so the state of the system is described by the vector $\mathbf{x} = (p, a, b)$,

Fig. 2. A minimal system of networks used in the example.

where p is the external input, and a and b are the internal states of networks A and B, respectively. The magnitude vector $\mathbf{g}$ for this system is $(0.5, 0.25, 0.25)$, so that half of each network's input consists of the external input, a quarter is its own state, and a quarter the peer's state. γ is set to 0.6. Let us assume that the system is in the resting state (p_0, a_0, b_0), which is also the initial memory trace in both networks A and B.

We want to present this system with a sequence of three different symbols (activity patterns) $p_1 p_2 p_3$, and expect the system to recall it later, after specifying the initial symbol p_1 only. We also want the system to learn the reverse sequence consisting of the same symbols, i.e. $p_3 p_2 p_1$. Context-sensitivity will be demonstrated if, after the presentation of any given symbol in the first sequence, the system's internal state is different than after the presentation of the same symbol in the second sequence. This is equivalent to a cognitive system being able to distinguish between, e.g. "John loves Mary" and "Mary loves John".

3.1 Training

Assuming that the system is in state (p_0, a_0, b_0), we change the external input from p_0 to p_1. The overlap between the the new state (p_1, a_0, b_0) and the memory trace (p_0, a_0, b_0) is $m[(p_1, a_0, b_0), (p_0, a_0, b_0)] = 0.5 < \gamma$, so the learning rule is invoked asynchronously in each of the networks. The order in which the networks change their state is irrelevant to our consideration, so for clarity, let us assume that network A always changes state first. A new pattern a_1 is generated in network A, and a the new trace (p_1, a_1, b_0) is added to its memory. A similar process happens in network B, except here the trace remembered is (p_1, a_1, b_1). A similar sequence of state changes happens when p_2 and p_3 are presented to the network. They are summarised in Table 1.

Table 1. Summary of state transitions for the input sequence $p_1 p_2 p_3$.

Time	Input	Current State	Updated Network	New State	New Trace in the Updated Network
t_1	p_1	$a_0 b_0$	A	$a_1 b_0$	(p_1, a_1, b_0)
t_2		$a_1 b_0$	B	$a_1 b_1$	(p_1, a_1, b_1)
t_3	p_2	$a_1 b_1$	A	$a_2 b_1$	(p_2, a_2, b_1)
t_4		$a_2 b_1$	B	$a_2 b_2$	(p_2, a_2, b_2)
t_5	p_3	$a_2 b_2$	A	$a_3 b_2$	(p_3, a_3, b_2)
t_6		$a_3 b_2$	B	$a_3 b_3$	(p_3, a_3, b_3)

Now, starting from the resting state (p_0, a_0, b_0) again, we present the network with the sequence $p_3p_2p_1$. Again, after the presentation of p_3 the highest two overlaps in network A are $m[(p_3, a_0, b_0), (p_0, a_0, b_0)] = 0.5$ and $m[(p_3, a_0, b_0), (p_3, a_3, b_2)] = 0.5$. Both are below γ, so a new internal state a_4 is generated. Network B will now have only one highest overlap equal to $m[(p_3, a_4, b_0), (p_3, a_3, b_3)] = 0.5$, so b_4 will too be generated. Table 2 lists the state transitions. Note how the

Table 2. Summary of state transitions for the input sequence $p_3p_2p_1$.

Time	Input	Current State	Updated Network	New State	New Trace in the Updated Network
t_1	p_3	a_0b_0	A	a_4b_0	(p_3, a_4, b_0)
t_2		a_4b_0	B	a_4b_4	(p_3, a_4, b_4)
t_3	p_2	a_4b_4	A	a_5b_4	(p_2, a_5, b_4)
t_4		a_5b_4	B	a_5b_5	(p_2, a_5, b_5)
t_5	p_1	a_5b_5	A	a_6b_5	(p_1, a_6, b_5)
t_6		a_6b_5	B	a_6b_6	(p_1, a_6, b_6)

internal network's state is different for the two sequences, demonstrating the fact that the system treats the symbols differently depending on the situation in which they occur.

3.2 Recall

At this point the memory traces stored in networks A and B are as shown in Table 3. Now we want to see the recall of the trained sequences from the initial symbols. Starting from the resting state (p_0, a_0, b_0), we present pattern p_1 to the system. In network A, the highest overlap will be $m[(p_1, a_0, b_0), (p_1, a_1, b_0)] = 0.75 > \gamma$, so the network will change state to a_1. In network B $m[(p_1, a_1, b_0), (p_1, a_1, b_1)] = 0.75 > \gamma$, and the network will change state to b_1.

At this point, we remove the external input signal, which is equivalent to supplying a null input vector $(0, ..., 0)$ in the underlying binary representation. In network A, the highest overlaps will be $m[(0, a_1, b_1), (p_1, a_1, b_0)] = m[(0, a_1, b_1),$

Table 3. Memory traces in networks A and B after training.

Network A	Network B
(p_0, a_0, b_0)	(p_0, a_0, b_0)
(p_1, a_1, b_0)	(p_1, a_1, b_1)
(p_2, a_2, b_1)	(p_2, a_2, b_2)
(p_3, a_3, b_2)	(p_3, a_3, b_3)
(p_3, a_4, b_0)	(p_3, a_4, b_4)
(p_2, a_5, b_4)	(p_2, a_5, b_5)
(p_1, a_6, b_5)	(p_1, a_6, b_6)

$(p_2, a_2, b_1)] = 0.25$. For the purpose of pure recall we do not apply the learning rule, but watch what the network comes up with when left to its own devices. We observe that patterns a_1 and a_2 are in competition, but since fatigue will be affecting a_1, a_2 will eventually win. A similar process will take place in network B, leading to the overall state $(0, a_2, b_2)$ corresponding to the input pattern p_2. From there, the situation will repeat itself, and eventually the system will arrive at state $(0, a_3, b_3)$ which corresponds to input pattern p_3.

A similar process will take place when, starting from (p_0, a_0, b_0), the system is presented with pattern p_3 and then 0. The system's state will move to $(0, a_4, b_4)$, then $(0, a_5, b_5)$, and then finally to $(0, a_6, b_6)$, which corresponds to the input sequence $p_3 p_2 p_1$.

4   Discussion

The example given above demonstrates the general principles of the model's operation. In particular, it shows that, thanks to the ability to develop internal representations, a system of the kind discussed in this report is capable of remembering sequences of symbols. An important observation made here is that there exists a level of abstraction which permits a relatively accurate analysis of the model without considering individual units and synapses. In fact, the larger-scale components of the system (i.e. the networks A and B in the example) can be treated as single processing elements operating on symbolic values. The construction of the elements does not compromise the biological plausibility of the model, since it relies purely on Hebbian learning (triggered by postsynaptic change) and competitive inhibition. In fact, the level of abstraction allows the study of associative systems without committing oneself to a particular neurological paradigm, whether it is based on spikes or the average cell activity.

Equally important is the ease and efficiency with which symbolic operations can be implemented on a computer. Larger and more sophisticated systems can be studied this way, giving an opportunity to tackle more processing-intensive tasks.

References

1. Daniel J. Amit. *Modeling brain function : the world of attractor neural networks.* Cambridge University Press, 1989.
2. Michael A. Arbib, Péter Érdi, and János Szentágothai. *Neural Organization: structure, function, and dynamics.* The MIT Press, Cambridge, MA, 1998.
3. John A. Barnden. Encoding complex symbolic data structures with some unusual connectionist techniques. In John A. Barnden and J. B. Pollack, editors, *High Level Connectionist Models*, volume 1 of *Advances in Connectionist and Neural Computation Theory.* Ablex, Norwood, N.J., 1991.
4. David J. Chalmers. Syntactic transformations on distributed representations. *Connection Science*, 2:53–62, 1990.
5. Lonnie Chrisman. Learning recursive distributed representations for holistic computation. *Connection Science*, 3(4):345–366, 1991.

6. Jeffrey L. Elman. Distributed representations, simple recurrent networks, and grammatical structure. *Machine Learning*, 7:195–225, 1991.
7. J. J. Hopfield. Neural networks and physical systems with emergent collective computational abilities. In *Proceedings of the National Academy of Sciences*, volume 79, pages 2554–2558, 1982.
8. Lokendra Shastri and Venkat Ajjanagadde. From simple associations to systematic reasoning: A connectionist representation of rules, variables and dynamic bindings using temporal synchrony. *Behavioral and Brain Sciences*, 16:417–494, 1993.

Spiking Associative Memory and Scene Segmentation by Synchronization of Cortical Activity

Andreas Knoblauch and Günther Palm

Abteilung Neuroinformatik, Fakultät für Informatik, Universität Ulm,
Oberer Eselsberg, D-89069 Ulm, Germany
Tel: (0049)-731-50-24151; FAX: (0049)-731-50-24156;
(knoblauch,palm)@neuro.informatik.uni-ulm.de

Abstract. For the recognition of objects there are a number of computational requirements that go beyond the detection of simple geometric features like oriented lines. When there are several partially occluded objects present in a visual scene one has to have an internal knowledge about the object to be identified, e.g. using associative memories.
We have studied the bidirectional dynamical interaction of two areas, where the lower area is modelled to match area V1 in greater detail and the higher area uses Hebbian learning to form an associative memory for a number of geometric shapes. Both areas are modelled with simple spiking neuron models, and questions of "binding" by spike-synchronisation and of the effects of Hebbian learning in various synaptic connections (including the long-range cortico-cortical projections) are studied.
Presenting a superposition of three stimulus objects corresponding to learned assemblies, we found generally two states of activity: (i) relatively slow and unordered activity, synchronized only within small regions, and (ii) faster oscillations, synchronized over larger regions. The neuron groups representing one stimulus tended to be simultaneously in either the slow or the fast state. At each particular time, only one assembly was found to be in the fast state. Activation of the three assemblies switched within a few hundred milliseconds.

Keywords: associative memory, cell assemblies, visual cortex, oscillations, synchronization, attention, binding problem, scene segmentation

1 Introduction

In biological neural systems information is processed in a parallel and distributed manner where different aspects of a single entity are represented at different locations using population codes. This may lead to the so-called *binding problem* [20], if a natural scene or situation contains several such entities at the same time. As a solution, the temporal correlation hypothesis was suggested, where coding is extended to the temporal domain [20,6,30]. More exactly, distributed sets of neurons signalling various elementary features of objects as a spatial activation pattern synchronize their spikes when they respond to the same object,

S. Wermter et al. (Eds.): Emergent Neural Computational Architectures, LNAI 2036, pp. 407–427, 2001.
© Springer-Verlag Berlin Heidelberg 2001

and are desynchronized when they respond to different objects. This idea goes back to the Hebbian notion of cell assemblies [11] where neurons show that their stimuli belong together in the outside world by spiking together and Hebbian synapses between two neurons increase their positive connection strength when they spike together. Thus "belonging together" implies "firing together" implies "wiring together". This mechanism leads to the formation of cell assemblies [24] and it should show up in the correlation between the spikes generated by two or more neurons (cf. [2,1]). Ubiquitous experimental findings of synchronized oscillations in the gamma range (30-60 Hz) has lead to some extensions of this idea that use these fast oscillations as a reference clock for representing different entities at different phases of the oscillation period (e.g. [28]). But there exists only little experimental evidence for this kind of phase coding scheme. Moreover, experiments using several stimuli simultaneously revealed evidence against phase-coding. Recordings from two different neuron groups representing two different entities (e.g. two moving bars in opposite direction) showed consistently uncorrelated activity for the two neuron groups leading to flat correlograms [8, 18], while phase-coding models would predict modulated correlograms with non-zero time lag of the central peak. In this work we try to reconciliate the original temporal correlation hypothesis with experimental results.

First, we develop a new model of an associative memory [37,23] which can be interpreted as one or several "columns" of a "higher" non-sensory cortical area. We address the question how and where synchrony could be detected, i.e. how can the synchronized activation of two cell groups (presumably within 4-5ms) be read out by cortical circuits despite relatively high membrane constants (presumably within 10-20ms) and the presence of ubiquitous noise. For our model we demonstrate that by the interaction between a "Hebbian" excitatory and a particularly constructed inhibitory population, we obtain the capability of autonomous threshold control and fast pattern separation of associative memory patterns. Interestingly, we were able to derive an efficient technical version of an associative memory from our biological model [16] which is capable of completing a retrieval for n neurons in $O(\log^2 n)$ steps in parallel using *sparse coding* [23, 25].

In a second step this model is incorporated into a larger model of the orientation selective subsystem of the primary visual pathway that is capable of reproducing qualitatively most of the findings of more detailed models [38,31] which have been simulated in a non-oscillatory parameter regime (see [34] for analysis of transitions between oscillatory and non-oscillatory regimes).

In this model the primary visual area P is reciprocally connected to a central visual area C modelled as an associative memory containing representations for all the used stimuli (as in the unidirectional model of [35]). Fast oscillatory activity in P, synchronized over larger distances, is specifically induced by the reciprocal connection when the stimulus assembly in C gets activated. Using three stimuli at the same time, we found synchronized activity within neuron groups representing the same stimulus leading to modulated correlograms with a central peak, while activity of neuron groups representing different stimuli was uncorrelated with flat correlograms, as in the experiments.

Finally, we discuss the results of our simulations in the context of possible alternative interpretations of the temporal correlation hypothesis. We suggest that *fast* oscillatory or fluctuating activity may improve pattern retrieval and separation (cf. [34]), while feature binding may be solved on a slower time scale through a kind of self-generated attention switching as an emergent network property of our model.

2   Spiking Associative Memory

A variety of associative memory models has been introduced to demonstrate the capability of neural networks to learn and retrieve patterns (e.g. [37,23,13, 14]). Often these models use continuous variables that are interpreted as firing rates, and they exhibit some problems when addressed with a superposition of several patterns. Normally, this will result in an possibly incomplete activation of a superposition of stored patterns.

Wennekers and Palm demonstrated the pattern separation capability of a spiking variant of the Willshaw model [35]. They used only two populations of neurons, one excitatory spiking population that was autoassociatively connected according to the clipped Hebbian learning rule [11,37,23], and a single gradual inhibitory interneuron reciprocally coupled with the excitatory population. Connections from an input population coupled to the excitatory population and initiated a retrieval. The function of the inhibitory population was to control excitation so that only one of the addressed populations got activated.

However, this model was faced with two problems. First, it only works in a very restricted activity range. In particular, activation may explode when there are too many additional active elements in the address pattern, or when a superposition of several incomplete patterns is used as an address. Second, the temporal dimension was not really considered: introducing realistic synaptic and axonal delays, the pattern separation capability may vanish, since the indirect activation of inhibition is too slow. In the following we present a more elaborated model closely related to the algorithm used in the classical Willshaw model [37, 23] that overcomes these problems.

In the autoassociative Willshaw model, stored patterns are represented by fully connected subsets of neurons, also called assemblies. When addressing with a certain pattern, the result is that every neuron i can be assigned a counter value $c^{(i)}$ usually interpreted as membrane potential corresponding to the number of connected address neurons. The result of the retrieval is the set of neurons with counter values exceeding a global threshold Θ. For threshold setting, two strategies are used: (i) Either one knows that the address pattern is an incomplete version of a stored pattern, then the threshold equals the number of active units in the address pattern and we obtain all neurons of the stored pattern, plus eventually some false ones. (ii) Or when nothing is known about the address pattern, one has to adjust the threshold step by step until the number of active units in the resulting pattern has a desired value (cf. [36]). Note that the second strategy constrains the pattern sizes to be about the same for all patterns.

Here we want to consider a realisation of the associative memory by spiking neurons. To this end we model the membrane potential $x^{(i)}$ of each neuron as a linear differential equation that is driven by the synaptic inputs or equivalently by the counter values split up into three categories,

$$\frac{d}{dt}x^{(i)} = A \cdot c_H^{(i)} + B \cdot \left(c_A^{(i)} - \alpha \cdot c_\Sigma\right), \quad A, B > 0, \quad 0 < \alpha \le 1 \quad (1)$$

where $c_H^{(i)}$ is the number of spikes received (possibly heteroassociatively) from the address area, $c_A^{(i)}$ the number of spikes received (autoassociatively) from within the area, and c_Σ the total number of spikes that occurred in the addressed area. Parameters A and B determine strengths of inter-areal and intra-areal inputs respectively. Parameter α determines the so called *separation strength*. E.g. a separation strength of $\alpha = 0.9$ means that the autoassociative feedback is only excitatory for neurons connected to more than 90 percent of the neurons that have fired in the addressed area. For all other neurons, the autoassociative feedback is inhibitory.

To implement something analogous to the first threshold strategy described above, dx/dt should be positive for $c_A \approx c_\Sigma$ and negative for $c_A \ll c_\Sigma$. This is achieved by choosing $\alpha \approx 1$ and making the (intra-areal) autoassociative feedback much stronger than the (extra-areal) heteroassociative. This is also justified for biological models, since half of the synapses on a cortical pyramidal cell are made by axons belonging to local (autoassociatively coupled) neurons of the same area, while the other half results from remote areas [5]. Experiments revealed that a local cortical neuron group is connected to five to eight other cortical patches, what suggests that the local autoassociative feedback may be about five to eight times stronger than input from one extra-areal neuron group.

Now the choice of threshold Θ becomes independent of the size of the address pattern. This is because as long as all neurons that exhibited a spike belong to one assembly, for all other neurons of this assembly we have $c_A = c_\Sigma$, and therefore their potentials continue to increase towards the threshold. Further, miscarried retrievals (e.g. if neurons of different assemblies have already fired) exhibit a tendency to break down at an early time, because the condition $c_A \approx c_\Sigma$ is no longer fulfilled for any neuron. Thus we obtain autonomous threshold control and autonomous detection of failed retrievals.

For a direct biological implementation of the spike counter model, consider the network structure of Figure 1. The network consists of two areas. The addressing area R makes input connections in the addressed area C, which consists of one excitatory population C, and two inhibitory populations C^S (*separating* inhibition) and C^T (*terminating* inhibition). Note that only population C^S receives input from the addressing area.

Considering only the two populations C and C^S, the idea is to make C^S behave as a twin-population of C. Therefore, the input connection $R \rightarrow C^S$ is the same as $R \rightarrow C$, the heteroassociative connection $C \rightarrow C^S$ is analog to

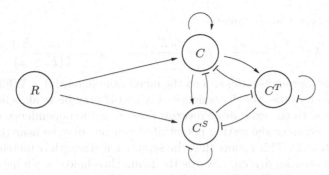

Fig. 1. Network structure of the spiking associative memory. Excitatory synaptic connections are denoted by →, inhibitory ones by ⊣.

the autoassociative connection of C, and the two unspecific inhibitory feedbacks $C^S \to C^S$ and $C^S \to C$ are also the same. The result is, that in C and C^S always corresponding neuron populations get activated. So, information about the spike counter c_A is conveyed through the excitatory autoassociative connections of population C, while information about the spike counter c_Σ can be conveyed through the unspecific (e.g. full) connections from population C^S onto C. This property is even independent of many features of the neuron model (including synaptic and axonal delays). The only requirement is that excitatory and inhibitory inputs superimpose in an approximately subtractive manner on a time scale according to the retrieval length (in our model 4–5ms).

We suggest that excitatory and inhibitory conductances play the role of the spike counters. In our model, excitatory and inhibitory conductances g_e and g_i superimpose linearly (equations 5–6). Therefore we can write

$$g_e \approx K \cdot c_H + L \cdot c_A, \quad g_i \approx M \cdot c_\Sigma \tag{2}$$

for relative synaptic strengths K, L and M, if the spike counts represent spikes in a short time window (dependant on the decay constants of the synaptic conductances). Below, we will find that a retrieval is done in 4–5ms, and therefore the approximation is justified. Neglecting noise processing and assuming an essential subtractive superposition of EPSPs and IPSPs, we obtain from equation 7 the following approximation.

$$\tau_x \frac{d}{dt} x \approx g_e(E_e - x) + g_i(E_i - x)$$
$$\approx (K \cdot c_H + L \cdot c_A)(E_e - x) + M \cdot c_\Sigma(E_i - x)$$
$$= K(E_e - x)c_H + L(E_e - x)\left(c_A - \frac{M(x - E_i)}{L(E_e - x)}c_\Sigma\right) \tag{3}$$

This is equation 1 for parameters

$$A = \frac{K(E_e - x)}{\tau_x}, \quad B = \frac{L(E_e - x)}{\tau_x}, \quad \alpha = \frac{M(x - E_i)}{L(E_e - x)}. \tag{4}$$

Note that the parameters depend on the membrane potential x. While for biological parameter-values, the dependence of A and B may be weak, since x is small in comparison to the excitatory reversal potential, the dependence of α may be much stronger, since the resting potential of neurons may be near the inhibitory reversal potential. This means that the separation strength α increases with the membrane potential driving towards the firing threshold. So we have essentially implemented our technically spike counter model using an excitatory and an inhibitory neuron population where afferents for the inhibitory one are copies of the excitatory one.

Since our model relies on a balanced excitation and inhibition, it is not always guaranteed that the excitation can be controlled. Furthermore, it may even be useful to choose the separation strength to be not stationary in order to work in different regimes (cf. [24], chapter 12). In an unbalanced regime of weak separation strength, the inhibition produced by C^S may be much too weak to control excitation. So to be sure that finally inhibition prevails, there may be a need for the second inhibitory population C^T to be activated at the end of the retrieval. This can simply be achieved by making the afferents (and efferents) of C^T unspecific. Using e.g. a full connection $C \to C^T$ results in a synchronous activation of C^T and those neurons of an addressed assembly that get no extra-areal input (e.g. because the address pattern was incomplete).

Figure 2A-C demonstrates the typical behaviour of our spiking associative memory. For addressing R we activated subsequentially two assemblies A_{R1} and A_{R2} heteroassociativley corresponding to the assemblies A_{C1} and A_{C2} stored in area C (Fig. 2A). A_{R2} was activated $t_{\text{diff}} = 2$ms after A_{R1}, and the temporal dispersion of activation times of individual neurons was a Gaussian with standard deviation of 1.5ms. Figure 2B shows the activation of the two addressed assemblies in C. While the first assembly gets activated completely, the second assembly is mostly suppressed with the most activated neurons lying in the overlap with the first assembly. Figure 2C shows the activation of the three neuron populations in C. The populations C and C^S show a symmetric behaviour, while population C^T gets activated at the end of the retrieval.

We compared pattern retrieval and separation quality for several variants of our model [15,16]. Some results for the base model SSI (Specific Separating Inhibition) as described above and a variant NSI (No Separating Inhibition) with inactivated C^S are shown in Figure 2D,E. We simulated for two conditions. Either with low or with high memory load (quotient of learned synapses to total number of synapses). The plots show the normalized separation quality s_N [16] over the average spike time difference t_{diff} of the two addressing assemblies in R (we have $s_N = 0$ if all addressed assemblies get equally activated, and $s_N = 1$ if exclusively *one* addressed assembly gets activated).

Even for low memory load (Fig. 2D) the differences between the models are revealed. While the model variant with inactivated population C^S exhibits

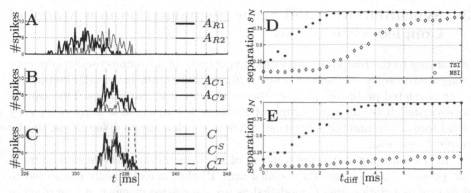

Fig. 2. Pattern retrieval and separation in the biological associative memory model when addressed with a superposition of two patterns. Sizes of populations C and C^S were 1600 neurons, size of C^T was 400. Average pattern size was 160. **A-C**: Typical retrieval in the base model (SSI) when addressed with two superimposed pattern (average spike time difference $t_{\text{diff}} = 2$ms). **A**: Spike activity in the address area R. Two patterns A_{R1} (thick line) and A_{R2} (thin line) with temporal overlap are activated. **B**: Spike activity in population C. Only assembly A_{C1} (thick line) addressed by the earlier pattern A_{R1} gets completely activated, while A_{C2} (thin line) is suppressed. **C**: Spike activities in the three populations C, C^S and C^T. Activity of C and C^S is symmetric, while C^T is activated at the end of the retrieval. **D-E**: Quality of pattern separation s_N over average spike time difference t_{diff} for two overlaying address patterns. The results for the model SSI ($*$) as described in the text and a variant NSI ($\diamond$) without C^S are shown. **D**: Quality s_N for low memory load (only 10 stored patterns). **E**: Quality s_N for high memory load (25% of the synapes for heteroassociation from R, and 40% for autoassociation within C were formed).

pattern separation only for relatively high values of t_{diff}, the complete model accomplishes pattern separation also for small values of t_{diff}. The differences increase, when the memory load is increased (Fig. 2E). Here, the variant without C^S show generally bad results even for large values of t_{diff} because even addressing with single patterns leads to low quality retrievals. In contrast the complete model still shows acceptable results. Simulations of further variants have demonstrated that there is no need for a one-to-one relation for the assemblies in C and C^S. Excitatory afferents of C^S may even be unspecific [15,16] as long as C^S gets extra-areal input as C.

Note that depending on the operation mode of connected associative memories, relatively small values of s_N for a single retrieval are sufficient. E.g. in an oscillatory regime with two reverberating associative memories, the separation is improved from one to the next iteration. This behaviour is known as *iterative retrieval* in technical associative memories [29,32]. Our results presented in the next section suggest indeed a similar function of synchronous oscillations found in the visual system.

3   Synchronization and Scene Segmentation in a Model of Coupled Visual Areas

According to the temporal correlation hypothesis, synchronous oscillations were suggested to solve the *binding problem* [6,30]. This problem actually consists of a set of related problems and may appear in quite different shapes. Two aspects investigated here are the binding of features belonging to a single entity by synchronization, and the separation of features belonging to different entities by desynchronization.

In the following we present a model of the orientation selective subsystem of the primary visual cortex (here called area P) that is reciprocally connected to a central association area C modelled as an associative memory as described in the previous chapter. We tested our model with single objects as well as with a superposition of several stimulus objects (triangle, ellipse and rectangle) which are represented in the retinal area R connected unidirectionally to area P. Here the representation of each stimulus in P is the activation of a certain set of orientation selective patches. And the binding problem is both to identify the patches that belong to the same representation, and to identify corresponding representations in the two areas P and C. Our goal here is, to present a model that reproduces the corresponding experimental findings of stimulus-specific correlation and decorrelation of neural activity [8,18].

3.1   Network Model

Now the network consists of three areas (R,P and C) each composed of several neuron populations. Figure 3A shows the connection scheme of neuron populations. In area R (retina) input patterns corresponding to stimulus objects in the visual field are represented in a 100×100 bitmap. Area P (primary visual cortex) consists of 100×100 excitatory spiking neurons (population P) and 100×100 inhibitory gradual neurons (population P^I). Connections from R and inside P are modelled corresponding to the subsystem of orientation selective columns in the primary visual cortex, resulting in the patchy stimulus representations shown in Figure 4B. At a certain location in P, only the neurons with orientation preferences best matching the active pattern in R get strongly activated, while neighboring cells are suppressed by recurrent inhibition.

Area C (central visual area) is modelled as the SSI-variant of our spiking associative memory described in the last chapter. Populations C and C^S both have sizes 40×40, while population C^T has size 20×20. The stored patterns are topographic random representations of the stimuli and additional pure random patterns to fill up the memory matrixes to about 0.05. As can be seen in Figure 4C, the original stimulus shapes are only rawly preserved in the stored patterns.

Figure 3B shows the arrangement of orientation selective columns in area P. For computational reasons we confined orientations to four discrete values $0°$, $45°$, $90°$ and $135°$. Further the arrangement is composed of the regular alignment of 10×10 identical blocks of size 10×10. Each such block corresponds to roughly 1mm^2 cortex and consists of four subblocks with centers (depicted as small black circles) around which orientation preference rotates for 180°. The exact symmetric layout of our model was inspired by a model of Grinvald (cited as personal communication in [12]).

Each neuron in population P receives topographically orientation tuned gradual input from a local neighbourhood in R. Figure 3Ca plots the synaptic strength kernel for afferents of a neuron sensitive for horizontal orientations. The kernel can be computed using a two-dimensional Gaussian. The Gaussian is properly rotated for neurons with other preferences.

The internal excitatory connection from and to population P couples specifically neurons which have similar orientation preferences *and* additionally are near neighboured or collinear aligned. This is as expected from Hebbian correlation learning during the presence of stimuli rich of contours. Figure 3Cb shows exemplarily the kernel for afferents of a neuron P_x (cross x in Fig.3A).

Connections involving inhibitory neurons of P^I are modelled as non-specific with respect to orientation selectivity, therefore synaptic strength depends exclusively on neuron distance. For a neuron in P^I, Figure 3Cc shows the strength of afferents from P. The kernel is identical for all neurons and can be computed using a two dimensional symmetric Gauss function. Each neuron in P^I inhibits itself and the corresponding neuron in P.

To overcome the problem of artificial stimulus-locked synchronicity even between distant patches, we introduce spatially and temporally correlated noise

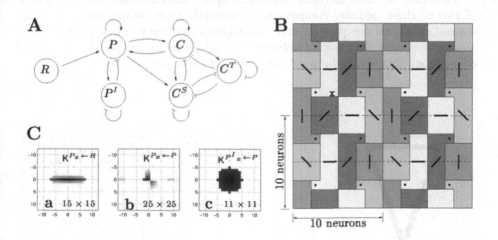

Fig. 3. Structure of the model of reciprocally coupled areas. **A**: Network of the different neuron populations. The network consists of three areas. The retinal area R, the primary visual area P, and the central association area C. **B**: Arrangement of orientation selective columns in area P. Only a 20 × 20 clip is shown, whereas area P is composed of 100 identical blocks of size 10 × 10. The black points mark centers around which orientation preference rotates for 180°. **C**: Plots of afferent kernels of the neurons at the location corresponding to the cross (x) shown in B. For the excitatory neuron, afferents from R (a) and from P (b) are shown, while for the inhibitory neuron afferents from P are depicted (c). Numbers indicate kernel sizes.

as ubiquitously found in experiments (e.g. [3]). The result is simply that distant activated patches show no more synchronization artefacts, i.e. correlograms remain flat as in experiments (e.g. [6]).

Figure 4 plots the representations of the triangle stimulus in the different areas R,P and C. Projections of the patchy representations in P onto area C are used to define topographical random patterns for area C which were stored in the associative memory. In our simulations, the average size of stored patterns was 80 neurons for the stimulus and the random patterns.

For the reciprocal connections between areas P and C, we modelled essentially a topographically confined heteroassociation. As patterns in P, we used the spike rate patterns elicited by a stimulus in P after a threshold operation. A neuron was defined as belonging to the pattern, if its spike rate was greater than 18 spikes/sec (see Fig.4B). Neurons in P project onto neurons in C and C^S within a neighbourhood (diameter 13) around the topographic corresponding location in area C. For neurons in C, feedback projections onto neurons in P were similarly topographic (diameter 33). As in the Willshaw-model of heteroassociative memory, synapses between two neurons in P and C were present, if both neurons belong to at least one pair of corresponding patterns in P and C. The synaptic strength for a connection between two neurons is proportional to the highest spike rate of the neuron in P, exhibited when stimulated with a pattern that both neurons represent. Note that the reciprocal connection is symmetric with respect to synaptic strength.

Generally we used distance dependent spike transmission delays composed of two or three additive components: a synaptic base component, a distance dependent axonal component, and sometimes a random component. For intra-areal connections delays were between 0.8 and 2.8 ms, while for the inter-areal connections delays were between 3.8 and 9.8 ms (see discussion).

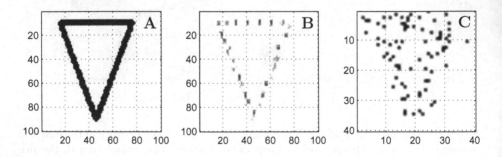

Fig. 4. Representations of the triangle stimulus in the three areas of the model. **A**: Stimulus representation in area R. **B**: Representation in area P is patchy due to the orientation selective modules. **C**: Representation in area C is a topographic random pattern. Original shape is rawly preserved.

3.2   Simulation Results

In the next subsections we present some simulation results for different network and stimulus configurations. First, we investigate the role of feedback from higher area C to primary area P using a single stimulus by comparing results from simulations with and without feedback. And after this we investigate scene segmentation in our complete model using three overlapping stimuli. For each network and stimulus configuration, we used the same parameter set (neuron parameters, connection strengths, delays), and simulated the model for 2000 ms with stationary stimulus configuration. Recorded spike activity from the excitatory populations P and C over the complete simulated time window was used for computation of average firing rates and correlation histograms.

Single Stimulus Condition: For the single stimulus condition we used the triangle stimulus (Fig. 4) and simulated our model with and without the feedback from area C to P.

Figure 5 shows the simultaneously recorded (summed) spike activity in populations P and C. The plots on the left (Fig. 5A-D) show the results for the model without feedback $P \leftarrow C$, while the plots on the right (Fig. 5E-H) show the results for the complete model. Without feedback, activity in areas C and P is relatively slow and not very regular. Activity within patches of P is obviously synchronized, but in a quite irregular manner, while activity of different patches seems to be rather independent. In contrast, the behaviour gets more ordered for the complete model. Now both C and P show regular fast oscillatory behaviour, and different patches in P seem to be synchronized. Even for many activated neurons, oscillations are reflected in the membrane potential (Fig. 5H, bottom neuron). However, other activated neurons exhibit visible membrane potential oscillations only seldom and spike activity is rather uncorrelated (Fig. 5H, upper two neurons).

Results of spike rate and correlation analysis are shown in Figure 6. For the model without $P \leftarrow C$, the rate pattern in P (Fig. 6A,B) reflects the stimulus representation (see Fig. 4) of the triangle stimulus. However, the patches apparently are delimited in a rather blurred way. Spike rates seem to be distributed unimodally with most neurons resting near zero activity and only few neurons reaching up to 40 spikes/sec. In contrast, the rate distribution in C (data not shown) was clearly bimodal. The activated neurons were only the ones belonging to the triangle assembly and the spike rates were narrowly distributed around rates of $25 - 30$ spikes/sec. For the complete model (Fig. 6D,E), demarcation of patches in the rate pattern is sharper because of increased inhibitory activation due to the feedback. The spike rates increased for many neurons, and now also the rate distribution in P is clearly bimodal with many neurons still resting at a low rate, and a second population of neurons with rates broadly distributed above 20 spikes/sec. The situation in C was similar to the case without feedback, but activity was slightly faster.

To analyse temporal interaction between different patches as well as between the two areas, we computed auto and cross correlation histograms (ACHs/CCHs)

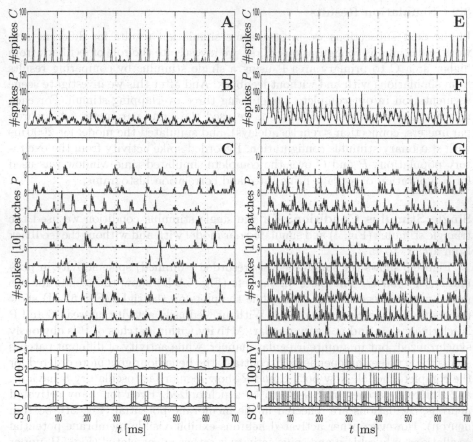

Fig. 5. Spike trains for single stimulus condition. Left plots (A-D) correspond to the model with inactivated feedback from area C. Right plots (E-H) correspond to the complete model. **A/E**: Summed spike activity in population C. **B/F**: Summed spike activity in population P. **C/G**: Summed spike records of 10 different activated patches representing the triangle stimulus in population P. **D/H**: Potentials and spikes of single units taken from three activated patches of P.

on the patch and population level (detailed data in [15,17]). For the model without feedback $P \leftarrow C$, the ACH of population P exhibited only a weak central peak indicating non-oscillatory weak global synchronization. Correspondingly, the CCHs for different patches remained flat on average. They exhibited stronger peaks near zero time lag only for neighboured patches, while for distant patches, correlograms remained essentially flat with small peaks at random time lag (see Fig. 6C). ACH of population C exhibited a strong central peak indicating the fast retrieval processes in the associative memory, but only weak side peaks indicating relatively irregular oscillations. As expected, the CCH for activity in P and C remained flat indicating uncorrelated activity in the two areas. The results for the complete model revealed fast oscillatory synchronization (about

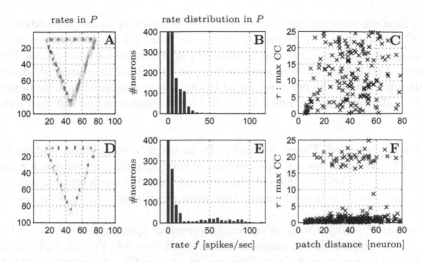

Fig. 6. Analysis of spike activity in P for the single stimulus condition. Plots in upper row (A-C) correspond to the model with inactivated feedback from area C. Plots in lower row (D-F) corresponds to the complete model. **A/D**: Spike rate pattern for population P. **B/E**: Histogram of spike rate distribution in population P. **C/F**: Correlation maximum (maximized over time lag τ) over patch distance for pairs of activated patches in P.

50 Hz) in the range of the feedback from area C. All correlograms for intra-areal activity exhibited now strong *central* peaks with side peaks (see Fig. 6F). The correlation strength between patches was increased and depended only weakly on patch distance. In the CCH for activity in populations P and C, the central peak was shifted for about half a period indicating antiphasic oscillatory activity. This behaviour is typical for the mutual activation of two reciprocally connected areas by *one* activity wave running iteratively forth and back as reported earlier in simulation studies [28,4]. However, in experiments inter-areal oscillatory synchronization apparently occurs predominantly at zero phase [6, 9,7,10]. We will discuss this discrepancy below after presenting the simulation results for overlayed stimuli.

In summary, the function of the feedback $P \leftarrow C$ as suggested by our simulation results for the single stimulus condition may be the following. Considering spike rates, activity is specifically enhanced for neurons representing relevant features because of additional excitation from the higher area. For the same reason, activity is suppressed by indirectly evoked inhibition for many neurons with RF-properties matching less precisely the stimulus configuration. Together, this leads to a sharper activation pattern of an assembly representing the scene in the visual field. Considering temporal relations, the range of synchronized spike activity is extended by feedback from the higher area. While without feedback, synchronization is restricted to a patch according to the local intra-areal connections of area P, activating feedback extends synchronization to the whole area.

This happens because the locally induced synchronicity in the associative memory C is transmitted to the primary area P by converging/diverging connections. More generally, we suggest that the range of synchronization is determined by the extent of (i) locally induced synchronicity in the higher area, and (ii) divergent feedback from the higher area. In addition, the sychronization is accompanied with oscillatory activity, much faster and more regular as without feedback.

Superposition of Three Stimuli: To investigate scene segmentation in the complete model, we used a superposition of three stimuli: the triangle as before, a rectangle and an ellipse. Except the different stimulus configuration, simulation conditions and the model parameters were exactly the same as in the single stimulus conditions.

Figure 7 shows the simultaneously recorded spike activity in populations P and C, similar to the single stimulus condition (cf. Fig. 5). Spike activity is depicted separately for the different (sub)assemblies representing the three stimuli (triangle, ellipse and rectangle). The recordings of one assembly in C (Fig. 7A, e.g. the bottom row for the rectangle assembly) show periods of fast oscillatory activity lasting for a few hundred milliseconds alternating with longer periods of essential silence. Comparing the three recordings from C, one obtains that only one assembly is in the oscillatory state at a time. This results from the structure of the associative memory in area C preferring separated activation of the three addressed assemblies (see previous section).

Activity of the three (sub)assemblies in area P reveals similar results (Fig. 7B-D) as in C. The recordings show periods of *fast* and precise (small phase jitter) oscillatory activity lasting again for a few hundred milliseconds alternating with periods of relatively *slow* und unordered activity. While the *"slow" state* and the *"fast" state* can clearly be distinguished on assembly and patch level, for many single units the distinction is much harder (Fig. 7D, upper two recordings).

Corresponding assemblies in P and C are always at the same time in the fast or in the slow state. Moreover, looking at the recordings of one assembly (e.g. the triangle), the behaviour in the fast state is very similar to the results from the single stimulus condition in the complete model (cf. Fig. 5E-H), while the behaviour in the slow state is very similar to the results from the single stimulus condition in the model without feedback $P \leftarrow C$ (cf. Fig. 5A-D).

Figure 8 shows the analysis of spike recordings. Anlalysis of mean firing rates shows a superposition of the assemblies representing the three stimuli (Fig. 8A–D). While the three patchy assemblies in P can clearly be distinguished (Fig.8A), the superposition of the three assemblies in C results seemingly in a random activity pattern (Fig.8C, cf. Fig. 4). Patches representing different stimuli apparently are activated differently. At least the patches representing the triangle seem to be weaker than the other patches. However, the spike rate histogram for P (Fig.8B) shows a bimodal distribution with most neurons having small spike rates. The average firing rate of the activated neurons lies well below 50 spikes/sec (the frequency of the fast oscillations), and the fastest neurons reach no more than 75 spikes/sec. In area C the results are qualitatively similar

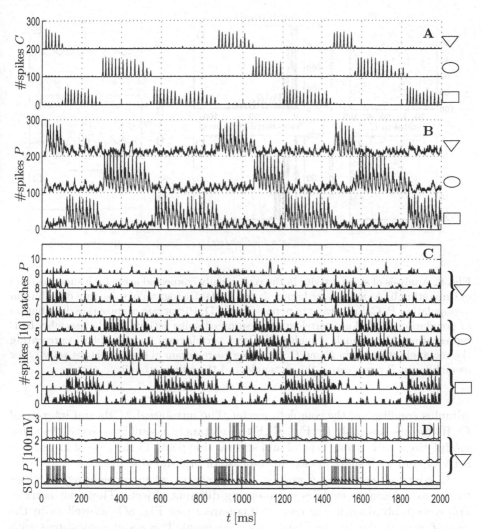

Fig. 7. Spike trains for superposition of three stimuli in the complete model. **A**: Summed spike activity in population C for the triangle (top row), ellipse (middle row) and rectangle assembly (bottom row). **B**: Summed spike activity in population P for the triangle (top row), ellipse (middle row) and rectangle assembly (bottom row). **C**: Summed spike records of 10 different activated patches representing the triangle (rows 6–9), the ellipse (rows 3–5) and the rectangle stimulus (rows 0-2) in population P. **D**: Potentials and spikes of single units taken from three activated patches of P representing the triangle stimulus.

with the average of the activated neurons firing at about $20 - 25$ spikes/sec and only few neurons reaching up to 40 spikes/sec.

Some results of correlation analysis are shown in figure 8E-G (cf. [15,17]). Looking at only one assembly, the results were generally very similar to the single

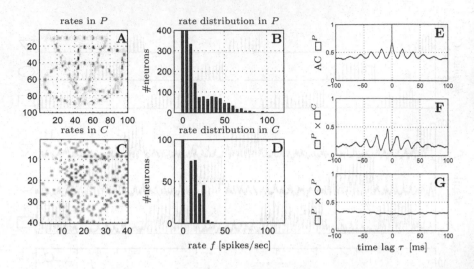

Fig. 8. Analysis of spike activity in the reciprocally connected areas for a superposition of three stimuli. **A**: Spike rates in population P. **B**: Histogram of spike rate distribution in population P. **C**: Spike rates in population C. **D**: Histogram of spike rate distribution in population C. **E**: Auto correlogram of summed spike activity for the assembly in P representing the rectangle stimulus. **F**: Cross correlogram for assemblies representing the rectangle stimulus in P and C. **G**: Cross correlogram of summed spike activity for assemblies in P representing the rectangle and the ellipse stimuli.

stimulus condition of the complete model. The intra-areal ACHs and inter-areal CCHs for one assembly (in Fig. 8E,F the rectangle assembly) showed the same modulations with strong central and several side peaks in the case of intra-areal activity, and with shifted peaks in the case of inter-areal activity.

In contrast, correlations were generally absent when considering activities from two different assemblies representing different objects. These flat correlograms were obtained in the case of intra-areal (see Fig. 8G) as well as in the case of inter-areal activities (data not shown here). This result is consistent with findings of specifically modulated and unmodulated correlograms in experiments using multiple stimulus objects [8,18], while earlier proposed phase-coding models (e.g. [28]) would predict shifted peaks.

In summary, activities of assemblies representing the same entity are synchronized with modulated correlograms, while assemblies representing different entities exhibit uncorrelated activity with flat correlograms. Scene segmentation in our model is accomplished through a kind of self-generated attention switching in a sequence from the one to the next present pattern. The assembly representing the most salient object is first attended resulting in a state of fast and precise activation induced by reciprocal activity exchange between the two areas. After some time the assembly gets fatigued due to habituation and the assembly representing the next salient object switches from a slow, asynchronous state to a fast synchronous state. Here, the binding problem is solved through self-generated attention switching on a time scale slower than the one suggested by phase-coding models.

According to our results in the last chapter concerning the structure of the associative memory, the scene segmentation capability of our model has vanished when we inactivated population C^S in area C. The result was that all assemblies were in the fast synchronous state at the same time, the superposition catastrophe. Varying the time constants and strength of neuron habituation (see equation 10), we were able to control the time scale on which the binding through self-generated attention happened.

4 Discussion

Starting from retrieval strategies in the simple binary associative memory [37, 23], we proposed a new model of spiking associative memory similar to the model of [35], but incorporating two populations of inhibitory neurons (separating and terminating inhibition) differently involved in the retrieval of stored patterns. While one inhibitory population receives extra-areal input in the same way as the excitatory population, the other receives only intra-areal input. The result is a quasi-autonomous threshold control and failure detection. Capability of fast and very robust pattern separation is kept even for high memory loads up to 0.4. However, then separation for address patterns strongly overlapping in time is only partially accomplished and must be improved by iterative retrieval (similar to [29] and [32]). Our experiments show that a good performance of iterative retrieval does not require the inhibition to be specific [15,16].

We then investigated the interactions of two reciprocally connected visual areas, the primary visual area P and the central visual area C. For area P we presented a model of the orientation selective subsystem of V1, while area C was modelled as spiking associative memory as described above. To test for the function of the reciprocal inter-areal connection, we simulated the model with and without feedback $P \leftarrow C$ using a single stimulus. Without feedback we found in P only slow and unordered oscillations at about $20-25$ Hz, synchronized only between neighbouring neuron groups. In contrast, the oscillations in the complete model became faster ($50-60$ Hz) and more coherent, synchronized at zero phase within the whole range of feedback from C.

Scene segmentation was studied in this model using a superposition of three different stimulus objects. The neuron groups representing one object tended to be simultaneously in either a slow state (similar to the single stimulus condition with inactivated feedback) or a fast state (similar to the single stimulus condition of the complete model) At any particular time, only one assembly (of the three) was found to be in the fast state. After persistence for about 100 to 300 milliseconds, this assembly got fatigued due to habituation and left the fast state, whereby another assembly switched from the slow to the fast state. Cross correlations between cell groups coding features corresponding to a common object were modulated (with a peak at zero), while cross correlations remained flat if the cell groups coded features corresponding to different objects. These results are consistent with experimental findings of specifically modulated and unmodulated correlograms [8,18]. In contrast, earlier proposed *phase coding models* (e.g. [28]) predict modulated correlograms between two groups of neurons in both cases, with a central peak if the two groups code the same entity, and with a shifted peak if the two groups code different entities. In these models, simultaneously presented stimuli are segmented by serially activating the corresponding assemblies in an invariable order within one gamma-period.

In our model the prevalence of one of the three stimulus objects is switched on a time scale of 100 ms as opposed to < 10ms in phase coding models. This points to the involvement of attentional mechanisms in this switching of the interpretation which is to a large amount initiated by the higher associative cortical area. With this interpretation our network also reproduces findings concerning the influence of attention on single neurons in sensory areas (e.g. [27]).

The described switching of activated assemblies may be related to experimental findings of synchronization on different time scales, e.g. the T,C and H-peaks of correlograms in [22]. On the larger time scale corresponding to the C-peaks, we reproduced correlograms with centered peaks around zero even for inter-areal connections, and also the correlograms with different peak types at the same time (e.g. T and C) are reproduced by our model [17].

There still is one point in our simulations that is apparently inconsistent with physiological observations. In Fig. 8F we see the antiphasic oscillations between the two areas P and C, whereas zero phase seems to be most commonly observed even between different cortical areas [6,9,7,10]. In further simulations [17], we varied the inter-areal delays. Using only short delays, we obtained near zero phase synchronization, but the fast oscillations were absent and the self-generated attention switching was no longer present, because inhibition could no longer be bridged by the short delays. Using longer delays, we obtained indeed zero-phase lag and preserved the scene segmentation capability. This suggests that the problem could be resolved, for example, by introducing a wider distribution of time delays in cortico-cortical projections including very short and also quit long ones (cf. [33]), or by using other excitatory mechanisms on a larger time scale (e.g. bursting, NMDA channels).

We are aware that in our model the function of the fast oscillations is essentially to enable fast pattern separation and to improve the quality of representations by iterative retrieval [29,32], while the binding mechanism allows the simultaneous representation of several objects only on a slower time-scale of several hundred milliseconds. Results of psychological experiments (e.g. [26]) suggest both the existence of a rather long time-window for experience of complex situations (> 0.5 sec) and the need for a higher temporal resolution than this (30 − 40 ms). This points to a functional role of processes faster than the attention switching in our model, i.e. oscillations or fluctuations within an enhanced state lasting for some hundred milliseconds may still carry more detailed information about spatio-temporal relations in the outside world.

Appendix: Neuron Model

For simulations we chose a one-point spiking neuron model at an intermediate level of complexity, comparable with the so-called PTNR10 model of [19], that produces relatively realistic input-output dynamics.

The postsynaptic excitatory and inhibitory membrane conductances g_e and g_i are determined by the excitatory and inhibitory presynaptic input in_e and in_i. The dynamics of g_e (g_i analogous) is defined by

$$\tau_e \frac{d}{dt} g_e(t) = -g_e(t) + in_e(t), \tag{5}$$

$$in_e = \sum_{p \in P_{e,s}} w_p \sum_{s \in S_p} \delta(t - s - d_p) + \sum_{p \in P_{e,r}} w_p \, y(t - d_p). \tag{6}$$

Here $P_{e,s}$ contains the excitatory *spiking* neurons presynaptically coupled to the considered neuron. Analogously, $P_{e,g}$ contains the coupled *gradual* neurons. w_p is the synaptic strength, and d_p the total delay of the connection from neuron p. $\delta(t)$ is the Dirac impulse, so an arriving spike causes an instantaneous increase of the conductance followed by an exponential decrease according to the time constant τ_e (5 ms) or τ_i (7 ms). $y_p(t)$ is the output (mean firing rate) of gradual neuron p. The evolution of the transmembrane potential x is governed by the following low-pass equation.

$$\tau_x \frac{d}{dt} x = -x + [g_e(E_e - x) + g_i(E_i - x)] \frac{g_e + g_0}{g_e + g_i + g_0} + X_N \tag{7}$$

Without any input the membrane potential x decays exponentially with time constant τ_x (10ms). Excitatory or inhibitory conductances drive x towards the excitatory or inhibitory reversal potential E_e (80 mV) or E_i (−10 mV). A further divisive effect of inhibition may be modelled by appropriate values of parameter g_0 (100). The noise potential X_N is defined as

$$X_N(t) = \sigma_N \, N(t)(1 + \sigma_M \, in_{e,r}(t)), \tag{8}$$

where $N(t)$ is random with expectation zero, while σ_N and σ_M determine the noise power. Values of $\sigma_M > 0$ cause an modulatory influence of the gradual excitatory synaptic input $in_{e,r}$, defined as the last sum in equation 6. Sometimes $N(t)$ is spatio-temporally correlated. For the spiking model the output variable $y(t)$ indicates the generation of a spike. This happens as soon as the membrane potential x exceeds the threshold $\Theta(t)$, where for the last spike time $s(t-)$

$$\Theta(t-) = \begin{cases} \infty & , \quad t - s(t-) \leq R_a \\ \Theta_\infty \cdot (1 + \frac{R_r}{t - s(t-) - R_a}) + h(t-) & , \quad \text{otherwise.} \end{cases} \tag{9}$$

Refraction is modelled by introducing absolute and relative refractory periods R_a (2 ms) and R_r (4 ms), Θ_∞ is the asymptotic threshold value for $t \to \infty$. The dynamics of habituation $h(t)$ is described by

$$\frac{d}{dt} h(t) = -\frac{1}{\tau_h} h(t) + H \sum_{s \in S} \delta(t - s), \tag{10}$$

where S contains the spike times of the neuron. So each spike causes an instantaneous increase of $h(t)$ by H ($3\,\mathrm{mV}$ for C/C^S) followed by an exponential decay according to τ_h ($150\,\mathrm{ms}$).

References

[1] A.M.H.J. Aertsen, M. Erb, and G. Palm. Dynamics of functional coupling in the cerebral cortex: an attempt at a model-based interpretation. *Physica D*, 75:103–128, 1994.

[2] A.M.H.J. Aertsen, G.L. Gerstein, M.K. Habib, and G. Palm. Dynamics of neuronal firing correlation: Modulation of "Effective Connectivity". *Journal of Neurophysiology*, 61(5):900–917, 1989.

[3] A. Arieli, A. Sterkin, A. Grinvald, and A. Aertsen. Dynamics of ongoing activity: Explanation of the large variability in evoked cortical responses. *Science*, 273:1868–1871, 1996.

[4] A. Bibbig. *Oszillationen, Synchronisation, Mustertrennung und Hebb'sches Lernen in Netzwerken aus erregenden und hemmenden Neuronen (in German)*. PhD thesis, Department of Neural Information Processing, University of Ulm, Germany, 2000.

[5] V. Braitenberg and A. Schüz. *Anatomy of the cortex. Statistics and geometry.* Springer-Verlag, Berlin, 1991.

[6] R. Eckhorn, R. Bauer, W. Jordan, M. Brosch, W. Kruse, M. Munk, and H.J. Reitboeck. Coherent Oscillations: A mechanism of feature linking in the visual cortex? *Biol. Cybern.*, 60:121–130, 1988.

[7] A.K. Engel, P. König, A.K. Kreiter, and W. Singer. Interhemispheric synchronization of oscillatory neuronal responses in cat visual cortex. *Science*, 252:1177–1179, 1991.

[8] A.K. Engel, P. König, and W. Singer. Direct physiological evidence for scene segmentation by temporal coding. *Proc. Natl. Acad. Sci. USA*, 88:9136–9140, 1991.

[9] A.K. Engel, A.K. Kreiter, P. König, and W. Singer. Synchronization of oscillatory neuronal responses between striate and extrastriate visual cortical areas of the cat. *Proc. Natl. Acad. Sci. USA*, 88:6048–6052, 1991.

[10] A. Frien, R. Eckhorn, R. Bauer, T. Woelbern, and H. Kehr. Stimulus-specific fast oscillations at zero phase between visual areas V1 and V2 of awake monkey. *NeuroReport*, 5(17):2273–2277, 1994.

[11] D.O. Hebb. *The organization of behavior. A neuropsychological theory.* Wiley, New York, 1949.

[12] K.P. Hoffmann and C. Wehrhahn. Zentrale Sehsysteme (in German). In J. Dudel, R. Menzel, and R.F. Schmidt, editors, *Neurowissenschaft (in German)*, chapter 18, pages 405–426. Springer-Verlag, Berlin/New York, 1996.

[13] J.J. Hopfield. Neural networks and physical systems with emergent collective computational abilities. *Proceedings of the National Academy of Science, USA*, 79:2554–2558, 1982.

[14] J.J. Hopfield. Neurons with graded response have collective computational properties like those of two-state neurons. *Proceedings of the National Academy of Science, USA*, 81(10):3088–3092, 1984.

[15] A. Knoblauch. Assoziativspeicher aus spikenden Neuronen und Synchronisation im visuellen Kortex (in German). *Diploma thesis, Department of Neural Information Processing, University of Ulm, Germany*, 1999.

[16] A. Knoblauch and G. Palm. Pattern separation and synchronization in spiking associative memories and visual areas. *submitted to Neural Networks*, 2000.

[17] A. Knoblauch and G. Palm. Synchronization and scene segmentation in a realistic model of reciprocally connected visual areas. *in preparation for Biological Cybernetics*, 2000.

[18] A.K. Kreiter and W. Singer. Stimulus-dependent synchronization of neuronal responses in the visual cortex of the awake macaque monkey. *J. of Neurophys.*, 16(7):2381–2396, 1996.

[19] R.J. MacGregor. *Neural and Brain Modeling*. Academic Press, San Diego, 1987.

[20] C.v.d. Malsburg. Am I thinking assemblies? In G. Palm and A. Aertsen, editors, *Brain Theory*, pages 161–176. Springer-Verlag, Berlin/Heidelberg, 1986.

[21] R. Miller, editor. *Time and the Brain. Conceptual Advances in Brain research.* Harwood Academic Publishers, Amsterdam, 2000.

[22] L.G. Nowak and J. Bullier. Cross correlograms for neuronal spike trains. Different types of temporal correlation in neocortex, their origin and significance. In Miller [21], chapter 2, pages 53–96.

[23] G. Palm. On associative memories. *Biological Cybernetics*, 36:19–31, 1980.

[24] G. Palm. *Neural Assemblies. An Alternative Approach to Artificial Intelligence.* Springer, Berlin, 1982.

[25] G. Palm. Computing with neural networks. *Science*, 235:1227–1228, 1987.

[26] E. Pöppel. Temporal mechanisms in perception. *International Review of Neurobiology*, 37:185–202, 1994.

[27] J.H. Reynolds and R. Desimone. The role of neural mechanisms of attention in solving the binding problem. *Neuron*, 24:19–29, 1999.

[28] R. Ritz, W. Gerstner, U. Fuentes, and J.L. van Hemmen. A biologically motivated and analytically soluble model of collective oscillations in the cortex. II. Applications to binding and pattern segmentation. *Biol. Cybern.*, 71:349–358, 1994.

[29] F. Schwenker, F.T. Sommer, and G. Palm. Iterative Retrieval of Sparsely Coded Associative Memory Patterns. *Neural Networks*, 9:445–455, 1996.

[30] W. Singer and C.M. Gray. Visual feature integration and the temporal correlation hypothesis. *Annu.Rev.Neurosci.*, 18:555–586, 1995.

[31] D.C. Somers, S.B. Nelson, and M. Sur. An emergent model of orientation selectivity in cat visual cortical simple cells. *J.Neurosci.*, 15:5448–5465, 1995.

[32] F.T. Sommer and G. Palm. Improved bidirectional retrieval of sparse patterns stored by hebbian learning. *Neural Networks*, 12:281–297, 1999.

[33] H. Swadlow. Information flow along neocortical axons. In Miller [21], chapter 4, pages 131–155.

[34] T. Wennekers. *Synchronisation und Assoziation in Neuronalen Netzen (in German)*. Shaker Verlag, Aachen, 1999.

[35] T. Wennekers and G. Palm. On the relation between neural modelling and experimental neuroscience. *Theory in Bioscience*, 116:273–289, 1997.

[36] T. Wennekers, F.T. Sommer, and G. Palm. Iterative retrieval in associative memories by threshold control of different neural models. In H.J. Herrmann, D.W. Wolf, and E. Pöppel, editors, *Supercomputing in Brain Research: From Tomography To Neural Networks*, pages 301–319. World Scientific, Singapore, 1995.

[37] D.J. Willshaw, O.P. Buneman, and H.C. Longuet-Higgins. Non-holographic associative memory. *Nature*, 222:960–962, 1969.

[38] F. Wörgötter and C. Koch. A detailed model of the primary visual pathway in the cat: Comparison of afferent excitatory and intracortical inhibitory connection schemes for orientation selectivity. *J.Neurosci.*, 11:1959–1979, 1991.

A Familiarity Discrimination Algorithm Inspired by Computations of the Perirhinal Cortex

Rafal Bogacz[1], Malcolm W. Brown[2], and Christophe Giraud-Carrier[1]

[1] Dept. of Computer Science, University of Bristol, Bristol BS8 1UB, UK
{bogacz,cgc}@cs.bris.ac.uk
[2] Dept. of Anatomy, University of Bristol, Bristol BS8 1TD, UK
M.W.Brown@bristol.ac.uk

Abstract. Familiarity discrimination, i.e. the ability to recognise previously experienced objects is important to the survival of animals, but it may also find practical applications in information technology. This paper describes the *Fam*iliarity discrimination based on *E*nergy algorithm (FamE) inspired by the computations of the perirhinal cortex - the area of the brain involved in familiarity discrimination. In FamE the information about occurrences of familiar records is encoded in the weights of a neural network. Using the network, FamE can discriminate whether a given record belongs to the set of familiar ones, but cannot retrieve the record. With this restriction, the network achieves much higher storage capacity for familiarity discrimination than other neural networks achieve for recall. Therefore, for a given number of familiar records, the description of the weights of the network occupies much less space in memory than the database containing the records itself. Furthermore, FamE can still classify a record as familiar even if it differs in a substantial proportion of its bits from its previous representation. FamE is also very fast. Preliminary results of simulations demonstrate that the algorithm may be applied to real-world problems.

1 Introduction

Animal and humans possess the ability to discriminate familiarity for an impressive number objects. Human subjects, after seeing thousands of different pictures once, can still recognise the individual pictures as familiar [21]. This ability to recognise previously seen objects is important to the survival of animals, but it may also find practical applications in information technology. For example, the ability to determine whether a web site has been visited or not is very useful to a software agent searching the web. Familiarity discrimination is analogous to checking whether a record belongs to a certain set (i.e., the set of familiar records). From this point of view there exists a wide range of possible applications. One example could be a program controlling a security camera at the entrance of a building - it does not need to identify a person (e.g., retrieve its name), but only discriminate whether this person belongs to the set of persons authorised to enter the building. Another example could be checking whether a record is stored in a database without actually searching the database.

S. Wermter et al. (Eds.): Emergent Neural Computational Architectures, LNAI 2036, pp. 428–441, 2001.

To solve problems related to familiarity discrimination, current methods create a database of familiar records and use it to check whether a given record is contained in the database. Storing the database allows retrieval of information so this approach is not specialised for familiarity discrimination and hence is far from optimal. It is used because nowadays disk space is cheap and database searching is relatively fast. However, in the case when a user does not look for an exact match in the database but for something similar, database searching is much slower because one cannot use standard searching techniques such as indexing or hashing.

This paper describes the recently developed algorithm, *Fam*iliarity discrimination based on *E*nergy (FamE) [5] and shows its applicability. In FamE the information about occurrences of familiar records is encoded in the weights of a neural network. Using the network, FamE can discriminate whether a given record belongs to the set of familiar ones, although it cannot retrieve the record. For example, it cannot retrieve the record after being given only partial information (e.g., a key). With this restriction, the network achieves much higher storage capacity for familiarity discrimination than other neural networks achieve for recall. Therefore, for a given number of familiar records, the description of the weights of the network occupies much less space in memory than the database containing the records itself. Furthermore, FamE can still classify a record as familiar even if it differs in a substantial proportion of its bits from its previous representation. FamE is also very fast.

FamE is inspired by the computations performed by the perirhinal cortex. Work in amnesic patients and in animals has established that discrimination of the relative familiarity or novelty of visual stimuli is dependent on the perirhinal cortex [1,2,8,17]. Damage to the perirhinal cortex results in impairments in recognition memory tasks that rely on discrimination of the relative familiarity of objects [16]. Within the monkey's perirhinal cortex, ~25% of neurons respond strongly to the sight of novel objects but respond only weakly or briefly when these objects are seen again [8,22]. We have created a model of the perirhinal cortex which is consistent with many experimental observations [6]. Since the perirhinal network has the properties mentioned in the previous paragraph for FamE, the perirhinal cortex alone may rapidly discriminate the familiarity of many more stimuli than current neural network models indicate could be recalled (recollected) by all the remaining areas of the cerebral cortex. This efficiency and speed of detecting novelty provides an evolutionary advantage, thereby giving a reason for the existence of a familiarity discrimination network in addition to networks used for recollection.

FamE differs in important respects from the artificial neural networks used for familiarity discrimination in industrial applications [19,10]. In these approaches familiarity discrimination is regarded as detecting typical patterns of device behaviour, since atypical (i.e., novel) patterns may be a sign of malfunction. Hence such models assume that familiar patterns create clusters in representation space (and the synaptic weights of their neurons often encode prototypes of familiar patterns, e.g., see [10]). In contrast, the model outlined here does not require any assumptions concerning the distribution of patterns and it discriminates whether a particular pattern was presented previously rather than whether the pattern is typical. The information processing of FamE is somewhat similar to that in a novelty detector [14,15], but the novelty detector is an abstract model of a single neuron, with correspondingly limited storage capacity. The proposed model is a network of neurons constructed as to have a very large storage capacity.

The description of the algorithm is given in Section 2. In section 3 the storage capacity for familiarity discrimination is investigated. Section 4 discusses implementation issues which must be considered before the application of FamE. Section 5 shows how the algorithm performs on real data. Finally, section 6 discusses the relation of FamE to other techniques.

2 Algorithm Description

FamE stores information about the familiar patterns in the weights of a Hopfield network. The Hopfield network provides a simple model of associative memory [12]. It is a fully connected recurrent neural net consisting of N neurons, whose activations are denoted by x_i. The active state of a neuron is represented by 1, and the inactive state by -1. The patterns stored by the network are denoted by ξ^μ and the number of these patterns by P. The weight of the connection between neurons j and i is denoted by w_{ij} and computed according to Hebb rule [12]:

$$w_{ij} = \begin{cases} \dfrac{1}{N} \displaystyle\sum_{\mu=1}^{P} \xi_i^\mu \xi_j^\mu & \text{if } i \neq j \\ 0 & \text{otherwise} \end{cases} \tag{1}$$

The energy of the Hopfield network is defined by [12]:

$$E(x) = -\frac{1}{2} \sum_{i=1}^{N} x_i \sum_{j=1}^{N} x_j w_{ij} \tag{2}$$

The value of the energy function is usually lower for stored patterns and higher for other patterns [3]. Therefore, the value of the energy may be used for familiarity discrimination, which in this context corresponds to checking whether a pattern is stored in the Hopfield network [5]. Normally, the Hopfield network is used for retrieval of information by updating one-by-one the activities of the neurons (a process called relaxation). In FamE, the neurons do not perform any computations (i.e., there is no relaxation), but familiarity discrimination is achieved by checking the value of the energy function after delivery of a pattern. In other words, the discrimination is done not by the Hopfield network itself, but by an external entity which sets up the activations of the neurons according to a discriminated pattern and calculates the network's energy for this pattern.

We have already showed that the average value of the energy for stored patterns is $-N/2$, while for novel patterns it is 0 [5]. Therefore, by taking as a threshold the middle value of $-N/4$, we can define a familiarity discrimination criterion, namely, if $E < -N/4$, then the pattern is classified as familiar, and as novel otherwise.

In [6] we showed that the neural network designed to mimic neuronal activity in the perirhinal cortex performs similar computations during familiarity discrimination. The energy of the Hopfield network is an artificial function whose value is calculated by a double summation (see Equation 2). The model perirhinal network effectively calculates a similar function also by a double summation implemented by two layers of neurons - the first layer performs the first summation and the second layer the second summation. For details see [6].

3   Storage Capacity

Using signal-to-noise analysis we have established that the FamE algorithm using a network of N neurons can discriminate familiarity with 99% reliability for $0.023N^2$ uncorrelated patterns [5]. This capacity is much greater than the standard capacity of the Hopfield network for retrieval, namely $0.145N$ [3]. If a higher reliability is required, the familiarity discrimination capacity decreases slightly but it is still of order N^2, e.g., for an error probability of 10^{-4} the capacity is about $0.009N^2$ and for an error probability of 10^{-6}, about $0.006N^2$.

If the human perirhinal cortical network operates on similar principles, its theoretical capacity may be estimated on the assumption that it contains $\sim 10^7$ pyramidal neurons [13], 25% of which discriminate familiarity, each with $\sim 10^4$ synapses. With a probability of error of 10^{-6}, the perirhinal network could store $\sim 10^8$ patterns, each with up to 0.25×10^7 bits. A pile of books storing these patterns would be ~ 7000 km high (equal to the Earth's radius). The speed of searching this database is also impressive. It would take 20 ms for light to traverse the pile, while discriminating familiarity in the model of the perirhinal cortex takes only ~ 10 ms.

FamE demonstrates generalisation and is resistant to disruption by noise - a pattern will still be classified as familiar even if it differs in a substantial proportion of its bits from its previous representation. More precisely, a pattern will be classified as familiar if the Hamming distance (number of different bits) between the pattern and one of the stored patterns is small. Therefore, before applying the methods one should ensure that data is represented in such a way that similar pieces of information (from the point of view of the user) have similar representations in Hamming space.

The above considerations concerning capacity assume that the weights of the network are represented as real numbers. When in the Hopfield network the weights are replaced by binary values, i.e., positive weights by 1 and negative weights by -1, then the capacity drops from $0.145N$ to about $0.1N$, i.e., it decreases by a factor of 0.69 [11]. A similar decrease is observed in the case of FamE. For example, for a probability of error of 1%, the capacity decreases from $0.023N^2$ to $0.016N^2$. After the conversion of the weights to binary values, the average value of the energy for stored patterns is not $-N/2$ anymore. Hence, the energy discrimination threshold (below which patterns are classified as familiar)

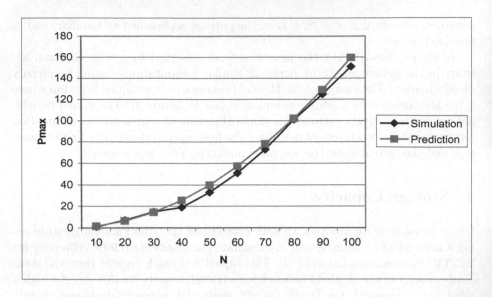

Fig. 1. Comparison of the simulated familiarity discrimination capacity of the network with binary weights, with theoretical predictions

cannot be taken as $-N/4$. The threshold should be found empirically. First, one should compute what the average energy value is for patterns stored in the network (by delivering stored patterns one by one to the network and calculating the energy without modifying the weights). Then, one should compute the average energy for a number of patterns which are not stored in the network: the discrimination threshold should be taken as the mean of these two averages.

Figure 1 compares the simulated familiarity discrimination capacity of the network with binary weights, with theoretical predictions. For each number of inputs N, and for each number of stored patterns P, the behaviour of the network was tested on 1000 random patterns. Among these patterns, 500 were stored patterns and 500 were random patterns for which the absolute value of the correlation with each stored pattern was less than 0.5. For each number of inputs N, P_{max} is taken as the maximum number of stored patterns for which the error rate is $\leq 1\%$.

In the case of traditional databases, a database storing K records of length N bits occupies NK bits of memory (e.g., on the disk). The number of bits occupied on average by one record may be defined as the memory occupied by the database divided by the number of records, that is, $NK/K = N$. In the case of a network with N neurons and binary weights, the database occupies $N^2/2$ bits, because each weight is represented by one bit and the weights are symmetrical, so only half of them need be stored. Assuming a probability of error of 1%, the number of records for which FamE may discriminate familiarity is $0.023N^2 \times 0.69$. Hence, the number of bits occupied on average by one record is equal to $(N^2/2)/(0.023N^2 \times 0.69) = 31.51$ bits (i.e., 3.94 bytes). In contrast to traditional databases, here the space occupied by a record is constant and does not depend on the number of bits in the record. For a probability of error of 10^{-4} the space occupied by a record is 80.52 bits (10.6 bytes) and for a probability of error of 10^{-6} the space is 131.75 bits (16.47 bytes). These data show that FamE is especially useful for databases containing large records. For example, for a database with records of 1.5KB, the weights of the network allowing FamE to discriminate familiarity with a probability of error of 10^{-6}, would occupy only about 1% of the space taken by the database with the records.

4 Implementation

Section 2 describes how the FamE algorithm works. However, before using it for practical applications, some implementation issues need to be considered.

Every Hopfield network has limited capacity for familiarity discrimination. The size of the network required by a particular application (i.e., the number of neurons) is determined by the number of bits in the records. The capacity of the network is determined by its size and required probability of error. Hence, the network capacity may differ from the capacity required by the application (e.g., the number of records in the database). If the capacity required is smaller, then one can use a network with sparse connections - we have showed that if connections are removed from the network, the storage capacity decreases in proportion to the number of connections removed [6]. If the capacity required is larger than can be achieved using a single network, then one can use a number of Hopfield networks and discriminate familiarity by checking the energy of every network one-by-one.

There are two types of errors the FamE algorithm may make: to classify a novel pattern as familiar (false-recognition) and to classify a familiar pattern as novel (non-recognition). Although, as we showed in [5] false-recognition errors cannot be avoided, non-recognition errors can be eliminated. There are two methods of eliminating or reducing non-recognition errors. First, in the case where many networks are used (to accommodate a database larger than the capacity of a single network), the patterns written to the first network and not recognised by it may be included in the set of patterns being written to the second network, and so on. Second, the number of non-recognition errors may also be reduced by using a different algorithm to modify weights. Instead of presenting all the patterns to the network and modifying the weights only once for each pattern according to Equation 1, one can present patterns in a number of epochs (i.e., show each pattern once and modify the weights, then show the patterns again and so on). One should start with all the weights initialised to zero and for each pattern modify the weights by an amount δ depending on how confident the network is of the novelty of the pattern:

$$\Delta w_{ij} = \frac{1}{N}\delta x_i x_j, \text{ where } \delta = \begin{cases} \dfrac{N/2 + E(x)}{N/2} & \text{for } E(x) > -N/2 \\ 0 & \text{otherwise} \end{cases} \quad (3)$$

The coefficient δ is large for patterns having high energy (and thus classified as novel). Hence for such patterns the magnitude of weight modification is high. On the other hand, δ is small for patterns having low energy, so for these patterns the weights are not changed significantly. Equation 3 is a generalisation of Equation 1, since when $\delta = 1$, they are equivalent. We have showed in [7] that this algorithm always finds values of weights resulting in the correct classification of stored patterns as familiar if such values exist. In practice this means that in many cases it eliminates non-recognition errors completely. In [7] we also proposed a biological mechanism which may implement a similar mechanism in the perirhinal cortex.

The results presented in section 3 concerned the case where patterns are not correlated. The condition of uncorrelation is likely to be satisfied by the patterns of neuronal activity for which the network of the perirhinal cortex discriminates familiarity because, as suggested in [4], the activity of sensory neurons encodes independent features.

The storage capacity of the Hopfield network decreases when the stored patterns are correlated [11]. A similar decrease is also observed in the case of FamE. For many real-world applications, the binary representations fo records are correlated. For example when the records contain text in ASCII code, then the most significant bit of each byte is equal to zero because in ASCII code because letters have code values less than 128 in ASCII. As demonstrated in section 5, the correlation in data decreases the capacity dramatically, hence normally some form of pre-processing will be required before FamE is used. In the remainder of this section a few such pre-processing techniques are suggested, but usually, specialised pre-processing will need to be developed for each application.

When the records in a particular application contain text, the simplest method for removing correlation is to code the text into smaller numbers of bits, or remove the most significant bit from each byte. We will refer to this method as *7-bit coding*. Another simple technique consists of using a code in which each alphanumerical symbol is assigned a random 8-bit code instead of its standard ASCII code. We will refer to this method as *random coding*. Since this code is generated randomly, it possesses much less structure than ASCII and the text records coded by random coding are less correlated than those coded in ASCII. However, these two simple operations can only go so far in reducing correlation.

The method most commonly used for removing correlation is Principal Component Analysis (PCA) [20]. Although PCA is normally used for data having continuous values, it may also be used for binary-valued data. The number of inputs to PCA may be set as the number of bits. The number of outputs should be smaller since PCA also removes redundancy. PCA transforms the records to vectors of real values and those real-valued vectors can be simply converted to binary records by replacing positive values by 1 and negative values by -1.

Removing correlation is one element of pre-processing, but one should also ensure that similar pieces of information (from the point of view of the user) have similar binary representations in Hamming space. This pre-processing is specific

Table 1. Meaning of Bits in Phonetic Coding

vowels			consonants				
lips pos.	tongue pos.		voicing	pos. of articulation		manner of articulation	
spread	front, middle	back, middle	unvoiced	bilabial, labiodental, alvedar	alvedar, palatal, velar, grottal	plosive, fricative	fricative

for each application, but here we show an example for record de-duplication - an application on which we are currently testing FamE. This work is in collaboration with Optima, a company that produces database management software for storing customer data for marketing purposes. These databases may contain several millions of records and the records are quite long. The databases also often contain duplicates of records, e.g., the same person can be included several times with a misspelled name or an address in a different format. The duplicates should be removed because sending marketing information to the same person twice increases the costs and is not good for the company image. Currently, during de-duplication, a special key for each record is generated and each key is compared with all the others. So de-duplication of a database containing K records requires $K^2/2$ comparisons, which is inefficient, especially considering the size of the databases. Hence, de-duplication is an appropriate application for FamE.

Since duplicates are often the result of misspelling, pre-processing should ensure that letters with similar pronunciation have similar representation. In different languages particular letters are pronounced in different ways. Furthermore, in many languages, pairs of letters represent only one sound (e.g., 'th' in English). For simplicity, we assume one sound per letter and Latin pronunciation.

We code each letter into one byte, where each bit represents a certain property of pronunciation. We refer to this coding as *phonetic coding*. The sounds of speech may be classified based on the way in which the air stream is modified by the vocal tract. The two basic groups of sounds are vowels and consonants. In our phonetic coding, we use the first 3 bits to encode vowels and the last 5 bits to encode consonants (i.e., for each vowel, the last 5 bits are set to 0, and for each consonant, the first 3 bits are set to 0).

The first bit of a vowel's code represents lips position during articulation (i.e., spread or round), and the second and third bits represent tongue position (i.e., front, middle or back). The first bit of a consonant's code (i.e., fourth bit of the phonetic code) represents voicing (i.e., voiced or unvoiced), the second and third bits represent the position of articulation (i.e., bilabial, labiodental, alvedar, palatal, velar or grottal), and the fourth and fifth bits represent the manner of articulation (i.e., plosive, fricative, nasal, lateral or aproximal) [18]. The detailed meaning of each bit is shown in Table 1.

Table 2. Comparison of Errors Made by FamE on Text with Different Pre-processing Methods

Number of patterns per network	447	175	106
Theoretical prediction of error	1%	10^{-4}	10^{-6}
Error without pre-processing	67.44%	62.16%	54.89%
Error with 7-bit coding	67.25%	51.74%	41.85%
Error with random coding	26.19%	9.49%	1.86%
Error with PCA	7.2%	0.37%	0.03%
Error with random coding and PCA	1.16%	0.05%	0.03%

In phonetic coding, the codes of letters with similar pronunciation, such as 't' and 'd', and 'm' and 'n', differ only in one bit, while the codes of letters with different pronunciations are very different. Other methods of pre-processing suitable to text are detailed in [9].

5   Simulation Results

Section 3 compares the storage capacity obtained in simulation with theoretical predictions for random, uncorrelated data. However, as we mentioned in the previous section in most applications the representation of records are correlated. This section shows how FamE performs for correlated data. We describe tests carried on two sets of data, one containing text and one made up of customer records.

We tested FamE on records containing text by taking a large amount of text, dividing it into parts of equal length, and building a database with records containing these parts of the text. As the text we chose part of The Holy Bible, Luke's Gospel, chapters 1-10. The text occupied 65390 bytes. Different methods of pre-processing were tested and the size of the records was chosen in such a way that each record after pre-processing occupied 21 bytes. Hence, the size of the network was 21x8 = 168 neurons. The number of networks used to accommodate the whole database depended on the choice of probability of error. The algorithm was tested for each network storing 447, 175 and 106 patterns - these numbers were selected because they yield theoretical predictions of error of 1%, 10^{-4} and 10^{-6}, respectively. To eliminate non-recognition errors, the weights were modified according to Equation 3 in two epochs, and any record still not recognised was written to the next network (see section 4). After writing information about the records to the network, the weights of the neurons were converted to binary values. Then, the performance of the algorithm was tested by checking the familiarity of all records written to the network and an equal number of records from the database which had not been written to the network. The errors were averaged over all the networks used to accommodate the database. The errors obtained for different methods of pre-processing are shown in Table 2.

Table 3. Comparison of Errors Made by FamE on Customer Database with Different Pre-processing Methods

Number of patterns per network	228	89	54
Theoretical prediction of error	1%	10^{-4}	10^{-6}
Error without pre-processing	52.09%	45.39%	41.96%
Error with random coding	47.46%	30.49%	20.77%
Error with phonetic coding	66.82%	54.59%	47.46%
Error with PCA	11.08%	0.51%	0.05%
Error with random coding and PCA	0.74%	0.0%	0.0%
Error with phonetic coding and PCA	5.63%	0.19%	0.03%

When FamE is applied to text without any pre-processing, it performs even worse than chance. When one observes the results for particular networks, one can see, that for the first network the error is about 50% (i.e., it performs at random) and for the following networks the error increases. This increase comes from the fact that records not recognised by one network are in the set for the next network. The non-recognised records are the ones which are particularly difficult for FamE, so they increase the error of the following networks.

When 7-bit coding is used (before pre-processing the records had a length of 24 bytes), the error is still very large. When random coding is used, the error decreases slightly but is still unacceptably large.

The error is strongly reduced by using PCA pre-processing. Before pre-processing the records had a length of 35 bytes. They were divided into 7 chunks of 5 bytes, and PCA was applied to each chunk separately to reduce the size of the PCA network. Each chunk was reduced to 3 bytes so the records were reduced to 3x7 = 21 bytes. When PCA was applied to text coded in ASCII, the error decreased to reasonable values (row 'Error with PCA' in Table 2). It is still far from the theoretical prediction because the correlation in ASCII code is very large. Although the outputs from PCA are always uncorrelated [20], the patterns delivered to the familiarity discrimination network after PCA pre-processing do not have to be, since they are created by converting real-valued PCA outputs to binary patterns. When one first applies random coding, and then PCA, the actual error approaches the predicted one (last row of Table 2).

The performance of FamE was also tested on the records from a customer database provided by Optima. For simplicity, only three fields, namely first name, last name and city, were selected. Different pre-processing methods were tested and the size of the records was chosen in such a way that each record occupied 15 bytes following pre-processing. Hence, the size of the network was 15x8 = 120 neurons. The algorithm was tested for each network storing 228, 89 and 54 patterns - corresponding to theoretical predictions of error of 1%, 10^{-4} and 10^{-6}, respectively. Non-recognition errors were avoided using the same method as in the previous experiment. The errors obtained for different pre-processing methods are shown in Table 3

Table 4. Energy Function for Records with Similar Pronunciations for Different Types of Coding

			Energy Value	
First Name	Last Name	City	Phonetic	Random
Vincent	Hopfield	Bristol	-600	-656
Fynsind	Habfield	Bryztol	-424	-4
James	Bond	London	-156	24
Anakin	Skywalker	Deathstar	-36	-40

The second, third and fourth rows of Table 3 show errors obtained with different types of coding without PCA. In these experiments, each recorded consisted of 15 bytes, where each group of 5 bytes encoded the first five letters of the first name, last name and city, respectively. Table 3 shows the error is very large without PCA.

The last three rows of Table 3 show the errors when PCA pre-processing was used. Before applying PCA, each record consisted of 21 bytes, where 7 bytes were used for each field of the record. The size of each record was reduced to 15 bytes using a single PCA network. The error obtained for random coding is lower than the one obtained with phonetic coding, because phonetic coding gives more correlated records. When one looks for an exact match, phonetic coding is the most appropriate, but in the case of de-duplication we want records with similar pronunciations to be classified as familiar as well.

We performed another experiment to check how random and phonetic codings work for records with similar pronunciations. We added another record to the Optima database: "Vincent Hopfieldi, Bristol". From each record, the first seven letters of the first name, last name and city were encoded using random coding and phonetic coding. The size of each record was then reduced from 21 to 10 bytes using PCA. The database consisted of 54 records and was written to a single Hopfield network of 100 neurons. Then, the energy of the network was checked for the stored record "Vincent Hopfiled", a misspelled version "Fynsind Habfield" and two other records not stored in the database. The values of the energy function are given in Table 4. none of the names in Table 4 were in the original Optima database.

Table 4 shows that for both types of coding, the energy is low for the stored record "Vincent Hopfield, Bristol", and is much closer to zero for records which have not been stored, namely "James Bond, London" and "Anakin Skywalker, Deathstar". However, the energy for the record "Fynsind Habfield, Bryztol" is low for phonetic coding and close to zero for random coding. That is because the representations of "Fynsind Habfield, Bryztol" and "Vincent Hopfield, Bristol" are very similar for phonetic coding, and completely different for random coding. Hence, the phonetic coding is more appropriate when one requires records with similar pronunciation to be classified as familiar.

After analysing the binary representations of "Fynsind Habfield, Bryztol" and "Vincent Hopfield, Bristol" created by phonetic coding and PCA, one can observe that the first bits are mostly the same, while the later bits are different. This is due to the fact that the first principal components are directions in which the data have the highest variance, so the first outputs from PCA carry the most information about the record [20]. The later outputs carry much less information and hence a small change in the record may change their values very much. This property is undesirable, because even very similar records may have many different bits in their representations. Furthermore, when the first outputs of PCA are transformed to binary values, much information is lost by conversion from real to binary numbers. These properties show that PCA, although very simple to implement, is not an ideal method of pre-processing for this application and it would be interesting to investigate other methods of feature extraction.

6   Discussion and Conclusion

It is interesting to compare FamE with other algorithms which may potentially be used for familiarity discrimination, e.g., hashing. Hashing is used in databases, where records are often divided into a large number B of buckets. A hashing function takes a record's key and produces an integer between 0 and B-1, determining in which bucket the record should be stored.

Let us consider the following algorithm for familiarity discrimination. Instead of storing records let us just store the values of the hashing function for these records. To check whether a record is stored in the database, we simply compute the value of the hashing function and check whether this value is stored in the database.

The above algorithm seems to be very naive and vulnerable to errors. However, with a good hashing function, whose outputs have a large enough number of bits, the probability of error is very small. For example, if the database has 10^6 records and the value of the hashing function is encoded on 5 bytes, the probability of error is less than $10-6$. Thus, great "compression" of information may be achieved, as in FamE. In addition, it is possible (but with very low probability) that a new record maps to the same value of the hashing function as one of the stored records. Hence, it would be classified as a stored record and the algorithm would make false-recognition errors. It is interesting that non-recognition errors will never be made by this hashing-based algorithm, analogously to the fact that non-recognition errors may be eliminated from FamE.

Familiarity discrimination using the hashing function is very vulnerable to noise, however. If two patterns differ even in a single bit, their values under a hashing function may be very different. On the other hand, FamE is very robust to noise, since it classifies a pattern as familiar even if it differs from the stored one in several bits.

The FamE algorithm is suitable for hardware implementation. Firstly, all the internal summations in Equation 2 may be done in parallel for each i. Hence, calculating energy could be done in just two processing steps. Secondly, FamE is very robust to damage - loss of connections between neurons or even of whole neurons causes only a decrease in capacity proportional to the damage [5].

If such fast implementation were available, FamE could be used for searching massive databases, when one does not require an exact match. Currently, searching databases for near-exact matches is slow because normal searching techniques such as indexing or hashing are not applicable. Large databases could be divided into parts and records from each part encoded in the weights of the Hopfield network thus creating a kind of "neural index". In order to find a record one can check the energy in each network and in this way identify whether a similar record is stored in the database and in which part. One may then search only the identified part using other techniques. The property of FamE that non-recognition errors can be eliminated guarantees that if the record exists in the database it will always be found. The false-recognition errors do not affect this application because, even if the algorithm falsely indicates that the record is stored in one part of the database, the further search of this part will show that it is not stored.

This paper discusses the Familiarity discrimination based on Energy (FamE) algorithm, inspired by the presumed computations of the perirhinal cortex. The algorithm allows fast and accurate familiarity discrimination with high storage capacity. The initial experiments demonstrate that FamE may be applied to real-world problems. Many new applications for FamE are likely to emerge in the future, especially due to the increasing sizes of Internet databases.

Acknowledgements. We are grateful to Steven Cole for useful comments and help in implementation, and to Optima for providing the database and support. This work is supported in part by ORS and MRC grants.

References

1. Aggleton, J.P. and Shaw, C. (1996). Amnesia and recognition memory: a re-analysis of psychometric data. *Neuropsychologia*, **34**:51-62.
2. Aggleton, J.P. and Brown, M.W. (1999). Episodic memory, amnesia and the hippocampal-anterior thalamic axis. *Behavioral Brain Science*, **22**:425-498.
3. Amit, D.J. (1989). *Modelling Brain Function*. Cambridge University Press, Cambridge, UK.
4. Barlow, H.B. (1989). Unsupervised Learning. *Neural Computation*, **1**:295-311.
5. Bogacz, R., Brown, M.W. and Giraud-Carrier, C. (1999). High capacity neural networks for familiarity discrimination. In *Proceedings of the International Conference on Artificial Neural Networks (ICANN'99)*, Edinburgh, UK, 773-776.
6. Bogacz, R., Brown, M.W, and Giraud-Carrier, C. (2000). Model of familiarity discrimination in the perirhinal cortex. Submitted.
7. Bogacz, R., Brown, M.W. and Giraud-Carrier, C. (2000). Frequency-based error back-propagation in a cortical network. To appear in *Proceedings of the International Joint Conference on Neural Network (IJCNN'00)*, Como, Italy.
8. Brown, M.W. and Xiang, J.Z. (1998). Recognition memory: Neuronal substrates of the judgement of prior occurrence. *Progress in Neurobiology*, **55**:149-189.

9. Cole, S. (2000). Record de-duplication using a familiarity discrimination neural network. Final Year Project, University of Bristol, Department of Computer Science.

10. Granger, E., Grossberg, S., Rubin, M.A. and Streilein, W.W. (1998). Familiarity discrimination of radar pulses. *Advances in Neural Information Processing Systems*, **11**:875-881.

11. Hertz, J., Krogh, A. and Palmer, R.G. (1991). *Introduction to the Theory of Neural Computation*. Addison Wesley.

12. Hopfield, J.J. (1982). Neural networks and physical systems with emergent collective computational abilities. *Proceedings of the National Academy of Science*, **79**:2554-2558.

13. Insausti, R., Juottonen, K., Soininen, H., Insausti, A.M., Partanen, K., Vainio, P., Laakso, M.P. and Pitkanen, A. (1998). MR volumetric analysis of the human entorhinal, perirhinal and temporopolar cortices. *American Journal of Neuroradiology*, **19**:659-671.

14. Kohonen, T., Oja, E. and Ruohonen, M. (1974). Adaptation of a linear system to a finite set of patterns occurring in an arbitrarily varying order. *Acta Polytechnic Scandinavian Electric Engineering*, **25**.

15. Kohonen, T. (1989). *Self-organisation and Associative Memory*. Springer-Verlag, Heidelberg, Third Edition.

16. Murray, E.A. (1996). What have ablation studies told us about the neural substrates of stimulus memory? *Seminars in Neurosciences*, **8**:13-22.

17. Murray, E.A. and Bussey, T.J. (1999). Perceptual-mnemonic functions of the perirhinal cortex. *Trends in Cognitive Science*, **3**:142-151.

18. Owens, F.J. (1993). *Signal Processing for Speech*. Macmillan Press, London.

19. Roberts, S. and Tarassenko, L. (1995). A probabilistic resource allocating networks for novelty detection. *Neural Computation*, **6**:270-284.

20. Sanger, T.D. (1989). Optimal unsupervised learning in a single-layer feedforward neural network. *Neural Networks*, **2**:59-473.

21. Standing, L. (1973). Learning 10,000 pictures. *The Quarterly Journal of Experimental Psychology*, **25**:207-222.

22. Xiang, J.Z. and Brown, M.W. (1998). Differential neuronal encoding of novelty, familiarity and recency in regions of the anterior temporal lobe. *Neuropharmacology*, **37**:657-676.

Linguistic Computation with State Space Trajectories

Hermann Moisl

Centre for Research in Linguistics, University of Newcastle upon Tyne

Abstract. This paper addresses the key question of this book by applying the chaotic dynamics found in biological brains to design of a strictly
sequential artificial neural network-based natural language understanding (NLU) system. The discussion is in three parts. The first part argues that, for NLU, two foundational principles of generative linguistics,
mainstream cognitive science, and much of artificial intelligence –that
natural language strings have complex syntactic structure processed by
structure-sensitive algorithms, and that this syntactic structure determines string semantics– are unnecessary, and that it is sufficient to process strings purely as symbol sequences. The second part then describes
neuroscientific work which identifies chaotic attractor trajectory in state
space as the fundamental principle of brain function at a level above
that of the individual neuron, and which indicates that sensory processing, and perhaps higher cognition more generally, are implemented by
cooperating attractor sequence processes. Finally, the third part sketches
a possible application of this neuroscientific work to design of an a sequential NLU system.

Introduction

The key question of this book is: 'What can we learn from cognitive neuroscience
and the brain for building new computational neural architectures?'. This paper
addresses that question in relation to natural language processing (NLP), and
the answer involves application of chaotic dynamics found in biological brains
to design of artificial neural network (ANN)-based string processing architectures. Specifically, it argues that, in designing and implementing NLP systems
for semantic interpretation of natural language strings, henceforth referred to as
natural language understanding (NLU) systems:

1. Two foundational principles of generative linguistics, mainstream cognitive
   science, and much of artificial intelligence (AI) and NLP –that NL strings
   have complex syntactic structure processed by structure sensitive algorithms,
   and that this syntactic structure determines string semantics– are unnecessary, and it is sufficient in principle for NLU purposes to process strings
   purely as symbol sequences, but there are substantial practical problems
   associated with sequential NLU.
2. Neuroscientific results support the principle of sequential NLU, and also
   provide potential solutions to the practical problems associated with it.

S. Wermter et al. (Eds.): Emergent Neural Computational Architectures, LNAI 2036, pp. 442–460, 2001.

3. It consequently makes sense to investigate the feasibility of sequential NLU by developing an NLU system based on the results in (2).

Two preliminary comments. The first is that NLU is here understood as language engineering –as the design and implementation of machines that process natural language for some purpose– and not as cognitive modeling. Secondly, given the current deluge of results on brain mechanisms generated by new imaging techniques, there are bound to be and in fact are controversies about their interpretation. Given the above construal of NLU, however, there is no need to engage in these controversies: it is legitimate to select and use brain mechanisms proposed by neuroscience purely on the grounds that they seem useful from an engineering point of view. The ideal, of course, is to engineer in accordance with the way the brain really does language, since it is the only known physical device that implements all human linguistic abilities and is thus the definitive model, but pending final understanding of the relevant mechanisms, there is no need to wait for the dust to settle.

1   The Need for Complex Syntactic Structure in NLU

This section is in two main parts. The first motivates the challenge to the need for complex syntactic structure in NLU, and the second presents the case for the principle of sequential mapping of NL strings to meanings.

1.1   Motivation

In the 1930s and 1940s, mathematical logicians formalized the intuitive notion of an effective procedure as a way of determining the class of functions that can be computed algorithmically. A variety of formalisms was proposed –recursive functions, lambda calculus, rewrite systems, artificial neural networks, automata– all of them equivalent in terms of the functions they can compute. Automata theory was to predominate in the sense that, on the one hand, it provided the theoretical basis for the architecture of most current computer technology, and, on the other, it is the standard computational formalism in numerous science and engineering disciplines. Automata theory was, moreover, soon applied to modeling of human intelligence, and has dominated thinking about human cognition ever since [43,40,20,39,80]. This approach to cognitive modeling attained apotheosis in the late 1970s, when Newell and Simon [61] proposed the Physical Symbol System Hypothesis (PSSH), where 'physical symbol system' is understood as a physical implementation of a mathematically stated effective procedure, a prime example of which is a programmed Turing Machine. The essence of the PSSH approach to cognitive modeling was set out by Fodor and Pylyshyn in 1988 [32]:

- There are representational primitives –symbols– called atomic representations.
- Being representational, symbols have semantic content, that is, each symbol denotes some aspect of the world.

- A representational state consists of one or more symbols, each with an associated semantics, 'in which (i) there is a distinction between structurally atomic and structurally molecular representations, (ii) structurally molecular representations have syntactic constituents that are themselves either structurally molecular or structurally atomic, and (iii) the semantic content of a representation is a function of the semantic contents of its syntactic parts, together with the syntactic structure'.
- Input-output mappings and the transformation of mental states 'are defined over the structural properties of mental representations. Because these have combinatorial structure, mental processes apply to them by virtue of their form'.

Therefore, 'if in principle syntactic relations can be made to parallel semantic relations, and if in principle you have a mechanism whose operation on expressions are sensitive to syntax, then it is in principle possible to construct a syntactically driven machine whose state transitions satisfy semantic criteria of coherence. The idea that the brain is such a machine is the foundational hypothesis of classical cognitive science'.

Modeling of the human language faculty has been paradigmatic for the PSSH-based approach in cognitive science. The discipline concerned with this language modeling, generative linguistics, has developed considerably since Chomsky's pioneering work in the mid-1950s, but at least one foundational principle has remained unchanged: that NL strings have a structure beyond the strictly temporal or spatial sequentiality of speech utterances and text. More specifically, they are held to have a compositional phrase structure of the sort described above by Fodor and Pylyshyn, or, slightly more formally, given a phrase structure grammar (or equivalently, an automaton) that generates a language L, the tree diagrams that represent the structure of sentences in L must allow simultaneous left and right phrasal nonterminal branching from parent nodes; this is what is meant by 'complex syntactic structure' throughout the current discussion. The notion that such phrase structure determines sentence meaning is, moreover, central to currently dominant approaches to NL semantics [55,76, 25,11,26]. In model theoretic semantics [63,54], for example, the link between syntactic structure and meaning is made via the principle of compositionality, which says that the meaning of a syntactically complex expression is a function of the meanings of its constituent words and of the syntactic rules by which they are combined, and are realized by grammars in which each syntactic rule is associated with a semantic rule that specifies the meaning of a constituent in terms of the meanings of its own immediate constituents (the 'rule-to-rule' hypothesis). A derivation according to such a grammar on the one hand generates a sentence with an associated syntactic structure, and on the other the meaning of the sentence; this explains how syntactic structure systematically determines the meanings of the sentences of a language.

The PSSH has, furthermore, been influential not only in the cognitive sciences but also in AI, here understood as an engineering discipline which seeks to design and implement machines that emulate (aspects of) human behaviour. AI has historically been closely associated with PSSH-based cognitive science [43,73]: in essence, cognitive science has proposed models, and AI has attempted to implement them. NLU, more particularly, has been closely tied to developments in generative grammar.

Now, since the early 1980s, perceived shortcomings of the PSSH-based approach to cognitive science has generated a variety of challenges to its dominance –by artificial neural network [58,2,21,22] and dynamical system theories [83,48, 69,10,84,38,62,4] as paradigms for cognitive modeling, and by a radical shift of emphasis away from the study of the innate 'higher' functions like reasoning and language to concentration on developmental issues like the evolution of cognition and the interrelationship of mind, brain, body, and external environment [47,23, 46,39,62]. At the same time, the main thrust of AI research has retreated from the grand aim of general, human-level machine intelligence and has concentrated on developing systems that work well in restricted domains, such as expert systems, handwriting and speech recognition, and robotics and vision in specific industrial applications [31,57]. In response, at least some of those AI researchers who still believe in the achievability of general machine intelligence have increasingly turned to the same approaches as their cognitive science colleagues [6,3,82,29,57]. In NLU specifically [42,7,1,27,24], there now exist practical natural language interfaces in domain-specific applications, primarily databases and expert systems, but to my knowledge no reliable, broad-coverage natural language understanding system currently exists, nor do the prospects for one look promising. The problem is not with syntax –after several decades' effort, parsing algorithms capable of supporting reasonably large-scale general NLU systems are available [67,1,77,66,5] – but with implementation of semantics and pragmatics [42,1,68,56,71]. For natural language understanding, linguistic expressions must be related to the system's awareness of, interaction with, and expectations of the world, and the main lesson of the NLU work done in the 1970s and early 1980s is that any NLU system needs, at the very least, to represent, manipulate, and update real-world knowledge efficiently. Various knowledge and belief representation and update mechanisms have been developed [49,1,57], such as logic formalisms, semantic networks, schemas, frames, scripts, and rules, and while these have been shown to work well in small-scale, carefully managed applications, none has so far been scaled up successfully to usefully large general NLU systems.

The motivation for proposing to dispense with compositional structure and structure-sensitive processing in NLU, therefore, is simply that it has reached an impasse [30] (see however reviews of [30] in Artificial Intelligence 80 (1996)). This is not to claim that the PSSH-based approach to NLU cannot succeed. If human language, and cognition more generally, are computable functions [64,65, 9], then a PSSH virtual machine that implements them must exist. The problem is that no one has found it, nor does it look like it will be found soon, and as such it seems sensible to try other approaches.

1.2   Sequential Mapping of Strings to Meanings

It should be uncontroversial to observe that the research agendas of generative linguistics and NLU differ: to judge by their extensive research literatures, linguistics is science and NLU is engineering. Like any other science, linguistics aims to state empirically adequate and maximally expressive theories to explain its domain of interest, whereas the aim of NLU is to design and construct physical machines that respond to natural language utterances or text in an acceptably human-like way. Being physical, every NLU system is bounded in all aspects of its operation –in the lengths of its input and output strings, in the number of strings in can process in its operational lifetime, and in the memory and processing time at its disposal. This means that the unbounded input and output tapes and memories of automata theory need not be a design consideration, and in particular that the I/O sets which an NLU device will be asked to process must be finite. Now, a fundamental result in automata theory is that any finite mapping can be implemented by a finite state automaton (FSA) [44,8], and to an FSA every string has the same structure: strict sequence. It follows that any NLU function can be implemented by an FSA, and that sequential processing of input strings is sufficient for the purpose.

That FSA architecture, and thus sequential processing, is sufficient for NLU has been known since the early years of generative linguistics. Indeed, Chomsky himself pointed it out [12,14,15]. Despite that, finite state NLU has been all but ignored until fairly recently [74,51,18,50,19,52,70]. The various reasons for the lack of interest in FSAs are valid from a generative linguistics point of view, but irrelevant for NLU. These are briefly dealt with in what follows (see [16] and [17] for closely related discussions).

- Trivial finiteness
  One might want to argue that to insist on the finiteness of NLU functions is an excessively theoretical point –that string length, I/O set sizes, and the number of processing states / memory size in a real-world application might be so large as to be unbounded for all but the most abstemiously theoretical purposes– and that this disqualifies FSAs in practice. This has some initial plausibility, but consider these two arguments. Firstly, how long are NL strings in daily speech intercourse and text production? There are almost certainly studies that provide maxima, minima, means, and standard deviations for string length across different languages, but speaker / reader intuition serves to make the required point here (for a discussion of sentence length bounds see [53]). In everyday fact, most strings are very short – a dozen words, perhaps, or occasionally two or three times that number. Few extend beyond, say, 100 words, and those that do become increasingly incomprehensible. For NLU purposes, therefore, we are not dealing with what one might call trivially bounded strings which are so long –perhaps 100,000,000 words– that they are unbounded for any practical purpose. And, secondly, while it is true that, even with the above practical string length constraint, a native speaker of some language will typically produce a very

large number of strings in a lifetime, one has to keep in mind that NLU
is engineering. Vocabulary size, string length, and the permitted range of
syntactic patterns are under the designer's control. A fully human level of
linguistic competence is very difficult to achieve, as the history of AI over
the past several decades has shown, but if one is prepared to work to a less
general specification the problem size can be scaled down by fiat.

– Weak generative capacity

A fundamental result of formal language and automata theory is that FSAs
cannot generate the language $a^n b^n$, that is, sets of strings in which a sequence of some symbol a is followed by exactly the same length sequence
of another symbol b, where n is any positive integer. It has further been
claimed (misleadingly, but that is another matter) that NLs contain strings
with the $a^n b^n$ pattern, and that FSAs are consequently unable to generate
the class of natural languages [13,41,63]. This argument assumes unbounded
n, which is legitimate given the aims of linguistic theory. But, in terms of
NLU as language engineering, n must be bounded; where n is bounded it is
prespecifiable, and in such a case the $a^n b^n$ pattern can be generated by an
FSA. The argument against finite state NLU from weak generative capacity
is thus irrelevant. An analogous argument applies to the cross-serial dependency pattern that was used to disqualify the class of context free grammars
/ pushdown automata, and by transitivity regular grammars / FSAs, as a
basis for NL modeling [41,79,63].

– Strong generative capacity

Complex syntactic structure is fundamental to the explanatory capacity of
linguistic syntactic theory because it allows intuitions and empirical findings
about natural language word order to be expressed in satisfying generalizations. Because they cannot deal with complex syntactic structures, FSAs are
effectively useless for linguistic modeling. Chomsky pointed out their impov-
erished explanatory capacity in the 1950s, and they have for that reason
rightly been ignored by generative linguists ever since. On the present view,
however, NLU is not linguistic modeling, but is interested in constructing
physical devices with some particular I/O behaviour. And, in view of the
foregoing discussion, there is no good theoretical reason to prefer any one
automaton class over the others in NLU design: a finite string set does not
imply the computational class of the automaton that generates it. In other
words, explanatory adequacy is irrelevant to the choice of computational
architecture for NLU purposes.

– Compositional semantics

As noted, compositionality explains how syntactic structure systematically
determines the meanings of the sentences of a language. An FSA can support a compositional semantics, but that semantics is explanatorily trivial.
Because an FSA treats the syntax of every string as strictly sequential, an
FSA-based compositional semantic theory can assert only one thing –that
the meanings of all sentences in all languages are sequentially concatenative.
This amounts to proposing a semantic theory which pairs each syntactically
legal string with a meaning, and simply lists the pairs. Because a list cap-

tures no generalizations, FSAs are of no interest for NL semantic theory. Once again, however, NLU is not primarily concerned with explanation, and this conclusion is consequently irrelevant. A compositional semantic theory that uses complex syntactic structure gives a particular explanation of how sentences map to meanings, but from a language engineering point of view a list would be equally valid in principle. NLU is, in short, not required to adopt compositionality.

There is, then, no theoretical obstacle to a finite state, and thus sequential, approach to design and implementation of NLP and more specifically NLU devices. In fact, the past decade has seen a marked increase of interest in a finite state approach to NLP in such areas as speech recognition, phonological and morphological analysis, syntactic parsing, and information extraction from text [42,16,59,75,17]. There is, however, a significant silence on semantics, for reasons we are about to come to.

In principle, design of a device that maps strings to meanings is easily formulated. Because NLs are finite sets for NLU purposes, it is possible to define a string-to-meaning function f of (string, meaning) pairs. Once the set exists, all that is required for implementation is an algorithm for table lookup: given a string as input, the device returns the associated meaning as output. An appropriately configured FSA is one possible algorithm. Each lexically distinct string drives the machine through a characteristic state sequence, and the final state, which uniquely identifies the string, is mapped to the associated meaning. As a generative linguistic account of the human language faculty, this is of course completely inadequate since it amounts to no more than a list and hence explains nothing, as noted. For NLU, however, the only issue is whether or not such an approach is feasible in a specific NLU domain. And there are in fact difficult problems with it; the two main ones are:

1. Meaning and its representation
   To be able to construct a (string, meaning) pair list, it is necessary to have a clear idea what 'strings' and 'meanings' are, and to have a way of representing them so that they can be processed. The ontology and representation of symbol strings is straightforward. For meanings they are anything but. There is a long history of philosophical debate about how the concept of linguistic meaning might be understood, and, at the moment, there is a range of possibilities with little agreement among them [Craig 1998]. Model-theoretic NL semantics, pioneered by Montague, is the currently-dominant approach within generative linguistics [63,54], but others, such as the 'use' theory of linguistic meaning propounded by Wittgenstein, are still being developed [45]. Because it relies crucially on the notion of complex syntactic structure in the generation of sentence meaning, model-theoretic semantics is unavailable to a sequential NLU system for practical purposes, and choice among remaining alternatives is not clear. Nor, assuming a choice has been made, is it clear how abstractly-characterized meanings should be represented.
2. List construction
   Assuming a suitable meaning representation has been adopted, it is possible

to compile (string, meaning) pair lists explicitly for small, domain-restricted applications. Extension to general, real world NLU would, however, clearly be extremely onerous and almost certainly impractical.

The neuroscientific results described in the next section offer a good starting point for solutions to these problems.

2 Meaning and Sequential Processing in the Brain

On the basis of his work on biological sensory systems, Walter Freeman proposes fundamental principles of brain function and how these implement essential aspects of human cognitive behaviour. These proposals relate directly to sequential NLU, in that they include both a coherent view of the nature of meaning, and an account of the biological implementation of meaning that features purely sequential brain dynamics, with no reference to complex syntactic structure. We look first of all at Freeman's view of meaning and then at its implementation (what follows is based on [81,33,34,35,36,37,38]). A caveat, however. What follows is a very brief summary of a comprehensive account of human behaviour and its biological basis developed over a lifetime's research. As such, it inevitably and grossly oversimplifies, and may also misconstrue or misrepresent Freeman's work; apologies, where appropriate, are offered in advance.

2.1 Meaning

Intentionality is fundamental to Freeman's view of meaning. This is not the intentionality of twentieth-century analytic philosophy and of PSSH-based cognitive science, where the term denotes the 'aboutness' of mental representations, and is used to designate the relation between mental states and objects or events in the world, whether real or imaginary. Rather, his understanding of the term is derived from that of the thirteenth-century philosopher Thomas Aquinas, for whom intentionality had to do with goal-directed action in the world, and modification of the self in response to that consequences of that action as a way of coming to understand the world and the place of the self in it. Analyzed in terms of intentionality in this sense, an organism's existence in the world over time is a sequence of intents, where an intent has three stages: (1) formulation of a goal to whose realization an action will be directed, (2) perception of the environment, execution of the intended action, and perception of consequences for the self, and (3) learning from these consequences relative to the goal. Intents in the sequence are not independent, but are linked to one another in that goal formulation, execution, and adaptation at any given point in the sequence occur in the context of, and are informed by, the organism's lifetime intentional history up to that point. Thus, a hungry cat formulates a goal of catching food. On the basis of previous intentional action, it knows that it can achieve that goal by hiding in tall grass, priming itself for a spring, and then pouncing on an animal smaller than itself. It has a visual perception of suitable prey and pounces, but because the smaller animal on this occasion is an urban rat that fights back, the cat experiences pain. From the perceptions of rat and of pain, the cat learns to pick on different small animals in future. This sort of intentionality is characteristic of all vertebrates, in that they observably act in furtherance of their survival in specific environments and modify their behaviour in response to the consequences of their actions. There are degrees of intentionality, however: a salamander has much simpler, shorter-term goals and capacities for action and learning than does a human.

Meaning arises from intentionality. It is the organism's learned awareness of the interrelationship of perceived states of the world, its goals, and its actions in pursuance of those goals, at any given point in its intentional evolution. In the above example, the visual perception of a rat comes to mean pain to the cat as a result of learning. Meaning is, moreover, organism-specific: to a bigger, tougher cat perception of a rat might mean something very different. And, again, there are degrees of meaning commensurate with the richness of the intentionality available to a species. A given odour can presumably mean only a limited range of things to a salamander –food, danger, safety– whereas, to a human, it can not only mean such things, but also, for example, a subtle blend of emotional experience.

2.2   The Implementation of Meaning

Freeman models the vertebrate brain as a hierarchy of interacting nonlinear dynamical systems whose global evolution implements intentionality. In humans, these dynamical systems correspond to the modules of the limbic system, the sensory cortices, and the areas of the neocortex associated with higher cognitive functions. All three are required to implement full human intentionality, but to keep the discussion tractable the focus will be on the first two only. This will be sufficient for present purposes: the interaction of limbic system and sensory cortices is the necessary basis for human intentionality, and the principles that Freeman proposes for its operation extend straightforwardly to interaction with the neocortex (on which see [38] chapters 5-7).

The rest of this subsection develops the implementation of meaning in three parts. The first looks at the methodology and results of Freeman's work on the olfactory system, which is the empirical basis for his proposals on general brain function. The second then outlines his interpretation of these results in terms of dynamical systems theory. And the third describes his proposals for extending the dynamical systems analysis of the olfactory system to other perceptual modalities such as vision, and to the integration of these modalities in the enthorinal cortex and the hippocampus so as to implement perceptually grounded meaning.

The Olfactory System. Freeman has concentrated on olfaction because it is the primary perceptual system in evolutionary terms, and, in his view, is fundamental to understanding of the other perceptual modalities.

The operation of the sensory modalities has two distinct aspects: sensation and perception (see also [80]). Sensation is the transduction of external signals to spatial patterns of neural activation through stimulation of neural receptors on the sensory surface, here the nose, whereas perception involves transformation of this spatial pattern within the brain. Freeman has been concerned with perception. He has studied the behaviour of the olfactory system, comprising the nasal receptors, olfactory bulb, and olfactory cortex, in rabbits and cats. This was done by surgically attaching an 8 x 8 array of electrodes to the olfactory bulb. The animals were then trained to respond to conditioned and unconditioned stimuli and, when training was complete, an EEG was used to record the responses of the bulb both to training odorants and to odorants not previously experienced. The main observations relevant to the current discussion were:

- At rest, that is, when no learned odorant was present for the animal to smell, the EEG showed continual background activity with an aperiodic waveform.
- With each inhalation of a learned odorant there was a burst of activity which ended at the onset of exhalation. The burst was an oscillation with a common waveform throughout the bulb, but whose amplitude varied from place to place in it. This amplitude modulation (AM) constitutes a bulb-wide spatial pattern of activation. There was a characteristic AM pattern associated with each learned stimulus in the sense that each presentation of a specific odorant produced an AM pattern that differed from those associated with other learned odorants.
- When the animal was presented with a stimulus that was unconditioned during training, or a stimulus not previously encountered, there was no burst.
- When an animal trained on some set of stimuli was further trained to add a conditioned stimulus to its existing repertoire, two things happened: a new AM pattern was produced when the new odorant was presented after training, and all existing AM patterns were modified to a greater or lesser degree.

The Olfactory System as a Dynamical System. Neural populations within the brain are instances of macroscopic ensembles in an extensive range of naturally-occurring complex systems that evolve over time; examples are ecologies, social groups, and weather systems. They share the following characteristics with such ensembles: (i) there are many semi-autonomous elements, (ii) each element interacts with many others, (iii) the input-output relations of the elements are nonlinear, and (iv) there is an energy source and sink. The behaviour of such natural systems over time is standardly modeled using dynamical systems theory. A brain consists of a very large number of elements with nonlinear input-output behaviour that interact by means of rich and recurrent interconnections, and has a metabolic energy source and sink. As such it is a natural step to extend the dynamical systems approach to the study of the brain as well, and that is what Freeman has done.

The olfactory bulb and cortex constitute a system of coupled oscillators whose nonlinearity and recurrence permits the full range of behaviours associated with nonlinear dynamical systems, including not only point and limit-cycle but also chaotic attractors:

- The aperiodic wave which the EEG detects in the absence of sensory stimulus is the olfactory bulb in a chaotic attractor. Each module on its own has only a point attractor and a limit cycle attractor, that is, an oscillation with a characteristic frequency. The chaotic activity results from the coupling of the modules by both forward and feedback connections. Their characteristic frequencies differ, and they cannot agree on a common frequency, and so oscillate aperiodically. That this activity results from the modules' interaction is clear: on the one hand, it does not arise from stimulation originating elsewhere in the brain, since it persists even when the olfactory system is surgically isolated from the brain, and on the other, if the olfactory modules are surgically separated, they immediately settle into their characteristic oscillation. This chaotic basal state keeps the olfactory system in a constant state of readiness, so that even a small perturbation of the right sort can move it quickly to another attractor.
- The AM patterns are chaotic attractors in the bulb's state space. In a trained system, the stimulus from a specific odorant increases the gain in the bulb, and causes a state transition from the basal chaotic attractor to an attractor associated with that stimulus. On exhalation, the stimulus from the transduction pattern is removed, the gain is reduced, and the bulb returns to its basal chaotic attractor. In other words, the bulb is destabilized by gain from external stimulation.
- The odorant-specific attractors, that is, the AM patterns, are formed during training by the development of assemblies in the bulb via Hebbian synaptic modification. The result, in a trained animal, is that the olfactory bulb has a landscape of chaotic attractors including a basal attractor to which the system reverts in the absence of stimulation, and one for each of the learned odorants. The change in all existing AM patterns when a new odorant is added to the animal's training results from crowding of the attractors in the landscape as a new one is added.

Perceptually-grounded Meaning. On the above understanding of meaning, perception of an odour on its own means nothing; indeed, training on unconditioned odorants does not result in formation of an AM pattern / attractor for the odorant. To mean something, an odorant has to correlate with perceptions from one or more other sensory modalities. How is this implemented in the brain? By perceptual integration in the enthorinal cortex. The enthorinal cortex is one of the modules of the limbic system, and is remarkable first and foremost for the large number of other brain areas with which it interacts. Of particular relevance here is the fact that all the sensory cortices send their output to, and receive feedback from, the enthorinal cortex (see Fig. 1).

This indicates that the enthorinal cortex is the place where perceptual integration is implemented. In support, Freeman notes that, by observing the sensory cortices of rabbits trained in various sorts of conditioned input –light,

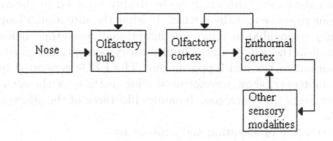

Fig. 1. Multisensory integration in the enthorinal cortex

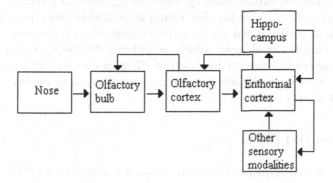

Fig. 2. Sequencing of multisensory perceptions via the interaction of enthorinal cortex and hippocampus

sound, touch– he was able to detect essentially the same dynamic behaviour as he observed in the olfactory system. The implication is that all the sensory systems use the same dynamics, and that, given this uniformity, the enthorinal cortex could in principle integrate its various perceptual inputs into multisensory perceptions or gestalts. Such a gestalt carries the meaning of a particular stimulus for the organism.

It needs, furthermore, to be kept in mind that intentionality is sequential. Individual sensory perceptions and integrated multisensory perceptions are embedded in a dynamic process which implements the organism's interaction with the world in real time. There are two issues to be addressed here (see Fig. 2):

– Implementation of sequencing

   The enthorinal cortex, which as we have just seen is proposed as the locus of multiperceptual integration, sends most of its output to the hippocampus, and via feedback connections also receives most of the hippocampal output, so that the two modules are in constant interaction. Now, the hippocampus has been shown experimentally to be deeply involved in the orientation of behaviour in space and time, which is what the intentional loop does. It is, therefore, reasonable to infer that intentional space-time behaviour is implemented in the sequencing of multisensory perceptions via the interaction of enthorinal cortex and hippocampus. The EEGs generated by this interaction, moreover, show waveforms similar to those of the sensory cortices, which suggests that here, too, dynamics like those of the olfactory cortex are at work.

− Implementation of attention and expectation

  Perception is an active process. Organisms do not simply respond to environmental stimuli, but fix their attention on aspects of the environment and have expectations of the environment for which they prepare. Attention and expectation are implemented by feedback signals that modulate response to the next sensory stimulus. For example, a multisensory perception in the enthorinal cortex is fed back to the various sensory cortices, thereby providing an additional input which, together with the next external sensory input, determines the cortices' output. Response to sensory input is thus not merely reactive, but is sensitive to the temporally prior context in which the current input occurs.

3   Sequential NLU

We have seen that sequential string processing is sufficient for NLU in principle, but that there are serious problems with it in practice. Freeman's proposals on how the brain implements intentionality and thus meaning both support the principle of sequential processing and offer an attractive approach to dealing with these problems. True, the proposals are based to a large extent on research into the role of the limbic system in sensory processing the results of which, when extrapolated to cognition more generally, are hypothetical and need to be empirically tested to gain acceptance in the brain and cognitive sciences. As noted in the Introduction, however, it is sufficient for NLU that they be plausible enough to be worth investigating relative to the engineering problem in question irrespective of their validity for brain function, and in my view they are. This section therefore sketches the relevance of Freeman's proposals to design of a sequential NLU architecture.

3.1   Sequential Processing

Freeman identifies attractor trajectory in state space as the fundamental principle of brain function at a level above that of the individual neuron. Each of the modules in sensory processing subsequent to transduction is driven by input through a sequence of attractors; there are no complex syntactic structures, only cooperating sequential processes. If that view is correct, then, since the brain in fact implements language, sequentiality in this sense must be sufficient. Freeman says little about language, and it may well turn out that, for this characteristically human and thus exceptional cognitive function, the brain does indeed use complex syntactic structure, in which case NLU would be well advised to use it as well. In the meantime, Freeman provides good cause for investigating sequentiality on its own.

3.2   Problems with Sequential NLU

Meaning Representation. However one understands 'meaning', it seems uncontroversial to say that the meaning of linguistic expressions has something to do with the way humans experience the material world. PSSH-based AI and NLU systems have formalized this relationship as a mapping from strings to states of the world, and have attempted to implement such mappings using explicitly-designed, system-internal representations of real-world domains of discourse. How best to represent (aspects of) the world and beliefs about the world is one of the main issues in PSSH-based AI, and a variety of formalisms for it exist, as noted earlier. It was also noted earlier, however, that this approach has only been moderately successful at best, and its relative lack of success is what has motivated alternative approaches to semantic interpretation and implementation over the last two decades or so, including the present one. The main trend in these alternative approaches has been based on a growing conviction of the importance of evolution in understanding cognition and the brain mechanisms that implement it [20] –that brains evolved as controllers for the bodies of creatures which had to survive in specific environments, and that the higher functions like language and reasoning are based on neural mechanisms developed for this purpose. As a consequence, recent 'embodied, embedded' cognitive science and AI have been much more interested than the PSSH tradition ever was in the interrelationship of mind, brain, body, and environment both as a means of explaining cognition and as a way of implementing cognitive functions (for example, [3,47,10,20,23,23,39,46,80]). With respect to meaning in particular [20], it is clear that, notwithstanding the concentration on linguistic meaning in the PSSH tradition, there is meaning apart from language –animals learn to behave meaningfully relative to their environments, and so do human babies before they have acquired language– and the idea that such meaning is the basis for linguistic meaning is at least plausible enough to merit further investigation. Consonant with this general trend, there has been a movement in AI and NLU away from attempting to map strings to explicitly designed representations which, inevitably, reflect a designer's analysis of what is significant in a task domain, and instead to relate language to system-internal states that are learned from interaction with the world without designer intervention. At its most ambitious, this approach aims to embed NLU systems in robotic agents that not only receive inputs from an environment via some combination of sensory transducers, but also interact with and change the environment by means of effectors. The aim is for the agents to develop internal states based on self-organization with the environment –'Concepts are thus the "system's own", and their meaning is no longer parasitic on the concepts of others (the system designer)' [28]– and are intended to learn the meanings of linguistic expressions from their use in environment-interactive situations.

Freeman's work is part of this trend in that it uses self-organizing integration of perceptions abstracted from sensory transducers as the mechanism of meaning implementation. What singles it out is that the proposed mechanism is based on empirical neuroscientific evidence; since the brain is the only mechanism currently known to implement linguistic meaning, the sensible approach is to attempt to mimic it.

Generation of (String, Meaning) Pair Lists. In Freeman's architecture, any given sensory input means the sequence of multisensory perceptions in whose generation it has participated. This includes linguistically relevant acoustic input sequences, and so, for a given string, there is a corresponding multisensory perceptual sequence or, put another way, a (string, meaning) pair. Because the architecture assumes availability of environmental inputs over time, there is no need to compile a (string, meaning) pair list explicitly. Such a list generates itself over time, as the system learns from its interactions with the environment.

Application to ANN-based NLU Design. Because Freeman is specific on the biological neural mechanisms involved in the implementation of intentionality and meaning, it is possible to carry these mechanisms directly over into ANN-based language engineering. Implementation of the architecture in Fig. 2 –including, crucially, the nonlinear dynamics among modules– involves a departure from the way that most ANN-based NLP has been done over the past decade or so [60], which, in essence, has been to implement explicitly-designed input-output mappings using feedforward or recurrent multilayer perceptrons and some variant of backpropagation. The required ANN architecture will have at least the following features:

- Processing units that output pulse trains in nonlinear response to input, not just numerical values which can be interpreted as representing an average firing rate as in currently-used ANN architectures, and that fire asynchronously relative to one another.
- Modules that contain excitatory and inhibitory neurons whose interaction generates oscillatory behaviour.
- Feedforward and feedback connection among modules to enable chaotic behaviour within the modules.
- An unsupervised, local learning mechanism such as the Hebbian to allow for self-organization within modules.

Conclusion

Motivated by the lack of progress in broad-coverage natural language understanding systems based on the processing of complex syntactic structure, the foregoing discussion has proposed sequential processing as an alternative approach to implementation of string-to-meaning mapping. This is theoretically justifiable, but has substantial practical problems associated with it. Freeman's work on the neural basis of intentionality and meaning both supports the principle of sequential processing for implementation of cognitive functions, including language understanding, and also provides potential solutions to the associated problems. As such, it constitutes an attractive basis for development of an ANN-based NLU system.

Clearly, all this amounts to the merest sketch of a possible approach to NLU design. On the one hand, as Freeman himself makes clear, the multisensory perceptions that are synthesized and sequenced by the limbic system are a necessary basis for the implementation of meaning, but not a sufficient one; full linguistic meaning must involve processing of limbic output by the neocortex, and nothing has been said about that here. On the other, none of the fundamental NLU issues, such as the problem of string ambiguity and the roles of memory and reasoning in linguistic comprehension, have been addressed. Whether or not a sequential NLU system based on Freeman's ideas will be able to deal with these issues better than existing approaches, and perhaps succeed in general NLU, remains to be seen.

References

1. Allen J. (1995) Natural Language Understanding. Benjamin / Cummings, Redwood City, California
2. Bechtel W., Abrahamsen A.(1991) Connectionism and the Mind. Basil Blackwell
3. Beer R. (1995) A Dynamical Systems Perspective on Environment Agent Interactions. Artificial Intelligence 72:173-215
4. Beer R.(2000) Dynamical Approaches to Cognitive Science. Trends in Cognitive Sciences 4(3):91-99
5. Briscoe T.(1996) Robust Parsing. In [24]
6. Brooks R.(1991) Intelligence without Representation. Artificial Intelligence 47:139-59
7. Carbonell J., Hayes P. (1992) Natural Language Understanding. In [78]
8. Carroll J., Long D. (1989) Theory of Finite Automata. Prentice-Hall, Englewood Cliffs, New Jersey
9. Chalmers D. (1996) Minds, Machines, and Mathematics. Psyche 2. http://psyche.cs.monash.edu.au/psyche-index-v2.html
10. Chiel H., Beer R. (1997) The Brain has a Body: Adaptive Behaviour Emerges from Interactions of nervous System, Body, and Environment. Trends in Neurosciences 20:553-7
11. Chierchia G., McConnell-Ginet S. (2000) Meaning and Grammar: an Introduction to Semantics. 2nd ed. MIT Press, Cambridge, MA
12. Chomsky N. (1957) Syntactic Structures;. Mouton, s'Gravenhage
13. Chomsky N. (1959) On Certain Formal Properties of Grammars. Information and Control 1:91-112
14. Chomsky N. (1961) On the Notion 'Rule of Grammar'. Proceedings of the 12th Symposium in Applied Mathematics, American Mathematical Society

15. Chomsky N., Miller G. (1965) Finitary Models of Language Users', Readings in Mathematical Psychology 2, ed. Bush R., Galanter E., Luce D. John Wiley, New York
16. Christiansen M. (1992) The (Non)-necessity of Recursion in Natural Language Processing. Proceedings of The 14th Annual Conference of the Cognitive Science Society, University of Indiana
17. Christiansen M., Chater N. (1999) Toward a Connectionist Model of Recursion in Human Linguistic Performance. Cognitive Science 23:157-205
18. Church K. (1980) On Memory Limitations in Natural Language Processing. TR MIT/CS/TR-45, Massachusetts Institute of Technology
19. Church K., Ejerhed E. (1983) Finite State Parsing. Papers from the Seventh Scandinavian Conference of Linguistics, University of Helsinki
20. Cisek P. (1999) Beyond the Computer Metaphor: Behaviour as Interaction. In [62]
21. Clark A. (1989) Microcognition: Philosophy, Cognitive Science, and Parallel Distributed Processing. MIT Press, Cambridge MA
22. Clark A. (1993) Associative Engines: Connectionism, Concepts, and Representational change. MIT Press, Cambridge MA
23. Clark A. (1997) Being There. Putting Brain, Body, and World Together Again. MIT Press, Cambridge MA
24. Cole R., Mariani J., Uszkoreit H., Zaenen A., Zue V. (1996) Survey of the State of the Art in Human Language Technology. Centre for Spoken Language Understanding, Oregon Graduate Institute of Science and Technology. http://cslu.cse.ogi.edu/HLTsurvey/
25. Craig E. (ed) (1998) Routledge Encyclopedia of Philosophy. Routledge, London
26. Cruse D. (2000) Meaning in Language: an Introduction to Semantics and Pragmatics. Oxford University Press, Oxford
27. Dale R., Moisl H., Somers H. (eds) (2000) Handbook of Natural Language Processing. Marcel Dekker, New York
28. Dorffner G., Prem E. (1993) Connectionism, Symbol Grounding, and Autonomous Agents. Proceedings of the 15th Annual Conference of the Cognitive Science Society
29. Dorffner G. (ed) Neural Networks and a New Artificial Intelligence. International Thomson Computer Press, London
30. Dreyfus H. (1992) What Computers Still Can't Do. 2nd ed. MIT Press, Cambridge MA
31. Finlay J., Dix A. (1996) An Introduction to Artificial Intelligence. UCL Press, London
32. Fodor J., Pylyshyn Z. (1988) Connectionism and Cognitive Architecture: a Critical Analysis. Cognition 28:3-71
33. Freeman W. (1991) The Physiology of Perception. Scientific American 264 (2):78-85
34. Freeman W. (1992) Tutorial in Neurobiology: from Single Neurons to Brain Chaos. International Journal of Bifurcation and Chaos 2:451-82
35. Freeman W. (1994) Chaos in the CNS: Theory and Practice. Flexibility and Constraint in Behavioral Systems, ed. Greenspan R., Kyriacou C. John Wiley, New York
36. Freeman W. (1994) Qualitative Overview of Population Neurodynamics. Neural Modeling and Neural Networks, ed. Ventriglia F. Pergamon, Oxford
37. Freeman W. (1999) Consciousness, Intentionality, and Causality. In [62]
38. Freeman W. (1999) How Brains Make Up their Minds. Weidenfeld and Nicholson, London

39. Freeman W., Nunez R. (1999) Restoring to Cognition the Forgotten Primacy of Action, Intention, and Emotion'. In [62]
40. Gardner H. (1985) The Mind's New Science. A History of the Cognitive Revolution. Basic Books, New York
41. Gazdar G., Pullum G. (1985) Computationally Relevant Properties of Natural Languages and their Grammars. New Generation Computing 3:273-306
42. Gazdar G., Mellish C. (1989) Natural Language Processing in LISP. Addison-Wesley, Wokingham, UK
43. Haugeland J. (1985) Artificial Intelligence: the Very Idea. MIT Press, Cambridge MA
44. Hopcroft J., Ullman J. (1979) Introduction to Automata Theory, Languages, and Computation. Addison Wesley, Wokingham, UK
45. Horwich P. (1998) Meaning. Clarendon Press, Oxford
46. Hurley S. (1998) Consciousness in Action. MIT Press, Cambridge MA
47. Johnson M. (1987) The Body in the Mind: the Bodily Basis of Meaning, Imagination, and Reason. University of Chicago Press, Chicago
48. Kelso J. A. S. (1995) Dynamic Patterns: the Self-organization of Brain and Behavior. MIT Press, Cambridge MA
49. Kramer B., Mylopoulos J. (1992) Knowledge Representation. In [78]
50. Krawer S., Tombe L. (1981) Transducers and Grammars as Theories of Language. Theoretical Linguistics 10:173-202
51. Langendoen D. (1975) Finite State Parsing of Phrase-structure Languages and the Status of Readjustment Rules in Grammar. Linguistic Inquiry 6:533-54
52. Langendoen D., Langsam Y. (1984) The Representation of Constituent Structure for Finite State Parsing. Proceedings of the Conference for Computational Linguistics, COLING 84
53. Langendoen D., Postal P. (1984b) The Vastness of Natural Languages. Blackwell, Oxford
54. Lappin S. (1996) The Handbook of Contemporary Semantic Theory. Blackwell, Oxford
55. Larson R., Segal G. (1995) Knowledge of Meaning. MIT Press, Cambridge, MA
56. Lochbaum K., Grosz B., Sidner C. (2000) Discourse Structure and Intention Recognition. In [27]
57. Luger G., Stubblefield W. (1998) Artificial Intelligence. Structures for Complex Problem Solving. Addison Wesley Longman, Harlow, UK
58. McClelland J., Rumelhart D. (1986) Parallel Distributed Processing. Explorations in the Microstructure of cognition. MIT Press, Cambridge MA
59. Moisl H. (1992) Connectionist Finite-state Natural Language Processing. Connection Science 2:67-91
60. Moisl H (2000) NLP Based on Artificial Neural Networks: Introduction. In [27]
61. Newell A., Simon H. (1976) Computer Science as Empirical Enquiry: Symbols and Search. Communications of the Association for Computing Machinery 19:113-26
62. Nunez R., Freeman W. (1999) Reclaiming Cognition. Imprint Academic, Thorverton, UK
63. Partee B., ter Meulen a., Wall R. (1990) Mathematical Methods in Linguistics. Kluwer, Boston
64. Penrose R. (1989) The Emperor's New Mind. Oxford University Press, Oxford
65. Penrose R. (1994) Shadows of the Mind. Oxford University Press, Oxford
66. Pereira F. (1996) Sentence modeling and parsing. In [24]
67. Petrick S. (1992) Parsing. In [78]

68. Poesio M. (2000) Semantic analysis. In [27]
69. Port R., van Gelder T. (1995) Mind as Motion. Explorations in the Dynamics of Cognition. MIT Press, Cambridge MA
70. Pulman S. (1986) Grammars, Parsers, and Memory Limitations. Language and Cognitive Processes 1:197-225
71. Pulman S. (1996) Semantics. In [24]
72. Pylyshyn Z. (1986) Computation and Cognition. MIT Press, Cambridge MA
73. Pylyshyn Z. (1992) Cognitive science. In [78]
74. Reich P. (1969) The Finiteness of Natural Language. Language 45:831-43
75. Roche E., Schabes Y. (1997) Finite-state Language Processing. MIT Press, Cambridge MA
76. Saeed J. (1997) Semantics. Blackwell, Oxford
77. Samuelsson C., Wiren M. (2000) Parsing Techniques. In [27]
78. Shapiro S. (1992) Encyclopedia of Artificial Intelligence. 2nd ed. John Wiley and Sons, New York
79. Shieber S. (1985) Evidence against the Context Freeness of Natural Language. Linguistics and Philosophy 8:333-43
80. Skarda C. (1999) The Perceptual Form of Life. In [62]
81. Skarda C., Freeman W. (1987) How Brains Make Chaos in order to Make Sense of the World. Behavioral and Brain Sciences 10 (2):161-95
82. Steels L., Brooks R. (1995) The Artificial Life Route to Artificial Intelligence. Lawrence Erlbaum, Hove, UK
83. Thelen E., Smith L. (1994) A Dynamic Systems Approach to the Development of Cognition and Action. MIT Press, Cambridge MA
84. van Gelder T. (1998) Cognitive Architecture: What choice do we have? In Constraining cognitive theories: issues and opinions, ed. Pylyshyn Z. Ablex, Norwood NJ

Robust Stimulus Encoding in Olfactory Processing: Hyperacuity and Efficient Signal Transmission

Tim Pearce[1], Paul Verschure[2], Joel White[3], and John Kauer[3]

[1] Department of Engineering, University of Leicester, University Road LE1 7RH,
United Kingdom. t.c.pearce@le.ac.uk
[2] Institute of Neuroinformatics, University/ETH Zürich, CH-8057 Zürich,
Switzerland. pfmjv@ini.phys.ethz.ch
[3] Department of Neuroscience, Tufts University Medical School, Boston, MA 02111,
USA. jwhite@opal.tufts.edu jkauer@opal.tufts.edu

Abstract. We investigate how efficient signal transmission and reconstruction can be achieved within the olfactory system. We consider a theoretical model of signal integration within the olfactory pathway that derives from its convergent architecture and results in increased sensitivity to chemical stimuli between the first and second stages of the system. This phenomenon of signal integration in the olfactory system is formalised as an instance of hyperacuity. By exploiting a large population of chemically sensitive microbeads, we demonstrate how such a signal integration technique can lead to real gains in sensitivity in machine olfaction. In a separate computational model of the early olfactory pathway that is driven by real-world chemosensor input, we investigate how spike-based signal and graded-potential signalling compares for supporting the accuracy of reconstruction of the chemical stimulus at later stages of neuronal processing.

1 Introduction

The olfactory system provides an ideal model to consider the issues of robust sensory signal transmission and efficient encoding/decoding within neural systems. It must overcome large shifts in operating conditions occurring over time, that together add up to a continual state of flux at its periphery, the olfactory epithelium. A key constraint is that the main sites for chemical transduction, Olfactory Receptor Neurons (ORNs), are in a rapid and continuous state of development and programmed apoptosis (at least in mammals) which differentiates them from all other sensory neurons within the nervous system [1]. This neurogenesis means that the total number of receptors innervating the first point of signal processing, the olfactory bulb, fluctuates over time as signals from degenerating ORNs cease and axons from large numbers of newly formed ORNs make their way to integration sites called glomeruli. How this is achieved is a fascinating and recently uncovered story of axonal guidance [2] but in the context of robust stimulus encoding we are primarily interested in the effect of this

S. Wermter et al. (Eds.): Emergent Neural Computational Architectures, LNAI 2036, pp. 461–479, 2001.

turnover of receptors on signal transmission. The key issue here is how the olfactory system manages to cope with changing numbers of receptors, yet still generate a consistent signal to support odour perception over time.

Another factor of crucial importance when considering robust signal processing in the olfactory system is evidence suggests that not only do the numbers of receptors change as a result of neurogenesis, but also shifts in the response characteristics of ORNs occur during the act of perception. Receptor adaptation or fatigue is a key factor here and is known to occur in ORNs as their response adapts strongly during exposure to high levels of specific chemicals or repeated exposure [3]. Since the olfactory system relies on an entire population of broadly-tuned chemosensors (of which there now appears to be around a thousand in mice, fewer in fish [4]) these shifts in tunings may play a fundamental role in determining the stability of the system as a whole, and so this provides another perspective from which to consider robust signal transmission and processing within the olfactory pathway.

Signals from ORNs must be transmitted over relatively large distances from the olfactory epithelium at the top of the nasal passages, through the cribiform plate, and into the olfactory bulb where they are integrated at the glomeruli. In general, action potentials are used within the nervous system to encode and represent the stimulus between the transducer and first site of processing. This encoding strategy possesses robust noise-resistant properties that result from its intermittent discretised nature [5] that will be considered later in this chapter. Astonishingly the system solves this transmission problem as well as improving sensitivity at the first stage of processing over and above that obtained at the receptor level. We will formalise this phenomenon as an instance of hyperacuity.

These shifts in operating conditions have a direct impact on the reliable processing of sensory information within the olfactory pathway since it must overcome constant change and external noise sources in order to maintain a robust capability for characterising and discriminating complex mixtures of molecular stimuli. Despite changes in both receptor numbers and their characteristics, odour perceptions are remarkably stable with time (subject to respiratory infection of course). The ability of the olfactory system to achieve robust performance in the face of such a high degree of change appears to derive principally from its neuronal architecture in combination with the signal encoding strategies employed, as we will discuss here.

We will consider two models of the early stages of the olfactory pathway that speak to the issue of how a robust signal representing the stimulus is transmitted to the first stages of processing and how the quality of the signal is maintained during this process. Specifically, the first model will provide a probabilistic interpretation of signal integration of receptor signals at glomeruli, which predicts a lowering of detection limits at the system level compared with individual receptors. This signal integration model will be tested experimentally by applying it to data obtained from real-world chemosensor microbeads that mimic key properties of olfactory receptors. In the second model, the issue of signal transmission within the early stages of the olfactory pathway will be addressed, by comparing an action-potential based model with one mediated by graded-voltage signals. This allows us to investigate under which operating conditions the signal integrity is maintained at each glomerulus.

A number of questions will be addressed using these models. For example, can signal integration at the glomerulus account for sensitivity enhancement observed in the biology? Is it reasonable to consider sensitivity enhancement within the olfactory system as an instance of hyperacuity, and if so then how should this be quantified? Also, how does spiking and graded signal encoding affect signal integrity within the early stages of olfactory processing?

Our models are simple, yet capture what we believe to be the key features of the early olfactory pathway; they are population-based, probabilistic, and spiking. This enables us to better understand the biology by making predictions about the performance of alternative coding and processing schemes that can be reasonably hypothesised, such as comparing spiking and graded-voltage signal transmission.

The models we consider are implemented as part of a biologically plausible artificial nose, which is driven by real-world optical microbead chemosensor input. The sensors have a number of properties similar to biological olfactory receptors and prove ideal for implementing functional models of the biological olfactory pathway. It is possible to operate the signal processing models and these sensing elements in combination and in real-time to comprise an artificial olfactory system [6]. Consequently, research in this area can inform us not only about neuronal information processing within biological systems but also on how to achieve better design in the field of machine olfaction.

After discussing the implications of our models for robust signal processing we will conclude this chapter with a discussion of how signal encoding and processing strategies within the olfactory system may inform more general architectures for computation that are based upon emerging results in neuroscience.

2   Receptor Convergence and Olfactory Hyperacuity

A marked feature of the mammalian olfactory system is the massive convergence of spiking receptor input from thousands of olfactory receptor neurons onto glomeruli, the first stage of processing in the olfactory bulb [4]. This convergence appears to be fundamental to the operation of the olfactory system since it is conserved across many species. This arrangement raises the question of how reliable odour encoding can be achieved in view of large numbers of discretised receptor inputs that converge onto the olfactory bulb?

We contend that one consequence of the massive convergence of sensory input [7] within the olfactory bulb is sensitivity enhancement. This arrangement is schematised in Fig. 1 where n receptors (n being in the order of 2-10 thousand in mammals) expressing the same receptor protein(s) generally converge onto two glomerular regions [8]. In the simplest scheme, we can consider the spike-trains generated by individual receptors as statistically independent Poisson processes (after Van Drongelen et al. [9]), where the probability of observing

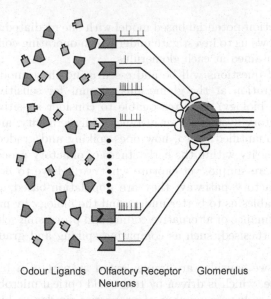

Odour Ligands   Olfactory Receptor   Glomerulus
                 Neurons

Fig. 1. A schematic of receptor convergence at the early stages of the mammalian olfactory pathway. Odour molecules are thought to interact with putative 7-transmembrane domain receptor proteins within the hair-like cilia of Olfactory Receptor Neurons (ORNs) leading ultimately to the generation of an action potential. The vigour of the cell response depends on both the suitability of the ligand to activate the second-messenger cascade signalling pathways and also the number of ligand-receptor interactions occurring at a specific cell. Action potentials produced by ORNs propagate over relatively large distances to reach the glomeruli of the olfactory bulb, which act as common sites for integration.

$(N = X)$ action potentials within a time-window, δt, is governed by the Poisson distribution

$$P_r(N = X) = \frac{\lambda_r^X}{X!}e^{\lambda_r} \tag{1}$$

where $\lambda_r = k_s \delta t$ and k_s is the mean firing rate expected for each stimulus, s. Since olfactory receptors probably have different ranges of tuning to particular stimuli, we would expect k_s to vary for a particular receptor over a given test-set of odorants. However, one effect of convergence of receptor input at the glomerulus might be to aggregate multiple spike-trains over a period of time. So while the statistics of spike generation at the receptor level may be governed by λ_r, at the glomerulus, $n\lambda_r$, spikes are expected on-average in time-window, δt. The spiking input to each glomerulus is considered as another Poisson process, but now with time-constant $\lambda_g = n\lambda_r$. The signal-to-noise ratio (SNR) enhancement of this convergent architecture is derived from the dispersion of the aggregated signal at the glomerulus, λ_g, compared with that of the individual receptor spike-trains, λ_r, so

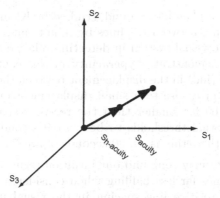

Fig. 2. A simplified representation of odour space in which each axis corresponds to a separate chemical component – the distance along each axis corresponds to its concentration. Since many simple odour compounds exist, this state space will be of high dimensionality and contain significant redundancy, since many permutations of odour concentration will never occur in the natural world. Any given point in odour space represents a complex mixture of volatile compounds with a unique fingerprint of relative concentrations, that elicits a specific perceived odour quality. The vectors demonstrates the just noticeable difference (jnd) in the stimulus that is required to either detect a difference from a single chemoreceptor, s_{acuity}, within the olfactory system or psychophysically as reported perceptually, $s_{\text{h-acuity}}$.

$$\text{SNR} = \frac{\sigma_g}{\sigma_r} = \left(\frac{\lambda_g}{\lambda_r}\right)^{1/2} = \left(\frac{n\lambda_r}{\lambda_r}\right)^{1/2} = \sqrt{n} \qquad (2)$$

and we expect an enhancement in sensitivity to follow $\sqrt{n}$, with increasing receptor numbers, n. This is a form of hyperacuity where the biology takes advantage of the statistics at the receptor level in order to generate overall system sensitivity that is greater than that of the underlying detectors.

There can be many forms of hyperacuity within a single sensory modality. For example, within the visual system three forms are commonly discussed; colour perception, vernier-style hyperacuity, and stereo-optic depth perception [10]. In each case the overall perceptual performance has been measured empirically using psychophysical experiments and then compared with theoretical physical limits imposed on the sensory system, such as receptor spacing or diffraction limits imposed by the optics of the eye. The point at which acuity becomes hyperacuity can be measured empirically when the overall psychophysical detection limits exceed those calculated from physical constraints placed on the sensory system or as measured electrophysiologically at the receptor level.

Probably the most widely studied example of hyperacuity in the visual system is during the perception of relative spacings in the visual field in two dimensions at the plane of fixation – so-called vernier-style hyperacuity. The effect can be measured empirically using a wide range of psychophysical experiments, a well known example being the estimation by an observer of relative spacing between four parallel lines on a plain background, studied by Klein and Levi (1985) [11]. By varying the spacing between the lines by minute amounts they managed to test the acuity of the visual system in detecting relative displacement shifts in the visual field. To demonstrate hyperacuity in this context requires the just noticeable difference (jnd) in the displacement between the lines, as reported by the observer, which gives rise to a small displacement of the projected image onto the retina, to be far smaller than the receptor spacing. Klein and Levi measured the perceptual thresholds to be *ca.* 0.9 seconds of arc, whereas the receptor spacing on the retina is *ca.* 30 seconds of arc [11].

While hyperacuity may confound our intuition regarding detection limits in the biology, it becomes far less baffling when considered in a statistical sense. For the example of relative line spacing in the visual field, the signals from many more than a single receptor can be called upon to solve the task. By recruiting the signals from a population of receptor cells it is possible to surpass the detection limits to which individual receptors are subject. Population coding and hyperacuity can be considered to be closely related phenomena.

Within the olfactory system, at first glance there appears to be two very separate forms of hyperacuity present. The first of these relates to sensitivity enhancement, formalised above. Here, ORNs expressing identical single receptor proteins (or combinations thereof) aggregate their signals at common sites, resulting in an overall jnd to a preferred compound that exceeds that of individual chemoreceptors. We may refer to this as a kind of concentration hyperacuity, which may be quantified by comparing reported detection thresholds obtained from psychophysical or electrophysiological experiments conducted using pure odour compounds with those thresholds observed at the individual receptor level using single-unit recordings. Another form of olfactory hyperacuity can be considered to arise from the combined action of a broadly tuned population of ORNs. By combining signals from many ORNs expressing different receptor proteins, the later stages of the olfactory pathway may enhance discrimination between similar complex odour stimuli over and above that achievable by any single chemoreceptor type. We may refer to this as a form of odour quality hyperacuity which may be quantified by comparing ORN single-unit recordings in response to paired odour stimuli for their ability to account for discrimination of the same odour pair as reported psychophysically or measured electrophysiologically.

Consideration of the underlying neuronal architecture uncovers just how closely related these two forms of hyperacuity might be. For example, overlapping receptor tunings also contribute to lower detection limits to single compounds and so it is not possible to attribute the phenomena of sensitivity enhancement to a single receptor type. Similarly, convergence of single receptor types also supports better multicomponent odour quality discrimination within the olfactory system and so single receptor types make significant contribution to encoding quality. Also note that in this context odour quality and quantity are not entirely separable since changes in concentration of some odour compounds are known to have marked effects on the perceived odour quality. A single definition of olfactory hyperacuity might suffice that avoids this awkward distinction. As a working definition consider

> Olfactory hyperacuity is demonstrated by the discrimination of two chemically different odour stimuli (which may vary in both quantity and quality) observed at the later stages of the olfactory pathway and measured either psychophysically or electrophysiologically, that cannot be accounted for by any single underlying chemosensor.

Notice that this definition is broad enough to encompass many olfactory scenarios – in particular changes in both odour intensity and quality. Fig. 2 shows an odour space representation, which provides an intuitive understanding of this definition for olfactory hyperacuity. Here, each point in odour space has an associated odour perception, the quality of which varies with changes in the relative concentration between the components of a complex odour mixture. The nearest excursion from a given point in odour space that produces a shift in the reported perception corresponds to the jnd in the stimuli that can be recognised – vector $s_{h-acuity}$. The excursion that demonstrates a statistically significant change in response at the single-unit level represents the jnd for that particular ORN – vector s^i_{acuity} for receptor class i. To demonstrate olfactory hyperacuity, the magnitude of jnd observed at the perceptual level must be far smaller than for any underlying receptor

$$|s_{h-acuity}|_2 \ll |s^i_{acuity}|_2 \text{ for all } i, \tag{3}$$

where $| \bullet |_2$ is the usual l-2 Euclidean norm.

Hyperacuity of this form has already been demonstrated within the olfactory system through electrophysiological measurements in both mammals and insects. Duchamp-Viret *et al.* (1989) measured sensitivity enhancement to a variety of single odour ligands at both the receptor and olfactory bulb level in the frog [12]. Their results show a clear lowering of the detection limits at the bulb level when compared with that observed for the underlying receptors. Interestingly, this effect is observed only when a large portion of the olfactory mucosa is exposed, compared with a punctate delivery to the receptor sheet. This provides further evidence that hyperacuity in the olfactory system relies upon the recruitment of a large population of receptors. Given that the convergence ratio of receptors onto glomeruli in the frog is similar to other small mammals and estimated to be *ca.* 1000 [12,13] the theoretical model of hyperacuity at the front-end of the olfactory system, represented by (2), predicts a lowering of the detection limit by a factor of *ca.* 32. In support of this prediction, Duchamp-Viret and co-workers observed between 1-2 orders of magnitude sensitivity enhancement in their measurements [12].

Similar measurements in the antennal lobe of cockroach have been made by Boeckh and co-workers demonstrating spectacular sensitivity enhancement to pheromone compounds - regularly between 1-4 orders of magnitude but also as high as six orders of magnitude between measurements taken at the antennae and the Macroglomerular Complex (MGC) of the antennal lobe [14]. These results are intriguing since although high convergence ratios of ORNs onto specialised glomeruli in the MGC of insects have been reported, these are nowhere close to the enormous convergence ratios required to support such extreme hyperacuity (10^{12}:1 receptor:glomeruli convergence ratios). Alternative mechanisms must be involved in enhancing sensitivity to such a high degree – one example of which might be noise-shaping [15].

The early stages of the olfactory pathway not only ensures efficient and robust signal transmission from the transduction sites to the first stage of processing, but clearly lower the overall system detection limits in the process. This provides only one example of how the convergent architecture of the olfactory pathway can teach us valuable lessons about robust signal processing within neural systems as it maintains high levels of sensitivity to relevant stimuli.

3   Hyperacuity in an Artificial Nose

Can we demonstrate such sensitivity enhancement within practical chemical sensing technology using the mechanism of hyperacuity in the olfactory system as a model? Even though artificial nose systems typically rely on arrays of widely tuned non-specific chemosensors, numbers of individual sensing elements are usually restricted overall, to reduce both system complexity and implementation costs. As a consequence it has not been possible to exploit the statistical properties of large numbers of chemosensor elements, simply due to lack of sensor numbers, as has been successfully exploited in the biological olfactory pathway to boost detection limits.

Optical microbead sensor technology, as depicted in Fig. 3a is ideally placed for such neuromorphic implementation. Enormous populations of microbeads may be deployed in a small area (the diameter being *ca.* 30 µm), from which the signal produced by each sensor element can be addressed individually. Only in such an arrangement can the issues of population coding in chemical sensing be addressed realistically. Individual microbead sensors are broadly-tuned to a wide range of organic compounds and so are reminiscent of the wide ranging but preferentially tuned responses observed in ORNs [16]. It is possible to effectively tune these devices by choosing different polymer/dye combinations so as to replicate to some extent the diversity of receptor types present in the biology. The devices are also small and low-power – a 3mW output laser-diode assembly can energise billions of optical microbeads in tandem, making it a useful chemical sensor technology in its own right [17].

Our aim, then, is to investigate the statistics across a population of identical optical microbeads in order to test for evidence of olfactory hyperacuity as demonstrated in the biology. This would provide both a practical method for sensitivity enhancement in chemical sensing instrumentation as well as add credence to the biological model discussed in Sect. 2.

For this work 201 optical microbeads with similar response characteristics were imaged on a glass slide as these were exposed to different dilutions of the saturated vapour headspace of a single chemical compound – toluene. Fabrication details for the microbeads have been reported elsewhere [18]. The fluorescence

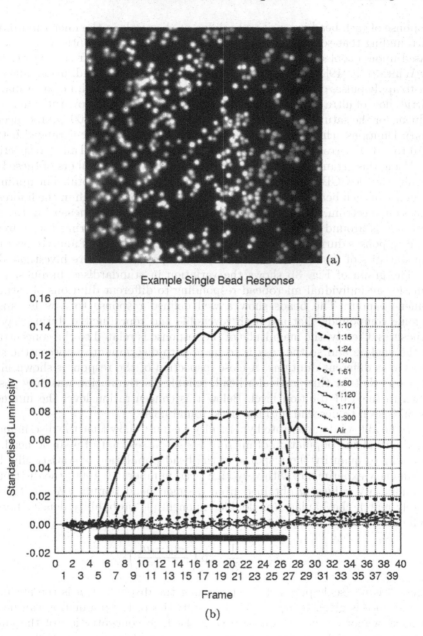

Fig. 3. (a) A grey-scale image showing 201 fluorescent microbeads responding to saturated toluene vapour at a single point in time. (b) A single bead response to different air dilutions of saturated toluene vapour over time, indicating concentration discrimination down to 1:61 dilution. The data were standardised by taking the average greyscale value for each individual bead and for each frame subtracting the greyscale value of the respective bead on the first frame of the sequence and finally dividing the result by the same first frame value. This results in a fractional pre-processing metric. The solid black line indicates odour exposure.

response of each bead is sensitive to different chemicals in the microenvironment surrounding that sensor and can be imaged using a simple optical arrangement based upon a cooled CID camera and microscope lens similar to that described by White *et al.* [19] Odour delivery was achieved using an air-dilution olfactometer to apply pulses of analyte vapour to region of beads being imaged. A make-up carrier flow of ultra zero grade air was controlled at varying flow-rates to act as a diluent for the saturated vapour headspace of toluene (*ca.* 3800 ppm concentration). Dilutions achieved using this odour delivery arrangement ranged between 1:10 to 1:300, corresponding to $154.8\,\mathrm{nmol\,ml^{-1}}$ and $5.2\,\mathrm{nmol\,ml^{-1}}$ respectively.

Using this arrangement it was possible to image large numbers of these beads within a single CCD-frame - in this experiment 201 beads in total. The luminosity response of each bead to the analyte (across a small bandwidth in the fluorescent emission spectrum, 10 nm) may be accurately assessed by measuring the grey-scale levels around localised points in the image. After storing the individual bead responses during exposure to different concentrations of analyte over time, the statistics of the responses across the bead population were investigated.

The graph of Fig. 3b shows the variation in standardised luminosity over time for an individual microbead responding to different dilutions of saturated toluene vapour. The beads show a well defined response to a wide range of organic compounds that are both reversible and reproducible during repeated exposures. Clearly, the magnitude of the response is related to the concentration of the analyte and so the task is to be able to discriminate between the single analyte at different dilutions. A close inspection of the responses shown in Fig. 3b shows that for this particular bead, reliable discrimination was only possible down to the 1:61 dilution level. Below this concentration level the luminosity signal is seen to descend into the background noise.

To be able to demonstrate hyperacuity we must show that the discrimination capability of the population of microbeads surpasses that of a single bead. The best way to quantify the response of the bead population is statistically. From preliminary experiments with very high bead numbers (> 1000) the distribution of luminosity values, y obtained from a single bead population was found to closely match the Laplace (or double exponential) probability density function (pdf) at any particular point in time

$$p(y) = \frac{1}{2\beta} \exp\left(\frac{-|y - \mu|}{\beta}\right) \tag{4}$$

where β is the scale parameter determining the dispersion, μ is the mean, and the variance is given by $\sigma_y^2 = 2\beta^2$. The statistics of the population can be used to make a more accurate assessment of the true concentration of the analyte using fundamental concepts from signal detection theory.

Given a single bead measurement, y, we can assign it to the most likely dilution class, H_i, by maximising the a posteriori probability $p(H_i|y)$. In the case of equally likely dilution classes, it is simple to show using Bayes rule that this is equivalent to maximising the conditional probability, and so our decision rule reduces to (for two classes, H_1 and H_0)

$$\text{if } p(y|H_1) > p(y|H_0) \quad \text{choose } H_1, \text{ otherwise choose } H_0 \tag{5}$$

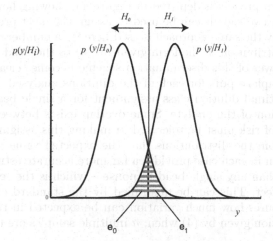

Fig. 4. The detection of two signals with distinct Probability Density Functions (pdfs). Using a threshold function it is possible to minimise the likelihood of making an error in assigning a hypothesis, H_0 or H_1, to observation y. The areas ϵ_0 and ϵ_1 represent the probability of making this error.

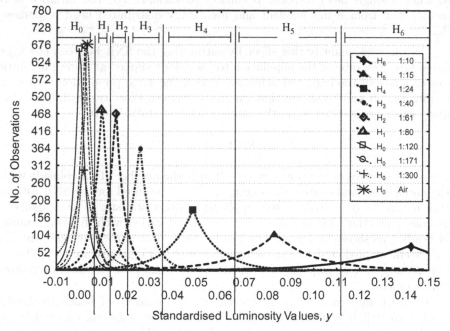

Fig. 5. Best fit Laplace distributions to the microbead population response to a variety of dilution categories. Distinct pdfs are observed down to 1:80 dilution taken at frame 25 where the bead luminosity response is maximum. The figure demonstrates how the expected value (the peak) of each distribution provide a far more accurate and robust measure of the actual dilution class - this being the basis of sensitivity enhancement in our artificial nose.

This decision process is depicted in Figure 4a, showing how the overlapping distributions of two signals can be used to assign the most probable class membership. Exactly the same approach is used here for a number of dilution classes. By fitting a distribution of the form given by (4) to the bead responses to each dilution the power of this discrimination scheme becomes clear. Figure 5 shows the multiple Laplace pdfs for each of the dilutions analysed. This can be used to make an optimal dilution class assignment for a single bead response to an unknown dilution of the analyte. Some overlap exists between the fitted pdfs, and so a level of risk must be tolerated in making this assignment. However, it is also clear from the distributions that the expected value (mean) across the bead population in each case provides a far more accurate estimator of the dilution category than any single bead response – which is the central issue in any hyperacuity effect. This can be quantified by the standard error in the mean, σ_μ which measures how much variation can be expected in the expected value of the distribution given by (4) when, n multiple samples are taken from a bead population.

$$\sigma_\mu = \frac{\sigma_y}{\sqrt{n}} = \frac{2\beta^2}{\sqrt{n}} \qquad (6)$$

so while a single bead response is subject to variance $2\beta^2$ the variance of the mean taken from n independent and identically distributed bead responses is $\frac{2\beta^2}{\sqrt{n}}$.

We can also quantify this effect by estimating the SNR between the aggregated bead response of the population to the odour applied at a specific dilution (an estimate of the expected value $\hat{\mu}_{Hi}$) with the same population responding to air, $\hat{\mu}_{H0}$. It can be shown that the SNR between these two estimates of the mean can be calculated using the Student's t-test statistic [20]

$$\hat{SNR} > t_{\alpha,\nu} \qquad (7)$$

where the degrees of freedom $\nu = 2n - 2$, n is the number of microbeads within the population, and α gives the significance level required for the SNR estimate (taken here as $\alpha = 0.05$). After applying a t-test statistic it was possible to estimate how the SNR of the aggregated signal varied with differing numbers of beads – as shown in Fig. 6. The results show clear agreement with the SNR enhancement predicted by the biological model and demonstrate how olfactory hyperacuity may be implemented within an artificial nose to achieve sensitivity enhancement.

Figure 6 may also be used to estimate the number of beads required to reach a particular system detection limit and so solve a specific odour detection problem. Assuming a minimum SNR of 3 for reasonable detection, it is clear from extrapolating the characteristics that the 1:171 toluene dilution odour detection task could be solved with ca. 500 concurrent bead measurements and the 1:300 task with ca. 1,300 bead measurements.

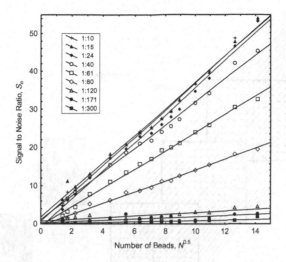

Fig. 6. Variation in the estimated SNR for a population of microbeads with different dilutions of analyte and bead numbers, $\sqrt{n}$.

4   Stimulus Encoding in the Olfactory System

Another issue of interest when considering robust signal processing within neural systems is how sensory signals are encoded in the CNS. It appears that a wide range of encoding strategies are employed by the biology in order to efficiently transmit sensory information under a wide range of conditions. Examples of coding strategies include graded potentials (usually over short distances), action potentials, rate codes, and specific temporal codings.

The key method of signal transmission between the first two stages of the olfactory pathway is known to be action potentials. Although these spiking signals are known to provide an efficient mechanism for long-distance transmission in the nervous system, how it can provide the signalling basis for reliable and accurate stimulus encoding is still debated [21,22]. One approach to address this issue is to investigate different stimulus encoding schemes within computational models. In this section, we investigate the mass action of sensory input to a simple olfactory model that is driven by optical chemosensors to understand its behaviour in what has been termed a high-input regime [23].

In this context realistic chemosensory input derived from optical microbead input confers advantages in terms of more natural statistics of sensory input than can be achieved with a small number of sensors or simulated input. Accordingly, we can investigate the behaviour of our model under a probabilistic, high-input regime akin to the biology.

We applied the data-set shown in Fig. 3b to a simple model of the front-end of the olfactory system. The layout for the model is shown in Fig. 7a where data from each individual bead is mapped onto a series of cell populations in

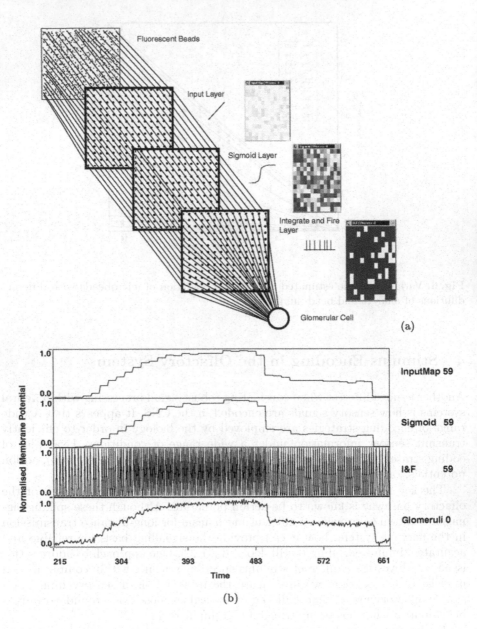

Fig. 7. (a) Architecture of a simple model of the early stages of the olfactory system, showing the different layers, including fluorescent bead population, input layer, sigmoid layer, integrate-and-fire layer, and a single glomerular cell entity. (b) Behaviour of the model over time at each successive stage of the system during a complete cycle of the stimulus – top to bottom: examples of the activity of a single simulated neuron belonging to the input layer responding to the bead response shown in Fig. 3b, sigmoid layer, and integrate-and-fire layer respectively. The bottom trace shows the stimulus after reconstruction from multiple spike trains at the level of the simulated glomerulus.

a 1-to-1 fashion. To address the stimulus encoding issues we consider here it is reasonable to connect a single simulated glomerular "cell" to a population of ostensibly identical fluorescent beads, to model the convergent architecture discussed in Section 2. In order to closely follow the biology of the system, the baseline response of the bead data was applied to a sigmoid layer in order to mimic the sigmoidal concentration dependence of transduction current within olfactory receptor neurons [24]. This also acts to auto-scale the data so that each bead response lies within the range $[0, 1]$.

Integrate-and-fire neurons were then used to generate a volley of spike trains that represent the stimulus using Poisson statistics as outlined in (1). This arrangement provides a simple yet reasonably accurate model of spike generation by olfactory receptors, which can be considered to produce spike trains with Poisson statistics where the mean firing frequency is sigmoidally dependent upon stimulus concentration. The convergence of receptor input in the biology has been represented in our model by the aggregation of spike trains from many integrate-and-fire cells at any one point in time at the glomerulus "cell". While the glomerulus does not exist as a cell entity in its own right (it comprises neuropil made up from axons from olfactory receptor neurons synapsing onto the dendrites of mitral, tufted, and periglomerular cells), we can use this in our model as a convenient site for integration that represents the excitatory effect of the receptor input on the olfactory bulb.

The behaviour of our model to real-world chemosensory input is shown in Fig. 7b. Here, a single cell in the input layer presents a complete cycle of the luminosity response of an individual microbead. This signal is then compressed by its corresponding cell within the sigmoid layer, and then transformed into a probabilistic spike train at the integrate-and-fire layer. By integrating a large number of such spike trains in both space and time (Fig. 7b, bottom), the glomerular cell is then able to reconstruct the stimulus. The key point to note here is that the signal can only be reconstructed accurately if spikes arrive at random points in time, and that at any point in time the glomerulus is receiving an accurate mean signal from the entire population of chemosensors. The system is therefore dependent upon the massively convergent rate-coded receptor input in order to accurately reconstruct the stimulus.

An important aspect related to how robustly the transmission scheme behaves under realistic input conditions, is how well the glomerulus is able to track the change in the stimulus over time using only discrete spiking input. This issue is central to the performance of any sensory system that must use the input provided from a population of receptors in order to make decisions about the stimulus, such as in the visual and auditory pathways. To estimate the SNR of the reconstructed stimulus, and so quantify the information available to any subsequent neuronal processing within the olfactory pathway, we conducted experiments on two models of the form shown in Fig. 7a in parallel. By comparing the signal produced from a single glomerulus in one model whilst being exposed to air, to an identical model during exposure to 1:10 dilution of saturated toluene vapour, an accurate estimate of the ability of the artificial sensory system to reconstruct the stimulus could be made.

To address the central issue related to robust signal encoding strategies we asked the question, does a rate-coded representation of the stimulus at the receptor level limit the signal quality that is recoverable at the glomerulus? To investigate this issue, we compared the results obtained from two models, one in which probabilistic spike trains representing the receptor input were integrated at the glomerular cell and another in which the graded (non-spiking) receptor input was transmitted directly to the glomerulus for integration.

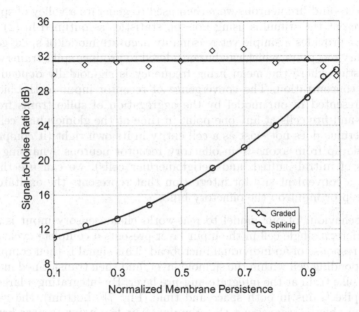

Fig. 8. A comparison of the signal-to-noise ratio obtained at the glomerulus cell of our neuronal model for two different encoding schemes – rate-coded spike trains and graded signal transmission.

The results comparing the SNR obtained at the glomerular cell for both operating regimes are shown in Fig. 8. The results demonstrate that direct transmission of graded receptor input gives rise to uniformly high SNR which is robust to variations in the charging time-constant at the glomerulus. For very short time constants the spiking rate-coded equivalent does very poorly, achieving an SNR at the glomerulus that is worse than that of any single receptor. However, for longer integration periods the SNR under the rate-coded regime approaches that achieved during graded signal transmission, showing that a similar efficiency of stimulus encoding can be achieved, but only within a specific range of temporal integration. The cost for adequately recovering a reasonable SNR at the glomerulus in the case of spiking rate-coded stimuli is, though, a far longer integration period, which slows the dynamics of the system as a whole and so limits the response time of the system. If the temporal dynamics of the odorant diffusion through the mucous layer of the olfactory epithelium and signal transduction dynamics at ORNs matched those of the glomeruli there would be no cost to pay since the system would act as a matched filter. Rate-coded stimulus encoding confers clear advantages in terms of its robust properties in the face of external noise sources, and so is preferable for reliable transmission over long distances such as between the epithelium and the olfactory bulb in the CNS.

These results indicate a clear trade-off between integration time and reconstructed signal quality during spike-based stimulus encoding. This represents an important aspect of how signal integrity can be maintained within a rate-coded regime which is fundamental to understanding the transmission of stimulus information throughout sensory systems.

5 Discussion and Summary

The focus of this chapter has been the robust transmission and reconstruction of sensory signals within neural systems. The olfactory pathway provides an excellent model system from which to address these issues. Two models of the early stages of this system have been presented; a mathematical model of sensitivity enhancement in the olfactory pathway and a computational model for comparing action- and graded-potential based signal transmission. Both models provide insight into aspects of robust signal processing and transmission. Massive convergence of receptor input coupled with population coding brings advantages to the olfactory system by way of fault-tolerance and sensitivity enhancement. Our computational model has demonstrated how spike based communication can be as efficient for signal transmission as graded-potential communication subject to temporal constraints.

How the CNS manages to transmit huge quantities of sensory information to the higher brain centres is a fascinating example of parallel processing. Individual sensory and bulbar neurons may only operate on a millisecond timescale, yet the entire olfactory system is able to make important decisions relating to the stimulus within a remarkable short period of time. Such a feat of information processing requires highly organised strategies to communicate and deal with sensory data in parallel. Investigating key organisation principles within the CNS, such as population coding provides a promising approach to replicating some of the robust properties of signal transmission whilst maintaining high bandwidth or sensory data.

An interesting aspect related to achieving such enormous bandwidth of sensory information that is suggested by our models, and merits further investigation is the matching of dynamics in the time domain of the perireceptor, signal transduction, transmission and processing stages. Further issues that might be investigated using a similar modelling approach are how stimulus and time is represented in the olfactory bulb and in particular the role of temporally complex mitral/tufted cell responses, and oscillations present within the system as a whole[25].

The authors wish to thank Keith Albert and David Walt (Chemistry Department, Tufts University, Boston, MA, USA) for fabricating the optical microbeads

478   T. Pearce et al.

used as part of this study. This work was funded by grants from the Royal Society (to TCP/PFMJV), SPP-SNF (to PFMJV), ONR, NIH and DARPA (to JSK).

References

[1] Singer M.S., Shepherd G.M., Greer C.A., Olfactory receptors guide axons, Nature **377** (1995) 19-20.
[2] Lin D.M., Ngai J., Development of the vertebrate main olfactory system, Curr. Opinion. Neurobiol. **9** (1999) 74-78.
[3] Dalton P., Psychophysical and behavioral characteristics of olfactory adaptation, Chem. Sens. **25** (2000) 487-492. .
[4] Christensen T.A., Heinbockel T., Hildebrand J.G., Olfactory information processing in the brain: encoding chemical and temporal features of odors, J. Neurobiol. **30** (1996) 82-91.
[5] Reike R., Warland D, de Ruyter van Steveninck R., and Bialek W., Spikes: Exploring the Neural Code, MIT Press: MA, USA, 1997.
[6] Pearce T.C., Verschure P.F.M.J.V. Olfaction: modeling and experimenting with an artificial nose, in NSF Report: Workshop on Neuromorphic Engineering Telluride, CO, USA. June 1998.
[7] Kauer, J.S., In: The Neurobiology of Taste and Smell. eds: T.E. Finger and W.L. Silver. John Wiley and Sons. pp. 205-231, 1987.
[8] Vassar R., Chao K.C., Sitcheran R., Nunez J.M., Vosshall L.B., Axel R., Topographic organization of sensory projections to the olfactory-bulb, Cell **79** (1994) 981-991.
[9] Van Drongelan W., Holley A., Doving K.B., Convergence in the olfactory system: quantitative aspects of odour sensing, J. Theor. Biol. **71** (1978) 39-48.
[10] Churchland P.S., Sejnowski T.J., The Computational Brain, MIT Press: Cambridge MA., 1992.
[11] Klein S.A., Levi D.M. Hyperacuity thresholds of 1 sec: theoretical predictions and empirical validation, J. Opt. Soc. Am. A **2** (1985) 1170-1190.
[12] Duchamp-Viret P., Duchamp A., Vigouroux M., Amplifying role of convergence in olfactory system: A comparative Study of receptor cell and second-order neuron sensitivities, J. Neurophysiol. **61** (1989) 1085-1095.
[13] Van Dronglelen W., Unitary recordings of near threshold response of receptor cells in the frog, J. Physiol. London, **277** (1978) 423-435.
[14] Boeckh J., Ernst K.D., Contribution of single unit analysis in insects to an understanding of olfactory function, J. Comp. Physiol. A **161** (1987) 549-565.
[15] Mar D.J., Chow C.C., Gerstner W., Adams R.W., Collines J.J., Noise shaping in populations of coupled model neurons, PNAS **96** (1999) 10450-10455.
[16] Dickinson T.A., White J., Kauer J.S., Walt D.R., A chemical-detecting system based on a cross-reactive optical sensor array, Nature **382** (1996) 697-700.
[17] Albert K.J., Lewis N.S., Schauer C.L., Sotzing, G.A., Stitzel S.E., Vaid T.P., Walt D.R., Cross-reactive chemical sensor arrays, Chem. Rev. **100** (2000) 2595-2626.
[18] Dickinson T.A., Michael K.L., Kauer J.S., Walt D.R., Convergent, self-encoded bead sensor arrays in the design of an artificial nose. Anal. Chem. **71** (1999) 2192-2198.

[19] White J., Kauer J.S., Dickinson T.A., Walt D.R., Rapid analyte recognition in a device based on optical sensors and the olfactory system, Anal. Chem. **68** (1996) 2191-2202.

[20] Sharaf M.A., Illman D.L., Kowalski B.R., Chemometrics, John Wiley & Sons: New York., 1986.

[21] Softky W.R., Koch C., The highly irregular firing of cortical cells is inconsistent with temporal integration of random EPSPs, J. Neurosci. **13** (1993) 334.

[22] Mainen Z.F., Sejnowksi T.J., Reliability of spike timing in neocortical neurons, Science **268** (1995) 1503.

[23] Shadlen M.N., Newsome W.T., The variable discharge of cortical neurons: implications for connectivity, computation, and information coding. J. Neuroscience **18** (1998) 3870.

[24] Lowe G., Gold G.H., Nonlinear amplification by calcium-dependent chloride channels in olfactory receptor cells, Nature **386** 1993 283-286.

[25] Laurent G., A systems perspective on early olfactory coding, Science **286** (1999) 723-728.

Finite-State Computation in Analog Neural Networks: Steps towards Biologically Plausible Models?

Mikel L. Forcada and Rafael C. Carrasco

Departament de Llenguatges i Sistemes Informàtics,
Universitat d'Alacant,
E-03071 Alacant, Spain.
{mlf,carrasco}@dlsi.ua.es

Abstract. Finite-state machines are the most pervasive models of computation, not only in theoretical computer science, but also in all of its applications to real-life problems, and constitute the best characterized computational model. On the other hand, neural networks —proposed almost sixty years ago by McCulloch and Pitts as a simplified model of nervous activity in living beings— have evolved into a great variety of so-called *artificial neural networks*. Artificial neural networks have become a very successful tool for modelling and problem solving because of their built-in learning capability, but most of the progress in this field has occurred with models that are very removed from the behaviour of real, i.e., biological neural networks. This paper surveys the work that has established a connection between finite-state machines and (mainly discrete-time recurrent) neural networks, and suggests possible ways to construct finite-state models in biologically plausible neural networks.

1 Introduction

Finite-state machines are the most pervasive models of computation, not only in theoretical computer science, but also in all of its applications to real-life problems (natural and formal language processing, pattern recognition, control, etc.), and constitute the best characterized computational model. On the other hand, neural networks, —proposed almost sixty years ago by McCulloch and Pitts [1] as a simplified model of nervous activity in living beings—, have evolved into a great variety of so-called *artificial neural networks*. Artificial neural networks have become a very successful tool for modelling and problem solving because of their built-in learning capability, but most of the progress in this field has occurred with models that are very removed from the behaviour of real, i.e., biological neural networks.

This paper surveys the work that has established a connection between finite-state machines and (mainly discrete-time recurrent) neural networks, and reviews possible ways to construct finite-state models in biologically plausible neural networks. The paper is organized as follows: section 2 describes the simultaneous inception of discrete-state discrete-time neural net models and finite-state machines; section 3 describes the relation between continuous-state discrete-time neural networks and finite-state machines; section 4 moves on to a more realistic model, namely, continuous-state continuous-time neural nets; spiking neurons as a biologically plausible continuous-time continuous-state model of finite-state computation is discussed in section 5. Finally, concluding remarks are presented in section 6.

2 The Early Days: Discrete-Time, Discrete-State Models

2.1 McCulloch-Pitts Nets

The fields of neural networks and finite-state computation started indeed simultaneously: when McCulloch and Pitts [1] formulated mathematically the behaviour of ensembles of neurons (after a number of simplifying assumptions such as the discretization of time and signals), they defined what we currently know as a finite-state machine (FSM). McCulloch & Pitts' simplified neurons work on *binary* signals and in discrete time; they receive one or more input signals and produce an output signal as follows:

- input and output signals can be *high* or *low*;
- inputs can be *excitatory* or *inhibitory*;
- the neuron has an integer activation threshold A;
- the neuron's output is high at time $t + 1$ if, at time t,
  - more than A excitatory inputs are high and
  - no inhibitory input is high

  and it is low otherwise.

McCulloch & Pitts' neuron may be easily shown to be equivalent to the current formulation of a *linear threshold unit*:

- input (u_i) and output (y) signals are 0 (*low*) or 1 (*high*);
- the neuron has real weights W_i, one for each input signal, and a real bias b;
- the neuron's output at time $t + 1$, $y[t + 1]$ is given by

$$y[t + 1] = \theta \left(\sum_i W_i u_i[t] + b \right) \tag{1}$$

where $\theta(x)$ is the step function[1] (time indices t, $t + 1$ may be dropped in some applications).

A McCulloch-Pitts net is a finite set of interconnected McCulloch-Pitts neurons. Some neurons receive external inputs; they are called *inputs to the net*. Other neurons compute the *outputs of the net*. The remaining neurons are called *hidden neurons*. Connections between neurons may form cycles (*i.e.*, the net may be *recurrent*). A McCulloch-Pitts net may be seen as a *discrete-time sequence*

---

[1] The step function is defined as follows: $\theta(x) = 1$ if $x \geq 0$, 0 otherwise.

processor which turns a sequence of binary input vectors $\mathbf{u}[t]$ into a sequence of binary output vectors $\mathbf{y}[t]$. The *state* of a McCulloch-Pitts at time t is the vector of the outputs at time t of its recurrent neurons (i.e., those involved in cycles, if any). Therefore, one may see a McCulloch-Pitts recurrent net as a finite-state machine[2] $M = (Q, \Sigma, \Gamma, \delta, \lambda, q_I)$, because

- The net may be found at any time t in a state from a finite set $Q = \{\text{high}, \text{low}\}^{n_X}$, where n_X is the number of recurrent units.
- The vector of inputs at time t takes values from a finite set $\Sigma = \{\text{high}, \text{low}\}^{n_U}$, where n_U is the number of inputs to the net. The set Σ is finite and may be seen as an alphabet.
- The vector of outputs at time t takes values from a finite set $\Gamma = \{\text{high}, \text{low}\}^{n_Y}$, where n_Y is the number of output signals going out from the network. This set is finite and may also be seen as an alphabet.
- The state of the net at time $t + 1$ is a function δ of inputs and states at time t which is defined by the architecture of the net.
- The output of the net at time $t + 1$ is a function λ of inputs and states at time t which is defined by the architecture of the net.
- The initial state $q_I \in Q$ is formed by the outputs of neurons and by the inputs to the net at time $t = 0$.

2.2   Regular Sets

Later, Kleene [3] formalized the sets of input sequences that led a McCulloch-Pitts network to a given state. He called them *regular events*; we currently know them as *regular sets* or *regular languages*. Nowadays, computer scientists relate regular sets to finite-state machines, not to neural nets [2, p. 28].

2.3   Constructing Finite-State Machines in McCulloch-Pitts Nets

Minsky [4] showed that any FSM can be simulated by a discrete-time recurrent neural net (DTRNN) using McCulloch-Pitts units; the construction used a number of neurons proportional to the number of states in the automaton; more recently, Alon et al. [5], Indyk [6], and Horne and Hush [7] have established better (sublinear) bounds on the number of discrete neurons necessary to simulate an finite state machine.

3   Relaxing the Discrete-State Restriction: Sigmoid Discrete-State Recurrent Neural Networks

Discrete neurons (taking values, for example, in $\{0, 1\}$) are a very rough model of neural activity, and, on the other hand, error functions for discrete networks are not continuous with respect to the values of weights, which is crucial for

---

[2] Such as Mealy and Moore machines, see [2, p. 42].

the application of learning algorithms such as those based in gradient descent. Researchers have therefore also shown interest in neural networks containing analog units with continuous, real-valued activation functions g such as the logistic sigmoid $g_L(x;\beta) = 1/(1+\exp(-\beta x))$, with β a positive number called the *gain* of the sigmoid. A logistic neuron has the form

$$y[t+1] = g_L\left(\sum_i W_i u_i[t] + b; \beta\right); \tag{2}$$

this relaxes the discrete-signal restriction. The logistic function has the following limiting properties: $\lim_{x\to\infty} g_L(x;\beta) = 1$ and $\lim_{x\to-\infty} g_L(x;\beta) = 0$, and $\lim_{\beta\to\infty} g_L(x;\beta) = \theta(x)$, where θ is the step function. The last property has the consequence that a logistic neuron with infinite gain is equivalent to a linear threshold unit. In summary, these are the advantages of relaxing the discrete-state restriction:

- Differentiability allows for gradient-descent learning
- Graded response gives a better model of some biological processes
- May, in principle, emulate discrete-state behaviour (the same as digital computers are built from analog units as transistors and diodes).

The use of analog units turns the state of the network into a real-valued vector. In principle, neural networks having neurons with real-valued states should be able to perform not only finite-state computation but also more advanced computational tasks.

3.1   Neural State Machines

A *neural state machine*, by analogy with a FSM $M = (Q, \Sigma, \Gamma, \delta, \lambda, q_I)$, is $N = (X, U, Y, \mathbf{f}, \mathbf{h}, \mathbf{x}_0)$ with

- $X = [0,1]^{n_X}$ (for n_X state neurons);
- $U = [0,1]^{n_U}$ (for n_U input signals);
- $Y = [0,1]^{n_Y}$ (for n_Y output neurons);
- $\mathbf{f} : X \times U \to X$ and $\mathbf{h} : X \times U \to Y$ are computed by feedforward neural nets;
- $\mathbf{x}_0 \in [0,1]^{n_X}$ is the initial state.

(the use of the $[0,1]$ interval is consistent with the use of the logistic sigmoid function, but other choices are possible). This construction is usually called a *discrete-time recurrent neural net* (DTRNN).

Similarly to FSM, we can divide neural-state machines in *neural Mealy machines* and *neural Moore machines* [2, p. 42]. In neural Mealy machines the output is a function of the previous state and the current input. Figure 1 shows a block diagram of a neural Mealy machine. Robinson and Fallside's [8] *recurrent error propagation nets*, used for speech recognition, are an example of a neural Mealy machine. In a neural Moore machine, the output is simply a function of the state just computed ($\mathbf{h} : X \to Y$), see figure 2. Elman's *simple recurrent net* [9] is an example of a neural Moore machine.

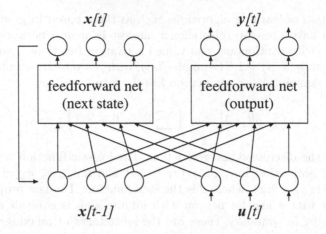

Fig. 1. A neural Mealy machine

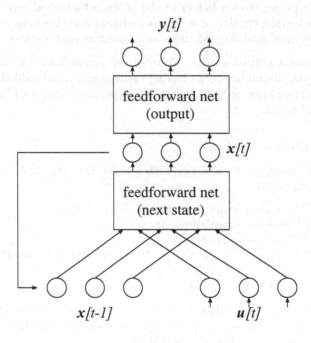

Fig. 2. A neural Moore machine

3.2   Learning Finite-State Behaviour in Sigmoid DTRNN

Under this intuitive assumption, a number of researchers set out to test whether
sigmoid DTRNN could learn FSM behaviour from samples [10,11,12,13,14,15,
16,17,18,19,20]. Sigmoid DTRNN may be trained to be FSM as follows:

- Define a *learning set* (input and output strings).
- Choose a suitable architecture (some DTRNN architectures are incapable of representing all FSM [21,22])
- Decide on an encoding for inputs.
- Define output targets for each symbol and establish tolerances.
- Initialize weights, biases and initial states adequately.
- Use a suitable learning algorithm to vary weights, biases, and, optionally, initial states until the net outputs the correct output string (within tolerance) for each input string in the *learning set*.
- If a *test set* has been set aside, check the network's *generalization* behaviour.
- If a FSM is needed, *extract* (see section 3.3) one from the dynamics of the network (even if dynamics is not finite-state!).

However, a number of open questions arise when training DTRNN to behave as FSM, among which

- How does one choose n_X? This introduces a definite inductive bias: too small a value may yield an incapable DTRNN; too big a value may hamper its generalization ability.
- Will the DTRNN exhibit finite-state behaviour? A continuous-state DTRNN has an infinite number of states available and therefore has no bias toward discrete-state behaviour.
- Will it indeed learn? A DTRNN may be computationally capable to perform a task but *learning* that task from examples may not be easy or even possible.
- As with all neural networks, learning may get trapped in undesirable local minima of the error function.

The results obtained so far by the researchers cited show that, indeed, DTRNN can learn FSM-like behaviour from samples, although some problems persist.

One such problem is called *instability*: after learning, FSM-like behaviour is observed only for short input sequences but degrades with sequence length; indeed, we will say that a DTRNN shows *stable* FSM behaviour when outputs are within tolerance of targets for input strings of *any* length. As will be explained later, DTRNN may be *constructed* to emulate finite-state machines; stable DTRNN constructed in that way have high weights, which in turn has the consequence that the error function has very small gradients; this may be part of the explanation why stable behaviour is hard to learn.

Another related problem occurs when the task to be learned has long-term dependencies, that is, when late outputs depend on very early inputs; when this is the case, gradient-descent algorithms have trouble relating late contributions to the error to small changes in the state of neurons in early stages of the processing; these problems have been studied in detail by Bengio et al. [23].

3.3   Extracting Finite-State Machines from Trained Networks

Once successfully trained, specialized algorithms may *extract* FSM from the dynamics of the DTRNN; some use a straightforward equipartition of neural state space followed by a branch-and-bound algorithm [12], or a clustering algorithm [10,16,20]. Very often, the finite-state automaton extracted behaves correctly for strings of any length, even better than the original DTRNN. But automaton extraction algorithms have been criticised [24,25] in the sense that FSM extraction may not reflect the actual computation performed by the DTRNN. More recently, Casey [26] has shown that DTRNN can indeed "organize their state space to mimic the states in the [...] state machine that can perform the computation" and be trained or programmed to behave as FSM. Also recently, Blair and Pollack [27] presented an increasing-precision dynamical analysis that identifies those DTRNNs that have actually learned to behave as FSM.

3.4 Programming Sigmoid DTRNN to Behave as Finite-State Machines

Finally, some researchers have set out to study whether it is possible to program a sigmoid-based DTRNN so that it behaves as a given FSM, that is, they have tried to formulate sets of rules for choosing the weights and initial states of the DTRNN based on the transition function and the output function of the corresponding FSM. In summary, to simulate a FSM in a sigmoid DTRNN one has to decide:

- How to encode the input symbols $\sigma_k \in \Sigma$ as vectors $\mathbf{u}_k \in U$. The usual choice is a (*one-hot*) encoding: $n_U = |\Sigma|$, $(\mathbf{u}_k)_i = \delta_{ik}$, where δ_{ik} is Kronecker's delta, defined as $\delta_{ik} = 1$ if $i = k$ and 0 otherwise.
- How to interpret outputs as symbols $\gamma_m \in \Gamma$. For example, nonempty disjoint regions $Y_m \subseteq Y$ (defined *e.g.* by a tolerance around suitable target values) may be assigned to each $\gamma_m \in \Gamma$.
- How many state neurons n_X to use (this will be discussed later).
- Which neural architecture to use.
- The values for weights.
- A value for the initial state $\mathbf{x}_0$ (for q_I).

As has been said in section 3.2, not all DTRNN architectures can emulate all FSM [21,22].

When does a DTRNN behave as a FSM? A DTRNN $N = (X, U, Y, \mathbf{f}, \mathbf{h}, \mathbf{x}_0)$ is said to behave as a FSM $M = (Q, \Sigma, \Gamma, \delta, \lambda, q_I)$ when (Casey 1996):

1. States $q_i \in Q$ are assigned nonempty disjoint regions $X_i \subseteq X$ such that the DTRNN N is said to be in state q_i at time t when $\mathbf{x}[t] \in X_i$.
2. The initial state of the DTRNN N, belongs to the region assigned to state q_I, that is, $\mathbf{x}_0 \in X_I$.
3. $\mathbf{f}_k(X_j) \subseteq X_i \quad \forall q_j \in Q, \sigma_k \in \Sigma : \delta(q_j, \sigma_k) = q_i$, where $\mathbf{f}_k(A) = \{\mathbf{f}(\mathbf{x}, \mathbf{u}_k) : \mathbf{x} \in A\}$ for short (see figure 3).
4. $\mathbf{h}_k(X_j) \subseteq Y_m \quad \forall q_j \in Q, \sigma_k \in \Sigma : \lambda(q_j, \sigma_k) = \gamma_m$, where $\mathbf{h}_k(A) = \{\mathbf{h}(\mathbf{x}, \mathbf{u}_k) : \mathbf{x} \in A\}$ for short (see figure 4).

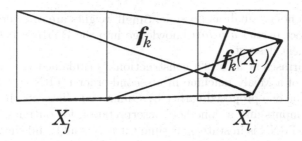

Fig. 3. Correctness of the next-state function of a FSM as computed by a DTRNN. The transition $\delta(q_j, \sigma_k) = q_i$ is illustrated

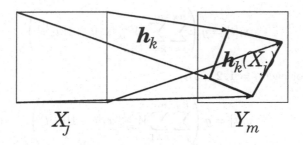

Fig. 4. Correctness of the output function of a FSM as computed by a DTRNN. The production of output $\lambda(q_j, \sigma_k) = \gamma_m$ is illustrated

Omlin and Giles [28] have proposed an algorithm for encoding deterministic finite-state automata (DFA, a class of FSM) in second-order recurrent neural networks which is based on a study of the fixed points of the sigmoid function. Alquézar and Sanfeliu [29] have generalized Minsky's [4] result to show that DFA may be encoded in Elman [9] nets with rational (not real) sigmoid transfer functions. Kremer [30] has recently shown that a single-layer first-order sigmoid DTRNN can represent the state transition function of any finite-state automaton. Frasconi et al. [31] have shown similar encodings for radial-basis-function DTRNN. All of these constructions use a number of hidden units proportional to the number of states in the FSM. More recently, Šíma [32] has shown that the behaviour of any discrete-state DTRNN may be stably emulated by a continuous-state DTRNN using activation functions in a very general class which includes sigmoid functions. In a more recent paper, Šíma and Wiedermann [33] show that any regular language may be more efficiently recognized by a DTRNN having threshold units. Combining both results, one concludes that sigmoid DTRNN can act as DFA accepting any regular language.

Recently, Carrasco et al. [22] (see also [34,35]) have expanded the current results on stable encoding of FSM on DTRNN to a larger family of sigmoids, a larger variety of DTRNN (including first- and second-order architectures), and a wider class of FSM architectures (DFA and Mealy and Moore FSM), by establishing a simplified procedure to prove the stability of a devised encoding and to obtain weights as small as possible. Small weights are of interest if encoding is used to inject partial *a priori* knowledge into the DTRNN before training it through gradient descent.

One of Carrasco et al.'s [22] constructions is explained here in more detail: the encoding of a Mealy machine in a second-order DTRNN; this construction is similar to the one proposed earlier by Omlin and Giles [28]. It uses a one-hot encoding for inputs and a "one-hot" interpretation for outputs, and $n_X = |Q|$ state units (DTRNN is in state q_i at time t if $x_i[t]$ is high and the $x_j[t], j \neq i$, are low). Then the initial state is chosen so that $x_I[0] = 1$ and all other $x_i[0] = 0$. A single-layer *second-order* neural net [12] is used for both δ and λ as follows:

– *Next state*: for each $i = 1, 2, \ldots n_X$,

$$x_i[t] = g \left(\sum_{j=1}^{n_X} \sum_{k=1}^{n_U} W_{ijk}^{xxu} x_j[t-1] u_k[t] \right) \tag{3}$$

– *Output*: for each $i = 1, 2, \ldots n_Y$,

$$y_i[t] = g \left(\sum_{j=1}^{n_X} \sum_{k=1}^{n_U} W_{ijk}^{yxu} x_j[t-1] u_k[t] \right), \tag{4}$$

where $g(x) = g_L(x; 1)$. The next-state weights are chosen as follows: $W_{ijk}^{xxu} = H$ if $\delta(q_j, \sigma_k) = q_i$ and $-H$ otherwise ($\sigma_k \in \Sigma$, $q_i, q_j \in Q$). The output weights are $W_{ijk}^{yxu} = H$ if $\lambda(q_j, \sigma_k) = \gamma_i$ and $-H$ otherwise ($\gamma_i \in \Gamma$). It may also be said that weights are $+1$ and -1 and that the gain of the sigmoid is $\beta = H$ (see section 3). Now, the key is to choose H high enough to ensure correct output for strings of *any* length; Carrasco et al. [22] obtain the lowest possible values for H by mathematical induction after a worst-case study; the results are shown in table 1. Encoding results such as this are very important: they show that

Table 1. Values of the weight parameter H for stable finite-state behaviour of a DTRNN as a function of the number of states $|Q|$ of the FSM (the value 2^+ indicates any value infinitesimally larger than 2)

$|Q|$	H
2	2^+
3	3.113
6	4.181
10	4.863
30	6.224

discrete-state behaviour may be obtained with continuous-state units having a finite sigmoid "gain" H (discrete-state behaviour is always guaranteed by an infinite gain, see section 3).

4   Relaxing the Discrete-Time Restriction

All of the encodings discussed are for finite-state machines in *discrete-time* recurrent neural networks which assume the existence of a non-neural external clock which times their behaviour and a non-neural storage or memory for the previous state of the network, which is needed to compute the next state from the inputs. However, real (biological) neural networks are physical systems that operate in continuous time and should contain, if involved in the emulation of finite-state behaviour, neural mechanisms for synchronization and for memory.

Clocks and memories are indeed present in digital computers that, on the one hand, show discrete-time behaviour, but, on the other, are built from analog transistors and diodes having continuous-time dynamics. Therefore, it should in principle be possible to build a more natural model of finite-state computation based on *continuous-time* recurrent neural networks (CTRNN). CTRNN are a class of networks whose inputs and outputs are functions of a continuous-time variable and whose neurons have a temporal response that is described as a differential equation in time (for an excellent review on CTRNN, see [36]). We are not aware of any attempt to describe the finite-state computational behaviour of CTRNN.

The continuous-time version of the sigmoid unit may be written as follows [37,38]:

$$\tau \frac{dy}{dt} = -y + g\left(\sum_i W_i u_i + b\right),\tag{5}$$

where τ is the unit's time constant (dynamics). The stationary (infinite-time) state is, obviously, that of a discrete-time or instantaneous neuron

$$y = g\left(\sum_i W_i u_i + b\right).\tag{6}$$

CTRNN may be constructed and trained [36] to process a *continuous input signal* into a *continuous output signal*:

$$\mathbf{u}(t) \to \mathbf{y}(t).\tag{7}$$

In particular, CTRNN may also be built to act as a memory (bistable device) or as a clock (oscillator). Therefore, CTRNN may emulate DTRNN (using CTRNN clocks and CTRNN memories), if inputs follow a precise chronogram (as they do in digital computers). As a corollary, CTRNN may be programmed to emulate FSM behaviour.

Nevertheless, a note regarding biological plausibility is in order. FSM behaviour is often a discrete-time simplification of continuous-time behaviour (e.g., speech, vision, etc.), which could be treated directly using CTRNN without constructing a discrete-time computational model. Indeed, while discrete-state discrete-time RNN were characterized in the 50's as finite-state machines, to our knowledge, the continuous-time continuous-state computational models represented by CTRNN have not been characterized theoretically.

5 Biologically Inspired Models: Spiking or Integrate-and-Fire Neurons

5.1 Spiking or Integrate-and-Fire Neurons

DTRNN and CTRNN are usually formulated in terms of sigmoid neurons transmitting amplitude-modulated signals, but most real neurons may be seen as using some kind of temporal encoding:

- trains of impulses (spikes) are sent,
- they are received and integrated by each neuron (possibly with some leakage),
- the neuron fires (sends an impulse) when the integral reaches a certain threshold.

These are the rudiments of *integrate-and-fire* or *spiking neuron* models. The computational capabilities of recurrent neural networks containing these biologically-motivated neuron models have yet to be explored. This opens a wide field for future interaction between theoretical computer scientists and neuroscientists.

5.2 Equivalence to Sigmoid Units and Implications

While integrate-and-fire or spiking models may be seen as very different from the sigmoid neurons discussed so far, they are not. The correspondence is, however, not straightforward. Biological neural systems perform visual processing (10 synaptic stages) in $\sim$ 100 ms [39]. But real neurons are very *slow*: firing frequencies are in the range of 100 Hz. Therefore, direct temporal encoding of analog variables such as the output of a continuous-state sigmoid unit are not biologically plausible. A variety of solutions have been proposed. One is called the *space-rate code*: analog variables in [0, 1] are encoded as the fraction of neurons in a pool of neurons which fire within a given time interval. Fast analog space-rate computation (e.g. of sigmoids) and temporal processing (e.g. bandpass filtering) has been demonstrated through simulation [40]. This suggests that finite-state machines may be indirectly encoded in integrate-and-fire networks by implementing a sigmoid DTRNN as a CTRNN and then converting the CTRNN into an integrate-and-fire network.

5.3 Direct Encoding of Finite-State Machines in Integrate-and-Fire Networks

A more direct, but related approach to finite-state behaviour in networks of spiking neurons has been recently proposed by Wennekers [41]. Wennekers' spiking neurons form *synfire chains*, that is, sequences of events in which a particular subset of integrate-and-fire neurons *synchronizedly fire* in response to the synchronized firing of another (or the same) subset of neurons after an approximately constant delay determined by the characteristics of the connections and the neurons themselves. Each state in the finite-state machine is associated with a particular subset of neurons and the network is said to be in a certain state when the corresponding subset is synchronizedly firing. State transitions are gated by special subsets which fire only when a certain external input is present and a certain state was firing one delay unit before.

6   Concluding Remarks

Our survey shows that

1. Very useful finite-state computation models stem from McCulloch & Pitts' [1] idealized discrete-signal discrete-time recurrent neural network models. These models are present in many successful theoretical and practical computational solutions.
2. Continuous-state discrete-time models are more convenient in learning settings and are shown to be able to stably emulate finite-state behaviour. These networks use non-neural (external) devices such as clocks and memories.
3. Continuous-time RNN may, in principle, emulate clocks, memories and, therefore, they should be capable of emulating FSM behaviour, although we are not aware of any work about this aspect.
4. Theoretical models of the computational capabilities of continuous-time continuous-state recurrent neural networks such as those available for discrete-time discrete-state RNN (i.e. finite-state machines) are still missing.
5. The study of more biologically plausible models of finite-state computation —e.g., integrate-and-fire neurons— has just started.

However, an important question remains unanswered: is finite-state computation still a relevant model of biological information processing or is it a computationally convenient simplification of this behaviour?

Acknowledgements. The authors thank Juan Antonio Pérez-Ortiz for comments on this manuscript and acknowledge the support of the Spanish Comisión Interministerial de Ciencia y Tecnologia through grant TIC97-0941.

References

1. W. S. McCulloch and W. H. Pitts. A logical calculus of the ideas immanent in nervous activity. *Bulletin of Mathematical Biophysics*, 5:115–133, 1943.
2. J. E. Hopcroft and J. D. Ullman. *Introduction to automata theory, languages, and computation.* Addison–Wesley, Reading, MA, 1979.
3. S.C. Kleene. Representation of events in nerve nets and finite automata. In C.E. Shannon and J. McCarthy, editors, *Automata Studies*, pages 3–42. Princeton University Press, Princeton, N.J., 1956.
4. M.L. Minsky. *Computation: Finite and Infinite Machines.* Prentice-Hall, Inc., Englewood Cliffs, NJ, 1967. Ch: Neural Networks. Automata Made up of Parts.

5. N. Alon, A. K. Dewdney, and T. J. Ott. Efficient simulation of finite automata by neural nets. *Journal of the Association of Computing Machinery*, 38(2):495–514, 1991.

6. P. Indyk. Optimal simulation of automata by neural nets. In *Proceedings of the 12th Annual Symposium on Theoretical Aspects of Computer Science*, pages 337–348, Berlin, 1995. Springer-Verlag.

7. B. G. Horne and D. R. Hush. Bounds on the complexity of recurrent neural network implementations of finite state machines. *Neural Networks*, 9(2):243–252, 1996.

8. Tony Robinson and Frank Fallside. A recurrent error propagation network speech recognition system. *Computer Speech and Language*, 5:259–274, 1991.

9. J. L. Elman. Finding structure in time. *Cognitive Science*, 14:179–211, 1990.

10. A. Cleeremans, D. Servan-Schreiber, and J. L. McClelland. Finite state automata and simple recurrent networks. *Neural Computation*, 1(3):372–381, 1989.

11. Jordan B. Pollack. The induction of dynamical recognizers. *Machine Learning*, 7:227–252, 1991.

12. C. L. Giles, C. B. Miller, D. Chen, H. H. Chen, G. Z. Sun, and Y. C. Lee. Learning and extracted finite state automata with second-order recurrent neural networks. *Neural Computation*, 4(3):393–405, 1992.

13. R. L. Watrous and G. M. Kuhn. Induction of finite-state languages using second-order recurrent networks. *Neural Computation*, 4(3):406–414, 1992.

14. Arun Maskara and Andrew Noetzel. Forcing simple recurrent neural networks to encode context. In *Proceedings of the 1992 Long Island Conference on Artificial Intelligence and Computer Graphics*, 1992.

15. A. Sanfeliu and R. Alquézar. Active grammatical inference: a new learning methodology. In Dov Dori and A. Bruckstein, editors, *Shape and Structure in Pattern Recognition*, Singapore, 1994. World Scientific. Proceedings of the IAPR International Workshop on Structural and Syntactic Pattern Recognition SSPR'94 (Nahariya, Israel).

16. P. Manolios and R. Fanelli. First order recurrent neural networks and deterministic finite state automata. *Neural Computation*, 6(6):1154–1172, 1994.

17. M. L. Forcada and R. C. Carrasco. Learning the initial state of a second-order recurrent neural network during regular-language inference. *Neural Computation*, 7(5):923–930, 1995.

18. Peter Tiňo and Jozef Sajda. Learning and extracting initial Mealy automata with a modular neural network model. *Neural Computation*, 7(4), July 1995.

19. R. P. Ñeco and M. L. Forcada. Beyond Mealy machines: Learning translators with recurrent neural networks. In *Proceedings of the World Conference on Neural Networks '96*, pages 408–411, San Diego, California, September 15–18 1996.

20. Marco Gori, Marco Maggini, E. Martinelli, and G. Soda. Inductive inference from noisy examples using the hybrid finite state filter. *IEEE Transactions on Neural Networks*, 9(3):571–575, 1998.

21. M.W. Goudreau, C.L. Giles, S.T. Chakradhar, and D. Chen. First-order vs. second-order single layer recurrent neural networks. *IEEE Transactions on Neural Networks*, 5(3):511–513, 1994.

22. Rafael C. Carrasco, Mikel L. Forcada, M. Ángeles Valdés-Muñoz, and Ramón P. Ñeco. Stable encoding of finite-state machines in discrete-time recurrent neural nets with sigmoid units. *Neural Computation*, 12, 2000. In press.

23. Y. Bengio, P. Simard, and P. Frasconi. Learning long-term dependencies with gradient descent is difficult. *IEEE Transactions on Neural Networks*, 5(2):157–166, 1994.

24. J.F. Kolen and Jordan B. Pollack. The observer's paradox: apparent computational complexity in physical systems. *Journal of Experimental and Theoretical Artificial Intelligence*, 7:253–277, 1995.
25. J. F. Kolen. Fool's gold: Extracting finite state machines from recurrent network dynamics. In J. D. Cowan, G. Tesauro, , and J. Alspector, editors, *Advances in Neural Information Processing Systems 6*, pages 501–508, San Mateo, CA, 1994. Morgan Kaufmann.
26. M. Casey. The dynamics of discrete-time computation, with application to recurrent neural networks and finite state machine extraction. *Neural Computation*, 8(6):1135–1178, 1996.
27. A. Blair and J. B. Pollack. Analysis of dynamical recognizers. *Neural Computation*, 9(5):1127–1142, 1997.
28. C. W. Omlin and C. L. Giles. Constructing deterministic finite-state automata in recurrent neural networks. *Journal of the ACM*, 43(6):937–972, 1996.
29. R. Alquézar and A. Sanfeliu. An algebraic framework to represent finite state automata in single-layer recurrent neural networks. *Neural Computation*, 7(5):931–949, 1995.
30. Stefan C. Kremer. *A Theory of Grammatical Induction in the Connectionist Paradigm*. PhD thesis, Department of Computer Science, University of Alberta, Edmonton, Alberta, 1996.
31. Paolo Frasconi, Marco Gori, Marco Maggini, and Giovanni Soda. Representation of finite-state automata in recurrent radial basis function networks. *Machine Learning*, 23:5–32, 1996.
32. Jiří Šíma. Analog stable simulation of discrete neural networks. *Neural Network World*, 7:679–686, 1997.
33. Jiří Šíma and Jiří Wiedermann. Theory of neuromata. *Journal of the ACM*, 45(1):155–178, 1998.
34. Ramón P. Ñeco, Mikel L. Forcada, Rafael C. Carrasco, and M. Ángeles Valdés-Muñoz. Encoding of sequential translators in discrete-time recurrent neural nets. In *Proceedings of the European Symposium on Artificial Neural Networks ESANN'99*, pages 375–380, 1999.
35. Rafael C. Carrasco, Jose Oncina, and Mikel L. Forcada. Efficient encodings of finite automata in discrete-time recurrent neural networks. In *Proceedings of ICANN'99, International Conference on Artificial Neural Networks*, 1999. (in press).
36. B. A. Pearlmutter. Gradient calculations for dynamic recurrent neural networks: a survey. *IEEE Transactions on Neural Networks*, 6(5):1212–1228, 1995.
37. F. J. Pineda. Generalization of back-propagation to recurrent neural networks. *Physical Review Letters*, 59(19):2229–2232, 1987.
38. L.B. Almeida. Backpropagation in perceptrons with feedback. In R. Eckmiller and Ch. von der Malsburg, editors, *Neural Computers*, pages 199–208, Neuss 1987, 1988. Springer-Verlag, Berlin.
39. S. Thorpe, D. Fize, and C. Marlot. Speed of processing in the human visual system. *Nature*, 381:520–522, 1996.
40. Thomas Natschläger and Wolfgang Maass. Fast analog computation in networks of spiking neurons using unreliable synapses. In *Proceedings of ESANN'99, European Symposium on Artificial Neural Networks*, pages 417–422, 1999.
41. Thomas Wennekers. Synfire graphs: From spike patterns to automata of spiking neurons. Technical Report Ulmer Informatik-Berichte Nr. 98-08, Universität Ulm, Fakultät für Informatik, 1998.

An Investigation into the Role of Cortical Synaptic Depression in Auditory Processing

Sue L. Denham and Michael J. Denham

Centre for Neural and Adaptive Systems
School of Computing
University of Plymouth
Plymouth UK
sue@soc.plym.ac.uk, mike@soc.plym.ac.uk

Abstract. In comparison with the speed and precision associated with processing in the auditory periphery, the temporal response properties of neurones in primary auditory cortex can appear to be surprisingly sluggish. For example, much of the temporal fine structure is lost, best modulation frequencies are generally low and the effects of forward masking can be detected for a surprisingly long time. The purpose of this investigation was to explore whether depression at thalamocortical synapses could account for these observations. We show that a model that incorporated synaptic dynamics could indeed replicate these and other experimental results and in addition provided a novel explanation for some effects of subthreshold stimuli.

1   Introduction

There are many aspects of auditory perception, such as the growth of loudness with duration and the effects of masking, which indicate that the auditory system performs some sort of temporal integration in processing incoming acoustic signals. However, the auditory system is also capable of fine temporal resolution, as evidenced by gap detection, double click discrimination, and also in the short latency and lack of jitter of onset responses in cortex [24]. This has been termed the resolution-integration paradox, i.e. how is it possible for a system to integrate information over long periods while retaining fine temporal resolution. Most accounts which satisfy the integration criterion use long time constants and therefore fail to behave swiftly enough to explain fine temporal resolution, and vice versa.

The time constants typically associated with sub-cortical processing differ substantially from those in the cortex. In comparison with the speed and precision associated with processing in the auditory periphery, the temporal response properties of neurones primary auditory cortex (AI) can appear to be surprisingly sluggish. For example, in the thalamocortical transformation of incoming signals a great deal of the temporal fine structure is lost [4] best modulation frequencies measured in AI are generally below 15 Hz [20] and the effects of a masker on a probe tone can be detected up to 400 ms after masker offset [2]. The focus in this paper is therefore on the temporal response properties observed in AI. As yet there have been no models proposed which can satisfactorily explain the observed behaviour of neurones in AI. Explanations in terms of intracortical inhibitory circuits have been proposed but inhibition does not provide an adequate account, at least in the case of forward

S. Wermter et al. (Eds.): Emergent Neural Computational Architectures, LNAI 2036, pp. 494-506, 2001.

masking which is unaffected by the application of a GABA antagonist [2]. On the other hand, simple threshold neural models cannot replicate such behaviour without some form of inhibition or by means of very long time constants, which as discussed above, would then prevent the model from satisfying the requirements for good temporal resolution.

Recently it has become apparent that cortical synaptic dynamics may be an important factor affecting the behaviour of biological neurones 13,1,21,19]. When synapses are repeatedly activated they do not simply respond in the same way to each incoming impulse and synapses may develop a short-term depression or facilitation, depending on the nature of the pre- and postsynaptic cells, and on the characteristics of the particular synapse involved [21,19]. Within current neural network models synapses are generally modelled as simple gains and it is interesting to consider how models of cortical processing might be enhanced by the inclusion of a richer synaptic model. If synapses are not simply be viewed as passive weighting elements in neuronal circuits, but rather as dynamical systems in their own right, then perhaps many of the response properties observed in AI might be explained in a relatively simple way. To explore this hypothesis, a model of cortical synaptic depression was used to investigate the computational properties of a neurone model that included dynamic synapses. This model was found to account for a wide range of experimental observations, including those outlined above.

2   The Dynamic Synapse Model

The dynamic synapse model we use here was presented in [22] and shown to replicate experimental results reported on synaptic depression. In fact, this model of the dynamics of neurotransmitter release had already been proposed much earlier by Grossberg [6], who derived a set of psychological postulates to explain the excitatory transients in transmitter release after a rest period and subsequent synaptic depression, which had been observed experimentally by Eccles [5]. This model of synaptic depression has been further developed and used by Grossberg in more recent years, for example to explain a number of important perceptual features involving the visual cortex. In the area of auditory modelling, a very similar model was also developed by Meddis [15] to describe transduction in cochlear inner hair cells.

The dynamic synapse model characterises the synapse by defining a "resource", e.g. the amount of neurotransmitter in the synapse, a proportion of which can be in one of three states: *available, effective, inactive*. The dynamical behaviour of the proportions of the resource that are in each of these states is determined by the system of three coupled differential equations below:

$$\frac{dx}{dt} = g.y(t).I(t) - \alpha.x(t)$$

$$\frac{dy}{dt} = \beta.w(t) - g.y(t).I(t)$$

$$\frac{dw}{dt} = \alpha.x(t) - \beta.w(t)$$

where $x(t)$ is the amount of *effective* resource, e.g. activated neurotransmitter within the synaptic cleft, as a proportion of the total resource, $y(t)$ is the amount of *available* resource, e.g. free neurotransmitter in the synapse, and $w(t)$ is the amount of *inactive* resource, e.g. neurotransmitter being reprocessed.

The input signal $I(t)$ represents the occurrence of a presynaptic action potential (AP) and is set equal to one at the time of arrival of the AP and for a small period of time δ thereafter, and otherwise is set equal to 0. The constant β determines the rate at which the inactive resource $w(t)$, is released to the pool of available resource on a continuing basis, and α represents the rate at which effective resource $x(t)$ becomes rapidly inactive again. The instantaneous efficacy of the synapse is determined by the variable g, which can be interpreted as the fraction of available resource released as a result of the occurrence of the presynaptic AP. It takes a value in the range zero to one.

The key idea behind the model is that there is a fixed amount K of total resource available at the synapse, a proportion $g.y(t)$ of which is activated in response to presynaptic activity, rapidly becomes inactive, and is then subsequently made available again through reprocessing. Thus, if the synapse is very active the amount of available resource $y(t)$ is rapidly reduced. There must then follow a period during which the synapse can recover in order to respond fully once more. This process appears to replicate the experimentally observed characteristics of synaptic depression, for example as reported in [14,22].

The EPSP at the synapse, $e(t)$, is computed from $x(t)$ using the following equation for the passive membrane mechanism [22]:

$$\tau_{EPSP} \cdot \frac{de}{dt} = \gamma.x(t) - e(t)$$

The neurone model used is described by the following system of equations, which has been adapted from a model described in [12]:

$$\tau_E \frac{dE}{dt} = -E(t) + V(t) + G_K(t).(E_K - E(t))$$

$$s(t) = 1, \text{if } E(t) \geq \theta(t), \text{else } s(t) = 0$$

$$\tau_{G_K} \frac{dG_K}{dt} = -G_K(t) + \eta.s(t)$$

$$\tau_\theta \frac{d\theta}{dt} = -(\theta(t) - \theta_0) + s(t)$$

where, $E(t)$ is the variation of the neurone's membrane potential relative to its resting potential, $V(t)$ is the driving input found by summing all the synaptic EPSPs, $G_K(t)$ is the potassium conductance, divided by the sum of all the voltage-dependent ionic membrane conductances, E_K is the potassium equilibrium potential of the membrane relative to the membrane resting potential, $\theta(t)$ is the firing threshold potential, θ_0 is the resting threshold, $s(t)$ is the variable which denotes firing of the cell, τ_E, τ_{EPSP}, τ_θ and τ_{GK} are time constants, and γ, χ and η are constant parameters.

3   Simulation Results

Not all cortical synapses are depressing; for example, synapses between cortical pyramidal neurones and bi-tufted GABAergic interneurons synapses are strongly facilitating [19]. However, thalamocortical synapses appear to be depressing; they are mediated by non-NMDA excitatory amino acids, depress rapidly and remain desensitised for some time [21]. In the simulations that follow, it can be seen that the response characteristics of the model neurone, when the dynamic synapse model is included turn out to be very similar to that found in primary auditory cortex.

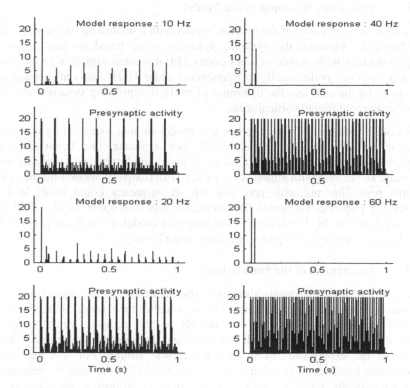

Fig. 1. Simulation of the transmission of signals between thalamic relay and cortical pyramidal cells. The model qualitatively replicates the behaviour of paired pyramidal cells in AI, which showed almost no response except at signal onset when stimuli exceeded 20 Hz [4].

3.1   Loss of Temporal Fine Structure

Differences between the response properties of thalamic and cortical neurones were described in [4]. Activity in thalamic relay cells and subsequent activity in paired pyramidal cells in AI was recorded, and it was found that even when thalamic activity was clearly synchronised to the stimulus up to 200 Hz, the paired cortical cell was unable to follow the details of the signal beyond about 20 Hz. The plots in figure 1 show the response of the model to spike trains generated to resemble typical thalamic activity in response to stimuli of the frequencies indicated. Total activity for 20 presentations is plotted both for the presynaptic spike trains and the model response. The model behaviour resembles that found experimentally. The model responds to details of the stimuli occurring at 10 Hz and to a lesser extent to details at 20 Hz, but for higher stimulus frequencies, the model only responds strongly at the onset of the signal. The reason for this is that at high frequencies successive presynaptic spikes arrive before the synapse has time to recover. This causes a strong depression of the synapse, resulting in the generation of very small postsynaptic EPSPs that are insufficient to raise the cell membrane potential above the firing threshold.

3.2   Frequency Response of the Model

The frequency response of the neurone model with a depressing synapse is illustrated in figure 2. Although the synaptic dynamics were tuned to match those found experimentally in the somatosensory cortex [14] it is interesting to note that the model clearly responds preferentially to frequencies under 10 Hz, as is also found in AI. It seems to be the case that the dynamics of cortical depressing synapses may be quite similar across different cortical areas.

For comparison the response of a neurone model without a depressing synapse is also shown. Clearly such a model could not replicate the behaviour observed experimentally without the addition of delayed inhibitory inputs which somehow increase in strength with stimulus frequency. Alternatively, modelling the synapse as a low pass filter but with very low cut-off frequency could result in a similar frequency response, but would fail simultaneously to account for the short response latency found in AI. The benefit of the proposed model is that it can account both for the low pass frequency response and short onset latency.

3.3   Limitations of the Simulations

For many of the experiments simulated, the nature of the thalamocortical signals is unknown, which makes it difficult to know whether the stimuli used as inputs to the model are realistic. However, the details of the acoustic stimuli used in the experiments are generally well documented and therefore it is desirable to be able to simulate the experiments using similar acoustic stimuli. For this reason a well-documented and tested peripheral model, DSAM [17], was used to generate signals characteristically found in auditory nerve fibre recordings in response to acoustic stimuli. The problem with this approach is that the rest of the subcortical auditory system has not been similarly modelled. Therefore, in the following simulations the output from the peripheral model is reprocessed to ensure that the firing rate remains below about 200 Hz by enforcing a reasonable refractory period. Clearly this ignores the computations which occur in the rest of the auditory system. However, the model can replicate a number of experiments and this would almost certainly be improved upon by more accurately modelling the thalamic-cortical signals. While recognising that this simplification likely to result in a poor approximation of actual thalamic relay cell activity, it is difficult at this stage to do much better, and has the added benefit of making the simulations tractable.

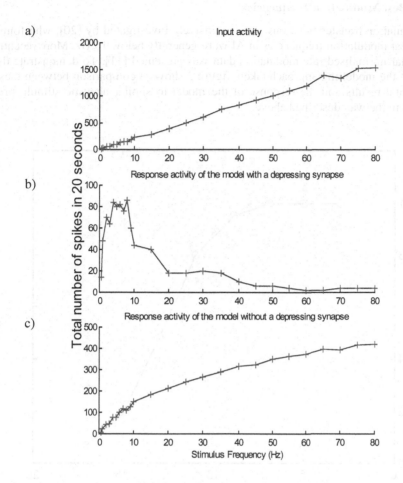

Fig. 2. Frequency response of the model. The response of the neuron model with and without a depressing synapse to an incoming spike train of the frequency indicated. Plots show a) total input activity; and the number of times the cell fired during the 20-second period b) with and c) without synaptic depression.

For the remainder of the simulations, the acoustic signals specified are processed by the DSAM model that includes an outer-middle ear transfer function, a gammatone filterbank, and Meddis' inner hair cell model. A simple stochastic spike generator model is used and a convergence of twenty inner hair cells to one auditory nerve fibre assumed. The spike trains are then processed to ensure that refractory periods are generally greater than 20ms. However, when more than one spike occurs simultaneously, as is possible with a combinations of twenty spike trains per channel, then the refractory period is allowed to decrease in proportion to the extent of the coincidence. This has the benefit of not destroying the enhanced onset response generated by the inner hair cells.

3.4   Best Modulation Frequencies

Rate modulation transfer functions were extensively investigated by [20], who found that the best modulation frequencies in AI were generally below 15 Hz. More recently very similar normalised rate modulation data was presented [11]. To demonstrate the validity of the modelling approach taken, figure 3 shows a comparison between these experimental results and the response of the model to similar acoustic stimuli, pre-processed in the way described above.

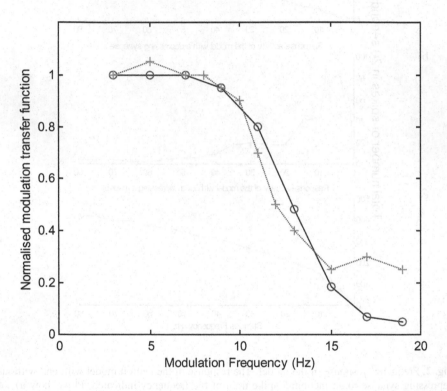

Fig. 3. Response to repeated tones at the given repetition rates; model 'o__o' and experimental results '+...+' [11]. Normalised repetition rate transfer functions are found using a stimulus consisting of 6 tones pulses at the repetition rate indicated and then calculating the mean response to the last 5 tones in the sequence divided by the response to the first tone; each tone has a duration of 25ms.

3.5   The Time Course of Forward Masking

Although there are undoubtedly a number of factors that contribute to the phenomenon of forward masking, it is clear that the depression of thalamocortical synapses must contribute to the total effect. Explanations for forward masking have also been sought in terms of lateral or forward inhibition. However, it has been shown that masking continues to exist even in the presence of a $GABA_A$ antagonist and therefore even if inhibitory inputs have some part to play they cannot provide a full account [2]. Both cortical forward masking and that evidenced behaviourally have been shown to last far longer than explainable in terms of peripheral adaptation [18]. The model clearly provides a mechanism for forward masking, since synapses that have been previously activated require time to replenish their transmitter stores and respond less strongly when depleted. The time course of synaptic recovery appears to be consistent with the time course of cortical forward masking. The tonotopic distribution of masking is also consistent with a model of forward masking in terms of the depression of thalamocortical synapses since it has been shown that masking is closely related to the receptive fields of cortical neurons [2,3]. Figure 4 shows the transmitter depletion at synapses across the tonotopic axis in response to masking stimuli at the intensities indicated.

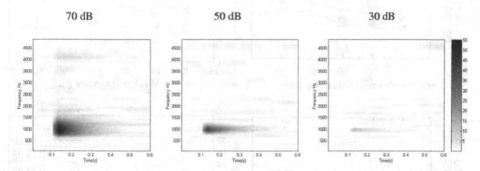

Fig. 4. Distribution and time course of transmitter depletion at synapses across the tonotopic axis in response to a 1000 Hz masker of 30 ms duration at the intensities indicated. Greyscale indicates the percentage depletion relative to that at the start of the stimulus. These results are very similar to Brosch and Schreiner's plots of the time course and distribution of masking [2].

An important aspect of this model is that it demonstrates that cortical forward masking could be dependent on presynaptic rather than postsynaptic activity. This offers a simple explanation for the puzzling experimental observation that masking is sometimes detected even in response to maskers that do not actually activate the target cell [2]. If masking is a result of transmitter depletion of thalamocortical synapses, then it would be quite possible for such synapses to become depleted by thalamic activity even though there is insufficient incoming activity to actually cause the cortical cell to fire, which is how the response to the masker was determined. Since these synapses would nevertheless be depleted, the probe tone could therefore be masked by the 'sub-threshold' masker.

3.6   The Effect of Masker Duration

In psychophysical experiments it has been shown that the degree of masking is affected by the duration of the masker and masking increases with masker duration (Kidd and Feth 1982). This was also found to be the case by Brosch and Schreiner [2] in their recordings in AI. However, the sensitivity to duration was observed even when the AI cell responded only at the onset of the masker, and although the effect of masker duration was noted, it was not suggested how this could occur. The model investigated here suggests a simple explanation, i.e. as long as there is some tonic incoming activity during the masker, then transmitter depletion at the thalamocortical synapses will be related to masker duration. Therefore, if as we hypothesise, the degree of masking is related to the degree of transmitter depletion at thalamocortical synapses, then the sensitivity to masker duration follows.

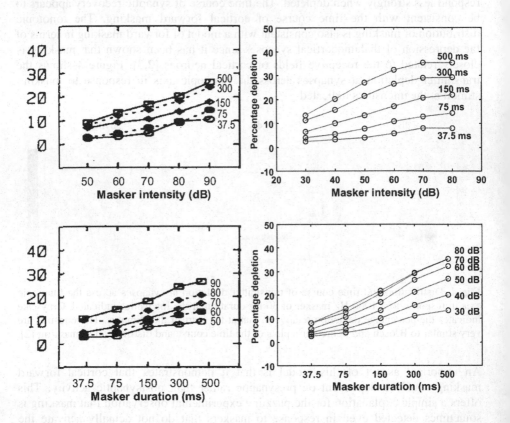

Fig. 5. The effect of masker duration. For comparison, experimental results relating masker duration and masker intensity to probe threshold shifts [10] are shown on the left and the percentage transmitter depletion in the model in response to similar stimuli is plotted on the right.

Brosch and Schreiner [2] did not include any detailed results on masker duration, so in figure 5, a comparison between the results in [10] and the model's response is shown. However, it should be noted that although these results are qualitatively the same, it is not clear how the degree of transmitter depletion in the model can be directly related to the probe threshold shifts plotted in [10].

3.7   Disruption of Synchronisation Responses by Subthreshold Stimuli

In a recent paper [16], Nelken suggested that his experiments showed a correlate of comodulation masking release. Activity was record in AI in response to noise modulated at 10 Hz, and was found to synchronise to each noise pulse as expected. However, when a very soft, even subthreshold, continuous pure tone with frequency corresponding to the cell's best frequency, was added to the noise, then this synchronisation was disrupted. In contrast, when the pure tone was added to an unmodulated noise then the response to the noise alone was indistinguishable from that to the noise plus tone. Nelken suggested that the cortex might therefore be able to detect masked sounds by means of their disruption of the more powerful masker.

Once again a simple explanation of Nelken's results is suggested by the model, which can easily replicate the experimentally observed behaviour as long as there is some tonic thalamic activity in response to the pure tone. Because the activity in response to the pure tone continues through the silent gaps between the noise pulses, this prevents the recovery of the synapses between noise pulses and so the synchronised response is disrupted. This explanation is also consistent with Nelken's unpublished observations that the synchronised response to the noise alone was far more reliably obtained when the noise was trapezoidally modulated, than when sine wave modulation was used. Figure 6 shows Nelken's experimental results and the model's responses to similar stimuli.

4   Discussion

In this paper it has been shown how a model neurone which incorporates dynamic synapses responds to a number of different stimuli. The results seem to indicate that synaptic depression at thalamocortical synapses may explain a number of aspects of the response properties of neurones in AI.

The nonlinearity of the dynamic synapse model allows it to behave in many situations like a low pass filter whilst also retaining a fast onset response. In response to repeated stimulation much above 10 Hz, synaptic depletion prevents the cell from responding except at the onset of the stimulus. However, the synaptic dynamics are not slow and after a period of rest the synapse can respond with a large EPSP to the onset of a new stimulus, which can result in a response of short latency. Since the reliability of a depressing synapse also appears to be related to amount of available transmitter [19], an aspect not included in this model, this means that after a period of rest such synapses will tend to respond very reliably as well. This behaviour is therefore consistent with the generation of onset responses of short latency and with little jitter. Although the cell tends to respond only at the onset of stimuli, important processing can continue to occur in the synapses throughout the duration of the stimulus. This allows the cell to exhibit sensitivity to stimulus duration, even when only active at stimulus onset. In addition, some of the apparent nonlinearities of responses measured in AI, such as the influence of subthreshold stimuli or interactions between different components of a complex stimulus [16], could be explained in this way.

Since synaptic depression operates at thalamocortical synapses which are the route through which sensory signals must pass in order to get to cortex, it seems likely that the dynamics of depressing synapses have a major role to play in sensory processing. Synaptic depression appears to result in a relatively infrequent sampling of the sensory inputs by cortex, where such information is presumably integrated with ongoing cognitive processes. This bears a remarkable similarity to Viemeister's 'multiple looks' model that was formulated in order to explain temporal processing in auditory perception and to resolve the resolution-integration paradox [23]. In this model, it is envisaged that 'looks' or samples from a short time constant process are stored in memory and can be accessed and processed selectively depending on the task.

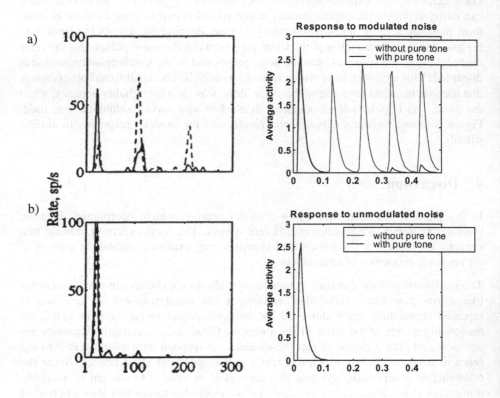

Fig. 6. a) Response of neurones in AI [16], left column, and the model, right column, to a wideband noise stimulus trapezoidally modulated at 10 Hz, without '---' and with '___' a continuous pure tone. b) Experimental and model responses when the noise is unmodulated.

The frequency response of the model bears a strong relationship to speech modulation transfer functions, with frequencies around 4 to 6 Hz being the dominant frequency of the envelope of speech signals. Syllables in speech are generally, although not always, distinguished by an amplitude peak preceded and closed by amplitude trough [9]. Therefore, when the model is stimulated by a speech signal, it has a tendency to fire at the onsets of syllables within the signal. Synaptic depression may therefore give rise to a syllable-like segmentation of speech signals within AI. Such segmentation could occur in parallel across the tonotopic axis, independently within each frequency channel. This suggestion is consistent with the experimentally observed response to species-specific calls by neurones in AI, which tend to fire primarily at the onset of segments or syllables within calls, irrespective of the characteristic frequency of the neurone [4,25,26]. One effect this would have is to increase temporal synchrony across the tonotopic axis thereby promoting the grouping of related frequency components of a call. Synchronous activity is likely to be important for the effective transmission of signals to further processing centres which integrate information across frequency channels.

In experiments in which species-specific calls were manipulated [26], it was also shown that speeding up of slowing down the signal, or reversing it all resulted in reduced responses. We suggest that the reasons for this differ between manipulations and that the behaviour of the model can help to explain these results. In the first case when the signal is slowed down, activity is still generated in response to syllable onsets, but since these occur at a slower rate, the total amount of activity per second decreases. In the second case, when the signal is speeded up, synaptic depression would prevent synchronisation to syllable onsets as effectively as for the control case. Finally, reversing the signal results in a reduced response, not because of a change in timing of the stimulus but because of the change in the nature of the transients in the signal. As shown in experiments on onset latency, sharp transients with abrupt rises are far more effective in generating responses than those with slow rise times [7]. Reversing the speech signal means that the transients generally become less abrupt and therefore generate reduced activity. It seems reasonable to suppose that communication sounds have evolved to optimise their detection by cortex, and that the communication sounds that are used are those which are most salient within AI. Hence, the similarity between the modulation transfer functions of speech signals and of those measured in auditory cortex. Interestingly, although derived very differently, the behaviour of the model is very similar to the RASTA filter and which was found to markedly improve speech recognition in noise [8].

5 Conclusions

By taking synaptic dynamics into account in modelling these experiments, it has been possible to account for a number of previously unexplained results in a fairly straightforward way. On the basis of these investigations it is suggested that the dynamics of thalamocortical synapses may help to explain the temporal integration observed in AI and in auditory perception.

References

1. Abbott LF, Varela JA, Sen K, Nelson SB, "Synaptic depression and cortical gain control", *Science,* Vol. 275, 220-224, 1997.
2. Brosch M, Schreiner CE, "Time course of forward masking tuning curves in cat primary auditory cortex", *J.Neurophys.*, 1997.
3. Calford MB, Semple MN, "Monaural inhibition in cat auditory cortex", *J. Neurophys.*, 73(5), 1876-1891, 1995.

4. Creutzfeldt O, Hellweg FC, Schreiner C, "Thalamocortical transformation of responses to complex auditory stimuli", *Exp. Brain Res.*, 39, 7-104, 1980.
5. Eccles JC. *The physiology of synapses*. New York: Academic Press, 1964.
6. Grossberg S, "On the production and release of chemical transmitters and related topics in cellular control ", *J. Theor. Biol.*, 22, 325-364, 1969.
7. Heil P, "Auditory cortical onset responses revisited. I. First-spike timing', *J.Neurophys.*, 2616-2541, 1997.
8. Hermansky H, Morgan N, "RASTA processing of speech", *IEEE Trans. On Speech and Audio Processing*, 2(4), 578-589, 1994.
9. Jusczyk PW, "The discovery of spoken language", MIT Press, 1997.
10. Kidd G, Feth LL, "Effects of masker duration in pure-tone forward masking", *J. Acoust. Soc. Am.*, Vol. 72 Num. 5, 1384-1386, 1982.
11. Kilgard MP, Merzenich MM, "Plasticity of temporal information processing in the primary auditory cortex", *Nature Neuroscience*, 1(8), 727-731, 1998.
12. McGregor RJ, *Neural and Brain Modelling*, Academic Press, 1989.
13. Markram H, Lubke J, Frotscher M, Sakmann B, "Regulation of synaptic efficacy by coincidence of postsynaptic APs and EPSPs", *Science*, Vol. 275, 213-215, 1997.
14. Markram H, Tsodyks M, "Redistribution of synaptic efficacy between neocortical pyramidal neurons", *Nature*, 382, 807-810, 1996.
15. Meddis R, "Simulation of mechanical to neural transduction in the auditory receptor", *J. Acoust. Soc. Am.*, 79(3), 702-711, 1986.
16. Nelken I, Yosef OB, "Processing of complex sounds in cat primary auditory cortex ", *Proc. Nato ASI on Computational Hearing*, 19-24, 1998.
17. O'Mard LP, Hewitt MJ, Meddis R, "DSAM :   Development System for Auditory Modelling", http://www.essex.ac.uk/psychology/hearinglab/lutear/home.html
18. Relkin EM, Smith RL, "Forward masking of the compound action potential: Thresholds for the detection of the N_1 peak", *Hear. Res.*, 53, 131-140, 1991.
19. Reyes A, Lujan R, Rozov A, Burnashev N, Somogyi P, Sakmann B, "Target-cell-specific facilitation and depression in neocortical circuits", *Nature Neuroscience*, 1(4), 279-285, 1998.
20. Schreiner CE, Urbas JV, "Representation of amplitude modulation in the auditory cortex of the cat. I. Comparison between cortical fields", *Hear. Res.*, 32, 49-64, 1988.
21. Thomson AM, Deuchars J, "Temporal and spatial properties of local circuits in neocortex", *TINS*, 17(3), 119-126, 1994.
22. Tsodyks MV, Markram H, "The neural code between neocortical pyramidal neurons depends on neurotransmitter release probability", *Proc. Natl. Acad. Sci, USA*, 94, 719-723, 1997.
23. Viemeister NF, Wakefield GH, "Temporal integration and multiple looks", *J. Acoust. Soc. Am.*, 90(2),858-865, 1991.
24. Viemeister NF, Plack CJ, "Time analysis", in Yost WA, Popper AN, Fay RR (Ed.s), *Human Psychophysics*, 1993.
25. Wang X, " What is the neural code of species-specific communication sounds in the auditory cortex?", *Proc. 11th Int. Symp. on Hearing*, Grantham, UK, 456-462, 1997.
26. Wang X, Merzenich MM, Beitel R, Schreiner CE, "Representation of a species-specific vocalisation in the primary auditory cortex of the common marmoset: Temporal and spectral characteristics", *J.Neurophys.*, 74(6), 2685-2706, 1995.

The Role of Memory, Anxiety, and Hebbian Learning in Hippocampal Function: Novel Explorations in Computational Neuroscience and Robotics

John F. Kazer and Amanda J.C. Sharkey

Dept. Computer Science, University of Sheffield, UK

Abstract. In this paper we aimed to show how memory and anxiety functions of the hippocampus could be combined computationally, using a simulation designed to investigate novelty detection and generalised anxiety disorder. We discuss data covering a wide range of hippocampal function, from episodic memory and navigation through novelty detection and anxiety.

The main conclusion to be drawn from the experiments performed upon the simulation is that, given the assumptions made about hippocampal neurophysiology, it provides a coherent prediction for a cause of GAD. That is, it predicts that GAD is caused by increased positive feedback in the loop involving hippocampal-mediated novelty detection and noradrenaline regulation. The act of showing that the computational simulation combines the computational nature of memory and anxiety provides a testable prediction of their compatibility. The clear novelty detection mechanism employed to combine them provides an excellent basis for deriving new simulations and neurological experiments to further develop our model.

1 Introduction

In this paper, we shall describe a computational simulation which seeks to reconcile data describing memory and anxiety functions of the hippocampus (see also [22] [21]). In doing so, a simple property of Hebbian learning enabling novelty detection within a neural network will be used to control a simulated mobile robot .

Our model uses Hebbian learning theory which forms the basis for many simulations and models of neocortical function. We shall concentrate upon the hippocampus. Popular computational simulations have covered auto-association (or episodic memory) [18] [30] [31] [32] and navigation [5]. However, a derived property, novelty detection, has also been investigated [7] [8] [29] [33]. We shall extend the previous uses of novelty within the model presented below. The derivation results from the use of Hebbian learning within computational models.

Why is an extension to the role of novelty detection interesting? It potentially leads to a mechanism for explaining a role for the hippocampus in an even

S. Wermter et al. (Eds.): Emergent Neural Computational Architectures, LNAI 2036, pp. 507–521, 2001.

wider context. Anxiety has been linked many times to the hippocampus (see section 1.3). No computational model, however, has been proposed by which the hippocampus could potentially be involved in the combination of anxiety and memory.

This paper will therefore describe an example by which a common computational mechanism built around novelty detection could result in a role for the hippocampus in memory and anxiety and indeed navigation. The example revolves around the use of a novelty detection mechanism based upon Hebbian learning to investigate generalised anxiety disorder (GAD). Experiments performed using the simulation will cover the role of individual and average unit responses to novel and familiar information and the subsequent changes in simulated anxiety.

Several issues need to be clarified before the simulation can be explored, justifying the proposed combination of (some) memory and anxiety functions within the hippocampus. In particular, because a mechanism, novelty detection, is suggested to form a computational link between memory and anxiety, there should be substantial anatomical and information processing overlaps between each of these. The following sections will examine anatomical and information processing aspects of memory, novelty and anxiety with respect to the hippocampus.

1.1   Episodic Memory and Place Cells

The hippocampus appears to be required for contextual, spatial or episodic memory tasks [6] [37]. An important role in navigation for the hippocampus cannot be ruled out [31]. Therefore, the combination of these diverse data indicate a wide ranging function for the hippocampus. This tallies well with its myriad anatomical associations with neo-cortical, limbic and brain-stem regions [19] [20].

1.2   Novelty Detection

Several lines of evidence suggest that the hippocampus has a role to play in novelty detection. These include cell recording [39], protein labelling [38] and the release of neuromodulators [23]. They reveal changes in activity or neuromodulator release within the hippocampus in response to novelty. The two key points with regard to this paper, however, are the computational nature of novelty detection and the fact that it may provide a link between the many possible hippocampal functions (outlined in section 1.1 above and section 1.3 below, with respect to anxiety). In computational terms, how might individual unit interactions lead to a computational mechanism which unites memory and anxiety? Alternatively, in which ways may memory and anxiety be linked functionally?

1.3   Anxiety

The study of anxiety in the context of the current paper is important for two related reasons. Firstly, it leads to very common psychiatric disorders (such as generalised anxiety disorder) and secondly because a computational simulation (as opposed to theory) has not been developed before. It is hoped that such an approach will lead to new ways of thinking about the disorder. For example, better understanding of the information processed during differing types of anxiety increase should lead to better diagnosis and targeting of treatments.

The essential symptoms of generalised anxiety disorder (GAD) are long-term excessive apprehension and worry about the future that has a vague and non-specific focus [2]. The vagueness is particularly marked in comparison with the specific triggers underlying panic and fear. It also reflects overlaps within treatment, for example depression is often linked to GAD. Blanchard *et al* [4], using lesion and drug studies, have suggested that GAD-like anxiety in animals is linked to hippocampal-dependent spatial anxiety stimuli, such as the uncertain location of predators.

Indeed, the links of the hippocampus with many regions associated with anxiety further reinforce the association. For example, the locus coeruleus [23], raphe nuclei [10] and amygdala [25]. Also, the theta rhythm, apparently so central to hippocampal functioning, has been linked to anxiety regulation [27] [40].

Finally, in their extensive reviews, Gray and McNaughton [11] [12] concluded firstly that anxiolytics (more than any other drug) have virtually the same effect upon anxiety-related behaviour as do lesions to the septo-hippocampal system. Secondly that control of the theta rhythm governs the nature of anxiety. Only anxiolytic drugs (either systemic or locally applied) exert control over the theta rhythm in the same way that stimulation or lesions do. The septo-hippocampal system is integral to generating and using the theta rhythm.

2   The Model and Simulated Robot

The computational model is distinct from the behavioural platform, a simulated robot (Nomadic Technologies v2.6.7). The robot provides information for learning and a method for making behaviourally explicit any changes in anxiety. A sensory pre-processing system provides organised input from the robots' sonar sensors to a neural network model of the septo-hippocampal system. The neural network is similar to that described by Hasselmo *et al* [16] [17], with a major addition being the presence of a parameter simulating some aspects of noradrenaline function.

2.1   The Hippocampus Network Architecture

The network architecture is shown in figure 1a [1]. The input vector approximates to the general sensory information from the entorhinal cortex and dentate gyrus [19]. It is provided by the pre-processor (section 2.2). Layers CA3 and CA1 represent the CA3 and CA1 regions in the hippocampus. The weights connecting these layers perform Hebbian learning of the input vectors provided by the pre-processor. The perforant path is represented by the copying of the input vector

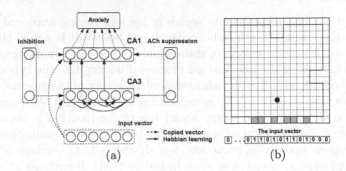

(a)   (b)

Fig. 1. (a) The neural network architecture. **(b)** The environment explored by the robot. The robot is shown by a black circle. The diagram illustrates the obstacles (enclosed by the bold lines), the classification squares used by the pre-processor and an example input vector for this location (bottom). The robot has detected regions of the bottom wall. The pre-processor selects which squares in its representation the regions are in (shaded squares) and sets the corresponding values to 1. All other values remain set to 0. Translation of the representation into an input vector starts at the top left and ends bottom right.

onto both CA3 and CA1. This input is applied at every learning cycle. Each input vector corresponds to a single robot location.

The septal region projects diffusely to the hippocampus and the neo-cortex. A major aspect of these projections is acetylcholine (ACh). In the model, septal ACh input performs synaptic suppression. Regulation of network processing in this manner has been described in detail by Hasselmo *et al* [16] [17] and will be briefly mentioned here.

2.2 Pre-processing

The nature of the pre-processing system is essential to the workings of the model. The input provided by it is not designed to directly model *in vivo* processes but to produce organised information. Organisation (of some type) is assumed to be an essential part of cognitive processing. The major feature of the organisation used is that input from locations which appear similar to the robot will be represented by similar vectors. This means that overlap between the current vector and those previously learnt will indicate familiarity, that the robot has already visited a similar location. The overlap which arises during exploration forms the basis for classification of locations as novel (low overlap) or familiar (high overlap) and therefore determines whether anxiety increases or decreases.

A method for allocentric representation of an environment sensed by sonar has been proposed by Lee [24], a version of which is used here. The environment representation is divided into a grid (figure 1b), stored within the pre-processor. The squares are not used for navigation but simply for classifying regions of the environment. Each detected region of an obstacle is positioned within the allocentric grid system by a path integration mechanism. Any given part of an obstacle will appear in the same grid square regardless of which position it was detected from. This forms the basis for producing organised information. The sensors are limited in range and scope so that at each location a sparse local representation is produced. The input vector is a simple transformation of this representation, with each of the 256 squares being represented by an element of the vector most of which will be set to 0 (bottom of figure 1b). Layers CA3 and CA1 also consist of 256 units because the input vector is copied onto them (figure 1a). There are thus three levels of description; the environment, a grid representation and the input vector.

2.3   The Hippocampus Model

The level of anxiety is determined by the classification of input vectors as novel or familiar (see section 2.4 for details). A high level of noradrenaline is also associated with the detection of novelty (section 1.3). In order to provide an account of the mechanism by which noradrenaline might lead to GAD, we need to examine the process by which inputs are classified as novel or familiar in more detail.

In the model, the novelty/familiarity classification depends upon the unit activation rate. The most important property of the network which affects activation rate is inhibition (equation 1). Inhibition regulates the classification of locations by limiting unit activation rate. Both the parameter NA (representing noradrenaline) and synaptic suppression affect inhibition but in different ways. Changes in NA (but not suppression) at one location effect the next. One effect of NA within the hippocampus may be to increase the level of inhibition [3] [15]. Hasselmo *et al* [16] [17] have also demonstrated how the faster physiological effects of acetylcholine (ACh) could be modelled as as synaptic suppression. Its effects upon the equations used here are similar (although simplified).

The following equations describe how unit activation and learning take place within the network. The effect of NA upon inhibition is shown by equation 1. The way in which ACh affects the learning of input vectors is also important for the classification process, as described below.

$$I = \iota \times (ave_i + ave_r) \times (1 + NA) \times ((1 - ACh) + 0.6) \tag{1}$$

ave_i is the average total input, ave_r average layer output and ι is a constant (0.24). NA represents noradrenaline.

$$ACh = ACh_{max} \times [1 - e^{(3 \times (\frac{T_{act}}{(N_{act}/3)} - 1.5))}] \tag{2}$$

ACh_{max} (0.9) is the initial and maximum value of ACh, which thereafter falls towards 0 (negative values of ACh are treated as 0). There are two values for ACh, one each for layers CA3 and CA1. T_{act} is the total layer output and N_{act} the number of units producing output. Unlike the equation used by Hasselmo *et al*, the average layer activity is normalised here in order to compensate for N_{act}, which varies. The activation of unit i (a_i) is calculated by equation 3.

$$a_i = \tanh[a_{i(t-1)} + p_i - (a_{i(t-1)} \times \lambda) - I + \sum_{j=1}^{256}(w_{ij} \times h_j \times \kappa \times (1 - ACh))] \tag{3}$$

p_i is the input vector element value multiplied by 0.1 (i.e. 0 or 0.1) and h_j is the CA3 unit activation value. κ (0.5) and λ (0.2) are constants and w_{ij} the weight from a_j to a_i. Unit activation is limited between 0 and 1. If the activation is less than a threshold, 0.1, no output is produced although the unit retains its activity. The following form of Hebbian learning is used.

$$\frac{\delta w_{ij}}{\delta t} = \begin{cases} \eta \times a_i \times a_j \times ACh & \text{if } a_i > 0.2 \\ 0 & \text{otherwise} \end{cases} \qquad (4)$$

η is a constant learning rate, set at 0.125. $0 \geq w_{ij} \leq 0.5$ and are initially set to 0.

Both ACh [16] [17] and the θ rhythm reduce crosstalk. The θ rhythm is simulated by resetting all the units to 0 before a new input vector is learnt. Without a reduction in crosstalk, the network would not be able to perform effective classification of locations as novel or familiar.

2.4   Anxiety

Generalised anxiety disorder (GAD) has been defined as abnormally high anxiety [2] and as dependent upon hippocampal hyperactivity [27]. To date, no computational mechanism has been proposed which accounts for GAD. However, increased anxiety, noradrenaline and θ rhythm occurrence are all associated with the perception of anxiogenic stimuli, such as novelty [28] [26] [40]. It is possible that GAD is a pathological version of these normal associations. In this paper, the possibility is investigated computationally and provides an example of how memory and anxiety may occur within the hippocampus.

In the model, anxiety is initially set at 0.1, varies between 0.3 and 0 and is incremented or decremented by 0.02 depending upon whether the last input vector was classified as novel or familiar respectively. A measure of unit activation rate is used to determine if an input vector (and therefore location) should be classified as novel or familiar. The rate is dependent upon the overlap between the current input vector and those previously learnt [13] and is therefore a natural property of Hebbian learning. If the average unit activation in layer CA1 reaches a threshold, 0.3, within 10 learning cycles, the input vector is classified as familiar. Otherwise it is classified as novel.

In order to simulate variations in the θ rhythm frequency, the learning period varies in length, depending upon vector classification. Learning lasts for 10 cycles if a vector has been classified as novel (i.e. fast θ) and 15 for a familiar vector (i.e. slow θ). If more vectors than normal are classified as novel, the simulated θ rhythm will be fast for longer. We suggest that this combination would be equivalent to hippocampal hyperactivity and therefore a possible cause of GAD (as defined by McNaughton [27]). It is assumed that use a more detailed model (e.g. [36]) would enable further explication of rhythm variations.

3   Experiments and Results

We have identified single unit responses to novelty as a key overlap within memory and anxiety theories of hippocampal function. Units respond differently to

new rather than familiar information and these responses are changed by learning. Novelty is an important part of anxiety responses in humans and rats. The
experiments below describe how differences in unit responses operate within the
simulation and how memory and anxiety may be combined.

3.1   Methodology

In each experiment, the simulated robot began at the centre of the world and
visited 900 locations, every third of which was learnt. The experiments consisted
of 10 averaged explorations of the worlds illustrated in figure 2, each using a
different initial weight connectivity. The same set of 10 connectivity patterns
were used each time. Anxiety was varied between 0 and 0.3.

Section 2 described the effects of NA regulation upon hippocampal network
processing and the default settings used for other parameters. In the following
experiments, three variations were used (see table 1). These encompassed a fixed
level of NA (i.e. static, or STAT), NA = anxiety (i.e. variable NA or NORM)
and NA = anxiety + 0.2 (HIGH). These shorthand labels will be used in the
subsequent descriptions.

Table 1. The three experimental conditions used.

STAT	Static: To illustrate the importance of a variable NA level and to act as a control to demonstrate the effects of NA across multiple locations. NA = 0.1
NORM	Variable: The standard set of parameters, which simulate a normal level of anxiety. NA = anxiety
HIGH	An increased level of NA feedback which seeks to simulate the disorder, GAD. NA = anxiety + 0.2

The STAT condition is important because it highlights the effects variable NA
has as series of locations or input vectors are processed. Normally (i.e. the NORM
condition) NA changes made at location A will affect location B and possibly
C, D etc. also. This is an important property and underlies the mechanism by
which NA affects the novelty/familiarity classification and its role in anxiety. In
the STAT condition, the positive feedback loop created by variable NA and novel
information does not occur. However, it is proposed that the HIGH condition
used in section 3.4 will lead to a larger positive feedback effect and therefore
higher anxiety.

3.2   Average Unit Output

Comparisons of the average output with the amount of novelty detected and
locations visited (i.e. time) were made. As the amount of novelty is dependent
upon average output within the simulation, figure 3a acts as an illustration of

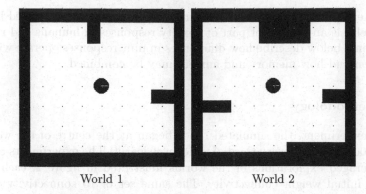

World 1   World 2

Fig. 2. The worlds explored by the simulated robot, illustrated at its start position in the middle. A simple measure of similarity showed that, in terms of the units active during exploration (and therefore total world regions detected) world 2 encompassed 65% of world 1.

simulation behaviour. A second set of data will be examined in section 3.3, detailing which units (correct or incorrect) were responsible for changes in novelty detection.

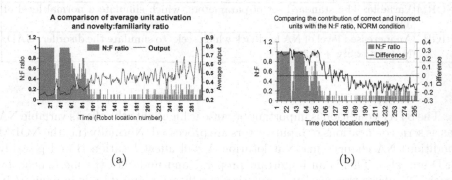

(a)   (b)

Fig. 3. (a) The relationship between average output and classification of input patterns (locations) as novel or familiar averaged over 10 explorations of world 1. The novel:familiar ratio was calculated by dividing the number of novel locations at each point in the exploration by 10. Therefore, if all exploration examples returned a novel pattern at a particular point (e.g. the 23rd location visited) a value of 1 would be entered upon the graph. The NORM condition was used in world 1. (b) Comparing the differences from the overall average unit output of correct and wrong units. If the contribution of average correct units is greater than that of wrong units, then the overall average was predominantly determined by correct units. Therefore, any change in anxiety level was due to learning between correct units. Any value for the difference above zero indicates this situation. See text for details.

Average output was calculated by averaging the output of all the active units in layer CA1 at the end of each learning episode. The average output of the active units rises as the duration of exploration increases. The regression values of 0.74 and 0.68 for average output in worlds 1 and 2 (data not shown) respectively demonstrate that there is a strong positive correlation between time and average unit activation. The rise is reflected by falls in novelty detection (figure 3a) as defined in the simulation.

3.3 Catastrophic Remembering

The average output increased during exploration. Figure 3a demonstrated that novelty falls with time as average output increased. This section examines average output (i.e. the average of those units that are active following learning at a specific location) in more detail. A closer look is necessary in order to confirm that any fall in novelty detection is due to correct units (i.e. those activated by the input vector). If this is not the case, catastrophic remembering [34] [35] is the cause of increased familiarity. Any classification of locations as familiar would therefore not be directly a result of the familiarity of the location, in terms of the unique configuration of the world visible to the robot there. It would be a function of the number of times different parts of the visible location (in terms of the grid squares used to portion out the world and form input vectors) have been seen previously, from any location. That is, although the location itself may be novel, the constituent parts are not. Indeed, in extreme cases, just one element of the input vector that has been involved in extensive previous learning may be sufficient to very quickly activate many incorrect units and therefore trigger familiarity. A limit of 0.5 was placed upon the weight size in the simulation but this only acts to scale the problem.

Figure 3b illustrates how the contribution of correct and incorrect units changes as the amount of novelty changes. The key value is the difference between the average output value and the average output of correct and incorrect units. The curve representing difference is derived from the following equation:

$$difference = (average_{correct} - average_{output}) - (average_{output} - average_{wrong})$$

Therefore, a positive value of difference indicates that the overall average unit output value was dominated by correct units. A negative value means that the average output was dominated by incorrect units. The graph in figure 3b illustrates this by revealing the locations at which the difference was positive or negative. It is clear that the amount of novelty falls to zero at least twice (at locations 30 and 95) before the difference became negative. Therefore, the initial familiarity increases experienced by the simulation during exploration were due to learning amongst correct units. Later familiarity (post location 110) was predominantly due to catastrophic remembering. However, the simulation was still able to detect new locations as novel following learning of 300 locations. If the same set of weights were subsequently used during training in a different world, novelty was detected (data not shown).

3.4 Anxiety

Section 2 described the role of NA within the simulation and how it is intended to correspond to noradrenaline *in vivo*. The core effect of NA is to increase inhibition and therefore raise the possibility that subsequent locations will be classed as novel. Novelty falls with time as shown by the results presented above. The results of the experiments described below will show how anxiety changes and how it is related to novelty and therefore memory.

There are two principle parts to this section, involving measurement of anxiety change during exploration of the two worlds in figure 2. The first part uses the STAT and NORM conditions, the second uses the NORM and HIGH conditions (see table 1) in order to test whether an increase in NA provides an explanation for a role of noradrenaline in GAD.

Normal variation in anxiety level over time. The aim of the set of experiments below were to determine whether the model could provide a computational account of the role of the SH system and noradrenaline in anxiety, using the STAT and NORM conditions. It is known that hippocampal activity (possibly in terms of an increased theta frequency) is raised during anxiety [12] [27]. The experiments will be discussed with respect to the novelty detection mechanisms and catastrophic remembering problems presented above.

In order to test the significance of the results, a Students t-test was performed upon values of summed anxiety level before it first fell to zero. The measure therefore captures both the amount of anxiety and time span of high anxiety.

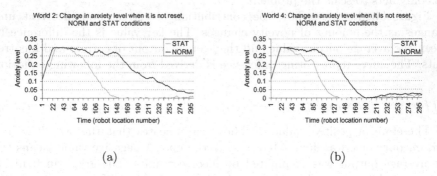

(a)   (b)

Fig. 4. Anxiety change for the NORM and STAT conditions during training, (a) world 1 and (b) world 2.

Figure 4 reveals that anxiety does indeed fall as expected (i.e. the result parallels the fall over time of novelty shown in figure 3a). Table 2 demonstrates that the experiments comparing the NORM and STAT conditions indicate that there is at least only a 5% chance that a sample will be incorrectly classed. The

Table 2. A measure of the total anxiety level before it first fell to zero was used to make comparisons between conditions. A paired, one-tailed Students t-test was used, n = 10 in all cases, in which the results for STAT and HIGH conditions were compared to the NORM condition. * indicates a significant difference from the NORM condition (p < 0.05).

World	NORM	HIGH	STAT
1	56.7, 15.2	68, 9.4*	27.1, 2.9*
2	28.6, 2.8	50.8, 8.7*	21, 2*

average values of the anxiety measure shown in table 2 demonstrate that the NORM condition causes higher levels of anxiety than the STAT condition.

As figure 4 suggests, therefore, the NORM condition generates significantly higher levels of anxiety than the STAT condition. The conclusion to be drawn from this experiment is that a variable value of NA does have a significant effect upon network information processing. Therefore feedback control by noradrenaline could form a mechanism for anxiety *in vivo*.

High noradrenaline feedback. The experiments presented in this section mirror those in the previous section in every way other than using the NORM and HIGH conditions rather than NORM and STAT. In the HIGH condition, NA = anxiety + 0.2. Therefore the effect of NA upon network processing is an increase over NORM for an equivalent level of anxiety. The aim is to determine whether the HIGH condition forms a plausible computational model for GAD, supporting the theory that GAD may be caused by higher than normal noradrenaline release into the hippocampus.

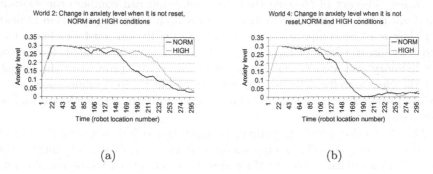

(a)   (b)

Fig. 5. Anxiety change for the HIGH and NORM conditions during training, (a) world 1 and (b) world 2.

The graphs in figure 5 indicate that the levels of anxiety are raised for the HIGH condition over NORM (which represents the same data used in the previous section). In each world and whether either resetting is used or not, the level of anxiety for HIGH remains higher for longer than in the NORM condition. Table 2 indicates that the difference between HIGH and NORM conditions is a significant increase at the 5% level.

These results therefore provide support for the prediction that abnormal increases in noradrenaline can lead to GAD. In a recent paper, McNaughton [27] discussed evidence that suggested hippocampal hyperactivity causes GAD. One possible explanation for this (none were proposed by McNaughton) is that there is an increased occurrence and frequency of theta during GAD. Theta has been recorded as increased in frequency during or following novelty detection and increased anxiety [40]. Unfortunately, the nature of the theta rhythm within the current simulation takes these changes into account as a core assumption. More realistic simulation of the hippocampus using oscillatory networks (relying upon a balance of excitation and inhibition, such as those described by Traub and Miles [36]) may be required to categorically state the relationship between NA increase and theta frequency changes.

However, the results in this section demonstrate that the simulation can provide a computational explanation for GAD which combines memory and anxiety data.

4   Discussion

In this paper we aimed to show how the postulated memory and anxiety functions of the hippocampus could be combined computationally, using a simulation designed to investigate novelty detection and GAD. One conclusion to be drawn from the experiments performed is that, given the assumptions made about hippocampal neurophysiology, the model provides a coherent prediction for a cause of GAD. That is, it predicts that GAD is caused by increased positive feedback in the loop involving hippocampal-mediated novelty detection and noradrenaline regulation. The act of showing that the simulation can combine the computational nature of memory and anxiety to provide a testable prediction indicates that they may be compatible. Of course, the prediction must be shown to be correct before theories of hippocampal function may be expanded to encompass memory and anxiety simultaneously. One certain product of the model is that the clear novelty detection mechanism employed to combine memory and anxiety provides an excellent basis for developing both the simulation and new neurological experiments.

The simulation provides a direct computational link between anxiety and Hebbian-learning models of hippocampal learning. Hebbian learning dictates that novel information should induce slow unit activation, whilst familiar information causes faster activation. In the simulation, the novelty detection mechanism based upon Hebbian learning and the subsequent rate of unit activation determined the change in anxiety level (see section 2.4 and equation 1). The change in anxiety level set the level of NA which inhibited the rate of unit activation. Therefore a positive feedback system was created, whereby novelty increased anxiety and NA, which tended (via inhibition of unit activation rate) to increase the subsequent likelihood of novelty detection.

However, this only provides one way in which to expand current models and simulations of the hippocampus. Further work will need to encompass even more physiological data. For example, single cell responses to novelty and familiarity, an expanded role for noradrenaline and an important role for 5-HT.

Pyramidal cells appear to react differently to novel and familiar information than suggested by conventional Hebbian learning theory [39]. Some cells in the hippocampus and surrounding neo-cortex have been found which react strongly (although with a rapid habituation) to novel information.

There are at least two possibilities. Firstly, the cells found so far with the unexpected properties are not related directly to novelty and familiarity at all. As it is difficult to specify the properties of single cells in terms of concrete information, such as place, it would seem reasonable to assume that correlating their properties with an abstraction such as novelty is more prone to error. Secondly, far more complex processes may be occurring within the hippocampus than encompassed by current Hebbian theory.

For example, the enhancement of pyramidal cell activity via noradrenaline has been demonstrated (e.g. [14] [23]). Hasselmo et al [15] have proposed a simulation in which noradrenaline suppresses units carrying erroneous information (e.g. that caused by interference) whilst potentiating meaningful unit activation. A key issue, however, is that Hasselmo et als simulation did not include learning. It would be extremely interesting to discover how their system might be linked with Hebbian learning. Alternatively, Hebbian theory should be altered to satisfy the new criteria, perhaps via extended intra-cellular mechanisms such as those suggested by Denham and McCabe et al [9].

We believe purely Hebbian learning models have reached their limit in terms of computational explanation with the work of O'Reilly and Rudy [30]. Linking of the successful simulation presented here, combining memory and anxiety, with more detailed simulation of Hebbian learning and neuromodulator effects, should be the next step.

References

1. D.G. Amaral and M.P. Witter. The three-dimensional organization of the hippocampal formation: A review of anatomical data. *Neurosci*, 31:571–591, 1989.
2. American Psychiatric Association. *Diagnostic and Statistical Manual of Mental Disorders (DSM4)*, 4th edition, 1994.
3. D.E. Bergles, V.A. Doze, D.V. Madison, and S.J. Smith. Excitatory actions of norepinephrine on multiple classes of hippocampal ca1 interneurons. *J Neurosci*, 16:572–585, 1996.
4. D.C. Blanchard and R.J. Blanchard. Effects of ethanol, benzodiazepines and serotonin compounds on ethopharmacological models of anxiety. In N. McNaughton and G. Andrews, editors, *Anxiety*, pages 117–134. University of Otago Press, 1990.

5. N. Burgess, J.G. Donnett, K.J. Jeffery, and J. O'Keefe. Robotic and neuronal simulation of the hippocampus and rat navigation. *The Royal Soc Phil Trans: Bio Sci B*, 352:1535–1543, 1997.

6. T.J. Bussey, J.L. Muir, and J.P. Aggleton. Functionally dissociating aspects of event memory: The effects of combined perirhinal and postrhinal cortex lesions on object and place memory in the rat. *J Neurosci*, 19:495–502, 1999.

7. G.A. Carpenter and S. Grossberg. The art of adaptive pattern recognition by a self-organizing neural network. *Computer*, pages 77–88, 1988.

8. M.J. Denham and S.L. McCabe. Biological temporal sequence processing and its application in robot control. In *Proc UKACC Int Conf on Control*, volume 427, pages 1266–1271. IEE, 1996.

9. M.J. Denham and S.L. McCabe. A model of predictive learning in the rat hippocampal principal cells during spatial activity. In *Proc WCNN*, 1998.

10. F.G. Graeff, F.S. Guimaraes, T.G. De Andrade, and J.F. Deakin. Role of 5-ht in stress, anxiety and depression. *Pharm Biochem Behav*, 54:129–141, 1996.

11. J.A. Gray. *The neuropsychology of anxiety*. Oxford Psychology Series. OUP, 1982.

12. J.A. Gray and N. McNaughton. *The Neuropsychology of anxiety: An enquiry into the functions of the septo-hippocampal system*. Oxford University Press, 2nd edition, In press.

13. R.W. Hamming. Error detecting and error correcting codes. *Bell Sys Tech J*, 2:147–160, 1950.

14. C.W. Harley. Noradrenergic long-term potentiation in the dentate gyrus. *Adv in Pharm*, 42:952–956, 1998.

15. M.E. Hasselmo, C. Linster, M. Patil, D. Ma, and M. Cekic. Noradrenergic suppression of synaptic transmission may influence cortical signal-to-noise ratio. *J Neurophys*, 77:3326–3339, 1997.

16. M.E. Hasselmo and E. Schnell. Laminar selectivity of the cholinergic suppression of synaptic transmission in rat hippocampal region ca1: Computational modeling and brain slice physiology. *J Neurosci*, 14:3898–3914, 1994.

17. M.E. Hasselmo, E. Schnell, and E. Barkai. Dynamics of learning and recall at excitatory recurrent synapses and cholinergic modulation in rat hippocampal region ca3. *J Neurosci*, 15:5249–5262, 1995.

18. M.E. Hasselmo, B.P. Wyble, and G.V. Wallenstein. Encoding and retrieval of episodic memories: Role of cholinergic and gabaergic modulation in the hippocampus. *Hippocampus*, 6:693–708, 1996.

19. R. Insausti, D.G. Amaral, and W.M. Cowan. The entorhinal cortex of the monkey: II. cortical afferents. *J Comp Neurol*, 264:356–395, 1987.

20. R. Insausti, D.G. Amaral, and W.M. Cowan. The entorhinal cortex of the monkey: III. subcortical afferents. *J Comp Neurol*, 264:396–408, 1987.

21. J.F. Kazer and A.J.C. Sharkey. The septo-hippocampal system and anxiety: A robot simulation. In *ICANN99: Proceedings of the 9th Int. Conf. Artificial Neural Networks*, pages 389–394, 1999.

22. J.F. Kazer and A.J.C. Sharkey. Towards an emotional robot: Simulating hippocampal-mediated anxiety. In Humphries G.W. Heinke D. and Olsen A., editors, *Connectionist Models in Cognitive Neuroscience*, Perspectives in Neural Computing, pages 102–111. Springer-Verlag, 1999.

23. V. Kitchigina, A. Vankov, C. Harley, and S.J. Sara. Novelty-elicited, noradrenaline-dependent enhancement of excitability in the dentate gyrus. *Eur J Neurosci*, 9:41–47, 1997.

24. D.C. Lee. *The map-building and exploration strategies of a simple sonar-equipped mobile robot: An experimental, quantitative evaluation*. CUP, 1996.

25. S. Maren, S.G. Anagnostaras, and M.S. Fanselow. The startled seahorse: Is the hippocampus necessary for contextual fear conditioning? *TICS*, 2:39–42, 1998.
26. K. Mason, D.J. Heal, and S.C. Stanford. The anxiogenic agents, yohimbine and fg 7142 disrupt the noradrenergic response to novelty. *Pharm Biochem Behav*, 60:321–327, 1998.
27. N. McNaughton. Cognitive dysfunction resulting from hippocampal hyperactivity - a possible cause of anxiety disorder? *Pharm Biochem Behav*, 56:603–611, 1997.
28. R. McQuade, D. Creton, and S.C. Stanford. Effect of novel environmental stimuli on rat behaviour and central noradrenaline function measured by in vivo microdialysis. *Psychopharm*, 1999.
29. J. Metcalfe. Recognition failure and the composite memory trace in charm. *Psych Rev*, 98:529–553, 1991.
30. R.C. O'Reilly and J.W. Rudy. Conjunctive representations in learning and memory: Principles of cortical and hippocampal function. Technical Report 99-01, Institute of Cognitive Science, Dept. Psychology, University of Colorado, Boulder, CO 80309, 1999.
31. A.D. Redish. *Beyond the cognitive map: Contributions to a computational neuroscience theory of rodent navigation*. PhD thesis, Computer Science, Carnegie Mellon University, 1997.
32. E.T. Rolls. A theory of hippocampal function in memory. *Hippocampus*, 6:601–621, 1996.
33. N.A. Schmajuk, Y.W. Lam, and J.A. Gray. Latent inhibition: A neural network approach. *J Exp Psych*, 22:321–349, 1996.
34. N.E. Sharkey and A.J.C. Sharkey. Adaptive generalisation. *Art Int Rev*, 7:313–328, 1993.
35. N.E. Sharkey and A.J.C. Sharkey. An analysis of catastrophic interference. *Conn Sci*, 7:301–329, 1995.
36. R.D. Traub and R. Miles. *Neuronal Networks of the Hippocampus*. CUP New York, 1991.
37. E. Tulving and H.J. Markowitsch. Memory beyond the hippocampus. *Curr Op Neurobiol*, 7:209–216, 1997.
38. H. Wan, J.P. Aggleton, and M.W. Brown. Different contributions of the hippocampus and perirhinal cortex to recognition memory. *J Neurosci*, 19:1142–1148, 1999.
39. J.Z. Xiang and M.W. Brown. Differential neuronal encoding of novelty, familiarity and recency in regions of the anterior temporal lobe. *Neuropharm*, 37:657–676, 1998.
40. J. Yamamoto. Relationship between hippocampal theta-wave frequence and emotional behaviors produced with stresses or psychotropic drugs. *Jpn J Pharm*, 76:125–127, 1998.

Using a Time-Delay Actor-Critic Neural Architecture with Dopamine-Like Reinforcement Signal for Learning in Autonomous Robots

Andrés Pérez-Uribe

Parallelism and Artificial Intelligence Group
Department of Informatics
University of Fribourg, Switzerland
Andres.PerezUribe@unifr.ch, http://www-iiuf.unifr.ch/~aperezu

Abstract. Neuroscientists have identified a neural substrate of prediction and reward in experiments with primates. The so-called *dopamine neurons* have been shown to code an error in the temporal prediction of rewards. Similarly, artificial systems can "learn to predict" by the so-called *temporal-difference* (TD) methods. Based on the general resemblance between the *effective reinforcement* term of TD models and the response of dopamine neurons, neuroscientists have developed a TD-learning time-delay actor-critic neural model and compared its performance with the behavior of monkeys in the laboratory. We have used such a neural network model to learn to predict variable-delay rewards in a robot spatial choice task similar to the one used by neuroscientists with primates. Such architecture implementing TD-learning appears as a promising mechanism for robotic systems that learn from simple human teaching signals in the real world.

Keywords: Learning robots, time-delay neural networks, actor-critic architecture, TD-learning, dopamine neurons, reinforcement learning, human teaching signals.

1   Introduction

Adaptive organisms have to deal with a continually changing environment. In order to survive, they must be able to predict or anticipate future events by generalizing the consequences of their behavioral responses to similar situations experimented in the past. Predictions result after combining the mechanisms of *search* and *memory*, and give adaptive organisms the time to prepare their behavioral reactions. Similarly, when a learning robot or an intelligent agent faces a particular situation, it may select one of a finite set of possible actions, and then, remember the actions that worked best and forget the inadequate actions [20]. The use of search and memory gives rise to what is called *trial-and-error* learning.

Neuroscientists have identified a neural substrate of prediction and reward in experiments with primates. The so-called *dopamine neurons* have been shown

S. Wermter et al. (Eds.): Emergent Neural Computational Architectures, LNAI 2036, pp. 522–533, 2001.

to code an error in the temporal prediction of rewards [15]. Similarly, artificial systems can "learn to predict" by the so-called *temporal-difference* (TD) methods [19]. Temporal-difference methods are computational models that have been developed to solve a wide range of reinforcement learning problems including game learning and adaptive control tasks [20]. In biological applications they have been used to replicate foraging behavior of bumblebees [8], learning of sequential movements [17], etc. The system learns to predict rewards by trial-and-error, based on the idea that "learning is driven by changes in the expectations about future salient events such as rewards and punishments" [15]. Such changes are detected by computing temporal differences of the estimations of rewards.

Based on the general resemblance between the TD-computation of the changes of expectations about rewards and the response of dopamine neurons, Suri and Schultz [18] developed a particular TD model they related to the behavior of monkeys in the laboratory. In this paper, we utilize such a neural network model to learn to predict variable-delay rewards in a robot spatial choice task similar to the one used by neuroscientists with primates. In Section 2, we describe the neural substrate of prediction and reward, identified by neuroscientists. In Section 3, we present the machine learning technique for learning with reinforcement signals. Sections 4 and 5 describe the neural architecture developed by neuroscientists and their experiments with primates. Section 6 describes our robotics experimental setup, Section 7 delineates the experimental results, and finally, Section 8 presents some concluding remarks.

2   Neural Substrate of Prediction and Reward

One clear connection between prediction and reward results from conditioning experiments [15]. Classical conditioning has a special place in the study of learning, since it has permitted to understand the rules that enable us to associate two events [16]. The fundamental rule is *temporal contiguity*: the conditioned stimulus (CS) precedes by some critical interval a second, reinforcing event, the unconditioned stimulus (US). The second rule is *contingency*: the conditioned stimulus predicts the occurrence of the unconditioned one.

It has been found that there is an optimal interval between a particular CS and a US that allows an animal to learn that the CS predicts the US. It is typically between 200 milliseconds and 1 second, even though in some cases the optimal interval may be longer. Such interval is surprisingly similar in all species for similar types of associative learning [16].

The so-called *dopamine neurons* have been shown to be "excellent feature detectors of the *goodness* of environmental events relative to learned predictions about those events" [15]. They emit a positive signal if a desired event is better than predicted; no signal, if a desired event occurs as predicted; and a negative signal, if the desired event is worse than predicted [15]. Thus, dopamine neurons appear to code an error in the temporal prediction of reward [18]: it has been found that they are activated by unpredicted rewards, by rewards during the first few stimulus-reward pairings, and when a reward occurs earlier or later than

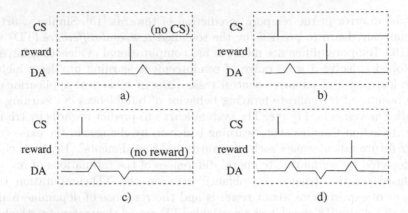

Fig. 1. Dopamine neuron activity. a) The dopamine neurons (DA) are activated by an unpredicted reward. b) After learning, the dopamine neurons are activated by the reward-predicting stimulus (CS) and not by the predicted reward. c) After learning, the dopamine neurons are activated by the reward-predicting stimulus (CS), but then experience a depression if the predicted reward fails to occur. d) After learning, the dopamine neurons code an error in the temporal prediction of reward (see [15] and [18]).

predicted (Figure 1). However, the activation following the reward decreases after repeated pairings and finally disappears.

The activation of the dopamine neurons may improve the selection of behaviors by providing advanced reward information before the behavior occurs, and may contribute to learning by modifying synaptic transmission [13], or may even play a general role in associative learning by 'regulating attention to the external environment and readiness to respond to unexpected events' [11].

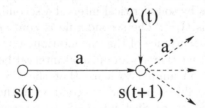

Fig. 2. Discrete time representation of a reinforcement learning problem.

3 Machine Learning with Reinforcement Signals

In machine learning applications, reinforcement learning tasks are generally treated in discrete time steps, that is, the problems are modeled as a Markov decision process or MDP (Figure 2). At each time step t, the learning system receives some representation of the environment's *state* $s(t)$, it *tries* an action a, and one step later it is *reinforced* by receiving a scalar evaluation $\lambda(t)$[1] (i.e., a reward or a punishment) and finds itself in a new state $s(t + 1)$. To solve a reinforcement learning task, the system attempts to maximize the total amount of reward it receives in the long run [20]. To achieve this, the system tries to minimize the so called *temporal-difference error* (also known as the effective reinforcement signal), computed as the difference between predictions at successive time steps.

In most machine learning applications, an automated trainer delivers a reward (or a punishment) to the learning system when it reaches a particular state (i.e., the reward is instantly available at time t+1). For example, in an obstacle avoidance learning task, a punishment signal may be delivered to the learning robot whenever it finds itself in a situation where one of its sensors is activated more than a certain threshold (which may correspond to the detection of an obstacle) [9]). However, in certain applications, the reward (or the punishment) may be provided by an external device or a human. In such cases, the reward signal may not be delivered (and received) at time $t + 1$ (immediate reward) but after a variable lapse of time.

One way to solve this problem is to use a time-delay neural network (TDNN) architecture [6], by implementing a kind of memory to represent the stimulus through time until the occurrence of a reward.

4   A Neural Network Model with Dopamine-Like Reinforcement Signal

Based on the general resemblance between the *effective reinforcement signal* of TD models and the response of dopamine neurons, Suri and Schultz [18] developed a TD model that was tested in a series of spatial delayed and non-delayed choice tasks, and related the performances of the model to the behavior of monkeys in the laboratory.

The model consists of an actor-critic architecture (Figure 3) which resembles the general architecture of the basal ganglia [1,13]. Both components receive the stimuli ($e_1(t)$ and $e_2(t)$ in the figure) coded as functions of time. The actor computes a weighted sum of the input stimuli to select an output action (modifiable weights are shown as heavy dots in the figure). A winner-take-all rule prevents it from performing two actions at the same time. The adaptation of the synaptic weights in the actor enables it to learn to associate stimuli with behavioral actions. The critic component computes a weighted sum of a temporal representation of the stimuli to estimate the reward predictions ($P_1(t)$ and $P_2(t)$ in Figure 3a). Every stimulus l is represented as a series of

---

[1] We use $\lambda(t)$ to refer to the external reinforcement signal, or 'primary reinforcement' and $r(t)$ to refer to the temporal difference error or 'effective reinforcement' to be consistent with Suri and Schultz [18], but $r(t)$ and $\delta(t)$ are more commonly used to refer to such signals, in the machine learning literature.

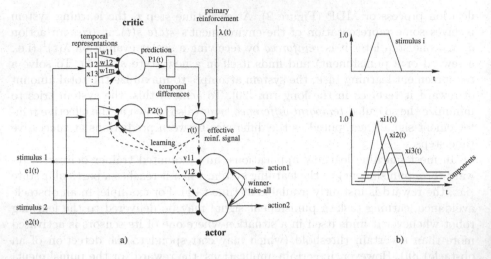

Fig. 3. a) Actor-critic architecture. The dots represent adaptive weight connections. The dashed lines indicate the influence of the effective reinforcement signal in those adaptive weights. b) Temporal stimulus representation.

components x_{lm} of different durations (Figure 3b). Each of these components influences the reward prediction signal according to its own adaptive weight w_{lm}. This form of temporal stimulus representation[2] allows the critic to associate a particular reward to the corresponding stimuli, even if the duration of the stimulus-reward interval is not null and variable. Indeed, what the critic attempts to do is to learn the exact duration of the stimulus-reward interval [18].

The temporal representation of the stimulus is computed as follows:
   a) At the onset of the stimulus:

$$x_{l1} = 1, x_{l,m \neq 1} = 0 \tag{1}$$

   b) k steps after the onset of the stimulus:

$$x_{l,m \leq k} = 0, x_{l,m > k} \rho^{m-1} , \tag{2}$$

where l is the stimulus, m is the component of its corresponding temporal representation, and ρ is a constant value in the range [0 1].

The critic generates the *effective reinforcement* signal as follows:

$$r(t) = d + \lambda(t) + \gamma P(t) - P(t-1) , \tag{3}$$

where, d is a constant value and accounts for the baseline firing rate of dopamine neurons ($d = 0$ in our implementation), $\lambda(t)$ is the primary reinforcement, γ is

---
[2] Virtually nothing is known about how the brain represents a stimulus for substantial periods of time [15].

a discount factor, $P(t)$ and $P(t-1)$ are the reward predictions at successive time steps, and $P(t) = \sum_l P_l(t)$ (i.e., the sum of the reward prediction for every stimulus l). The effective reinforcement is used to modify the adaptive synapses of the actor and critic components. The change in the critic weights w_{lm} is proportional to prediction errors (an underlying mechanism in engineering [20] and psychology [12]), and to the changes of the temporal stimulus representation [5, 18], as follows:

$$w_{lm}(t) = w_{lm}(t-1) + \eta_c(r(t) - d)[x_{lm}(t-1) - \gamma x_{lm}(t)]_+ , \qquad (4)$$

where, $[]_+$ indicates that a negative number is set to zero and a positive value remains unchanged, η_c is the learning rate of the critic.

The change in the actor weights v_{nl} is proportional to reward prediction errors, an action trace $\bar{a}_n(t)$, and the stimulus trace $\bar{e}_l(t)$, as follows:

$$v_{nl}(t) = v_{nl}(t-1) + \eta_a(r(t) - d)\bar{a}_n(t)\bar{e}_l(t) , \qquad (5)$$

where, η_a is the learning rate of the actor. The action and stimulus traces serve as a record of the occurrence of a particular action or the taking of an action. They are computed as follows:

$$\bar{a}_n(t) = min(1, a_n(t) + \delta\bar{a}_n(t-1)), and \qquad (6)$$

$$\bar{e}_l(t) = min(1, \bar{e}_l(t-2) + \delta\bar{e}_l(t-1)) \qquad (7)$$

The action selection is based on the weighted sum of the stimuli traces and a uniformly distributed source of noise $\sigma_n(t) \in [0, \sigma]$:

$$a_n(t) = \sum_l v_{nl}\bar{e}_l(t) - \sigma_n(t) \qquad (8)$$

Below is a table with the values of the most important parameters of the learning model. Most of them were set following the work by Suri and Schultz [18].

d	γ	η_c	η_a	δ	ρ	σ
0.0	0.98	0.08	0.1	0.96	0.94	0.5

In our implementation, we set the number of components of the temporal representation such that the model was able to associate a reward delivered up to 3 seconds after the selection of an action.

5   Neuroscientists' Experiments with Primates

In a series of experiments with primates, Schultz and colleagues [14] attached electrodes to the brains of monkeys to record electrical activity in dopamine-secreting neurons. The monkeys were trained to press a lever in response to a pattern of light to receive a reward (a squirt of juice).

A trial was initiated when the animal kept its right hand on a resting key. Illumination of an instruction light above a left or right lever indicated the target of reaching. A trigger light determined the time when the resting key should be released and the lever be touched. In one of several experiments with monkeys, Schultz and his colleagues [14] studied the dopamine neuron responses in a spatial choice task. In this task, the instruction and trigger lights appeared simultaneously, and both lights extinguished upon lever touch. The left and right instructions lights alternated semi-randomly. The monkey was rewarded for releasing the resting key and pressing the corresponding lever. It was not rewarded for incorrect actions like withholding of movement, premature resting key release or lever press, or incorrect lever choice [18].

In the following section, we present an autonomous robot experiment that replicates the experiments with primates described above. Basically, we used the neural network model presented in Section 4 to control a robot in the spatial choice task.

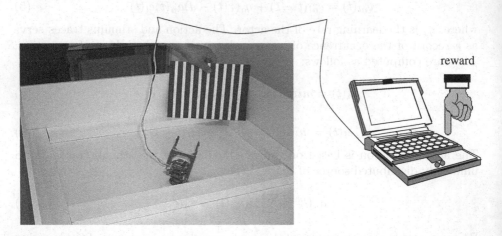

Fig. 4. Experimental setup. A Khepera robot with a linear vision system is placed in a workspace to replicate a primate learning task. The robot is connected to a host machine that controls its behavior through a neural network model and transmits human evaluating signals (rewards). The robot is intended to react to an instruction pattern.

6   Robotics Experimental Setup

In our experiments, an autonomous mobile robot is placed in a box of 75 cm × 45 cm with white walls (Figure 4). We have used the Khepera mobile robot [7] with a K213 vision turret developed by K-Team [2] (Figure 5). It contains a linear image sensor and a global light intensity detector. The image sensor is an array of 64 × 1 cells, giving a linear image of 64 pixels with 256 grey-levels each.

The view-angle is of about 36 degrees. The optics was designed to bring into focus objects at a distance of 5 to 50 cms in front of the robot. The image rate is 2 Hz.

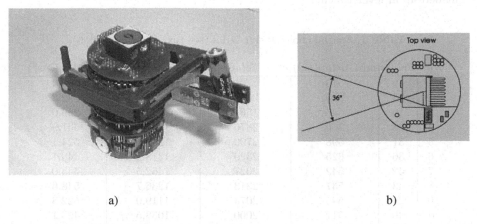

a)   b)

Fig. 5. a) The Khepera robot [3]. b) The K213 vision turret [2].

We programmed our robot [2,3] to turn around itself covering an angle of 180 degrees. While the robot moves, it takes a snapshot of its environment and transmits the 64 pixel image to a host machine (we used a serial link at 38400 Baud, but we could have used a radio link) that implements the neural network model. A pre-processing phase is executed to take the linear image of the environment and compute the number of "drastic" changes of activation between neighboring sensor cells (i.e., to detect a pattern like the one shown at the right side of the arena in Figure 4). If more than four "drastic" changes of activation between neighboring sensor cells are found, we apply a stimulus of amplitude 1.0 to the right input of the model (stimulus 1), when the robot is turning to the right, or to the left input, when the robot is turning to the left (stimulus 2).

The neural network model (see Section 4) operates at steps of 100 ms to activate one of three possible actions, which correspond to the fact of *perceiving a pattern while turning to the right* (action 1), *perceiving a pattern while turning to the left* (action 2), and *perceiving no pattern at all* (action 3). In Figure 3, we show only two actions, for the sake of simplicity. When the model chooses the correct action, an experimenter provides a reward by pushing a button on the host machine. Otherwise, the system starts a new trial after 3 seconds of waiting for the reward. Robot actions are only allowed if a stimulus is presented (indeed, in the experiments with monkeys, neuroscientists pretrained the animals to press levers when stimuli were presented, and an ongoing action is maintained during three successive time-steps, given that the time to press a lever in the real experiments was about 300 ms [18]).

In our experiments, the mobile robot stops turning around itself when it selects one of the actions. This was introduced to permit the experimenter to remove the stimulus pattern and place it somewhere else before starting a new trial. In the monkey experiments, the instruction lights automatically extinguished upon lever touch.

Table 1. Spatial choice task results (see text for explanation).

Run	Trials	min. delay(ms)	max. delay(ms)	mean delay(ms)	std. dev.(ms^2)
1	49	510	1701	1018.0	439.4
2	54	800	2000	1600.0	292.3
3	16	545	1564	1112.8	299.0
4	38	565	2876	1290.8	601.9
5	31	598	2799	1299.0	524.1
6	36	625	2329	1313.9	448.1
7	42	542	2627	1452.5	545.0
8	43	537	2313	1236.7	548.6
9	27	545	2073	1119.0	522.8
10	40	517	2000	1058.5	423.1

7   Experimental Results

During a trial, a pattern of 10 black columns and 10 white columns is presented to the robot at a distance of about 35 cms (Figure 4). The trial evolves as explained above. In average, the robot learned to respond correctly to 9 of the 10 last presentations of the pattern, after 37 trials. The average delay between the selected action and the corresponding reward (given by the experimenter) was 1250.2 ms. The maximum delay was 2876 ms, and the minimum delay was 510 ms. In these experiments, we set the baseline firing rate of dopamine neurons to zero ($d = 0$) instead of $d = 0.5$ as in Suri and Schultz [18]. Kakade and Dayan [4] have recently presented an interpretation of an additional role for the dopamine neuron activity in guiding exploration by means of *dopamine bonus* terms, which are computationally similar to the d term.

In Table 1, we present the results of 10 runs of the robot spatial choice learning task. In the second column, we present the total number of trials needed to respond correctly to 9 of the last 10 presentations of the pattern. Columns 3 to 5 show the minimum, the average, and the maximum delay between an action and the corresponding reward for each run. Column 6 shows the standard deviation of such delay for each run.

Figure 6 illustrates the behavior of the dopamine-like reinforcement signal during a trial in which the reward occurs later than predicted: the effective reinforcement signal (i.e., the dopamine-like response) signals the prediction of a reward at the onset of the stimulus, but then decreases at the predicted time of reward. Finally, it is increased at the new time of reward. It corresponds to the dopamine neuron activity presented in Figure 1d.

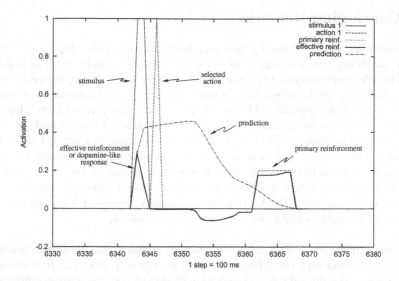

Fig. 6. Dopamine-like reinforcement signal during a learning trial in the spatial choice task.

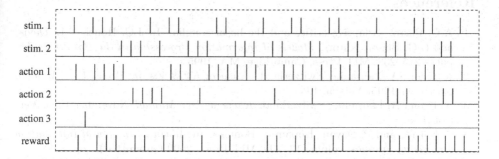

Fig. 7. Typical experimental run. The onset of stimulus 1 corresponds to the presentation of a pattern while the robot was turning to the right. Stimulus 2, while it was turning to the left. The robot "chooses" action 1 when it "thinks" it has perceived the pattern while turning to the right, and action 2, when it "thinks" it has perceived the pattern while turning to the left; finally, action 3, when it "thinks" it has not "perceived" the pattern.

Figure 7 represents a typical experimental run: the robot learned to respond correctly to 10 successive presentations of the pattern, after 42 trials. The first two rows show the activation of the two stimuli at every trial of the run. Rows 3 to 5 show the selected action by the robot, and finally, row 6 shows when the system was rewarded. The figure shows how the system learns very fast to discard action 3, but then, needs several tens of trials to learn to distinguish between 'pattern to the right' and 'pattern to the left' (i.e., when to choose action 1 and action 2).

8 Concluding Remarks

In a previous work, we used a hebbian learning model developed by neuroscientists studying conditioning in bees [8] to control a mobile robot in a foraging task [10]. We were motivated by the promising idea of using dopamine-like neural signals to modulate learning. In this work, we show that a time-delay actor-critic architecture implementing a TD model (i.e., with a dopamine-like reinforcement signal) enables an autonomous robot to learn a spatial choice task by a trial-and-error mechanism coupled with an externally provided reward delivered at arbitrary times. The approach appears as a promising mechanism for robotic systems that learn from simple human teaching signals in the real world. However, it should be noted that longer stimulus-reward intervals require a larger number of components in the temporal stimulus representation, thus a larger number of synaptic weights to be learned by the critic.

Acknowledgments. The author wishes to thank Prof. Wolfram Schultz for helpful comments and Dr. Thierry Dagaeff for his help with the first versions of the C-code of the neural network model with dopamine-like reinforcement signal. This work was supported by the Swiss National Science Foundation.

References

1. A.G. Barto. Adaptive critics and the basal ganglia. In J.C. Houck, J.L. Davis, and D.G. Beiser, editors, *Models of Information Processing in the Basal Ganglia*, pages 215–232. MIT Press, Cambridge, MA, 1995.
2. K-Team SA, Lausanne, Switzerland. *Khepera K213 Vision Turret User Manual*, November 1995. Version 1.0.
3. K-Team SA, Lausanne, Switzerland. *Khepera User Manual*, November 1995. Version 4.06.
4. S. Kakade and P. Dayan. Dopamine Bonuses. In *Advances in Neural Information Processing Systems*, volume 13. The MIT Press, 2000 (submitted).
5. A.H. Klopf. A neuronal model of classical conditioning. *Psychobiology*, 16(2):85–125, 1988.
6. K. J. Lang and G. E. Hinton. A time-delay neural network architecture for speech recognition. Technical Report CMU-DS-88-152, Dept. of Computer Science, Carnegie Mellon University, Pittsburgh, PA, December 1988.
7. F. Mondada, E. Franzi, and P. Ienne. Mobile robot miniaturization: A tool for investigating in control algorithms. In *Proceedings of the Third International Symposium on Experimental Robotics*, Kyoto, Japan, 1993.
8. P.R. Montague, P. Dayan, C. Person, and T.J. Sejnowski. Bee Foraging in uncertain environments using predictive hebbian learning. *Nature*, 377:725–728, October 26 1995.
9. A. Pérez-Uribe. *Structure-Adaptable Digital Neural Networks*, chapter 6. A Neurocontroller Architecture for Autonomous Robots, pages 95–116. Swiss Federal Institute of Technology-Lausanne, Ph.D Thesis 2052, 1999.
10. A. Pérez-Uribe and B. Hirsbrunner. Learning and Foraging in Robot-bees. In Meyer, Berthoz, Floreano, Roitblat, and Wilson, editors, *SAB2000 Proceedings Supplement Book*, pages 185–194, Honolulu, 2000. International Society for Adaptive Behavior.

11. P. Redgrave, T.J. Prescott, and K. Gurney. Is the Short Latency Dopamine Burst Too Short to Signal Reward Error. *Trends in Neurosciences*, 22:146–151, 1999.
12. R.A. Rescorla and A.R. Wagner. A theory of Pavlovian conditioning: variations in the effectiveness of reinforcement and non-reinforcement. In A.H. Black and W.F. Prokasy, editors, *Classical Conditioning II: Current Research and Theory*. Appleton-Century-Crofts, New York, 1972.
13. W. Schultz. Predictive reward signal of dopamine neurons. *Journal of Neurophysiology*, 80:1–27, 14 March 1998.
14. W. Schultz, P. Apicella, and T. Ljungberg. Responses of monkey dopamine neurons to reward and conditioned stimuli during successive steps of learning a delayed response task. *J. Neuroscience*, 13(3):900–913, 1993.
15. W. Schultz, P. Dayan, and P. Read Montague. A Neural Substrate of Prediction and Reward. *Science*, 275:1593–1599, 14 March 1997.
16. L.R. Squire and E.R. Kandel. *Memory: From Minds to Molecules*. Scientific American Library, New York, November 1998.
17. R.E. Suri and W. Schultz. Learning of sequential movements by neural network model with dopamine-like reinforcement signal. *Exp Brain Res*, 121:350–354, 1998.
18. R.E. Suri and W. Schultz. A Neural Network Model With Dopamine-Like Reinforcement Signal That Learns a Spatial Delayed Responde Task. *Neuroscience*, 91(3):871–890, 1999.
19. R.S. Sutton. Learning to predict by the methods of Temporal Differences. *Machine Learning*, 3:9–44, 1988.
20. R.S. Sutton and A.G. Barto. *Reinforcement Learning: An Introduction*. The MIT Press, 1998.

Connectionist Propositional Logic
A Simple Correlation Matrix Memory Based Reasoning System

Daniel Kustrin and Jim Austin

Dept. Computer Science, University of York, UK

Abstract. A novel purely connectionist implementation of propositional logic is constructed by combining Correlation Matrix Memory operations, tensor products and simple control circuits. The implementation is highly modular and expandable and in its present form it not only allows forward rule chaining but also implements *is a* hierarchy traversal which results in interesting behavior even in its simplest form.

1 Introduction

Correlation Matrix Memories (CMMs) are simple binary weighted feed forward neural networks which have been applied to a large number of real world applications in domains as diverse as molecular matching [10], image segmentation [1] and financial forecasting [6]. Their well understood semantics, complexity and storage properties makes them an ideal vehicle for preliminary research into connectionist reasoning systems since their operation characteristics are known and a number of efficient implementations (both in software and in hardware) exist [5]. In this paper CMMs have been employed as execution engines for a propositional logic implementation as well as for implementation of an *is_a* hierarchy. The actual system has been implemented as an interactive interpreter in which only the parser and proposition-to-code translation was not built from connectionist components: all actual processing is performed on a purely connectionist architecture. In its operation it is most similar to Smolenski's system [4] but in contrast to their architecture CPL is a purely binary connectionist system with defined semantics and integrated *is_a* hierarchy. Additionally it is also highly modular and thus scalable. CPL builds on initial CMM in reasoning theoretical papers [2,3] and represents a step in evolution towards connectionist higher order logic implementations.

The paper addresses the question of how complex data may be presented in a real neural system. The approach taken here is purely connectionist and, as such, could be used to show some principles of data organisation and storage in the human cortex. It must be stressed, however, that just because the implementation is plausible, it would not constitute a mode of how the human brain processes data.

The following section introduces the basic CMM operations, atom binding and coding procedures, Architectures section presents the actual implementation while execution section contains an annotated sample transcript from an interactive session with the system.

S. Wermter et al. (Eds.): Emergent Neural Computational Architectures, LNAI 2036, pp. 534–546, 2001.
© Springer-Verlag Berlin Heidelberg 2001

2   Operational Semantics

Correlation Matrix Memories are equivalent to single layer, binary weighted neural networks, but use have simple teaching and recall algorithms. Their power lies in application of thresholding and their ability to process overlapped input vectors.

Teaching of a matrix M given input vector I and output vector O is performed as

$$M' = M \bigvee OI^{\mathrm{T}} \tag{1}$$

and recall is simply

$$O = \Theta_\theta (MI) \tag{2}$$

where Θ_θ is a transfer function performing a hard threshold at level θ. Thresholding is a very powerful technique which allows extraction of correct associations even in presence of some noise; by manipulating threshold level it is possible to extract not only the matching association but also outputs which are "close" to it. Since CMMs work only on fixed weight binary vectors it is necessary to convert all input/output values into appropriate form. For propositional logic random binary vectors are generated for each term; these vectors are not necessarily orthogonal but are ensured to be sparse (few bits set) which allows accurate and almost error-free recalls. Error properties will not be discussed in this paper, see [9] for further information on CMM error properties. It is interesting to notice that it has been shown that binary CMMs exhibit lower crass-talk then non-binary single layer feed-forward neural networks for same associative tasks, which provides additional validation for use of CMMs [8].

The implemented connectionist propositional logic (CPL) is a slight extension of the traditional propositional logic since it recognises two types: propositions and atoms. Each proposition is formed by binding of two atoms and all rules (logic operations) are performed exclusively on propositions. Atoms were introduced to allow for more complex behavior and are a direct consequence of introduction of a *is_a* hierarchy. Additional level of complexity is introduced by having two modes of operation: learning of axioms and resolution. Before presenting each logic operation and its semantics it is necessary to define CPL syntax.

2.1   Syntax

The alphabet of the CPL is defined as

$$\Sigma = \left\{ \mathcal{A}, \mathcal{P}, \neg, \implies, \xrightarrow{is\text{-}a}, (,), \wedge, \otimes \right\} \tag{3}$$

where

$$\mathcal{A} = \{\top, \bot, \mathsf{any}, \mathsf{none}, \mathsf{a}_1, \mathsf{a}_2, \dots \} \tag{4}$$

$$\mathcal{P} = \{\mathsf{p}_1, \mathsf{p}_2, \dots \} \tag{5}$$

$$\mathsf{p}_i = \mathsf{a}_j \otimes \mathsf{a}_k \tag{6}$$

$$\text{Sentence} \rightarrow \text{AxiomSentence}$$
$$\text{Sentence} \rightarrow \text{ResolveSentence}$$
$$\text{AxiomSentence} \rightarrow \textbf{AXIOM} \ \text{AxiomTail}$$
$$\text{AxiomTail} \rightarrow \text{IsaTerm}$$
$$\text{AxiomTail} \rightarrow \text{AxiomExp}$$
$$\Longrightarrow \text{AxiomExp}$$
$$\text{AxiomExp} \rightarrow \text{BindTerm}$$
$$\text{AxiomExp} \rightarrow (\ \text{AxiomExp}\ )$$
$$\text{AxiomExp} \rightarrow \text{BindTerm} \wedge \text{AxiomExp}$$
$$\text{ResolveSentence} \rightarrow \textbf{RESOLVE} \ \text{ResolveExp}$$
$$\text{ResolveExp} \rightarrow (\ \text{ResolveExp}\ )$$
$$\text{ResolveExp} \rightarrow \text{ResTerm}$$
$$\text{ResolveExp} \rightarrow \neg \ \text{ResolveExp}$$
$$\text{ResolveExp} \rightarrow \text{ResTerm} \wedge \text{ResolveExp}$$
$$\text{ResolveExp} \rightarrow (\ \text{ResolveExp}\ ) \Longrightarrow$$
$$(\ \text{ResolveExp}\ )$$
$$\text{ResTerm} \rightarrow \text{BindTerm}$$
$$\text{ResTerm} \rightarrow \text{IsaTerm}$$
$$\text{IsaTerm} \rightarrow \text{Atom} \xrightarrow{is\text{-}a} \text{Atom}$$
$$\text{BindTerm} \rightarrow \text{Atom} \otimes \text{Atom}$$

Fig. 1. A grammar for well-formed formulae in CPL.

operator $\otimes$ is the binding operator. The truth values, $\top$ denoting truth and $\bot$ denoting false are special atomic constants as are any and none. It is immediately obvious that the disjunction connective is not part of the alphabet and that propositions, p_i, are constructed from atoms by application of the binding operator, $\otimes$. Due to the operational semantic of the CMMs (see below) it is possible to convert disjunctive expressions into CPL well-formed formulae. Any expression of the form:

$$p_i \vee p_j \vee \ldots \Longrightarrow p_q \tag{7}$$

can be expressed as

$$p_i \Longrightarrow p_q$$
$$p_j \Longrightarrow p_q$$
$$\vdots \Longrightarrow p_q \tag{8}$$

and any expression of the from

$$p_i \Longrightarrow p_j \vee p_k \vee \ldots \tag{9}$$

can be expressed as

$$p_i \Longrightarrow p_j$$
$$p_i \Longrightarrow p_k$$
$$p_i \Longrightarrow \vdots \tag{10}$$

These conversion forms allow rewriting of any disjunctive expression in CPL *wff*. Given the CPL alphabet it is necessary to present the grammar which shows the difference between atomic and propositional sentences as well as two modes of operation, see fig. 1.

Strictly speaking the grammar is ambiguous since it cannot be presented in LL(1) form. To solve this problem precedence ordering is defined (in the parser) as follows (from highest to lowest): $\otimes, \xrightarrow{is\text{-}a}, \neg, \wedge, \implies$.

2.2 Semantics

The easiest and most accurate way of describing CPL implementation is defining its semantics, given basic CMM and binary vector operations. The semantics of the CPL is closely tied to CMM operation and their properties. A CMM is a binary feed-forward neural network which accepts and produces fixed length, fixed weight (fixed desity of 1's) binary vectors. Consequently, each atom has to be converted into such a vector; in the implemented architecture this is done by assigning a random fixed length, fixed weight binary vector to each atom. Additionally, two out of four special atomic propositions (any and none) have special forms: any is represented by a vector with all bits set while none is a zero vector. The binding operator, $\otimes$, performs a outer product (tensor) between two atoms and produces a vector which given atom vector length of n and weight of k will be of length n^2 and weight k^2:

$$\mathsf{a}_i \otimes \mathsf{a}_j = \mathsf{a}_i \otimes \mathsf{a}_j = \begin{bmatrix} b_1^1 & b_1^2 & \cdots & b_1^n \\ b_2^1 & b_2^2 & \cdots & b_2^n \\ \vdots & \vdots & \ddots & \vdots \\ b_n^1 & b_n^2 & \cdots & b_n^n \end{bmatrix} \tag{11}$$

$$= \begin{bmatrix} b_1^1 & b_1^2 & \cdots & b_1^n \end{bmatrix} \tag{12}$$

where $b_l^k = \mathsf{a}_i^k \times \mathsf{a}_j^l$. Outer product is used for binding as it allows superposition of propositions without loss of association.

As stated above, a CMM is a type of a neural network it has two modes of operation: training and recall. These two modes are mapped on to axiom presentation and resolution. The training process can be defined as an operator action ($\mathfrak{T}$) between two sets of binary vectors $\mathbf{I} \in \mathbb{B}^n$ and $\mathbf{O} \in \mathbb{B}^n$, denoting relational mapping $\mathbf{I} \to \mathbf{O}$, and a CMM, M as follows:

$$\mathfrak{T}\left(\mathbb{B}^n, \mathbb{B}^n, \mathrm{M}^{n \times n}\right) \to \mathrm{M}^{n \times n} \tag{13}$$

$$\mathfrak{T}\left(\mathbf{I}, \mathbf{O}, M^i\right) = M^i \vee \mathbf{OI}^{\mathrm{T}} = M^{i+1}$$

given $M^0 = 0$. This equation shows that training is just a disjunction of outer products. Similarly, recall operator ($\mathfrak{R}$) takes a CMM, input vector and an integer (threshold) and produces an output vector:

$$\mathfrak{R}\left(\mathbb{B}^n, \mathrm{M}^{n \times n}, \mathbb{Z}\right) \to \mathbb{B}^n \tag{14}$$

$$\mathfrak{R}\left(\mathbf{I}, M, n\right) = \Theta_n\left(M\mathbf{I}\right) = \mathbf{O}$$

The threshold parameter can be varied to produce various effects but in this paper it is assumed to be set at the input vector weight at all times, see [3] for further information on application of thresholds in CMMs.

These three definitions (binding, training and recall) allow specification of the semantics of BindTerm, IsaTerm, AxiomTail implication and ResolveExpression implication terms in the grammar:

$$\text{Atom} \otimes \text{Atom} \quad : \quad (\mathbb{B}^n, \mathbb{B}^n) \to \mathbb{B}^{n^2} \tag{15}$$

$$\triangleq \quad \mathsf{a}_i \otimes \mathsf{a}_j \tag{16}$$

$$\text{Atom} \xrightarrow{is\text{-}a} \text{Atom} \quad : \quad (\mathbb{B}^n, \mathbb{B}^n)$$

$$\triangleq \quad \mathsf{a}_i \xrightarrow{is\text{-}a} \mathsf{a}_j \tag{17}$$

$$\triangleq \quad M_{is\text{-}a} \leftarrow \mathfrak{T}\,(\mathsf{a}_i, \mathsf{a}_j, M_{is\text{-}a})$$

$$\text{Atom} \xrightarrow{is\text{-}a} \text{Atom}$$

$$: \quad (\mathbb{B}^n, \mathbb{B}^n) \to \mathbb{B}^{n^2} \tag{18}$$

$$\triangleq \quad \mathsf{a}_i \xrightarrow{is\text{-}a} \mathsf{a}_j$$

$$\triangleq \quad \mathsf{a}_j = \mathfrak{R}\,(\mathsf{a}_i, M_{is\text{-}a}, \theta)$$

where $(\mathbb{B}^n, \mathbb{B}^n) \to \mathbb{B}^{n^2}$ (for example) is the type definition, first row is the grammar representation and the second row is the implementation definition. The above two forms of the *is-a* association relate to either training (axiom presentation), eq. 17, or recall (resolution), eq. 18. The resolution equation (eq. 18) returns a truth value which is a vector of length n^2 and weight k^2, see below. Similar equations can be constructed for the implication connective:

$$\text{BindTerm} \implies \text{BindTerm} \tag{19}$$

$$: \quad \left(\mathbb{B}^{n^2}, \mathbb{B}^{n^2}\right)$$

$$\triangleq \quad \mathsf{p}_i \implies \mathsf{p}_j$$

$$\triangleq \quad M_{rule} \leftarrow \mathfrak{T}\,(\mathsf{p}_i, \mathsf{p}_j, M_{rule})$$

$$\text{BindTerm} \implies \text{BindTerm} \tag{20}$$

$$: \quad \left(\mathbb{B}^{n^2}, \mathbb{B}^{n^2}\right) \to \mathbb{B}^{n^2}$$

$$\triangleq \quad \mathsf{p}_i \implies \mathsf{p}_j$$

$$\triangleq \quad \mathsf{p}_j = \mathfrak{R}\,(\mathsf{p}_i, M_{rule}, \theta)$$

As in the eq. 17 and eq. 18, two modes of operation exist: training and recall. In the recall (resolution) the return value is a truth token, see below. The CMM size in *is-a* case and rules case are different. The *is-a* uses matrix of size $n \times n$ and type $\mathbb{M}^{n \times n}$ while rules store uses a CMM of size $n^2 \times n^2$ and type $\mathbb{M}^{n^2 \times n^2}$.

The conjunction operator semantics are defined using bitwise 'or' operation ($\bigvee$) as follows:

$$\text{BindTerm} \wedge \text{BindTerm} \tag{21}$$

$$: \quad \left(\mathbb{B}^{n^2}, \mathbb{B}^{n^2}\right) \to \mathbb{B}^{n^2}$$

$$\triangleq \quad \mathsf{p}_i \wedge \mathsf{p}_j$$

$$\triangleq \quad \mathsf{p}_i \bigvee \mathsf{p}_j$$

As it can be seen from the above equations, the basic return value is a tensored vector of length n^2 and weight k^2. The four primitive atomic propositions $\top$, $\bot$, any and none are, although atomic, of the proposition size and weight since they operate on propositional level. The negation operator can, thus, be defined as

$$\neg \text{BindTerm} \quad : \quad \mathbb{B}^{n^2} \to \mathbb{B}^{n^2} \tag{22}$$

$$\triangleq \quad \neg \mathsf{p}_i$$

$$\triangleq \quad \begin{cases} \top & \text{if } \mathsf{p}_i = \bot \\ \bot & \text{otherwise} \end{cases}$$

This definition converts a proposition into a truth value. Although not ideal it is necessary since even under closed world assumption negation semantics is quite difficult. Alternative would be to denote $\neg \mathsf{p}_i$ as the set of all propositions excluding p_i, $\neg \mathsf{p}_i \triangleq \text{all} - \mathsf{p}_i$. This approach was not chosen since it doesn't map well to CMM architecture. The equality operation used in eq. 18 and eq. 20 can be defined as

$$\text{BindTerm} = \text{BindTerm} \tag{23}$$

$$: \quad \left(\mathbb{B}^{n^2}, \mathbb{B}^{n^2}\right) \to \mathbb{B}^{n^2}$$

$$\triangleq \quad \mathsf{p}_i = \mathsf{p}_j$$

$$\triangleq \quad \begin{cases} \top & \text{if } \mathsf{p}_i \text{ and } \mathsf{p}_j \text{ are identical} \\ \bot & \text{otherwise.} \end{cases}$$

2.3   Axioms and Resolution

CPL does not come with any predefined axioms. All axioms have to be presented to the architecture (trained). The resolution system is also very simple. Since both *is-a* and the rules sub-systems are built from the same building blocks they implement the same algorithm: chaining. In *is-a* mode they performs traversal up the *is-a* tree while in the rules sub-system mode they forward chain the rules.

The *is-a* hierarchy is a directed non-cyclical graph depicting both membership and subset relations. In traditional semantic networks there is a distinction between subset relations like $Cats \xrightarrow{Subset} Mammals$ and $Bob \xrightarrow{Member} Cats$ since they necessarily have different semantics, $Cats \subset Mammals$ and $Bob \in Cats$, respectively. In CPL this distinction is not made and a general *is-a* relation is used:

$Cat \xrightarrow{is\text{-}a} Mammal$ and $Bob \xrightarrow{is\text{-}a} Cat$. Not making this distinction allows simple application of the binding operator to atoms for formation of propositions. These propositions are then passed to the rule sub-module which uses them for rule searching. Traditionally it has been argued that using $is\text{-}a$ links instead of separate membership and subset links leads to inconsistencies [7]. Although the formal semantics of inheritance has not been developed for this system the problems outlined in [7] are not applicable since the propositions always have the form class $\otimes$ query atom in which only the class part is modifiable. The class part can only be changed to a superclass until the top of the inheritance graph has been reached. Since the $is\text{-}a$ hierarchy is assumed to be a directed non-cyclic graph termination is assured.

The rules sub-system accepts propositions and performs forward chaining on left-hand sides of the rules. Given an input rule $p_i \implies p_j$ it will search for all the rules matching p_i on the LHS. If none are found it will request modification of the rule LHS by the $is\text{-}a$ module. This request will ask for replacement of the p_i superclass by its superclass, if any exists, eg. if $p_i = a_j \otimes a_k$ and no rules are matched, the rule sub-system will request the super-class of a_j, say a_m, which will then be bound to a_k into a new proposition $p_n = a_m \otimes a_k$. This new proposition will then be used in another search. The searching process will end when the $is\text{-}a$ has reached the top of its hierarchy.

Although the architecture only uses forward chaining and a simple semantic network it is capable of performing resolution of a surprising flexibility.

3   Architecture

The connectionist architecture implementing the semantics of CPL has been built around correlation matrix memory libraries developed at York. The grammar and the semantics have been directly implemented in YACC and LEX on a Silicon Graphics Supercomputer. Although currently the architecture is simulated in software it is a truly connectionist artifact. Both $is\text{-}a$ and the rules sub-systems have been built using exactly the same basic building block presented in fig. 2 (figure is slightly simplified and shows only the recall connections and omits training/recall switching and input/output pre and post processing).

It is interesting to notice that all components used are implementable in purely connectionist form: the Correlation Matrix Memory is a simple binary feed-forward neural network, the Selector components can be constructed by using inhibitory neurons while the comparator is a simple bit-wise conjunction. The control input is used to "stop" the module from forward-chaining and keep it at the current search parameter. This is used, in the complete architecture, by the rules sub-module for control of $is\text{-}a$ hierarchy traversal. The selector component is used bi-directionally either as a input selector or output selector. Both variants are controlled by a control signal and the component semantics can be defined as follows, in input selection form:

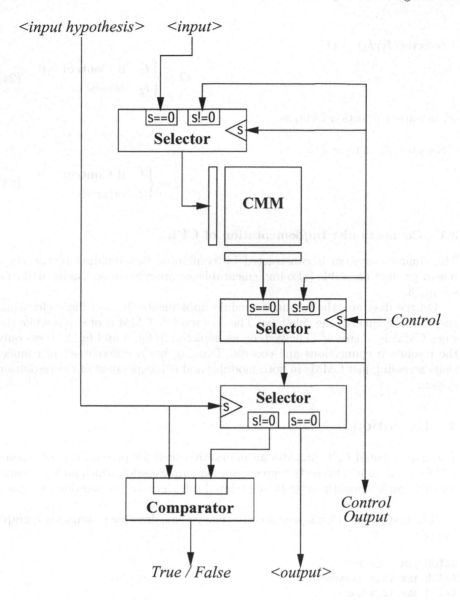

Fig. 2. Basic building block for the connectionist architecture implementing CPL. Diagram only shows the data paths for recall mode of operations.

Selector $(I_1, I_2) \to O$

$$O = \begin{cases} I_1 & \text{if } \mathbf{Control} = 0 \\ I_2 & \text{otherwise} \end{cases} \quad (24)$$

or, in output selection form, as

Selector $(I) \to O_i = I$

$$i = \begin{cases} 1 & \text{if } \mathbf{Control} = 0 \\ 2 & \text{otherwise} \end{cases} \quad (25)$$

3.1   Connectionist Implementation of CPL

The complete system is constructed by combining two building blocks via a tensor product ensemble (also implementable as a purely connectionist artifact), see fig. 3.

On the diagram, the left-hand module implements the *is-a* hierarchy while the RHS implements the rule-base. The *is-a* module CMM is of size n while the rules CMM is of size n^2. The system, as depicted in fig. 3 and fig. 2 shows only the resolution connections and control. Training has a different set of connections accessing just CMMs in both modules and is independent of the resolution system.

4   Execution

The implemented CPL provides an interactive shell for presentation of axioms and for resolution. This section presents an example session which involves learning of a simple semantic network and related rules and some resolution examples.

The first set of axioms presented to the systems describe a simple *is-a* graph (fig. 4):

```
AXIOM jim is_a man
AXIOM jim is_a professor
AXIOM dan is_a man
AXIOM dan is_a ra
AXIOM man is_a human
AXIOM professor is_a academic
AXIOM ra is_a academic
AXIOM academic is_a human
```

while the following rules define how these atoms are interrelated:

```
AXIOM human:any => mortal:true
AXIOM academic:any => mad:true
AXIOM professor:any => has_big_office:true
```

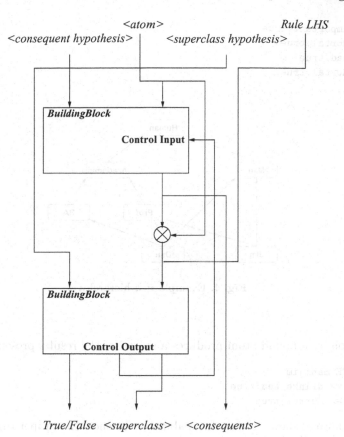

Fig. 3. Complete connectionist implementation of CPL. Only recall data paths are shown, as before.

```
AXIOM has_big_office:true => important:true
AXIOM man:jim => drinks_tea:true
AXIOM man:dan => drinks_coffee:true
```

Given the set of *is-a* axioms and the rule set it is possible to perform a variety of resolution queries. Resolving an atom searches both the *is-a* set and the rules set:

```
> RESOLVE jim
jim is_a man
jim is_a professor
jim is_a human
jim is_a academic
jim is_a human
jim => drinks_tea:true
jim => has_big_office:true
```

```
jim => important:true
jim => mortal:true
jim => mad:true
jim => mortal:true
```

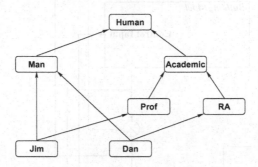

Fig. 4. Example *is-a* hierarchy.

Resolution on a bound atom produces a subset of the results presented above:

```
> RESOLVE man:jim
man:jim => drinks_tea:true
man:jim => mortal:true
```

These examples show the search capabilities. To show the proper rule resolution consider the following examples:

```
> RESOLVE (man:jim) => (has_big_office:true)
No.
> RESOLVE
    (professor:jim) => (has_big_office:true)
Yes.
```

This example shows how the system performs reasonable rule resolution, although jim is both man and professor the property of having a big office is only related to being a professor (via the rule `professor:any => has_big_office:true`). Hence the above example performs correctly. To test rule chaining consider the following result

```
> RESOLVE (professor:jim) =>  (important:true)
Yes.
```

which is the result of chaining rules `(professor:any) => (has_big_office:true)` and `(has_big_office:true) => (important:true)` which suggests that jim is important since he has a big office. Similarly, resolution

```
> RESOLVE (ra:dan) =>  (important:true)
No.
```

shows that dan is not important since he doesn't have a big office.

Since the architecture has no axioms predefined to produce proper *modus ponens* it is necessary to provide the axiom stating that truth implies truth:

```
AXIOM (true:true) =>
(true:true)
```

which allows for queries like

```
> RESOLVE (((professor:jim =>
              has_big_office:true) AND
             (has_big_office:true =>
              important:true)) =>
             (professor:jim =>
              important:true))
Yes.
```

5 Conclusion

This paper presented a pure binary connectionist reasoning system based on a propositional logic. The operational semantics were presented as were its modular structure and an example execution. Due to space restrictions this paper did not provide proof for the claims of its scalability or a thorough comparison with other similar systems.

Acknowledgements. This work is funded by EPSRC ROPA grant GR/L75559.

References

1. C Orovas J Austin. Cellular associative symbolic processing for pattern recognition. *MFCS '98 workshop on Grammer Learning*, pages 269–280, 1998.
2. J Austin. Correlation matrix memories for knowledge manipulation. In *International Conference on Neural Networks, Fuzzy Logic, and Soft Computing: Iizuka, Japan*, 1994.
3. J Austin. Distributed associative memories for high speed symbolic reasoning. *International Journal on Fuzzy Sets and Systems*, 82(2):223–233, 1995. Invited paper to the special issue on Connectionist and Hybrid Connectionist Systems for Approximate Reasoning.
4. C P Dolan and P Smolensky. Tensor product production system: a modular architecture and representation. *Connection Science*, (1):53–68, 1989.
5. J V Kennedy, J Austin, R Pack, and B Cass. C-NNAP: A parallel processing architecture for binary neural networks. In *International Conference on Neural Networks (ICANN 95)*, Perth, Australia, November 1995.
6. D Kustrin, J Austin, and A Sanders. Application of correlation memory matrices in high frequency asset allocation. In M Niranjan, editor, *Fifth International Conference on Artificial Neural Networks*. IEE, 1997.
7. D McDermott. Artificial intelligence meets natural stupidity. *SIGART Newsletter*, (57), 1976.

8. S O'Keefe. *Neural-Based Content Analysis of Document Images*. PhD thesis, Department of Computer Science, University of York, 1997.
9. M Turner and J Austin. Matching performance of binary correlation matrix memories. *Neural Networks*, 10(9):1637–1648, 1997.
10. M Turner and J Austin. A neural network technique for chemical graph matching. In M Niranjan, editor, *Proceedings of the Fifth International Conference on Artificial Neural Networks*. IEE, 1997.

Analysis and Synthesis of Agents That Learn from Distributed Dynamic Data Sources

Doina Caragea, Adrian Silvescu, and Vasant Honavar

Artificial Intelligence Research Laboratory, Department of Computer Science, Iowa
State University, Ames, Iowa 50011-1040, USA,
honavar@cs.iastate.edu,
http://www.cs.iastate.edu/~honavar/aigroup.html

Abstract. We propose a theoretical framework for specification and
analysis of a class of learning problems that arise in open-ended environments that contain multiple, distributed, dynamic data and knowledge
sources. We introduce a family of learning operators for precise specification of some existing solutions and to facilitate the design and analysis of
new algorithms for this class of problems. We state some properties of instance and hypothesis representations, and learning operators that make
exact learning possible in some settings. We also explore some relationships between models of learning using different subsets of the proposed
operators under certain assumptions.

1 Learning from Distributed Dynamic Data

Many practical knowledge discovery tasks (e.g., learning the behavior of complex computer systems from observations, computer-aided scientific discovery in
bioinformatics) present several new challenges in machine learning. The data
repositories in such applications tend to be very large, physically distributed,
often autonomously managed, and constantly growing over time (as new data
get added). Thus, there is a need for algorithms for learning from distributed
data by analysing the distributed data sets where they reside instead of shipping
large volumes of data across networks, in an incremental fashion, as the data
becomes available over time, without having to reprocess the already processed
data [5,14].

Although some incremental and distributed learning algorithms have been
proposed in the literature, most of them [9,4,15], do not guarantee generalization
accuracies that are provably close to those obtainable in the batch or centralized learning scenario. Some notable exceptions include parallel and distributed
versions [1,13,6,12] and incremental versions [3] of batch algorithms that preserve the underlying nature of the centralized algorithm. At present, with the
exception of some interesting results (e.g., *mistake bounds*) for the closely related
problem of *online learning* [7], a characterization of hypothesis classes that admit efficient exact or approximate distributed or incremental learning is lacking.
Yet from a practical standpoint, the design and implementation of such learning

S. Wermter et al. (Eds.): Emergent Neural Computational Architectures, LNAI 2036, pp. 547–559, 2001.
© Springer-Verlag Berlin Heidelberg 2001

agents is clearly of interest. Against this background, there is a need to address incremental and distributed learning problems in their full generality.

This paper presents some tentative steps towards a framework for specification, analysis, and synthesis of incremental and distributed learning agents. We define some learning and information extraction operators to formally model some existing learning algorithms. We explore some properties of instance and hypothesis representations, and learning operators that guarantee the existence of incremental and distributed learning algorithms with provable performance guarantees relative to their batch or centralized counterparts. We offer some examples to illustrate the use of this theoretical framework in designing new incremental and distributed learning algorithms.

2   Incremental Learning and Distributed Learning

A generic incremental learning scenario is shown in Fig. 1. In an *incremental learning scenario*, data sets $D_1, D_2, \cdots, D_n$ are assumed to become available to the learner at discrete instants in time $t_1, t_2, \cdots, t_n$. The learner starts with a (possibly null) initial hypothesis h_0 which constitutes the prior knowledge of the domain. We assume that the learner is typically unable to store the data in its raw form. Thus, it can only maintain and update its hypothesis base as new data becomes available. Thus, h_0 gets updated to h_1 on the basis of D_1, and h_1 gets updated to h_2 on the basis of data D_2, and so on.

In a *distributed learning* scenario, the data set is assumed to be physically distributed across multiple, possibly autonomous, data repositories $D_1, \cdots, D_n$. The learner can visit the repositories to gather the information necessary for generating knowledge (e.g., in the form of pattern classification rules) by processing the data where it is stored. Alternatively, the data repositories may transmit the information to the learner. In either case, we prohibit transport of raw data among different sites. A distributed learning scenario is shown in Fig. 2.

A number of variations on these basic incremental and distributed learning scenarios can be envisioned under different assumptions concerning where and when data processing is performed, what information is made available to the learner, etc. More generally, we can consider incremental learning from distributed data sources. Space does not permit a detailed discussion of such scenarios.

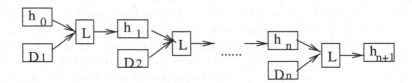

Fig. 1. Incremental Learning

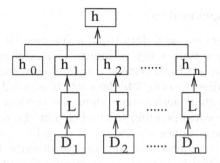

Fig. 2. Distributed Learning

3   Horizontal and Vertical Data Fragmentation

In many applications, the data set consists of a set of tuples where each tuple stores the values of relevant attributes. The distributed nature of such a data set can lead to at least two common types of data fragmentation: *horizontal fragmentation* wherein subsets of data tuples are stored at different sites; and *vertical fragmentation* wherein subtuples of data tuples are stored at different sites. Assume that a data set D is distributed among the sites $1, \cdots, n$ containing data set fragments $D_1, \cdots, D_n$. We assume that the individual data sets $D_1, \cdots, D_n$ collectively contain enough information to generate the complete dataset D. In many applications, it might be the case that the individual data sets are autonomously owned and maintained. Consequently, the access to the raw data may be limited and only summaries of the data (e.g., number of instances that match some criteria of interest) may be made available to the learner. Even in cases where access to raw data may not be limited, the large size of the data sets makes it infeasible to assemble the complete data set D at a central location.

3.1   Horizontal Fragmentation

In the distributed setting, the data is fragmented in such a manner that each site contains a set of data tuples. The union of all these sets constitutes the complete dataset. If the individual data sets (horizontal fragments) are denoted by $D_1, D_2, \cdots, D_n$, and the corresponding complete data set by D, then *Horizontally Distributed Data* (HDD) has the following property: $D_1 \cup D_2 ... \cup D_n = D$, where $\cup$ denotes set union. Hence, in this case, a distributed learning algorithm L_d is exact with respect to the hypothesis inferred by a learning algorithm L if it is the case that:

$$L_d(D_1, D_2, \cdots, D_n) = L(D_1 \cup D_2 \cup \cdots \cup D_n). \tag{1}$$

The challenge is to achieve this guarantee without providing L_d with simultaneous access to $D_1, \cdots D_n$.

Similarly, we can envision horizontal fragmentation of data in the incremental setting.

3.2   Vertical Fragmentation

In the distributed setting, each data tuple is fragmented into several subtuples each of which shares a unique key or index. Thus, different sites store *vertical* fragments of the data set. Each vertical fragment corresponds to a subset of the attributes that describe the complete data set. It is possible for some attributes to be shared (duplicated) across more than one vertical fragments, leading to overlap between the corresponding fragments. Let $A_1, A_2, \cdots, A_n$ indicate the set of attributes whose values are stored at sites $1, \cdots, n$ respectively, and let A denote the set of attributes that are used to describe the data tuples of the complete data set. Then in the case of *Vertically Distributed Data* (VDD), we have: $A_1 \cup A_2 \cdots \cup A_n = A$. Let $D_1, D_2, \cdots, D_n$, denote the fragments of the dataset stored at sites $1, \cdots, n$ respectively, and let D denote the complete data set. Let the ith tuple in a data fragment D_j be denoted as $t^i_{D_j}$. Let $t^i_{D_j}$.index denote the *unique index* associated with tuple $t^i_{D_j}$ and let $\times$ denote the *join* operation. Then the following properties hold for VDD:

1. $D_1 \times D_2 \times \cdots \times D_n = D$, and
2. $\forall\, D_j, D_k,\ t^i_{D_j}$.index $= t^i_{D_k}$.index.

Thus, the subtuples from the vertical data fragments stored at different sites can be put together using their unique index to form the corresponding data tuples of the complete dataset. It is possible to envision scenarios in which a vertically fragmented data set might lack unique indices. In such a case, it might be necessary to use combinations of attribute values to infer associations among tuples [1]. In what follows, we will assume the existence of unique indices in vertically fragmented distributed data sets.

In the case of vertically fragmented data, a distributed learning algorithm L_d is exact with respect to the hypothesis inferred by a learning algorithm L if it is the case that:

$$L_d(D_1, D_2, \cdots, D_n) = L(D_1 \times D_2 \times \cdots \times D_n). \qquad (2)$$

The challenge is to guarantee this without providing L_d with simultaneous access to $D_1, \cdots D_n$.

Similarly, we can envision vertical fragmentation of data over time in the incremental setting. This is of special relevance in applications where data representation may be augmented over time by the addition of new attributes (e.g., measurements obtained using novel experiments in an ongoing scientific project). It is possible to envision scenarios in which a vertically fragmented data set might lack unique indices. In such a case, it will be necessary to combine the attribute values to infer associations among tuples. We can also envision data sets that are both horizontally and vertically fragmented in space, time, or both.

4   Learning Operators

Let $\mathcal{X}$ be a sample space where from the examples are drawn, and let $\mathcal{D}$ be the set of all possible subsets of the sample space $\mathcal{X}$. We can assume that the subsets in $\mathcal{D}$ are obtained from $\mathcal{X}$ by sampling according to different probability distributions. Let $\mathcal{C}$ be the space of all possible functions that we may want to learn or approximate, and $\mathcal{H}$ the space of the functions that a learning agent can draw on in order to construct approximations of the functions in $\mathcal{C}$. In a typical inductive learning scenario, $\mathcal{H}$ is a set of hypotheses. However, in the analysis that follows, it is useful to allow $\mathcal{H}$ to include not only the hypotheses but also other functions defined over $\mathcal{D}$. Examples of such functions include those that compute statistical summaries of a given data set, select subsets of a data set, or in general, extract useful information from a given data set. In what follows, we define some learning operators.

A *learning operator* is specified by $L : \mathcal{D} \to \mathcal{H}$, where L denotes any inductive learning algorithm or information extraction algorithm. It takes as input a dataset D and returns a function h that satisfies some specified criterion with respect to the data set. For example, if L is a consistent learner, it outputs a hypothesis that is consistent with the data. In other scenarios, L might compute relevant statistics from D.

The *"inverse" learning operator* is specified by $L^{-1} : \mathcal{H} \to \mathcal{D}$. As opposed to the learning operator, it takes as input a function h and returns a dataset D that satisfies some specified criterion with respect to L and the data set D. For example, L^{-1} might output when given h, a data set which when provided as input to L, results in the output h.

The *selection operator* is specified by $Sel : \mathcal{D} \to \mathcal{D}$. It generates a new dataset D' based on an existing dataset D by selecting examples according to a specified criterion (e.g., by sampling D according to some desired probability distribution).

The *union operator* $\cup : \mathcal{D} \times \mathcal{D} \to \mathcal{D}$ takes as arguments two datasets D_1 and D_2 and outputs a new dataset $D = D_1 \cup D_2$. It may represent the standard set union, multi-set union, or any suitably well-defined operation.

The *augmentation operator* $L_A : \mathcal{H} \times \mathcal{D} \to \mathcal{H}$ augments or refines a function h by incorporating new data D according to some specified criterion. For instance, it may minimally modify a hypothesis so as to be consistent with new data.

The *combination operator* $L_C^n : \mathcal{H}^n \to \mathcal{H}$ produces a new function h by exploiting the information provided by the given functions $h_1, h_2, \cdots, h_n$.

This set of definitions is meant to be merely illustrative (and not exhaustive) with respect to the types of operators that might be useful in incremental and distributed learning settings. Because the augmentation and combination operators are more general in some sense than the learning operator, it is desirable to enforce some consistency conditions among these operators. Let $h_\emptyset := L(\phi)$. Then we would like the following equalities to hold:

1. $L_A(h_\emptyset, D) = L(D)$,
2. $L_A(h, \phi) = h$, and
3. $L_C^2(h_\emptyset, h) = h$.

As a consequence, the following less general equalities should also hold:

4. $L_A(h_\emptyset, \phi) = L(\phi) = h_\emptyset$, and
5. $L_C^2(h_\emptyset, h_\emptyset) = L(\phi) = h_\emptyset$.

The previous conditions basically ensure that the two more general operators L_A and L_C^2 behave nicely when provided an empty dataset as one of the inputs.

5   Incremental and Distributed Learning Criteria

Batch learning algorithms have been the subject of extensive experimental and theoretical analysis. Consequently, it is desirable to develop a theoretical framework which allows us to gain useful insights into the performance of distributed and incremental algorithms by relating them their batch or centralized counterparts.

In what follows, we define *exactness* of a distributed or incremental learning algorithm. This definition is intended to illustrate the sorts of analysis that are facilitated by the theoretical framework that is sketched out in this paper. A similar analysis can be performed with respect to other performance criteria that are motivated by the needs of specific applications.

Definition 1. *Given two datasets D_1 and D_2, and a learning algorithm L for a function class $\mathcal{H}$, we say that $D_1 \cup D_2$ is exact incremental-learnable and exact distributed-learnable with respect to the algorithm L if the conditions*

$$L_A(L(D_1), D_2) = L(D_1 \cup D_2), \text{ and} \tag{3}$$

$$L_C^2(L(D_1), L(D_2)) = L(D_1 \cup D_2) \tag{4}$$

hold (respectively).

Note that the union operator used above can have different meanings in different scenarios. In some cases, it may denote standard set union; in some others, it may stand for an operation that combines subtuples of a data tuple from different data sets based on the unique index that helps associate the data subtuples (see the discussion of horizontally and vertically fragmented data in the previous section for details).

The conditions for exact incremental and exact distributed learning can be easily generalized for the case where the complete data set is distributed among n data sets. In many real world problems involving sufficiently expressive concept classes, exact incremental or distributed learning may not be possible even in principle. In other instances, although possible in principle, it may not be feasible in practice for computational reasons. At present, a characterization of hypothesis classes that lend themselves to exact or approximate distributed or incremental learning is lacking. From a practical standpoint, the design and implementation of data and hypotheses representations that can support computationally efficient and scalable distributed and incremental learning algorithms is clearly of interest.

6   Models of Distributed and Incremental Learning

In this section we will explore relationships among some of the learning operators introduced above and define them in term of the others. Then we will explore some properties of instance and hypothesis representations and operators which guarantee the existence of exact incremental and exact distributed learning under certain assumptions. Provable equivalences among certain learning operators (or combinations thereof) can help to transform the algorithms developed in one setting (e.g. distributed learning) to another setting (e.g., incremental learning) under certain well-defined conditions.

Consider the learning operators defined as follows:

1) $L_A(h, D) := L(L^{-1}(h) \cup D)$
2) $L_C^2(h_1, h_2) := L(L^{-1}(h_1) \cup L^{-1}(h_2))$
3) $L_C^2(h_1, h_2) := L_A(L_A(h_\emptyset, L^{-1}(h_1)), L^{-1}(h_2))$
4) $L_C^2(h_1, h_2) := L_A(h_1, L^{-1}(h_2))$
5) $L_A(h, D) := L_C^2(h, L(D))$

Lemma 1. *Assume that $L(L^{-1}(h)) = h$. Then definition 3) is equivalent to definition 4).*

Proof. $L_A(L_A(h_\emptyset, L^{-1}(h_1)), L^{-1}(h_2)) = L_A(L(L^{-1}(h_1)) L^{-1}(h_2)) = L_A(h_1, L^{-1}(h_2))$.
The first equality follows from the consistency of L_A and the second one from the assumption made.

Theorem 1. *Assume that $L^{-1}(L(D)) = D$. Then the operators given by definition 1) and definition 2) satisfy the exact learning criteria. That is,*

(1) $L(D_1 \cup D_2) = L_A(L(D_1), D_2) \; \forall D_1, D_2 \in \mathcal{D}$ (in the incremental setting)
(2) $L(D_1 \cup D_2) = L_C^2(L(D_1), L(D_2)) \; \forall D_1, D_2 \in \mathcal{D}$ (in the distributed setting)

Proof. The proof of (1) above follows from the observation that $L_A(L(D_1), D_2) = L(L^{-1}(L(D_1)) \cup D_2) = L(D_1 \cup D_2)$. The proof of (2) above follows from the observation $L_C^2(L(D_1), L(D_2)) = L(L^{-1}(L(D_1)) \cup L^{-1}(L(D_2))) = L(D_1 \cup D_2)$.

Theorem 2. *Assume that $L^{-1}(L(D)) = D$ and that L_A satisfies the Exact Learning criteria. Then the operator given by definition 4) also satisfies the exact learning criteria in the distributed setting. That is,*
$L(D_1 \cup D_2) = L_C^2(L(D_1), L(D_2)) \; \forall D_1, D_2 \in \mathcal{D}$

Proof. $L_C^2(L(D_1), L(D_2)) = L_A(L(D_1), L^{-1}(L(D_2))) = L_A(L(D_1), D_2) = L(D_1 \cup D_2)$

Theorem 3. *Assume that L_C satisfies the exact learning criteria. Then the operator given by definition 5) also satisfies the exact learning criteria in the incremental setting. That is, $L(D_1 \cup D_2) = L_A(L(D_1), D_2) \; \forall D_1, D_2 \in \mathcal{D}$*

Proof. We have: $L_A(L(D_1), D_2) = L_C^2(L(D_1), (L(D_2))) = L(D_1 \cup D_2)$

The results of this section show how we can emulate some operators using other operators so as to guarantee exact learning under certain assumptions. The condition $L^{-1}(L(D)) = D$ is quite strong and is seldom met in practice. The next section explores the design of exact incremental and distributed learning algorithms under weaker assumptions. A more complete characterization of the necessary and sufficient conditions for exact or approximate distributed and incremental learning is a subject of our ongoing research.

7   Designing Exact Learning Agents

One approach to devising distributed or incremental algorithms based on an existing batch or centralized learning algorithm is by identifying the information requirements of the learner and designing efficient means of providing the necessary information to it in the distributed or incremental setting. This decomposition of the learning task into *information extraction* and *hypothesis generation* phases offers a general approach to adapting some of the existing learning algorithms to work in the distributed setting. The *hypothesis generation* component of the algorithm can be thought of as the *control* part of the algorithm, which triggers the execution of the *information extraction* part as needed. The execution of the two parts is typically interleaved in time. In this model of distributed learning, only the *information extraction* component has to effectively cope with the distributed nature of the data.

We illustrate this approach to design some incremental and distributed algorithms based on existing batch algorithms (e.g., instance based learning, decision tree learning, and support vector machines induction).

7.1   Incremental and Distributed K-Nearest Neighbor Classifiers

The k-nearest neighbor algorithm is an example of an instance based learning algorithm that can be easily transformed into an exact algorithm for learning from horizontally fragmented data in both incremental as well as distributed settings. The learning phase of a k-NN algorithm consists simply in storing the data and the information extraction is done during the classification phase. Thus, in the k-NN case we have incremental/distributed classification as opposed to the most algorithms where we have incremental/distributed learning, but centralized classification. The classification phase in a k-NN algorithm can be separated in two phases L_{extr} for information extraction phase, which returns the k closest neighbors of a given example x, and L_{proc} for information processing phase which will take a majority vote among the k closest neighbors. The representation of the result of L_{extr} part will be the same as the representation of the instances. In this case we can easily define an inverse operator L_{extr}^{-1} as being the identity. Therefore $L_{extr}^{-1}(L_{extr}(D)) = D'$, where D' contains the k closest points to the point x which should be classified. We will get the following sufficient conditions for exact incremental/distributed information extraction:

Incremental case:

$$L_{extr}(D_1 \cup D_2) = L_A(L_{extr}(D_1), D_2) := L_{extr}(L_{extr}^{-1}(L_{extr}(D_1)) \cup D_2). \quad (5)$$

Distributed case:

$$L_{extr}(D_1 \cup D_2) = L_C^2(L_{extr}(D_1), L_{extr}(D_2)) :=$$

$$L_{extr}(L_{extr}^{-1}(L_{extr}(D_1)) \cup L_{extr}^{-1}(L_{extr}(D_2))). \quad (6)$$

We call these properties *u-closure* properties. Suppose we have an algorithm $k\_nn(D, x)$ which compute the k closest neighbors of a new instance x in the set D (in the information extraction phase). The solution given by the k-NN algorithm for the instance x can be written as follows:

$$h(x) = \arg\max_{c \in C} |\{y | y \in k\_nn(x), \; h(y) = c\}|,$$

where C is the set of possible classes. Given n datasets $D_1, \cdots, D_N$ and an instance x to be classified, the information extraction algorithm works as follows: $k\_nn(k\_nn(D_1, x) \cup \cdots \cup k\_nn(D_n, x), x)$. Similarly, incremental information extraction (in the case of two data sets D_1 and D_2) works as follows: $k\_nn(k\_nn(D_1, x) \cup D_2, x)$.

Theorem 4. *The following two equalities hold for arbitrary data sets* $D_1, \cdots, D_n$*, and a new instance* x*:*

$$k\_nn(k\_nn(D_1, x) \cup \cdots \cup k\_nn(D_n, x), x) = k\_nn(D_1 \cup \cdots \cup D_n, x).$$

$$k\_nn(k\_nn(D_1, x) \cup D_2, x) = k\_nn(D_1 \cup D_2, x)$$

These two equalities guarantee the *u-closure* properties, and hence the resulting algorithms are exact.

7.2 Incremental and Distributed Induction of Support Vector Machines

Support Vectors Machines (SVM) [16] have proved to be a successful technique for batch learning. SVM summarizes the data in a very compact form by identifying the set of instances (the so-called *support vectors*) that specify the maximal margin hyperplane separating the two classes. We can devise an algorithm for exact learning of support vector machines from horizontally fragmented data in both incremental as well as distributed settings by decomposing the learning task into two phases: first extract some information about datasets using an algorithm L_{extr} for information extraction, and then process the information to generate a hypothesis using L_{proc} as needed. The information extracted by L_{extr} will have the same representation as the instances in the training set. More specifically, it will select a subset of the training set that is sufficient for exact learning. Taking again L_{extr}^{-1} to be the identity operator ($L_{extr}^{-1}(L_{extr}(D)) = L_{extr}(D)$), we can get similar conditions for incremental and distributed learning. Specifically, we can guarantee exact incremental and distributed learning if the following *u-closure* properties hold:

$$L_{extr}(D_1 \cup D_2) = L_A(L_{extr}(D_1), D_2) = L_{extr}(L_{extr}^{-1}(L_{extr}(D_1)) \cup D_2) \quad (7)$$

in the incremental setting, and

$$L_{extr}(D_1 \cup D_2) = L_C^2(L_{extr}(D_1), L_{extr}(D_2)) =$$

$$L_{extr}(L_{extr}^{-1}(L_{extr}(D_1)) \cup L_{extr}^{-1}(L_{extr}(D_2))) \quad (8)$$

in the distributed setting.

It can be easily seen that support vectors do not satisfy this property but the convex hulls of the instances that belong to the two classes do. This observation leads to exact incremental and distributed learning algorithms [3].

7.3 Distributed Decision Tree Algorithms

An approach to exact learning of decision trees from horizontally and vertically fragmented data is proposed in [12]. When data is horizontally distributed, examples for a particular value of a particular attribute are scattered at different locations. For finding the count of examples for a particular node in tree, all the sites are visited and count is accumulated. These counts are used by a decision tree construction algorithm [11] to find the best attribute among the set of examples being considered.

The formal description of the algorithm in the case of HDD is the following:

$$L_{HDD}(D_1, \cdots, D_n) := I_P(I_E(D_1), \cdots, I_E(D_n)). \quad (9)$$

where I_E is an operator which collects statistics about attributes at different sites and $L_C^n \equiv I_P$ combines the information gained and constructs the decision tree.

In vertically distributed datasets, each example has a unique index associated with it. Subtuples of an example are distributed among different datasets. However, they can be related to each other using their shared index. While constructing a branch of decision tree, the count of examples that satisfy the constraints on attribute values along the branch is found using unique indices to relate the subtuples of an example. To find the best attribute, a pass is made through all data sites to compute the count of examples.

The formal description of the algorithm in the case of VDD is as follows:

$$L_{VDD}(D_1, \cdots, D_n) := I_P(I_E(D_1), \cdots, I_E(D_n)). \quad (10)$$

In this case, I_E collects statistics and finds the best attribute at each location, and $L_C^n \equiv I_P$ uses this information to construct the tree. Specifically the statistics which are collected consist in counts for combinations of attribute values. As shown in [12], this approach yields exact and efficient algorithms for learning decision trees from horizontally, vertically, and both horizontally and vertically fragmented data sets in the distributed setting.

8   Additional Algorithms

This section explores the specification of some additional algorithms that have been proposed in the literature for distributed and incremental learning within the proposed theoretical framework.

8.1   Bagging, Boosting, and Stacked Generalization

Bagging [2] and Stacked Generalization [17] are hypotheses combination techniques used typically to improve the accuracy of a learning algorithm. Bagging works as follows:

$$L_{Bagging}(D) := Majority\_Vote(L_1(Sel(D)), \cdots, L_n(Sel(D))) \qquad (11)$$

where $L_1, \cdots, L_n$ are arbitrary learning algorithms, and the combination is done using a majority vote. It is possible to develop similar characterizations of stacked generalization, boosting, and related techniques. This opens up the possibility of approximate constructing incremental and distributed learning algorithms based on such techniques. The interested reader is refered to [8] for an example.

8.2   Meta-learning

Meta-learning [9] is a technique which combines independent classifiers generated from distributed databases into a single global classifier. The classifiers generated from individual databases are integrated so as to improve the overall predictive accuracy of the combined classifier. More specifically, the individual classifiers are used to generate predictions on a separate validation test. The predictions and the validation test are put together to form a new data set. The final classifier is obtained by a new learning process applied to this dataset. This process can be captured as follows:

$$L_{Meta}(D_1, \cdots, D_n, D_0) := L(L^{-1}(L_1(D_1)|D_0) \cup \cdots \cup L^{-1}(L_n(D_n)|D_0)) \quad (12)$$

where $D_1 \cup \cdots D_n = D$ and D_0 is the validation set and L is a learning operator.

8.3   DAGGER

Another strategy to combine multiple learned models can be found in [4]. The goal of DAGGER is to learn a single comprehensive model from distributed data sets. The key idea is to learn the hypotheses $h_1, \cdots, h_n$ from the datasets $D_1, \cdots, D_n$ and use them to guide sampling of a new set of informative examples from these datasets. Then the learning operator L generates a final model from the union of the informative examples as follows:

$$L_{DAGGER}(D_1, \cdots, D_n) := L(L^{-1}(L_1(D_1)) \cup \cdots \cup L^{-1}(L_n(D_n))), \qquad (13)$$

where the operators used are similar to those in meta-learning, and each hypothesis $h_i = L_i(D_i)$.

8.4   Incremental Tree Induction

Incremental tree induction (ITI) [15] is an algorithm that sequentially restructures a hypothesis in the form of a decision tree as new examples are encountered on the basis of some statistics that are maintained. Examples are processed one at a time. Hence, each new set contains only one example. Let n be the number of datasets (in this case, examples). We can describe this algorithm as follows:

$$L_{ITI}(D_1, \cdots D_n) := L_A(L_A(\cdots (L_A(h_\emptyset, D_1) \cdots, D_{n-1}), D_n), \qquad (14)$$

where L_A is the algorithm which updates the statistics and $h_\emptyset = L(\emptyset)$. This algorithm does not guarantee exact learning in the incremental setting.

9   Summary and Future Research

The results presented in this paper constitute some useful (albeit tentative) first steps toward the development of a theoretical framework for the specification and analysis of a class of learning problems that arise in open-ended, dynamic environments. Such learning tasks arise in many real-world applications involving knowledge acquisition from multiple, distributed, possibly autonomous data and knowledge sources. We have introduced a family of learning operators and illustrated their use to formally describe some existing approaches and to design new distributed and incremental learning algorithms with provable performance guarantees. We have identified some properties of instance and hypothesis representations and learning operators that make exact learning possible in some open-ended, dynamic environments under certain assumptions. Work in progress is aimed at the elucidation of the necessary and sufficient conditions that guarantee the existence of exact or approximate cumulative multi-agent learning systems in general, and different types of incremental and distributed learning agents in particular, in terms of the properties of instance and hypothesis representations and learning operators, communication operators, and knowledge requirements of agents. Also of interest are PAC-style and mistake-bound analysis of incremental and distributed learning in multi-agent learning systems under different assumptions. Long term goals of this research include: design theoretically well-founded multi-agent systems for learning from interaction with open-ended dynamic environments that include multiple data and knowledge sources (including other agents) and application of such multi-agent learning systems to large-scale data-driven knowledge discovery tasks in applications such as bioinformatics.

Acknowledgements. This research was funded in part by grants from the National Science Foundation (9982341, 9972653, 0087152), the John Deere Foundation, and Pioneer Hi-Bred, Inc. to Vasant Honavar.

References

[1] Bhatnagar, R., Srinivasan, S.: Pattern Discovery in Distributed Databases. Proceedings of AAAI, AAAI Press, 1997.

[2] Breiman, L.: Arcing Classifiers. Annals of Statistics, Volume 26, 1998.

[3] Caragea, D., Silvescu, A., Honavar, V.: Agents that Learn from Distributed Dynamic Data Sources. Proceedings of the Workshop on Learning Agents, Agents-00/ECML-00. Barcelona, Spain. June 2000.

[4] Davies, W., Edwards, P.: DAGGER: A New Approach to Combining Multiple Models Learned from Disjoint Subsets. Machine Learning 2000.

[5] Honavar, V., Miller, L., Wong, J.: Distributed Knowledge Networks. Proceedings of the IEEE Conference on Information Technology, Syracuse, NY, 1998.

[6] Kargupta, H., Park, B., Hershberger, D., Johnson, E.: Collective Data Mining: A New Perspective Toward Distributed Data Mining. Advances in Distributed and Parallel Knowledge Discovery, Eds: Hillol Kargupta and Philip Chan. AAAI Press. 2000.

[7] Littlestone, N.: The weighted majority algorithm. Information and Computation, 108:212-261, 1994.

[8] Polikar, R., Udpa, L., Udpa, S., Honavar, V.: Learn++: An Incremental Learning Algorithm for Multilayer Perceptron Networks. Proceedings of the IEEE Conference on Acoustics, Speech, and Signal Processing (ICASSP) 2000. Istanbul, Turkey. In press.

[9] Prodromidis, A.L., Chan, P.K.: Meta-learning in distributed data mining systems: Issues and Approaches. Book on Advances of Distributed Data Mining, editors Hillol Kargupta and Philip Chan, AAAI press, 2000.

[10] Provost, F., Hennessy, D.: Scaling Up: Distributed Machine Learning with Cooperation. Proceedings of the Fourteenth National Conference on Artificial Intelligence, 1996.

[11] Quinlan, J.R.: Induction of Decision Trees. Machine Learning, vol. 1, pp 81-106, 1986.

[12] Sharma, T., Silvescu, A., Andorf, C., Caragea, D., Honavar, V.: Algorithms for Learning from Distributed Data Sets. Tech. Rep. ISU-CS-TR 2000-10. Department of Computer Science, Iowa State University, Ames, IA, May 2000.

[13] Srivastava, A., Han, E.H., Kumar, V., Singh V.: Parallel Formulations of Decision-Tree Classification Algorithms. Data Mining and Knowledge Discovery: An International Journal, vol. 3, no. 3, pp 237-261, September 1999.

[14] Thrun, S., Faloutsos, C., Mitchell, M., Wasserman, L.: Automated Learning and Discovery: State-of-the-art and research topics in a rapidly growing field. AI Magazine, 1999.

[15] Utgoff, P.E., Berkman, N.C., Clouse, J.A.: Decision Tree Induction Based on Efficient Tree Restructuring. Machine Learning, 1997.

[16] Vapnik, V.: Statistical Learning Theory. Springer-Verlag, New York, 1998.

[17] Wolpert, D.H.: Stacked Generalization. Neural Networks, 5:241-259, 1992.

Connectionist Neuroimaging

Stephen José Hanson, Michiro Negishi, and Catherine Hanson[1]

Psychology Department
Rutgers University
Newark N.J. USA

Abstract. Connectionist modeling and neuroscience have little common ground or mutual influence. Despite impressive algorithms and analysis within connectionism and neural networks, there has been little influence on neuroscience, which remains primarily an empirical science. This chapter advocates two strategies to increase the interaction between neuroscience and neural networks: (1) focus on emergent properties in neural networks that are apparently "cognitive", (2) take neuroimaging data seriously and develop neural models of dynamics in the both spatial and temporal dimensions.

1 Introduction

In 1990 then President of the USA, George H W Bush, declared the "Decade of the Brain". This year the "Decade of the Brain" ended (although perhaps George W Bush, the son, will declare yet another decade), it is worth asking what happened during the Decade of the Brain and in particular what was the influence of neural computation on neuroscience. How did neuroscience help define or delineate aspects of neural computation during this last decade?

Neural Networks have become mainstream engineering tools (IEEE, 1998) helped stir a resurgence of statistical methods (esp. Bayesian methods). They have been incorporated into a large and diverse application base from medical to automotive and control applications. And Hollywood continues to believe "intelligence" is some property of a neural network. On the other hand, paradoxically, Neural Networks have had little or no effect on the larger mainstream neuroscience community or field. It is clear over the last decade, despite an increasing sophistication and development of neural network algorithms, that little has changed in the neuroscience field with respect to computation or the representational issues concerning neural tissue. Neuroscientists continue to focus on cell level mechanisms and generic properties of system level interaction.

Most people blame neural networks for this lack of impact on neuroscience and biological considerations of neural networks. Four reasons are often cited:

- They are not biologically plausible
- They do not scale well with large problems
- They are not new---just statistics
- They don't process symbols and humans do

---

[1] Also at Telcordia Technologies, Piscataway, New Jersey.

S. Wermter et al. (Eds.): Emergent Neural Computational Architectures, LNAI 2036, pp. 560-576, 2001.

I blame neuroscience. I see three reasons for this lack of connection:

- For nearly 100 years neuroscience has been essentially an empirical enterprise, one that has not easily embraced common abstract principles underlying common behavioral/physiological observations ("splitters" as opposed to "lumpers").
- System Neuroscience, which should have the greatest impact on computational approaches, in general is the most difficult level in which to obtain requisite data to constrain network models or provide common principles due to potential complexity of the multiple Neuron-Body problem.
- Most serious, is the level of analysis that Neuroscientists tend to cling to: the cellular level (or god help us the molecular level). This focus is notwithstanding the lack of fundamental identification of this anatomically distinct structure as also a unique unit of computation. It can be easily shown that computational regularity at the behavioral level does not force unique implementations at the neural level.

We suggest there are two strategies to encourage more connections between neuroscience and neural networks.

- One: Attempt to show emergent behavior that is more is similar to human COGNITIVE performance: Analyze the network representations to understand the nature of the interactions between learning and representations.
- Two: take neuroimaging data seriously and model it with neural networks (that embody dynamical systems), for example, as opposed to doing inferential statistics-- Neuroimaging data could be seen as spatio-temporal multivariate data, representing some time dynamical system distributed through a 3-d volume.

In the end we shall suggest it is also productive to look for ways to combine cognitively suggestive models with the data rich methods of neuroimaging such as EEG and fMRI.

2 Network Emergent Behavior: The Case of Symbol Learning

We argue it is useful to demonstrate Emergent behaviors in Networks that were
- not programmed,
- engineered or
- previously constrained
- by choice of architecture,
- learning rule or
- distributional properties of the data.

It is known that Recurrent Neural Networks can induce regular grammars from exposure to valid strings drawn from the grammar. However, it has been claimed that neural networks cannot learn symbols independent of rules (see Pinker).

A basic puzzle in the cognitive neurosciences (30) is how simple associationist learning which has been proposed to exist at the cellular and synaptic levels of a brain can be used to construct known properties of cognition which appear to require abstract reference, variable binding, and symbols. The ability of humans to parse sentences and to abstract knowledge from specific examples appears to be inconsistent with local associationist algorithms for knowledge representation (3, 8, 16, 20, 21, 22 but see 11). Part of the puzzle is how neuron-like elements could from simple signal processing properties emulate symbol-like behavior. Properties of symbols include the following (14).

> a set of arbitrary "physical tokens" scratches on paper, holes on a tape, events in a digital computer, manipulated on the basis of "explicit rules" that are   likewise physical tokens and strings of tokens. The rule-governed symbol-token manipulation is based purely on the shape of the symbol tokens (not their "meaning"), i.e., it is purely syntactic, and consists of "rulefully combining" and recombining symbol tokens. There are primitive atomic symbol tokens and composite symbol-token strings. The entire system and all its parts -- the atomic tokens, the composite tokens, the syntactic manipulations both actual and possible and the rules -- are all "semantically interpretable:" The syntax can be systematically assigned a meaning e.g., as standing for objects, as describing states of affairs).

As this definition implies a key element in the acquisition of symbolic structure involves a type of independence from the task the symbols are found in and the vocabulary they represent. Fundamental to this type of independence is the ability of the learning system to *factor* the generic nature of the task or rules it confronts with from the aspect of the symbols or vocabulary set which are arbitrarily bound to the input description or external referents of the task. In this report we describe a series of experiments with an associationist neural network that creates abstract structure that is context sensitive, hierarchical, and extensible.

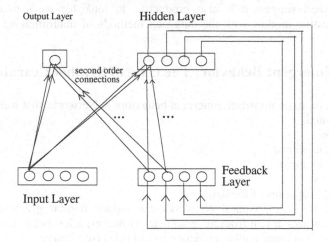

Fig. 1. The Recurrent Network Architecture used in the simulations. This is a simple neural network learning architecture that possesses a simple memory. All weights are subject to adaptation or learning, there are no fixed structures in the RNN prior or during learning.

Consider the simple problem of learning a grammar from valid, or positive set only sentences consisting of strings of symbols drawn randomly from an infinite population of such valid strings. This sort of learning might very well underlie the acquisition of language in children from exposure to grammatically correct sentences during normal discourse with their language community[2]. It is well known now that neural networks[3] can induce the structure of the FSM (Finite State Machine; for example see Fig. 2) only from presentation of strings drawn from the FSM (9, 25). In fact, recently (2) it has been shown that the underlying attractors of neural networks have no choice but to be in a one to one correspondence with the states of the state machine from which the strings are sampled. This surprising theorem is the precedence for proposing that a neural network embodying the underlying rules of a state machine in its attractor space could also learn to "factor" the input encoding or external symbols. In the present report, we employ a Recurrent Neural Network (RNN, see Figure 2) with a standard learning algorithm developed by Williams and Zipser, extended to second order connections (10). The network was trained with newly generated sentences until performance of the network met a learning criterion[4]. Input sentences were limited to 20 symbols long and were constructed from local binary encoding of each symbol.[5] All weights in the RNN were candidates for adaptation, no structures were fixed prior to or during learning.

Humans are known to gain a memorial advantage from exposure to strings drawn from a FSM over ones that would be constructed randomly (17, 18, 23, 24) as though they are extracting abstract knowledge of the grammar itself from exposure to strings drawn randomly from the FSM. A more stringent test of knowledge of a grammar

---

[2] Although controversial, language acquisition must surely involve the exposure of children to valid sentences in their language. Chomsky (3) and other linguists have stressed the importance of the a prior embodiment of the possible grammars in some form more generic than the exact target grammar. Although not the main point of the present report, it must surely be true that of the distribution of possible grammars, some learning bias must exist that helps guide the acquisition and selection of one grammar over another in the presence of data. What the nature of this learning bias is might be a more profitable avenue of research in language acquisition than the recent polarizations inherent in the nativist/empiricist dichotomy. (5, 16, 20, 22).

[3] Neural networks consist of simple analogue computing elements that operate in parallel over an input and output field. Recurrent Neural Networks are networks that have recurrent connections to their intermediate or "hidden" layers. Such recurrent connections implement a local memory of recent input/output and processing states of the network. Feedforward networks have only unidirectional connections and hence no mechanism for examining past inputs.

[4] That is, after each training sentence, the network was tested with randomly generated 1000 sentences and the training session was completed only when the network yielded below the low-threshold output node activity when sentences could not end and above high-threshold activity when they could end. Thresholds were initialized to 0.20 (high value threshold) and 0.17 (low value threshold) and were adapted using output values during the network was processing test sentences. The high threshold was modified to the minimum value yielded at the end of test sentences minus a margin (0.01) and the low threshold was modified to the high threshold minus another margin (0.02) during the test, and these thresholds were used for the next test and training sentences.

[5] Each word was represented as an activation value of 1.0 of a unique node in the input layer, while all other node activations were set to 0.0. The task for the network was to predict if the next word was END (in which case the output layer node activation was trained to become 1.0) or not (output should be 0.0). Note that when the FSM is at the end state, a sentence can either end or continue. Therefore at this state, the network is sometimes taught to predict the end a sentence and sometimes not. However the network eventually learns to yield higher output node activation when the sentence can end.

would be to expose the subjects to a FSM with one external symbol set and to see if the subjects transfer knowledge to a novel external symbol set. In principle in this type of task, it is impossible for the subjects to use the symbol set as a basis for generalization without noting the patterns that are commensurate with the properties of the FSM[6]. A version of this type of transfer is shown in Fig. 3. In this task new symbols are assigned randomly to the arcs, such that the external symbols are completely new. This task, which we call the Vocabulary Transfer Task, was used in the first simulation to train recurrent neural networks and to examine their ability to transfer over novel, unknown symbol sets.

Fig. 2. The SYMBOL transfer task. The figure shows two finite state machine representations, each of which has 3 states (1, 2, 3) with transitions indicated by arrows and legal transition symbols (A, B, ... , F) for each state. Note that this task involves no possible generalization from the transition symbol. Rather, all that is available are the state configuration geometries. The task explicitly forces the network to process the symbol set independently from the transition rule.

In this task, the network was trained on three regular grammars (the source grammar) which have the same syntactic structure (Fig. 2) defined on three unique sets of words and the effect of these prior training's to the training of the target grammar was measured as the network was trained with yet another new set of words. One of the indicators of such effect is the savings in terms of number of trials needed to meet the completion criterion. Fig. 3 shows the number of trials for both the source grammar trainings (vocabulary switching = 1, 2, ... , 9 in the figure) and the target grammar training (vocabulary switching = 10), averaged on 20 networks with different initial random weights. The result of vocabulary switching is in the first 9 cycles is a complete accommodation of the new symbol sets with near 100% savings. This accommodation represents the networks ability to create a data structure that is consistent with a number of independent vocabularies. More critically however there was 63% reduction in the number of required trainings for the new unseen vocabulary. This result is remarkable given the required independence of *syntax* and *vocabulary*. Apparently the RNN is able to partially factor the transition rules from its constituent symbol assignments after exposure to a diversity of vocabularies.

One obvious question that arises is whether the source of the *novel* transfer is due to a network *memory*. Our initial studies in this area showed that in fact local memory is

---

[6] Reber (24) showed that humans would significantly transfer in such a task, however his symbol sets allowed subjects to use similarity as a basis for transfer as they were composed of contiguous letters from the alphabet. However, recent reviews of the literature indicate that this type of transfer is common even across modalities (16).

important. We showed in a series of similar tasks that there was no significant savings in learning for feedforward networks that were exposed to rule learning contexts (e.g., "Penzias Problem") with subsequent permutation transfer[7]. At least from these preliminary studies it would suggest that *memory* in the network is an important component of the ability of a neural network to transfer its syntactic knowledge.

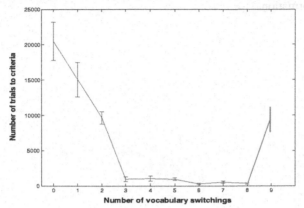

Fig. 3. The learning savings from subsequent relearning of the *symbol transfer* task. Each data point represents the average of 20 networks trained to criterion on the same grammar. The relearning cycle shows an immediate transfer to the novel symbol set which continues to improve to near perfect transfer through the ninth cycle (3 cycles of the 3 symbols sets) until the 10th cycle where a completely *novel* set is used with the same grammar. Over 60% of the original learning on the grammar independent of symbol set is saved.

How is the neural network accomplishing these abstractions? Note that in the vocabulary transfer task the Network as with a human subject has no possible way to transfer based on the external symbol set. It follows that the network must abstract away from the input encoding. In effect the network must find a way to buffer or recode the input in order to defer symbol binding until enough string sequences have been observed. If the network extracted the common syntactic structure, the hidden layer activities would be expected to represent the corresponding FSM states, regardless of vocabularies. This is, in fact, shown by linear discriminant analysis (LDA)[8]. After the network learned the first vocabulary, activity of the node was shown to be sensitive to FSM states (Fig. 4A). In this figure, different FSM states are

---

[7] Feedforward networks were trained on the Penzias task, a boolean counting task studied previously by Denker et al (4). A permutation task was defined which was similar to the vocabulary transfer task, but the feed-forward network showed only interference effects even with significant increases in the capacity of the network.

[8] LDA of the hidden unit states allows for a complete search of linear projections that are optimally consistent with organizations based on *state* from the FSM or by *vocabulary*. LDA was applied to the hidden unit activations over 20 Networks to find some a stable result for the preferred encoding of the input space. Evidence from the LDA for *state* representations would indicate that the RNN found a solution to the multiple vocabularies by referencing them hierarchically within each *state* based on context sensitivity within each vocabulary cluster.

represented by different clusters, while the different symbol sets are plotted with different graphic symbols. Note that these clusters represent attractors for the states in the FSM. Moreover if one starts a trajectory nearby one of the clusters it proceeds to a location nearby the cluster representing the appropriate state transition. Hence this space possesses context sensitivity in that coordinate positions encode both *state* and *trajectory* information.

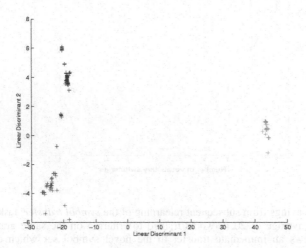

Fig. 4A. Linear Discriminant Analysis of hidden activities of networks that learned a single FSM/symbol set. Note that the different clusters represent different states while the "+" sign codes for the single symbol set.

After each of the three vocabularies were learned in three cycles, LDA of the hidden layer node activities with respect to FSM states (Fig. 4B) was contrasted with that with respect to vocabulary sets (Fig. 4C). The correct rate of discrimination clearly shows that the state space is organized by FSM states, since FSM states could be correctly classified by the former linear discriminants with the accuracy of 80% (SD=16, n=20) whereas vocabulary set could be classified correctly for only 45% (SD=9.7, n=20). Notice in both figures 4B and 4C relative to 4A that the symbol sets have spread out and occupy more of the hidden unit space with significant gaps between clusters of the same and different symbol sets. Moreover from Fig. 4B, one can also see that the vocabularies are hierarchically organized into states corresponding to the FSM. This hierarchical structure provides a super-structure for the accommodation of the already learned vocabularies and any new ones the RNN is asked to learn. It also can be seen from Fig. 4C that the hidden layer activations are also sensitive to , but not linearly separable by, vocabularies.

LDA after the test vocabulary is learned once also shows that the network state is predominantly organized by FSM states (Fig. 5), although the linear separation by FSM states of a small fraction of activities are compromised. This interference by the new vocabulary is not surprising considering that old vocabularies were not re-learned after the new vocabulary was learned. What is more interesting is the spatial location

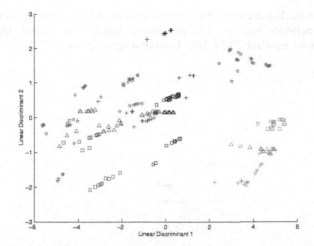

Fig. 4B. Linear discriminant analysis of the hidden units of the RNN that have learned three independent FSMs with three different symbol sets using state as the discriminant variable. Notice how the hidden unit space spreads out compared to Fig. 4A. Notice further that the space is organized by clusters corresponding to states which are internally differentiated by symbol sets (represented by different graphic symbols: +=ABC, triangle=DEF, square=GHI).

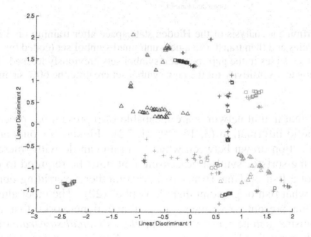

Fig. 4C. Linear Discriminant Analysis of hidden unit activities from networks trained on three independent FSM and three different symbol sets.   The LDA used the symbol set as the discriminant. Note the symbol sets are coded by 1 of 3 graphic codes same as in Fig. 4A and 5B.   In this case note that the discriminant function based on symbol set produces no simple spatial classification as in Fig. 4B which shows the same activations classified by the *state* of the FSM.

of the new vocabulary ("stars"). The hidden unit activity, again, clearly shows that *state* discriminant structure is dominant and organizes the symbol sets, the *fourth* vocabulary or unseen symbol set that the networks are exposed to, simply finds *empty* spots in the hidden units space to code it location relative to the existing state structure, hence indicating the strength of the present abstraction to encourage observance of the hierarchical *state* representation and its existing context sensitivity. In effect the network *bootstraps* from existing nearby vocabularies that would allow generalization to the same FSM that all symbol sets are using.

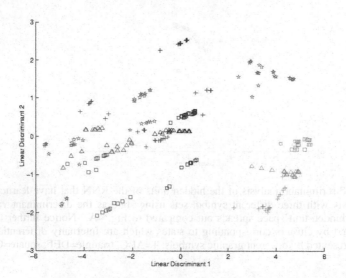

Fig. 5. Linear Discriminant analysis of the Hidden state space after training on 3 independent symbol sets for 3 cycles and then transfer to a new untrained symbol set (coded by *stars*). Note how the new symbol set slides in the gaps between symbol sets previously learned. Presumably this provides initial context sensitivity for the new symbol set creating the 60% savings.

It has been argued that neural networks are incapable of representing, processing, and generalizing symbolic information (3, 16, 20, 21, 22). Pinker, for one, argues there must be some distinction drawn between what the brain can do with "mere statistical information" and the sorts of symbol processing that must be required to understand that a "whale is not a fish or Tina Turner is a grandmother, overriding our statistical information about what fish or grandmothers look like" (20). The other alternative, as demonstrated by the present experiments is that neural networks that incorporate associative mechanisms can be sensitive to both *the statistical substrate* of the world and create data structures that have the property of following *a deterministic rule* and once learned can be used to override even large amounts of statistical evidence (e.g. from another FSM). Quite conveniently as demonstrated, these data structures can arise even when that rule has only been expressed implicitly by examples and learned by mere exposure to regularities in the data.

The next section focuses on Neuroimaging techniques and discusses some of the problems and some of the promise. This will help develop the idea of combining RNN for extraction of FSM properties of real-time cognition.

3   Taking Neuroimaging Seriously: New Tools for Neuroimaging Is Neuroimaging Just the 21st Century Phrenology?

Neuroimaging is an important new technology for he analysis and representation of cognitive processes and the neural tissue that supports them.   Nonetheless, these technologies run the danger of becoming the new Phrenology. Unfortunately, when it comes to studying cognitive processes, data analytic techniques commonly employed in neuroimaging limit and distort the hypotheses researchers can consider.

For example, several statistical problems exist within the assumptions of these analysis:

- Independence of voxels in space and time; clearly there is dependence structure in both time and space.
- Gaussian assumption unlikely to hold; hence stat tests will be inefficient and miss structure in low S/N environments.
- Contrastive testing is subject to linearity and minimal components assumptions
- Modularity metaphor --looking for local focal areas of computation--face and teapot areas!

Statistical thresholds that are chosen are enormously high, implying that either the underlying distribution is non-guassian or that locality of signal is preferred (10^{-5}, 10^{-20})!

We have looked at a number of these assumptions including recently showing that BOLD susceptibility in brain is non-gaussian (Hanson & Bly, (32)).   Recently we have looked specifically at the notion that the fMRI time series has a specific temporal dependence   (Murthy, Lange, Bly & Hanson (33)). We used two kinds of simple sensory tasks. The first task was a finger tapping task in a box car paradigm with 4 seconds of tapping and 4 seconds of no finger tapping. A GE 1.5T scanner was used to collect the data from a single subject. We also used an auditory task where a single tone was presented again to a single subject also in a boxcar presentation. Autoregressive time series models were used to model every voxel in the brain over time steps and the goodness of fit was collected. All goodness of fit above a criterion value (>95%) was re-projected back into the brain slice without further thresholding. In the figure 6 below we show two slices, one showing a standard SPM99 analysis of the boxcar for the finger tapping task (leftmost graphic) and the AR analysis showing the projection of an AR (3) model (rightmost graphic) which required no reference to the contrastive box car design, rather it created sensitivity in the brain slice due only to the temporal structure of the fMRI signal itself.

We have also used the time delay coefficients derived from the AR analysis which would indicate the sensitivity of the indicated tissue to the time dependence. Recently we have also shown that the time series is generally not stationary especially in areas that seem "activated".   In these cases and perhaps more generally it would be appropriate to consider ARIMA style models which can explicitly model nonstationarity.   We note in passing that a more general ARIMA model is in fact a Recurrent Neural Network which as we have shown previously has the useful property of extracting a unknown FSM from a time series.

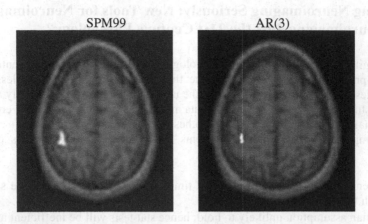

Fig. 6. Brain slices showing the "active" areas of the brain during a finger tapping task. Rightmost graphic shows a standard SPM99 analysis, while the leftmost graphic shows the AR model.

We have also used the time delay coefficients derived from the AR analysis which would indicate the sensitivity of the indicated tissue to the time dependence. Recently we have also shown that the time series is generally not stationary especially in areas that seem "activated". In these cases and perhaps more generally it would be appropriate to consider ARIMA style models which can explicitly model nonstationarity. We note in passing that a more general ARIMA model is in fact a Recurrent Neural Network which as we have shown previously has the useful property of extracting a unknown FSM from a time series.

4   Dynamics of Cognition: The Case of Event Perception and Signal Fusion (EEG, fMRI)

4.1 Event Perception: Perceiving and Encoding Events

Day to day experience is characterized, remembered, and communicated as a series of events. We think about driving to work, we remember having an argument with our spouse, and we tell a friend about our plans to attend the theatre next Saturday. Abbreviated phrases such as driving to work act as a type of shorthand notation for describing complex action sequences. Thus, our ability to communicate successfully with others using such labels as driving to work reflects a certain level of familiarity with the referenced activities that we share or presume to share with our intended audience.

How common is our knowledge about common events? Empirical work suggests that there is considerable consensus concerning the constituent actions of familiar events. For example, Bower, Black, and Turner (1) asked subjects to describe the typical actions involved in going to a restaurant, attending a lecture, getting up, grocery shopping, and visiting a doctor. They found that subjects showed considerable agreement about the composition of common events, many responses being offered by more than 70% of their subjects and very few being unique.

Familiarity with events may provide the basis for understanding and encoding new information. In a second experiment of their 1979 study, Bower et al. (1) asked subjects to parse prose stories centered on events such as visiting a doctor into smaller "parts." They found that subjects tended to choose similar points in the story as constituent boundaries. Agreement about event boundaries extends to online measures of parsing as well (e.g., Newtson (19); Hanson & Hirst (12)). For example, Hanson & Hirst (12) asked subjects to indicate the boundaries of events while viewing videotapes of common activities (e.g., playing a game) under various orientation instructions and found that subjects had little difficulty agreeing about the boundaries of such events.

4.2 Recurrent Nets and Schemata

Neisser's Perceptual Cycle

Neisser has suggested that perception is a cyclical activity in which
(1) memory in the form of schemata guides the exploration of the environment,
(2) exploration yields samples of available information, and
(3) data collected from the exploration process modifies the prevailing schema.
According to Neisser:
"The schema assures the continuity of perception over time in two different ways. Because schemata are anticipations, they are the medium by which the past affects the future; information already acquired determines what will be picked up next. (This is the underlying mechanism of memory, though that term is best restricted to cases in which time and a change of situation intervene between the formation of the schema and its use.) In addition, however, some schemata are temporal in their very nature. "One can anticipate temporal patterns as well as spatial ones. (p. 22-23)."

By focusing on the interaction of perception and memory, Neisser's "perceptual cycle" model offers a particularly fertile context for studying the processing of event information. However, because this is a processing model rather than a model of knowledge representation little emphasis is placed on the structure of schematized knowledge. Thus, it is not clear how turning ignition might be related to driving home or even what role the decomposition of events might play in generating the expectations purportedly used to guide sampling of available information. Germane to this issue is another that arises in relation to the proposed modification process. How does the prevailing schema change in response to the sampling process? In particular, what is the basis for the similarity between the ongoing situation and the schemata that are subsequently activated?

These questions lead to computational considerations in how one might implement a system which can represent schemata, their similarity and their dynamic properties in the presence of the ongoing stimulus situation. We discuss two possibilities, the first which has a classical status in the event-perception literature. This hypothesis concerning event structure, was first introduced by Schank & Ableson (28) and by Minsky, and is referred to by either "Scripts" or "Frames". This hypothesis basically asserts that the world can be clustered into simple categories that predict and organize the stimulus situations according to their goals and expectancies.

The second approach is a competing account which we will introduce here for the first time in the context of a Connectionist hypothesis concerning temporal processing of information. Temporal control in connectionist networks was introduced by Jordan (15, also see 26,27), as well as Elman (6,7) who first discussed the relationship between introducing recurrence into connectionist networks and temporal processing. Both Jordan and Elman were interested in the relationship between psychological constraints arising from phenomena involving recognition or production of serial order. Jordan's models focused on Lashley's general challenge to associationism which requires the addition of a memory to the connectionist network allowing context sensitive behavioral sequences. Similarly, Elman introduced a memory to the standard connectionist network in order to recognize simple grammars.

4.3 Categories and Temporal Control

In the present paper we introduce and focus on another important property of Recurrent connectionist networks, that of the evolution and dynamic control over the adjunct memory and the construction of recognition categories ("schemata") from the stimulus situation. The difference in our focus to the Jordan-Elman focus underscores our interest in the interaction between perception and memory. Our experiments and simulations our meant to elucidate the types and nature of interactions between a "memory" in the neural network and its dependence on evolving schemata (hidden layer) and present features of the situation. There is actually a closer relation of the present work to the general framework of Arbib who has used the term Schema as we use it here in this constructive sense as well as Hanson & Burr (11), who have emphasized the importance of representational and learning interactions in connectionist networks.

22 subjects asked to provide judgments of event change while watching a video tape of actors engaging in everyday events (eating in a restaurant, driving a car, working in an office.). One videotape showed two people playing a game of Monopoly and the other showed a woman in a restaurant drinking coffee and reads a newspaper. Subjects watched the videotapes under various orientations and pressed a response button whenever they believed a new event was beginning. In the present study, we used responses made when subjects had been oriented toward "small" events while viewing the tapes. This orientation produced the greatest number of perceived event boundaries.

There was high agreement between the 22 subjects on event boundaries and we chose cases where there was at least 75% agreement on the location of the event-second for an event boundary which produced 15 event-boundaries for averaging. Both EEG and fMRI were recorded simultaneously from a single subject making event boundaries judgments in the Restaurant tape, which was scripted each second with an actor who was drinking coffee and reading a newspaper in a restaurant and had been used previously in the behavioral study. ERP and fMRI were averaged before and after the 15 event boundaries and are time aligned (fMRI was sampled every 4 seconds and ERPs were taken every 1MS and then down-sampled for every second) and shown below 20 seconds before and 8 seconds after an event change.

The ERP shows positive active areas in visual areas and temporal lobe prior to event change evolving to a large negative wave starting from prefrontal areas and moving back towards temporal and visual areas. Simultaneous with EEG, we recorded from the same subject fMRI, this is also shown in figure 7, again time synchronized with the ERP averages. In the fMRI case, an arbitrary baseline was created from the first 4 scans as an average and then contrasted against the other 41 scans to produce a standard t-map. These t-maps were then treated as the ERPs and averaged within a window before and after an event change. Prior to an event change there is a large amount of distributed activity (these are at t values where p<.1 --low threshold case) early on, which "thins out" prior to an event change and seems also to lateralize towards the right hemisphere after an event change boundary.

Areas that were significantly active in what we termed the low threshold case (t values where p < 0.1) produced 10-12 areas which are shown in figure 8. These areas include Dorsal Lateral prefrontal cortex, anterior cingulate, precuneous and temporal lobe areas similar to what was seen in the ERP coherence maps. Also shown in figure 8 is a "meta analysis" of labels taken from the literature where these areas have been implicated in different tasks. Shown are at least 3 labels that have been used to characterize these areas from different studies or tasks in the neuroimaging literature.

EVENT BOUNDARY

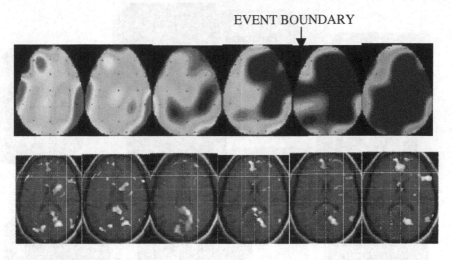

Fig. 7. ERP and fMRI showing concurrently collected and averaged activity 20 seconds before and 8 seconds after 15 judged event changes.

It is clear from the putative functions of these areas that it would be possible to characterize the event boundary judgments as a dynamical system that includes attentional , spatial orienting and detection functions (anterior cingulate, anterior parietal), which includes areas that have been implicated in so-called "attentional networks" (e.g. Posner). Such an event detection system would also necessarily seem to include planning for schema processing and some sort of short term buffer for comparison to well known schema that would be involved in memory and or language comprehension (temporal lobe). Finally, in terms of comparison and schema processing, functions such as imagery and spatial modeling (parietal lobule, precuneus) would be critical to such a dynamic schema detection circuit.

The most highly significant areas as indexed by the t-value was Anterior Cingulate and Dorsal Lateral Prefrontal Cortex. The subject also responded with a thumb movement to indicate a behavioral response to an event change which was most likely associated with SMA and primary motor cortex.

This potential circuit would not be apparent from the normative neuroimaging method which stresses single area functions rather than interactivity, distributed processing and systemic function.   Key to the proposal in this chapter is the concept that neuroimaging data is subject to dynamical systems analysis and in particular the type of sequential structure inherent in simple FSM as discussed earlier in this chapter. Without this kind of view, neuroimaging data are far too limited to reveal the likely complexity of commerce between brain areas.   Neural networks have considerable power and generalization scope that could be useful in analyzing and modeling neuroimaging time series, in particular we end on a proposal to fuse time rich signals (such as EEG) and space rich signals (such as fMRI) in order to create spatio-temporal signals that commensurate and sensitive to cognitive interactivity and system level brain function.

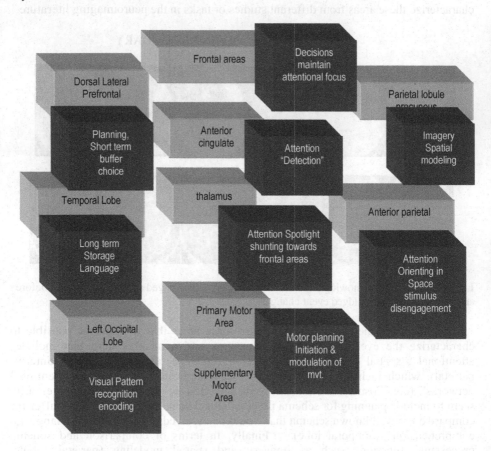

Fig. 8. SPM99 analysis of the event perception task. Using low threshold values.

4.4 Fusing EEG and fMRI for "Real Time" Cognitive Measurement

Because event perception tasks are productive in a cognitive sense and evokes lots of COGNITION at once: Perception, memory, sequence, grammar etc. Processes, system level and real-time processing are informative and potentially revealing. We claim it is important to understand the SYSTEM of brain areas that support COGNITIVE function as and interactive, distributed, time, dynamic structured neural network.   Specifically it is well known that the reconstruction problem using ERPs is intractable due to the lack of constraints on the position and the number of dipoles giving rise to scalp voltage potentials (ill-posed). We would constrain the ERP inverse equations with fMRI number and locations in a low threshold approach as described earlier.   Next one would solve the equations and iterate at that sampling rate of the fMRI image acquisition. Use the now solved for dipoles for initiation of the location estimation procedures from before (for example, in the kernel Gaussian method, the mean might be placed at the now identified dipole) and reestimate the locations of neural activity. Now use these new estimates to reseed the ERP inverse and iterate.   This kind of approach would allow the location stable measures in FMREI to be interpolated with the ERP estimators between fMRI sample points. In effect the fMRI would be augmented with a millisecond estimation of position between every sampled image. Further imposing of temporal regularizer would ensure that the ERP-fMRI estimator would remain smooth and stable between fMRI samples. The successful application of this new method would constitute the first demonstration of real time brain imaging.

5   Some Conclusions: CONNECTIONIST Neuroimaging

We need to look for emergent properties of networks that might guide the measurement for neuroscience.   Similar to the Grammar transfer task, the kinds of computations found were *METRIC* and similarity based. Neuroimaging Data can help constrain our modeling and provide us insights in to the complex spatio-temporal dynamical system of the brain.

References

1.   Bower, G.H., Black, J.B., & Turner, T.J. (1979). Scripts in memory for text. Cognitive Psychology, 11, 177-220.
2.   Casey, M, The dynamics of discrete-time computation with applications to recurrent neural networks and finite state machine extraction, Neural Computation, 8, 6 pp. 1135-1178, (1996).
3.   Chomsky. N, *Syntactic structures.* Mouton, The Hague (1957).
4.   Denker, D. Schwartz, B. Wittner, S. Solla, R. Howard, L. Jackel, J. Hopfield, Automatic learning, rule extraction and generalization, *Complex Systems,* 1 (5), 877-922 (1987).
5.   Elman, J.L., E. Bates, M. Johnson, A. Karmiloff-Smith, D. Parisi, K. Plunkett, *Rethinking Innatenes* (MIT Cambridge, 1996).
6.   Elman, J.L., Finding Structures in Time, *Cognitive Science,* 14 (1990*).*
7.   Elman, J.L. (1988). Finding structure in time. CRL Technical Report8801. Center for Research in Language, UCSD.

8.   Fodor, Z. Pylyshn, Connectionism and Cognitive Architecture: A critical analysis. In Pinker & Mehler (Eds.), *Connections and Symbols*, (MIT Cambridge, 1988).
9.   Giles, B. G. Horne, T. Lin, Learning a class of large finite state machines with a recurrent neural network, *Neural Networks,* **8,** (9), pp. 1359-1365 (1995).
10. Giles, L., Miller, C. B., Chen D. Chen, H. H. Sun G. Z., Lee, Y.C. (1992). Learning and Extracting Finite State Automata with Second-Order Recurrent Neural Networks, Neural Computation, (In Press)
11. Hanson S. & D. Burr What connectionist model learn: learning and representation in connectionist models, *Behavioral Brain Models,* **13,** (3), 471 (1990).
12. Hanson, C. & Hirst, W. (1989). On the representation of events: A study of orientation, recall, and recognition, Journal of Experimental Psychology: General, 118, pp. 124-150.
13. Hanson, S.J. & Burr, D. J. (1990). What Connectionist Models Learn: Learning and Representation in Connectionist Networks. Behavioral and Brain Sciences, 13, 3 pp. 477-518.
14. Harnad, S. The Symbol Grounding Problem , *Physica D* **42**: 335-346, (1990).
15. Jordan, *Serial Order: A parallel distributed processing approach,* ICS Technical Report (UCSD, 1986).
16. Marcus G., S. Viyayan, P. Bandi Rao, M. Vishton, *Science.* **283,** (1999).
17. Medin, D.L., & Schaffer, M.M. (1978). Context theory of classification learning. Psychol. Review, 85, 207-238.
18. Miller, G.A. and M. Stein, Grammarama I: Preliminary studies and analysis of protocols. Technical Report No. CS-2, Cambridge: Harvard University, CCS (1963).
19. Newtson, D. (1973). Attribution and the unit of perception of ongoing behavior. Journal of Personality and Social Psychology, 28, 28-38.
20. Pinker, S. Enhanced: Out of the Minds of Babes, *Science,* **283**, (1), 40-41, (1999).
21. Pinker, S. *How the Mind works.* (W.W. Norton & Co. 1997).
22. Pinker, S. *The Language Instinct* (Morrow & Co. 1994).
23. Reber, A. Implicit learning of artificial grammars. *Journal of Verbal Learning and Verbal Behavior,* **6,** 855-863 (1967).
24. Reber, A. Transfer of syntactic structure in synthetic languages. *Journal of Experimental Psychology,* **81,** 115-119, (1969).
25. Redington, J and N. Chater, Transfer in Artificial Grammar Learning: A reevaluation, *J. Exp. Psych: General,* **125:** 2, pp. 123-138. (1996).
26. Rumelhart, D., G. Hinton, and R. J. Williams, Learning representations by back-propagating errors, *Nature,* **323,** 9, (1986).
27. Rumelhart, D., Hinton, G. & Williams, R. (1986). Learning internal representations by error propagation. In D.E. Rumelhart D. and J.L. McClelland (Eds.) Parallel Distributed Processing I: Foundations. Cambridge, Mass: MIT Press.
28. Schank, R.C. (1982).Dynamic Memory: A theory of reminding and learning in computers and people. Cambridge: Cambridge University Press
29. Servan-Schreiber D., Cleeremans, A. & McClelland, J. (1988). Encoding sequential structure in simple recurrent networks. CMU Technical Report CS-88-183.
30. Special Issue on Cognitive Neuroscience, *Science,* **275,** 1580-1608 (1997).
31. Watrous, R. & Kuhn G. (1992). Induction of Finite -State Languages Using Second -Order Recurrent Networks, Neural Computation,
32. Hanson & Bly, 2000, The distribution of BOLD susceptibility in the Brain is Non-Gaussian
33. Murthy, Bly & Hanson, 1999, Identification of the fMRI Signal, Cognitive Neuroscience Society.

Author Index

Lecture Notes in Artificial Intelligence (LNAI)

Vol. 1919: M. Ojeda-Aciego, I.P. de Guzman, G. Brewka, L. Moniz Pereira (Eds.), Logics in Artificial Intelligence. Proceedings, 2000. XI, 407 pages. 2000.

Vol. 1925: J. Cussens, S. Džeroski (Eds.), Learning Language in Logic. X, 301 pages 2000.

Vol. 1930: J.A. Campbell, E. Roanes-Lozano (Eds.), Artificial Intelligence and Symbolic Computation. Proceedings, 2000. X, 253 pages. 2001.

Vol. 1932: Z.W. Raś, S. Ohsuga (Eds.), Foundations of Intelligent Systems. Proceedings, 2000. XII, 646 pages.

Vol. 1934: J.S. White (Ed.), Envisioning Machuine Translation in the Information Future. Proceedings, 2000. XV, 254 pages. 2000.

Vol. 1937: R. Dieng, O. Corby (Eds.), Knowledge Engineering and Knowledge Management. Proceedings, 2000. XIII, 457 pages. 2000.

Vol. 1952: M.C. Monard, J. Simão Sichman (Eds.), Advances in Artificial Intelligence. Proceedings, 2000. XV, 498 pages. 2000.

Vol. 1955: M. Parigot, A. Voronkov (Eds.), Logic for Programming and Automated Reasoning. Proceedings, 2000. XIII, 487 pages. 2000.

Vol. 1967: S. Arikawa, S. Morishita (Eds.), Discovery Science. Proceedings, 2000. XII, 332 pages. 2000.

Vol. 1968: H. Arimura, S. Jain, A. Sharma (Eds.), Algorithmic Learning Theory. Proceedings, 2000. XI, 335 pages. 2000.

Vol. 1972: A. Omicini, R. Tolksdorf, F. Zambonelli (Eds.), Engineering Societies in the Agents World. Proceedings, 2000. IX, 143 pages. 2000.

Vol. 1979: S. Moss, P. Davidsson (Eds.), Multi-Agent-Based Simulation. Proceedings, 2000. VIII, 267 pages. 2001.

Vol. 1991: F. Dignum, C. Sierra (Eds.), Agent Mediated Electronic Commerce. VIII, 241 pages. 2001.

Vol. 1994: J. Lind, Iterative Software Engineering for Multiagent Systems. XVII, 286 pages. 2001.

Vol. 2003: F. Dignum, U. Cortés (Eds.), Agent-Mediated Electronic Commerce III. XII, 193 pages. 2001.

Vol. 2007: J.F. Roddick, K. Hornsby (Eds.), Temporal, Spatial, and Spatio-Temporal Data Mining. Proceedings, 2000. VII, 165 pages. 2001.

Vol. 2014: M. Moortgat (Ed.), Logical Aspects of Computational Linguistics. Proceedings, 1998. X, 287 pages. 2001.

Vol. 2019: P. Stone, T. Balch, G. Kraetzschmar (Eds.), RoboCup 2000: Robot Soccer World Cup IV. XVII, 658 pages. 2001.

Vol. 2033: J. Liu, Y. Ye (Eds.), E-Commerce Agents. VI, 347 pages. 2001.

Vol. 2035: D. Cheung, G.J. Williams, Q. Li (Eds.), Advances in Knowledge Discovery and Data Mining – PAKDD 2001. Proceedings, 2001. XVIII, 596 pages. 2001.

Vol. 2036: S. Wermter, J. Austin, D. Willshaw (Eds.), Emergent Neural Computational Architectures Based on Neuroscience. X, 577 pages. 2001.

Vol. 2039: M. Schumacher, Objective Coordination in Multi-Agent System Engineering. XIV, 149 pages. 2001.

Vol. 2056: E. Stroulia, S. Matwin (Eds.), Advances in Artificial Intelligence. Proceedings, 2001. XII, 366 pages. 2001.

Vol. 2062: A. Nareyek, Constraint-Based Agents. XIV, 178 pages. 2001.

Vol. 2070: L. Monostori, J. Váncza, M. Ali (Eds.), Engineering of Intelligent Systems. Proceedings, 2001. XVIII, 951 pages. 2001.

Vol. 2080: D.W. Aha, I. Watson (Eds.), Case-Based Reasoning Research and Development. Proceedings, 2001. XII, 758 pages. 2001.

Vol. 2083: R. Goré, A. Leitsch, T. Nipkow (Eds.), Automated Reasoning. Proceedings, 2001. XV, 708 pages. 2001.

Vol. 2086: M. Luck, V. Mařík, O. Štěpánková, R. Trappl (Eds.), Multi-Agent Systems and Applications. Proceedings, 2001. X, 437 pages. 2001.

Vol. 2099: P. de Groote, G. Morrill, C. Retoré (Eds.), Logical Aspects of Computational Linguistics. Proceedings, 2001. VIII, 311 pages. 2001.

Vol. 2101: S. Quaglini, P. Barahona, S. Andreassen (Eds.), Artificial Intelligence in Medicine. Proceedings, 2001. XIV, 469 pages. 2001.

Vol. 2103: M. Hannebauer, J. Wendler, E. Pagello (Eds.), Balancing Reactivity and Social Deliberation in Multi-Agent Systems. VIII, 237 pages. 2001.

Vol. 2109: M. Bauer, P.J. Gmytrasiewicz, J. Vassileva (Eds.), User Modeling 2001. Proceedings, 2001. XIII, 318 pages. 2001.

Vol. 2111: D. Helmbold, B. Williamson (Eds.), Computational Learning Theory. Proceedings, 2001. IX, 631 pages. 2001.

Vol. 2116: V. Akman, P. Bouquet, R. Thomason, R.A. Young (Eds.), Modeling and Using Context. Proceedings, 2001. XII, 472 pages. 2001.

Vol. 2120: H.S. Delugach, G. Stumme (Eds.), Conceptual Structures: Broadening the Base. Proceedings, 2001. X, 377 pages. 2001.

Vol. 2123: P. Perner (Ed.), Machine Learning and Data Mining in Pattern Recognition. Proceedings, 2001. XI, 363 pages. 2001.

Lecture Notes in Computer Science